LIPID-MEDIATED SIGNALING

METHODS IN SIGNAL TRANSDUCTION SERIES

Joseph Eichberg, Jr., Series Editor

Published Titles

Lipid Second Messengers, Suzanne G. Laychock and Ronald P. Rubin

G Proteins: Techniques of Analysis, David R. Manning

Signaling Through Cell Adhesion Molecules, Jun-Lin Guan

G Protein-Coupled Receptors, Tatsuya Haga and Gabriel Berstein

Calcium Signaling, James W. Putney, Jr.

G Protein-Coupled Receptors: Structure, Function, and Ligand Screening,
    Tatsuya Haga and Shigeki Takeda

Calcium Signaling, Second Edition, James W. Putney, Jr.

Analysis of Growth Factor Signaling in Embryos, Malcolm Whitman and
    Amy K. Sater

Signal Transduction in the Retina, Steven J. Fliesler and Oleg G. Kisselev

Signaling by Toll-Like Receptors, Gregory W. Konat

LIPID-MEDIATED SIGNALING

Edited by

Eric J. Murphy
University of North Dakota

Thad A. Rosenberger
University of North Dakota

CRC Press
Taylor & Francis Group
Boca Raton   London   New York

CRC Press is an imprint of the
Taylor & Francis Group, an **informa** business

CRC Press
Taylor & Francis Group
6000 Broken Sound Parkway NW, Suite 300
Boca Raton, FL 33487-2742

First issued in paperback 2017

© 2010 by Taylor and Francis Group, LLC
CRC Press is an imprint of Taylor & Francis Group, an Informa business

No claim to original U.S. Government works

ISBN 13: 978-1-138-11396-1 (pbk)
ISBN 13: 978-0-8493-8141-6 (hbk)

Library of Congress Cataloging-in-Publication Data

Lipid-mediated signaling / editors, Eric J. Murphy and Thad A. Rosenberger.
    p. ; cm. -- (Methods in signal transduction series)
  Includes bibliographical references and index.
  ISBN 978-0-8493-8141-6 (hardcover : alk. paper) 1. Lipids--Physiological transport.
  2. Second messengers (Biochemistry) 3. Cellular signal transduction. I. Murphy, Eric
J. (Eric James), 1963- II. Rosenberger, Thad A. III. Title. IV. Series: Methods in signal
transduction.
  [DNLM: 1. Lipids--physiology--Laboratory Manuals. 2. Second Messenger
Systems--physiology--Laboratory Manuals. QU 25 L7642 2009]
  QP751.L5474 2009
  612'.01577--dc22
                                                    2009037296

Visit the Taylor & Francis Web site at
http://www.taylorandfrancis.com

and the CRC Press Web site at
http://www.crcpress.com

We dedicate the book to our professor and mentor, Lloyd A. Horrocks. Lloyd's training instilled in us our love for lipid metabolism and provided us a solid foundation on which we have built our careers. To this day, we carry on Lloyd's quest for understanding the biological significance of plasmalogens in the central nervous system and the potential for these unique, poorly understood phospholipids in signal transduction. Lloyd was a quiet man who we fondly remember for his encyclopedic knowledge of lipid metabolism, the methods used to study this vast field, and the application of these methods to problems. More importantly, we recall how he answered our questions with a question, forcing us to think more deeply about our own ideas, and to probe our own knowledge base to come up with answers. He pushed us to truly advance our training and provided a solid foundation in critical thinking. Upon his mentorship we base our careers and our methods for the training of the next generation of lipid biochemists. To you, Lloyd, we are truly indebted.

Contents

Part I Enzyme Systems Involved in Lipid-Mediated Signaling

Part II   Mass Spectrometry Methods for Lipid Analysis

Part III   Lipid-Mediated Regulation of Gene Expression

Part IV Methods to Assess Fatty Acid and Phospholipid Metabolism

Series Preface

The concept of signal transduction at the cellular level is now established as a cornerstone of the biological sciences. Cells sense and react to environmental cues by means of a vast panoply of signaling pathways and cascades. While the steady accretion of knowledge regarding signal transduction mechanisms is continuing to add layers of complexity, this greater depth of understanding has also provided remarkable insights into how healthy cells respond to extracellular and intracellular stimuli, and how these responses can malfunction in many disease states.

Central to advances in unraveling signal transduction is the development of new methods and refinement of existing ones. Progress in the field relies upon an integrated approach that utilizes techniques drawn from cell and molecular biology, biochemistry, genetics, immunology, and computational biology. The overall aim of this series is to collate and continually update the wealth of methodology now available for research into many aspects of signal transduction. Each volume is assembled by one or more editors who are leaders in their specialty. Their guiding principle is to recruit knowledgeable authors who will present procedures and protocols in a critical yet reader-friendly format. Our goal is to assure that each volume will be of maximum practical value to a broad audience, including students, seasoned investigators, and researchers who are new to the field.

Ten years ago the first volume in this series was published, which dealt with approaches to study lipid mediators in cell signaling. Since then, this field has continued to progress and expand. The vast quantity of accumulated new knowledge has been accompanied by a plethora of novel methodologies, as well as the improvement of older techniques that can be utilized to elucidate the role of lipids in signal transduction. The current volume provides a representative, yet quite broad, coverage of a variety of useful and important procedures.

The authors have divided their presentation into four sections. Part I outlines current methods for assaying the activity and expression of lipid-metabolizing enzymes important in conveying signals. Part II presents techniques that utilize mass spectrometry to profile phospholipids, sphingolipids, and markers of oxidative stress. Part III includes a thorough coverage of methodologies in the rapidly burgeoning area concerning the lipid-mediated regulation of transcription via nuclear receptors. Finally, Part IV includes chapters on techniques for fatty acid measurement and targeted cellular uptake, as well as details on procedures for investigating the kinetics of fatty acid metabolism *in vivo*. As a whole, both the quantity and the quality of information presented in this volume should be of high value to a wide range of researchers who wish to familiarize themselves with methodologies for studying lipid-mediated signal transduction.

Joe Eichberg

Preface

In 1999, *Lipid Second Messengers*, edited by Suzanne G. Laychock and Ronald P. Rubin, was published as part of the Methods in the Life Sciences—Methods in Signal Transduction series. This new addition represents an update to the previous publication and contains many new chapters that highlight the rapidly changing field of lipid-mediated signal transduction. This new edition is divided into four parts giving the reader an opportunity to examine a wide array of modern techniques used in the field. In Part I, methods used to assay and identify the expression of different enzymes involved in lipid-mediated signaling are described and includes chapters on phospholipases A_2, C, and D (Chapters 1 through 3) as well as a chapter on diacylglycerol kinase (Chapter 4), a key enzyme involved in regulating diacylglycerol levels in cells. This part also contains a chapter (Chapter 5) focused on the analysis of enzymes involved in the endocannabinoid system, giving readers the depth and breadth of the typical enzyme systems involved in lipid-mediated signaling. Part II highlights the use of mass spectrometry to assess the lipid content in various systems. Chapter 6 includes the use of shotgun lipidomics to examine phospholipid molecular species. Chapters 7 and 8 focus on techniques for the analysis of sphingolipids, isoprostane, and neuroprostane levels. Part III contains two chapters (Chapters 9 and 10) devoted to techniques used to determine the role of lipids in the activation of gene transcription. While Chapter 9 is quite extensive and covers many nuclear transcription factors, Chapter 10 highlights the use of advanced fluorescence methods to examine the interaction of proteins involved in lipid ligand–induced gene transcription. Finally, Part IV focuses on methods used to assess fatty acid uptake and metabolism. The use of classical chromatography techniques to measure fatty acids is presented in Chapter 11, which complements Chapter 6 on shotgun lipidomics. A chapter on fatty acid uptake and targeting (Chapter 12) is also found in this part, which complements chapters in many parts of this book, but also brings out many of the nuances important to measuring fatty acid uptake that are often overlooked. Chapter 13 focuses on using steady-state radiotracer kinetics to assess fatty acid incorporation and turnover *in vivo*, and outlines methods that are applicable to many different experimental paradigms.

By bringing together leading authorities in the field, this book describes a wide range of commonly used techniques used to assess lipid-mediated signal transduction. We hope that the depth and breadth of this new addition to the series will provide researchers additional insight into the field and a convenient compilation of methods to overcome research challenges.

Editors

Eric J. Murphy is an associate professor in the Department of Pharmacology, Physiology, and Therapeutics at the University of North Dakota, Grand Forks, in the School of Medicine and Health Sciences. His interest in lipid metabolism and signaling emerged at Hastings College, Nebraska, where he received his BA with majors in biology and history. His love for lipids was coalesced under the guidance of Lloyd A. Horrocks at The Ohio State University, Columbus. Professor Horrocks taught him many things and instilled in him the desire to examine the role of fatty acids and phospholipids in central nervous system function as well as an entrepreneurial spirit that he still carries to this day. After a short time in the pharmaceutical industry, Dr. Murphy entered the laboratory of Professor Friedhelm Schroeder at Texas A & M University, College Station, where he studied the impact of lipid-binding proteins on the uptake and metabolism of cholesterol and fatty acids. Dr. Murphy was a National Research Council senior fellow at the National Institute on Aging (NIA), Bethesda, Maryland, where he studied under the guidance of Dr. Stanley I. Rapoport. At the NIA, he began to appreciate the power of steady-state kinetic modeling to assess the uptake and turnover of fatty acids in the heart and brain, extending the breadth of his techniques used in the analysis of lipid metabolism. In 2000, he became an assistant professor at the University of North Dakota where he has risen through the ranks. In 2006, he became the editor-in-chief of *Lipids*, a journal of the American Oil Chemist's Society (AOCS), and is an active member of the AOCS, the American Society for Neurochemistry, and the International Society for Neurochemistry. In 2009, he was awarded the Thomas Clifford Faculty Research Achievement Award from the University of North Dakota. He currently serves on the board of directors of Unicrop OY, Helsinki, Finland, and since 2005, he has served as the chief scientific officer of Unicrop OY and Agragen, LLC, Cincinnati, Ohio. He has published over 75 peer-reviewed articles and given oral presentations at over 35 universities worldwide, and at national and international meetings.

Thad A. Rosenberger is an assistant professor in the Department of Pharmacology, Physiology, and Therapeutics at the University of North Dakota, Grand Forks, in the School of Medicine and Health Sciences. His research interests include understanding the extent by which alterations in lipid-mediated signal transduction contribute to the progression of injury associated with neuroinflammation. Dr. Rosenberger earned his doctoral degree from The Ohio State University, Columbus, in 1999 under the supervision of Lloyd A. Horrocks and completed a postdoctoral fellowship with Dr. Stanley I. Rapoport at the National Institute on Aging, Bethesda, Maryland, in 2004. He is currently a senior associate editor for *Lipids*, a journal of the American Oil Chemist's Society, and is an active member of the American Society for Neurochemistry, the International Society for Neurochemistry, and the Society for Neuroscience.

Contributors

Barbara P. Atshaves
Department of Physiology and
Pharmacology
College of Veterinary Medicine and
Biomedical Sciences
Texas A&M University
College Station, Texas

Olivier Barbier
Laboratory of Molecular Pharmacology
Center Hospitalier Universitaire de
Québec Research Center

and

Faculty of Pharmacy
Laval University
Quebec City, Quebec, Canada

Gwendolyn Barceló-Coblijn
Laboratory of Molecular Cell
Biomedicine
Department of Biology
Institut Universitari d'Investigacions en
Ciencies de la Salut
University of the Balearic Islands
Palma de Mallorca, Spain

Dhaval P. Bhatt
Department of Pharmacology,
Physiology, and Therapeutics
School of Medicine and Health
Sciences
University of North Dakota
Grand Forks, North Dakota

Jacek Bielawski
Department of Biochemistry and
Molecular Biology
Medical University of South Carolina
Charleston, South Carolina

Tiziana Bisogno
Istituto di Chimica Biomolecolare
Consiglio Nazionale delle Ricerche
Pozzuoli, Naples, Italy

Hua Cheng
Division of Bioorganic Chemistry and
Molecular Pharmacology
Department of Internal Medicine
Washington University School of
Medicine
St. Louis, Missouri

Luciano De Petrocellis
Istituto di Cibernetica
Consiglio Nazionale delle Ricerche
Pozzuoli, Naples, Italy

Vincenzo Di Marzo
Istituto di Chimica Biomolecolare
Consiglio Nazionale delle Ricerche
Pozzuoli, Naples, Italy

Ryan A. Frieler
Department of Pharmacology,
Physiology, and Therapeutics
School of Medicine and Health
Sciences
University of North Dakota
Grand Forks, North Dakota

Michael A. Frohman
Department of Pharmacology
Center for Developmental Genetics
Stony Brook University
Stony Brook, New York

Kaoru Goto
Department of Anatomy and Cell
    Biology
Yamagata University School of
    Medicine
Yamagata, Japan

Xianlin Han
Division of Bioorganic Chemistry and
    Molecular Pharmacology
Department of Internal Medicine
Washington University School of
    Medicine
St. Louis, Missouri

Yusuf A. Hannun
Department of Biochemistry and
    Molecular Biology
Medical University of South Carolina
Charleston, South Carolina

Heather A. Hostetler
Department of Physiology and
    Pharmacology
College of Veterinary Medicine and
    Biomedical Sciences
Texas A&M University
College Station, Texas

Yasukazu Hozumi
Department of Anatomy and Cell
    Biology
Yamagata University School of
    Medicine
Yamagata, Japan

Huan Huang
Department of Physiology and
    Pharmacology
College of Veterinary Medicine and
    Biomedical Sciences
Texas A&M University
College Station, Texas

Xuntian Jiang
Division of Bioorganic Chemistry and
    Molecular Pharmacology
Department of Internal Medicine
Washington University School of
    Medicine
St. Louis, Missouri

Jenny Kaeding
Laboratory of Molecular Pharmacology
Center Hospitalier Universitaire de
    Québec Research Center

and

Faculty of Pharmacy
Laval University
Quebec City, Quebec, Canada

Ann B. Kier
Department of Pathobiology
College of Veterinary Medicine and
    Biomedical Sciences
Texas A&M University
College Station, Texas

Hisatake Kondo
Division of Anatomy and Histology
Department of Rehabilitation
Faculty of Medical Science and Welfare
Tohoku Bunka Gakuen University
Sendai, Japan

Serena M. Lackman
Department of Pharmacology,
    Physiology, and Therapeutics
University of North Dakota
Grand Forks, North Dakota

Alberto M. Martelli
Department of Human Anatomical
    Sciences
University of Bologna
Bologna, Italy

Jane McHowat
Department of Pathology
St. Louis University School of Medicine
St. Louis, Missouri

Avery L. McIntosh
Department of Physiology and
  Pharmacology
College of Veterinary Medicine and
  Biomedical Sciences
Texas A&M University
College Station, Texas

Paul G. Millner
Department of Pharmacology,
  Physiology, and Therapeutics
University of North Dakota
Grand Forks, North Dakota

Ginger L. Milne
Division of Clinical Pharmacology
Vanderbilt University School of
  Medicine
Nashville, Tennessee

Jason D. Morrow (Deceased)
Division of Clinical Pharmacology
Vanderbilt University School of
  Medicine
Nashville, Tennessee

Cameron C. Murphy
Department of Pharmacology,
  Physiology, and Therapeutics
School of Medicine and Health
  Sciences
University of North Dakota
Grand Forks, North Dakota

Eric J. Murphy
Department of Pharmacology,
  Physiology, and Therapeutics
School of Medicine and Health
  Sciences
and
Department of Chemistry
University of North Dakota
Grand Forks, North Dakota

Tomoyuki Nakano
Department of Anatomy and Cell
  Biology
Yamagata University School of
  Medicine
Yamagata, Japan

Anca D. Petrescu
Department of Physiology and
  Pharmacology
College of Veterinary Medicine and
  Biomedical Sciences
Texas A&M University
College Station, Texas

Prerna Rastogi
Department of Pathology
St. Louis University School of Medicine
St. Louis, Missouri

Thad A. Rosenberger
Department of Pharmacology,
  Physiology, and Therapeutics
School of Medicine and Health
  Sciences
University of North Dakota
Grand Forks, North Dakota

Friedhelm Schroeder
Department of Physiology and
  Pharmacology
College of Veterinary Medicine and
  Biomedical Sciences
Texas A&M University
College Station, Texas

Janhavi Sharma
Department of Pathology
St. Louis University School of Medicine
St. Louis, Missouri

Wenjuan Su
Department of Pharmacology
Center for Developmental Genetics
Stony Brook University
Stony Brook, New York

Motohiro Tani
Department of Biochemistry and
  Molecular Biology
Medical University of South Carolina
Charleston, South Carolina

Jocelyn Trottier
Laboratory of Molecular Pharmacology
Center Hospitalier Universitaire de
  Québec Research Center

and

Faculty of Pharmacy
Laval University
Quebec City, Quebec, Canada

John A. Watt
Department of Anatomy and Cell
  Biology
School of Medicine and Health
  Sciences
University of North Dakota
Grand Forks, North Dakota

Youchun Zeng
Division of Bioorganic Chemistry and
  Molecular Pharmacology
Department of Internal Medicine
Washington University School of
  Medicine
St. Louis, Missouri

Part I

Enzyme Systems Involved in Lipid-Mediated Signaling

1 Methods for Measuring the Activity and Expression of Phospholipases A$_2$

Prerna Rastogi, Janhavi Sharma,
and Jane McHowat

CONTENTS

1.1 INTRODUCTION

Cellular membrane phospholipids are important structural elements that also serve as precursors for second messenger molecules that have important roles in cellular signaling processes. These second messenger molecules are generated by the action of intracellular or extracellular phospholipases. Phospholipase A_2 (PLA_2) enzymes are a ubiquitous group of esterases that cleave the *sn*-2 bond of phospholipids, leading to the stoichiometric release of a free fatty acid and a lysophospholipid. Free arachidonic acid serves as a precursor for eicosanoids such as prostaglandins and leukotrienes. Eicosanoids bind to specific G-protein-coupled receptors and participate in several inflammatory processes including atherosclerosis and asthma [1,2]. Lysophospholipids act as detergents and are, therefore, injurious to cells. They can further be acetylated to generate the platelet-activating factor (PAF), an important mediator in the recruitment of inflammatory cells to the endothelium [3]. Due to the diversity of PLA_2 enzymes and the role that each, or several, can play under specific conditions, it is of critical importance to characterize their diversity and their regulators.

1.2 CLASSIFICATION OF PHOSPHOLIPASES A_2

Mammalian PLA_2 comprise a family of distinct enzymes that can be divided into three main classes. These include secretory ($sPLA_2$), cytosolic ($cPLA_2$), and calcium-independent ($iPLA_2$) isoforms [4,5]. All three classes coexist in mammalian cells and may interact with each other. These classes have more recently been grouped according to their amino acid sequences and other conserved features. A brief classification is given in Table 1.1.

1.3 SECRETORY PHOSPHOLIPASE A_2

1.3.1 STRUCTURE, EXPRESSION, AND MECHANISM OF ACTION

The secretory PLA_2 enzymes have low molecular weights (14–18 kDa); require mM levels of calcium for catalysis; and possess a highly conserved calcium-binding loop (XCGXGG), zero to eight disulfide bonds, and a common catalytic site (DXCCXXHD) [6]. Histidine and aspartate residues form a His/Asp dyad at the active site. Activation of the enzyme occurs when a water molecule binds to histidine, which, in conjunction with aspartate and the calcium-binding loop, acts as a ligand cage for calcium. These enzymes do not have fatty acid selectivity but most group II subfamily members act preferentially on anionic phospholipids such as phosphatidylethanolamine/serine/glycerol compared to phosphatidylcholine (PtdCho) [7]. The interfacial binding to anionic phospholipid vesicles is in part due to the presence of several hydrophobic residues on the α-helical N-terminus, such as tryptophan. The indole side chain of tryptophan residues allows it to penetrate the lipid interface of the membranes, and this is particularly evident in group V $sPLA_2$ [8].

sPLA$_2$ IB contains a hydrophobic channel with the conserved active site, a disulfide bond between Cys11 and Cys77, and an additional 5 amino acid "pancreatic

TABLE 1.1

Classification of Phospholipases A$_2$

Class	Group/Alternate Names	Molecular Weight (kDa)	Calcium Requirement	Active Site Residue	Inhibitors
Secretory	I A, B	13–15	mM calcium	His/Asp	LY311727
phospholipase	II A–F	13–18	required for	dyad	LY315920
A$_2$ (sPLA$_2$)	III	15–18	hydrolysis		PGBx
	V	14			PX-18
	IX	14			PX-52
	X	14			
	XI A, B	12			
	XII	19			
	XIII	<10			
	XIV	13–19			
Cytosolic	IV A/cPLA$_2$α	85	µM calcium	Serine	AACOCF$_3$
phospholipase	IV B/cPLA$_2$β	114	required for		MAFP
A$_2$ (cPLA$_2$)	IV C/cPLA$_2$γ	61	translocation of		
	IV D/cPLA$_2$δ	92–93	enzyme to an		
	IV E/cPLA$_2$ε	100	intracellular		
	IV F/cPLA$_2$ζ	96	membrane		
Calcium-	VIA1,2/iPLA$_2$β	84–90	No calcium	Serine	BEL
independent	VI B/iPLA$_2$γ	88–91	required		AACOCF$_3$
phospholipase	VI C/iPLA$_2$δ	146			MAFP
A$_2$ (iPLA$_2$)	VI D/iPLA$_2$ε	53			
	VI E/iPLA$_2$ζ	57			
	VI F/iPLA$_2$η	28			
	VIIA,B/PAF-AH	45, 40			
	VIII A,B	26			

loop" in the middle [9]. It is secreted as an inactive zymogen by pancreatic acinar cells. The inactive N-terminal heptapeptide is then cleaved by trypsin to form the mature enzyme [9]. In addition to the pancreas, trace amounts of sPLA$_2$ IB also occur in the lung, kidney, and spleen.

Highly cationic sPLA$_2$ isoforms, including group IIA, IID, and V, are bound to cell surface heparan sulfate proteoglycans on the cytosolic side and may have access to certain compartments on the plasma membrane where substrate phospholipids are more sensitive to sPLA$_2$ [10]. The basic amino acid residues near the C-terminus impart this characteristic heparinoid binding [10]. The members of group II preferentially hydrolyze anionic phospholipids [11]. Group IIA sPLA$_2$ is both inducible and constitutively expressed. Group II sPLA$_2$, including II A and II D, possess a disulfide linkage between Cys50 and the C-terminal cysteine. An additional disulfide link between Cys87 and Cys93 occurs in the loop region of II C [12]. A unique 30 amino acid C-terminal extension is present in II F with the potential to dimerize with a cellular protein [13]. Although group V sPLA$_2$ lacks the typical disulfide linkages of both group I and II sPLA$_2$, its genes are clustered at chromosome 1p34–36 with those of the other members of group II, indicating a common ancestral origin [14].

Group III sPLA$_2$ is a 15 kDa protein with the characteristic calcium-binding loop and 10 cysteine residues. In humans, the catalytic site is flanked by N- and C-terminal regions, resulting in a 55 kDa protein that sets it apart from the other members of the sPLA$_2$ family. It is expressed in the human kidney, heart, skeletal muscle, and liver. Group III sPLA$_2$ demonstrates a 10-fold preference to hydrolyze anionic phospholipids compared to PtdCho [15].

Group X sPLA$_2$ has group I and II characteristic disulfide linkages and is secreted as a propeptide. It undergoes both proteolytic cleavage and N-glycosylation prior to catalysis [16]. Group X sPLA$_2$ hydrolyzes both phosphatidylethanolamine and phosphatidylcholine vesicles and is expressed in the gastrointestinal tract, human testis, spleen, thymus, and alveolar endothelial cells. Expression of group X sPLA$_2$ is elevated in colon carcinoma [17].

Although group XII sPLA$_2$ possesses the characteristic sPLA$_2$ central catalytic domain with a His/Asp dyad, the cysteine residues outside the catalytic site are at a different location. Calcium is required for activity, and it can hydrolyze anionic phospholipids with a greater than 10-fold preference compared to PtdCho. It has a robust expression in the kidney, heart, pancreas, and skeletal muscle [18].

1.3.2 FUNCTION

The wide range of sPLA$_2$ isoforms provides a challenge to distinguishing the function of each. However, with the availability of pharmacologic inhibitors, siRNA techniques, antibodies, and mouse models, additional data have now become available on the various roles played by each of these isoforms. Broadly, sPLA$_2$ enzymes participate primarily in arachidonic acid release and eicosanoid generation.

Group II A sPLA$_2$ is present in high concentrations at sites of inflammation including synovial fluid in rheumatoid arthritis, acute respiratory distress syndrome, inflammatory bowel disease, pancreatitis, and sepsis [19]. Balboa et al. demonstrated that in murine P388D1 macrophages, the generation of prostaglandin E$_2$ in response to long-term stimulation by lipopolysaccharide involves an initial activation of group IV A cPLA$_2$ that induces the expression of group V sPLA$_2$, which in turn induces both the expression of COX-2 and the production most of the arachidonic acid substrate [20].

As sPLA$_2$ promotes and amplifies transcellular eicosanoid biosynthesis in an autocrine or juxtacrine fashion, it can be regarded as a type of hormone [21]. Group II A sPLA$_2$ has a controversial role in tumorigenesis with evidence existing for it being both a tumor suppressor and a tumor promoter. The presence of sPLA$_2$ alters the intestinal crypt microenvironment and confers the ability to resist multiple adenoma formation in *Min* mice (dominant mutation in the homolog of APC gene) [22]. Conversely, in humans, high levels of sPLA$_2$ expression in familial adenomatous polyposis contribute to increased arachidonic acid and COX-2 levels and tumor formation [23]. The basic residues on sPLA$_2$ can bind to factor Xa and inhibit prothrombinase activity, which has an antithrombotic effect [24]. Additionally, sPLA$_2$-II A promotes the generation of antithrombotic prostaglandin I$_2$ by cultured endothelial cells [25].

The plasma-membrane-associated group V and X PLA$_2$ enzymes play a role in host defense against adenoviral infection [26]. Bronchial epithelial cells and lung fibroblasts that express groups V and X sPLA$_2$ showed marked resistance to adenovirus-mediated gene delivery. More recent experimental studies suggest that sPLA$_2$ enzymes are involved in traumatic and autoimmune-precipitated neurodegeneration and, therefore, a potential target for the treatment of nervous system disorders [27]. Thus, sPLA$_2$ has been described in a wide range of diseases that continues to grow, which suggests that the development of selective inhibitors can be invaluable but may also be associated with a high incidence of side effects.

1.4 CYTOSOLIC PHOSPHOLIPASE A$_2$

1.4.1 STRUCTURE, EXPRESSION, AND MECHANISM OF ACTION

Cytosolic phospholipase A$_2$ (cPLA$_2$) isoforms are members of group IV (Table 1.1) and are large molecular weight proteins (61–114 kDa). To date, four human cytosolic PLA$_2$ have been cloned and are designated as α, β, γ, and δ [28]. Additional enzymes have been assigned to this group based on the sequence similarities and include group IV B and IV C from human, group IV D from both human and murine [29,30], and group IV E and IV F from murine tissues [30]. cPLA$_2$ enzymes demonstrate a preference for choline phospholipids with arachidonic acid esterified at the *sn-2* position [31]. Enzymes in group IV A can also function as transacylases and lysophospholipases [32].

The cPLA$_2$ isoforms contain two catalytic domains A and B with the lipase consensus sequence, GXSGS, located within the catalytic domain A. They cleave membrane phospholipids utilizing a catalytic serine [49]. A C2 calcium lipid-binding domain is found in group IV A cPLA$_2$ containing about 120–130 amino acids and a typical type II topology of eight antiparallel β strands interconnected by six loops [54–56]. Two calcium ions bind to the distinct calcium-binding region on the loops, and the C2 domain binds to the neutral PtdCho [33,34]. The catalytic domain comprises a central mixed β-sheet surrounded by 14 β strands and 13 α helices. The nucleophilic Ser228 attacks the *sn-2* ester-linked fatty acid while Asp549 activates the catalytic center. Ser228, Asp549, and Asp200 are the critical residues that are preserved in group IV B and IV C cPLA$_2$. The catalytic dyad of Ser228 and Asp549 is placed at the bottom of the active site channel lined by hydrophobic residues [35]. After calcium induces translocation of the group IV A cPLA$_2$ to an intracellular phospholipid bilayer, the individual substrate molecules bind to the narrow cleft of the active site bringing the *sn-2* ester bond in close proximity to Ser228. The phosphate head group is stabilized by Arg200. This leads to the formation of an enzyme–substrate complex. Asp549 removes a proton by nucleophilic attack on the *sn-2* ester. The serine-acyl intermediate is formed when the proton is transferred to the lysophospholipid. Hydrolysis of the acyl intermediate leads to either dissociation of the enzyme from the membrane interface or binding to another phospholipid molecule, repeating the cycle. The potential rate-limiting step is the hydrolysis of the acyl-enzyme intermediate [35]. Another implication of this sequential reaction is that it allows for a more targeted delivery of PLA$_2$-enzyme-bound arachidonate to other downstream elements required for eicosanoid generation, thereby increasing metabolic efficiency [36]. Calcium is not required for hydrolytic activity but for translocation from the cytosol to intracellular membranes, primarily those of the nucleus and the endoplasmic reticulum [37]. The group IV C cPLA$_2$ contains an isoprenylation site at the C-terminus instead of the C2 domain and is predominantly membrane bound [38].

Cytosolic phospholipases are constitutively expressed in most human tissues. Following stimulation by proinflammatory cytokines, there is an increased release of eicosanoids in various organs [39]. The gene for human group IV A cPLA$_2$ is expressed on chromosome 1 with 18 exons and binding sites for NF-κB, GRE, AP-1, and AP-2. The typical housekeeping gene promoters, such as a TATA box, are absent. Instead, a binding site for a polypyrimidine sequence is responsible for a low constitutive expression [40]. Additionally, Wick et al. have demonstrated that oncogenic forms of Ras increase the transcription of cPLA$_2$ in normal lung epithelial cells through activation of the mitogen-activated protein kinase (MAPK) family. They identified a lung Kruppel-like factor that binds to a critical region of the cPLA$_2$ promoter and is responsible for the induction of cPLA$_2$ [41]. Cowan et al. identified a novel AAGGAG motif 30 nucleotides downstream that is bound by a TATA-box binding protein (TBP), which is critical for basal transcriptional activity of the human group IV A cPLA$_2$ promoter [42]. Group IV B cPLA$_2$ is expressed in the pancreas, brain, heart, and liver while group IV C cPLA$_2$ is found in the skeletal muscle [43]. Chiba et al. cloned group IV D cPLA$_2$ from a human keratinocyte cDNA library. The recombinant protein demonstrated a sixfold increase in specific activity toward linoleic acid, a prominent fatty acid derivative found in psoriatic skin [29].

1.4.2 FUNCTION

Group IV A cPLA$_2$ has a robust expression both in the mouse and the human. The importance of group IV A cPLA$_2$-mediated arachidonic acid and lysophospholipid generation is underscored by the fact that in cells generated from group IV A cPLA$_2$ null mice, there is a significant reduction in the amount of these lipid mediators released in response to injurious stimuli [44]. In response to an increase in intracellular calcium, the enzyme undergoes phosphorylation at Ser505 by the action of mitogen-activated protein kinases (ERK or p38MAPK), leading to an increase in catalytic activity. Both phosphorylation and intracellular calcium increases are required for enzyme activity [45].

In airway inflammation during bronchial asthma, group IV A cPLA$_2$ appears to be an important regulator of arachidonic acid generation and subsequent eicosanoid formation. In group IV A cPLA$_2$ knockout mice, the airway response to ovalbumin-induced anaphylaxis and bronchial reactivity to methacholine were significantly reduced [46]. Fujishima et al. have also demonstrated the critical importance of group IV A cPLA$_2$ in the immediate-phase generation of LTC$_4$, LTB$_4$, and PGD$_2$ in response to signaling through c-kit or FcεRI in bone marrow mast cells derived from mice with group IV A cPLA$_2$ disruption [47]. A potential application of a group IV A cPLA$_2$ inhibitor could thus be feasible in acute lung injury. Adult respiratory distress syndrome (ARDS) has a high mortality rate and has no effective treatment. In group IV A cPLA$_2$ null mice, lipopolysaccharide/zymosan-induced lung injury (a model for ARDS) failed to produce increased lung permeability, neutrophil sequestration, and compromised gas exchange, suggesting the importance of cPLA$_2$α in ARDS [48].

Group IV A cPLA$_2$ null mice show ulcerative lesions in their gastrointestinal tracts, indicating the importance of cPLA$_2\alpha$ in PGE$_2$ production [49]. A role for group IV A cPLA$_2$ has also been demonstrated in the onset of the parturition signaling cascade. In knockout mice, it appears that gestation is infrequent and accompanied by a delay in the onset of labor, along with a small litter size and more dead pups, compared to wild type mice [50]. Therefore, a possible role of group IV A cPLA$_2$ in ovulation and implantation has been suggested in conjunction with COX-2 and PGE$_2$ [50]. Finally, in a model of cerebral ischemia–reperfusion injury using middle cerebral artery occlusion in group IV A cPLA$_2$ knockout mice, the size of the infarct is much smaller when compared to wild type mice [51]. Thus, there is substantial evidence that cPLA$_2$ is a key enzyme in the generation of eicosanoids that can be inflammatory or protective, depending on the cell type or organ studied.

1.5 CALCIUM-INDEPENDENT PHOSPHOLIPASE A$_2$

1.5.1 Structure, Expression, and Mechanism of Action

The gene for classic group VI A iPLA$_2$ has 16 exons, which leads to the possibility of the formation of several splice variants [52]. Group VI A iPLA$_2$ is an 85 kDa protein with eight N-terminal ankyrin repeat sequences. The catalytic domain has a consensus lipase motif of GXSXG with Ser465 present in the catalytic center. The splice variant iPLA$_2$-VI A-2 has an additional 54 amino acid, proline-rich region bearing a sequence similarity to Smad proteins. An additional glycine-rich, nucleotide-binding motif (GXGXXG) occurs just prior to the catalytic site while a calmodulin-binding domain is located near the C-terminus. Unlike cPLA$_2$ and sPLA$_2$, the presence of calcium leads to the formation of the calmodulin–iPLA$_2$ complex and, subsequently, inactivates the enzyme [53]. Group VI B iPLA$_2$ contains a C-terminal peroxisomal localization sequence [54]. The N-terminus sequence contains several serine and threonine residues but lacks the ankyrin repeats seen in group VI B iPLA$_2$. Several putative protein kinase A and C phosphorylation sites and a consensus sequence for a MAPK phosphorylation site are also present. Group VI C–F are novel enzymes possessing iPLA$_2$ activity. Group VI C (neuropathy target esterase, NTE) is expressed in human neurons, and its esterase domain slowly hydrolyzes the *sn*-2 position fatty acid of PtdCho and plays a role in membrane homeostasis [55]. The other three group VI enzymes (D–F) hydrolyze arachidonic acid at the *sn*-2 position in the absence of calcium. Additionally, they possess high triacylglycerol lipase and acylglycerol transacylase activities [56].

This class of enzymes is ubiquitously expressed in a wide variety of cells and tissues and can be preferentially distributed in the membrane fraction. Serine is used for catalysis, and, similar to the hydrolytic action of cytosolic PLA$_2$, two sequential nucleophilic displacement reactions are involved leading to the formation of an acyl-enzyme intermediate and a lysophospholipid. This is rapidly hydrolyzed to release the free fatty acid. Myocardial iPLA$_2$ demonstrates unique characteristics when compared to other PLA$_2$ isoforms described previously, including selectivity for plasmalogen phospholipids and resistance to inhibition by methyl arachidonyl fluorophosphonate (MAFP). Activation of myocardial iPLA$_2$ results in the production of lysoplasmenylcholine and arachidonic acid, both of which can change the electro-physiologic properties of the myocardium [57,58].

1.5.2 FUNCTION

One of the major roles for $iPLA_2$ is arachidonic acid release and subsequent eico-sanoid formation. In human bladder microvascular endothelial cells, inhibition of $iPLA_2$ using bromoenol lactone (BEL) blocks the generation of arachidonic-acid-derived PGI_2 production [1]. In thrombin-stimulated endothelial cells, PAF produc-tion requires $iPLA_2$ activation but is thought to occur through a Co-A-independent transacylase remodeling pathway rather than a result of direct membrane alkylacyl glycerophosphocholine (PakCho) hydrolysis [59]. Inhibition of $iPLA_2$ by either BEL or $iPLA_2\beta$-specific antisense oligonucleotide decreases zymosan-stimulated PGE_2 production in P388D$_1$ macrophages [60], and in HEK 293 cells transfected with $iPLA_2$ β, ionophore-stimulated arachidonic acid release is augmented [61].

Zhang et al. have demonstrated that BEL treatment prior to chemotherapeutic agents such as cisplatin or vincristine reduced apoptosis by 30%–50% in HEK 293 and human kidney carcinoma cells (Caki-1) [62]. Doxorubicin, an anticancer drug, is associated with myocardial toxicity by virtue of its ability to inhibit $iPLA_2$. Since membrane-associated $iPLA_2$ represents the majority of myocardial PLA_2 activity, its inhibition by anthracyclines would critically impair the ability of cardiomyocytes to repair oxidized phospholipids leading to chronic cardiotoxicity of the anthracy-clines [63]. Additionally, in renal proximal tubule cells, inhibition of $iPLA_2$ by BEL reduced the cisplatin-induced annexin V binding and other markers of apoptosis including chromatin condensation and caspase 3 activity [64].

In isolated rat mesenteric arteries, BEL pretreatment inhibited the acetylcho-line-induced relaxation of phenylephrine constriction, indicating the importance of $iPLA_2$-produced arachidonic acid in this process. Calcium-independent PLA_2 also plays a role in the maintenance of the steady state of lysophosphatidylcholine levels in monocytes and macrophages of the intestinal tract. Inhibition of $iPLA_2$ results in a marked decrease in the levels of this metabolite and subsequent lysozyme secre-tion [65]. Inhibition of $cPLA_2$ results in a complete abrogation of the arachidonate mobilization response, but has no effect on lysozyme secretion, thereby identifying $iPLA_2$-mediated lysophosphatidylcholine production as a necessary component of the molecular machinery leading to lysozyme secretion in U937 cells [66].

Ramanadham et al. reported that RNA-specific silencing of $iPLA_2$ expression in INS-1 cells (rat insulinoma cells) significantly reduced insulin-secretory responses of INS-1 cells to glucose. Additionally, they demonstrated that the impaired glucose tolerance was due to insufficient insulin secretion rather than decreased insulin sen-sitivity [67]. Inhibition of $iPLA_2$ by BEL or antisense RNA inhibited both mitogen-induced peripheral blood lymphocyte and Jurkat T cell proliferation and provides evidence that $iPLA_2$ plays a key role in the lymphocyte proliferative response [68]. Gross and coworkers have also provided evidence that ischemia activates group VI A $iPLA_2$ in myocardium and that group VI A $iPLA_2$-mediated hydrolysis of membrane phospholipids can induce lethal malignant ventricular tachyarrhythmias during acute cardiac ischemia [69]. Thus, the majority of studies have highlighted multiple functions for iPLA$_2$ enzymes including cell signaling, secretion, and membrane phospholipid repair.

1.5.3 PLATELET-ACTIVATING FACTOR ACETYLHYDROLASE

The platelet-activating factor acetylhydrolases (PAF-AH) are two groups of serine iPLA$_2$ that selectively hydrolyze the acetyl group from the *sn*-2 position of PAF. Group VII A is secreted into the plasma and associates with lipoproteins whereas group VII B and VIII are intracellular enzymes. Group VII A PLA$_2$ is a 45 kDa secreted protein containing a lipase consensus motif GXS273XG [70], and the catalytic Ser/His/Asp forms a classic hydrolase triad. It also hydrolyzes phospholipids with oxidized fatty acyl groups (up to nine carbons in length) from the *sn*-2 position of PtdCho or phosphatidylethanolamine and can hydrolyze water-soluble phospholipids [71]. Its ability to hydrolyze oxidized phospholipids in low-density lipoprotein (LDL) particles makes it a potential target in atherosclerosis. Upregulation of PAF-AH activity is seen in the differentiation of immature bone marrow mast cells to mature mast cells [72]. Upregulation of group VII A PAF-AH in cells leads to increased transfer of acetate from PAF to endogenous acceptor lipids. Therefore, this enzyme may protect cells from oxidative damage and minimize the pathophysiologic effects of these metabolites [73].

Group VII B PAF-AH is a 40 kDa intracellular monomeric protein that hydrolyzes *sn*-2 acyl chains with <5 carbons [74]. High levels of expression are found in the liver and kidney. In response to oxidative stress, the enzyme undergoes a reversible translocation to the endoplasmic reticulum. Group VIII PLA$_2$ consists of two catalytic (either homomeric or heteromeric) 30 kDa subunits (α1 and α2) and one regulatory 45 kDa subunit [75,76]. The active site has a Ser/His/Asp triad with a 30 amino acid sequence located downstream from the active serine site that exhibits significant homology to the first transmembrane region of the PAF receptor. It therefore exhibits substrate specificity for PAF analogs but not for other short chain or oxidized phospholipids in contrast to group VII enzymes, which show broader specificity toward these substrates [75,76].

1.6 INHIBITION OF PHOSPHOLIPASE A$_2$ ACTIVITY

Since several PLA$_2$ enzymes hydrolyze membrane phospholipids utilizing the same catalytic site, the development of selective pharmacologic inhibitors for a specific isoform has proved difficult. Several pharmacologic inhibitors have been developed with the goal of inhibiting a specific isoform or group that have later been found to be less specific than at first thought and may also inhibit other enzymes involved in phospholipid metabolism. Historically, the PLA$_2$ inhibitors used were antimalarial drugs, aminoglycosides, and polyamines. These substances disrupted the phospholipid substrate or calcium ion levels and prevented the PLA$_2$ enzyme from acting on the membrane [77]. Several inhibitors, or groups of inhibitors, have now been developed that are selective for a class of PLA$_2$ enzymes. However, as more information becomes available, these may not be as selective as originally proposed. Due to the ambiguity of results obtained with pharmacologic inhibitors, researchers have tried alternate approaches such as siRNA or overexpression techniques and the use of transgenic and knockout mouse models to decrease or increase expression of a specific PLA$_2$ isoform in cells. In the next section, we discuss the development of pharmacologic inhibitors for each class of PLA$_2$ enzymes, followed by a brief discussion on the molecular biology techniques that have been employed in an effort to delineate the role of PLA$_2$ enzymes.

1.6.1 PHARMACOLOGIC INHIBITORS DESIGNED TO INHIBIT sPLA$_2$

One of the most widely used and best characterized sPLA$_2$ inhibitors developed recently has been 3-(3-acetamide-1-benzyl-2-ethyl-indolyl-5-oxy) propane phosphonic acid (LY311727) [77]. It has been described to inhibit both group II A and group V sPLA$_2$ and resides in the hydrophobic channel of sPLA$_2$, resulting in structural changes that bring the inhibitor in direct contact with the active site. A subsequently described analog of LY311727, [[3-(aminooxoacetyl)-2-ethyl-1-(phenylmethyl)-1H-indol-4-yl]oxy] acetate (LY315920) displays a 40-fold selectivity for group II A, non-pancreatic, sPLA$_2$ (IC$_{50}$ = 9 nM) when compared to group I B, pancreatic, sPLA$_2$. LY315920 was found to be inactive against cytosolic PLA$_2$ and has been proposed to represent a new class of anti-inflammatory drugs that is currently under clinical investigation.

PGBx, a prostaglandin oligomer, belongs to a family of compounds that selectively inhibit sPLA$_2$ and block arachidonic acid release from neutrophils [78]. This group of compounds has at least two fatty acid moieties with one unsaturated double bond, one organic group, and one active group or any salt form or ionized form thereof. PX-18 is a recent addition to this group and inhibits human sPLA$_2$ with an IC$_{50}$ of <1 μM but does not inhibit recombinant cPLA$_2$, endothelial cell cPLA$_2$, or iPLA$_2$ [78,79].

Recently, CHEC-9, a small peptide fragment of diffusible survival evasion peptide (DSEP), has been proposed to be an "uncompetitive" inhibitor of sPLA$_2$, presumably binding to the enzyme–substrate complex with its efficacy dependent on the concentration of both enzyme and substrate in the reaction medium. A subcutaneous injection of CHEC-9 interrupts the inflammatory cascade and promotes anti-inflammatory and neuron survival effects in cerebral cortical lesions in rats [80].

1.6.2 PHARMACOLOGIC INHIBITORS DESIGNED TO INHIBIT cPLA$_2$

Two mechanism-based inhibitors that were developed originally to inhibit cPLA$_2$ are arachidonyl trifluoromethyl ketone (AACOCF$_3$) and methyl arachidonyl fluorophosphonate (MAFP). Both these compounds were subsequently found to inhibit iPLA$_2$ at similar concentrations [81]. However, this is not surprising since these compounds act at the catalytic site which, as described above, is similar in both iPLA$_2$ and cPLA$_2$ classes. Both inhibitors possess an arachidonyl tail that is coupled to a serine reactive group. AACOCF$_3$ is a tight-binding, reversible inhibitor that forms a stable hemiketal with the active site serine residues in both cPLA$_2$ and iPLA$_2$ enzymes. MAFP binds irreversibly to inhibit both cPLA$_2$ and iPLA$_2$, possibly by phosphorylating the active site serine residues [82]. Neither of these inhibitors have any effect on sPLA$_2$ activity. Recent effort has been focused on developing analogs of MAFP and AACOCF$_3$ that are more selective for either cPLA$_2$ or iPLA$_2$.

Previous studies from our laboratory have demonstrated that pretreatment of endothelial cells with MAFP leads to an increase in basal and thrombin-stimulated PAF production as a direct result of inhibition of PAF-AH activity [83]. This is simultaneously accompanied by increased expression of cell surface adhesion molecules including P-selectin and increased adherence of neutrophils to the endothelial monolayer. The most selective cPLA$_2$ inhibitor described to date is pyrrolidine-1. This inhibitor has been reported to inhibit group IV A cPLA$_2\alpha$ with an IC$_{50}$ of 0.07 μM using (16:0, 20:4) PtdCho as a substrate by a protein–protein interaction. In contrast, the IC$_{50}$ for group IV C cPLA$_2$ (a calcium-independent isoform) is 1.2 μM, thus pyrrolidine-1 demonstrates almost a 20-fold preference for group IV A cPLA$_2$ [84].

1.6.3 PHARMACOLOGIC INHIBITORS DESIGNED TO INHIBIT iPLA$_2$

BEL appears to be the most isoform-specific inhibitor developed to date. It demonstrates 100-fold selectivity for iPLA$_2$ when compared to cPLA$_2$ and sPLA$_2$ isoforms [85]. In addition, Jenkins et al. reported that separation of racemic BEL into its R and S enantiomers demonstrated a 10-fold selectivity of S-BEL for iPLA$_2\beta$ (group VI A1) and of R-BEL for iPLA$_2\gamma$ (group VI B) [86]. This study highlights the use of chiral mechanism–based inhibitors to further discriminate between PLA$_2$ isoforms. However, the selectivity of BEL is not absolute, since it inhibits phosphatidate phosphohydrolase, an enzyme that converts phosphatidic acid to diacylglycerol (IC$_{50}$ = 8 μM) [87].

1.6.4 ALTERNATIVES TO PHARMACOLOGIC INHIBITION

Studies designed to elucidate the role of specific PLA$_2$ isoforms under various conditions have increasingly involved the use of molecular biology techniques to support and verify the previous findings obtained with pharmacologic inhibitors. Several studies have used antisense RNA for a single PLA$_2$ isoform to modulate cellular responses. For example, in airway epithelial cells, antisense knockdown of cPLA$_2$ downregulated the cytokine-induced release of arachidonic acid [88]. Su et al. have shown that knockdown of iPLA$_2$ results in a delay in hormone-induced differentiation of adipocytes [89]. However, several considerations must be taken when using antisense techniques. It is important to determine that the mRNA, the protein, and the activity attributed to the isoform in question are all decreased or eliminated from the cell system prior to experimental intervention. Additionally, care must be taken that off-targets are not affected by the siRNA treatment and that a decrease in one PLA$_2$ isoform does not result in a compensatory increase in another to maintain cellular function or integrity.

As discussed above, there have been several knockout or transgenic mouse models developed to study a specific PLA$_2$ isoform. However, as with antisense RNA studies, care must be taken to exclude any possible alteration in total PLA$_2$ activity that may occur as a result of upregulation of an alternate isoform. Additionally, depending on the physiologic/pathologic function being studied, a mouse model may not be entirely suitable as a representation of the role of PLA_2 in man. For example, Gross and coworkers have determined that the mouse myocardium contains extremely low levels of $iPLA_2$ activity. They demonstrated that the mouse was a species-specific knockdown of the human pathologic phenotype of ischemia-induced ventricular arrhythmias that results from activation of $iPLA_2$ and the resultant increase in the release of free fatty acids and lysophospholipids. Indeed, the development of a cardiac myocyte-specific $iPLA_2\beta$ transgenic mouse resulted in ischemia-induced tachyarrhythmias, similar to those observed in human ischemic heart [68].

The use of pharmacologic inhibitors, siRNA techniques, and transgenic or knockout animals to manipulate the activity of PLA_2 and subsequent phospholipid metabolite production continues to be an area of active research. While the possibility of a single drug target seems remote at this point, the redundancy of different PLA_2 classes and isoforms within various species and even in different organ systems remains to be elucidated. However, preliminary studies in animal models and in vitro systems can still provide extremely useful insights into the therapeutic potential of PLA_2 modulation.

1.7 MEASUREMENT OF PHOSPHOLIPASE A$_2$ ACTIVITY

Since all PLA_2 isoforms essentially catalyze the same enzymatic reaction, the attribution of activity to an individual isoform and the development of specific PLA_2 inhibitors have remained difficult. These parameters are complicated by the presence of multiple PLA_2 isoforms present in cells and tissues. As a consequence, the development of assays that can be interpreted as specific for a single PLA_2 isoform has built on the subtle differences between each. Several variables that can be manipulated in an assay system include the calcium concentration, the phospholipid substrate, and the inclusion of selective inhibitors in the assay system.

Various chromogenic substrates have been used to assay phospholipase activity in different cells. For example a continuous, rapid method for measuring PLA_2 activity in brain cytosolic fractions using pyrene-labeled PtdCho as a substrate was proposed by Yarger et al. [90]. This method bypasses the hazards associated with the use of radioactive materials and can be rapid and easily performed in a continuous manner. The substrate, 1-hexadecanoyl-2-(1-pyrenedecanol)-*sn*-glycero-3-phosphocholine (b-py-C10-HPC) (Molecular Probes, Eugene, Oregon), was added to brain cytosol (obtained from decapitated adult female CB-57 mice). The assay was initiated by the addition of 2 mM substrate in ethanol. Fluorescence readings were taken every minute over a 10 min period using a Perkin Elmer L5 fluorimeter, with the excitation wavelength set at 345 nm and the emission wavelength set at 377 nm. Fluorescence intensity was converted to molar concentration using a pyrenedecanoic acid (py-C10) standard curve. To determine the specificity of this assay, BEL was incubated at various concentrations with the cytosolic protein (50 mg) for 10 min at 30°C in a darkened sample chamber. Following the incubation, the reaction was initiated by the addition of 4 mL of 1 mM substrate in ethanol for a final substrate concentration of 2 mM. The product was dried under a continuous stream of N_2 and then dissolved in 100 mL chloroform. Subsequently, 50 μL of the sample in chloroform was spotted on an activated (plate maintained at 110°C for 2 h) silica gel G thin layer chromatography plate (Analtech, Newark, Delaware) and developed in a solvent system composed of petroleum ether:diethyl ether:acetic acid, 110:90:3.8, v:v:v. Free fatty acids, diacylglycerides, and phospholipids were separated, respective bands scraped, and the fluorescence intensities measured in ethanol. Fluorescence intensities were converted to pmol using a standard curve.

Group-specific assays for purified isoforms were previously described by Yang et al., but these would not allow for the presence of multiple PLA_2 isoforms within a cell or tissue type [91]. Subsequently, Lucas and Dennis have designed group-specific assays using selective inhibitors to distinguish PLA_2 activity in biological samples. These rely on the inclusion of a low percentage of inhibitors (see Table 1.1) in the assay system and manipulation of the substrate composition to select for a class of PLA_2 enzymes. For example, the authors utilize the preference of $sPLA_2$ for phosphatidylethanolamine substrates and the preference of $cPLA_2$ for arachidonylated substrates in their assay system. The most specific assay is clearly the inclusion of EDTA in the assay buffer to measure $iPLA_2$ activity. The calcium requirements of $sPLA_2$ and $cPLA_2$ render these isoforms inactive in the presence of a calcium chelator and hence any activity measured can be reasonably attributed to $iPLA_2$. An exception to this general rule is group IV C $cPLA_2$ or $cPLA_2\gamma$. This isoform is a calcium-independent, membrane-associated enzyme that has been classified as a $cPLA_2$ due to its sequence homology to other group members. However, this isoform is unique among $iPLA_2$ members as it is resistant to BEL (Table 1.1), thus inclusion of BEL in the assay system in the presence of EDTA would effectively remove any contribution of calcium-independent PLA_2 activity and thus be able to measure $cPLA_2\gamma$.

In addition to the problem of assigning measured activity to a particular class or group of PLA_2 isoforms, there can be a dramatic difference in measured PLA_2 activity dependent upon the amount and molecular species of substrate and the assay conditions used, including temperature, time, and the amount of protein used as the source of PLA_2 enzyme.

To address this, we previously measured PLA_2 activity in human umbilical artery endothelial cells (HUAEC) using several previously published methods that were designed to selectively measure $iPLA_2$ and/or $cPLA_2$ activities. We found that

- The absolute value for PLA_2 activity could vary considerably, depending on the molecular species of substrate used, concentration of substrate, and incubation time and temperature.
- Regardless of the assay system, membrane-associated PLA_2 activity was consistently greater than cytosolic activity [59].
- Both membrane and cytosolic PLA_2 activities were greater in the presence of EGTA than in the presence of 1 mM Ca^{2+}, suggesting that the majority of endothelial cell PLA_2 activity is Ca^{2+}-independent.
- There was a significant decrease in measured PLA_2 specific activity when comparing activity measured after 5 and 30 min in the same assay system.

- Longer incubation times were associated with a loss in reaction linearity, presumably due to the accumulation of free fatty acid or degradation of the enzyme.
- Maximal reaction velocities are consistently achieved with substrate concentrations greater than $50\,\mu M$, and this may explain why we observed higher PLA_2 activity with $100\,\mu M$ substrate compared to $30\,\mu M$ substrate in our comparison studies.

Finally, we have measured PLA_2 activity in the same assay system with phospholipid substrates that differ at the *sn*-1 fatty acid linkage to the glycerol backbone. Once again, using the endothelial cell protein as our source of enzyme, we observed measurable PLA_2 activity in both the cytosolic and membrane fractions using plasmenylcholine (PlsCho), phosphatidylcholine (PtdCho), and alkylacyl glycerophosphocholine (PakCho), with the highest activity measured in the membrane fraction in the absence of Ca^{2+} and using PlsCho as the substrate. Importantly, in endothelial cells stimulated with thrombin, increased membrane-associated $iPLA_2$ activity was observed with PlsCho substrate only. Further analysis of membrane phospholipid hydrolysis in thrombin-stimulated endothelial cells revealed selective PLA_2-catalyzed hydrolysis of plasmalogen phospholipids, demonstrating a distinct preference for plasmalogen substrates in this cell type [59].

The techniques currently being used in our laboratory include immunoblotting (Protocol A), immunostaining (Protocol B), and activity measurements using radiolabeled phospholipid substrate to detect the presence of PLA_2 in various types of cells (Protocols C and D). Detailed protocols for each are presented below.

Protocol A: Immunoblotting for the presence of $iPLA_2$ isoforms

1. Homogenize cells (or tissue) and suspend in lysis buffer containing HEPES $20\,mM$ (pH 7.6), sucrose $250\,mM$, dithiothreitol $2\,mM$, EDTA $2\,mM$, EGTA $2\,mM$, β-glycerophosphate $10\,mM$, sodium orthovanadate $1\,mM$, phenylmethylsulfonyl fluoride $2\,mM$, leupeptin $20\,\mu g/mL$, aprotinin $10\,\mu g/mL$, and pepstatin A $5\,\mu g/mL$.
2. Sonicate the cell suspension on ice for six bursts of $10\,s$ and centrifuge at $10,000 \times g$ at $4°C$ for $20\,min$ to remove cellular debris and nuclei.
3. Separate the cytosolic and membrane subcellular fractions by centrifuging the supernatant at $100,000 \times g$ for $60\,min$.
4. Resuspend the pellet in lysis buffer and centrifuge at $100,000 \times g$ for $60\,min$ twice to minimize contamination of the membrane fraction with cytosolic protein.
5. Resuspend the final pellet in lysis buffer containing 0.1% Triton X-100.
6. Mix the protein (cytosol or membrane) with an equal volume of SDS sample buffer and heat at $95°C$ for $5\,min$ prior to loading onto a 10% polyacrylamide gel.
7. Separate the protein by SDS/PAGE at $200\,V$ for $35\,min$ and transfer electrophoretically to PVDF membranes (Bio-Rad, Richmond, California) at $100\,V$ for $1\,h$.

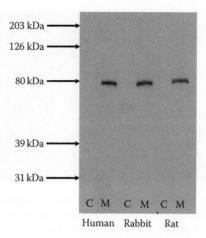

FIGURE 1.1   Immunoblot analysis of iPLA$_2$ β in cytosolic (C) and membrane (M) fractions isolated from human, rabbit, and rat cardiac myocytes. Proteins were separated by SDS-polyacrylamide gel electrophoresis and transferred to polyvinylidene difluoride membranes. Membranes were probed with anti-iPLA$_2$β antibody (1 in 2000 dilution; Cayman Chemical, Ann Arbor, Michigan) and horseradish peroxidase–linked anti-rabbit antibody (1 in 20,000 dilution; Amersham, Arlington Heights, Illinois). Immunoblots were detected with enhanced chemiluminescence and exposure to film for up to 5 min. Specific iPLA$_2$β protein in the membrane fraction migrated at approximately 80 kDa compared with molecular weight markers. No iPLA$_2$β was detectable in the cytosolic fractions at comparable amounts of protein.

8. Block the nonspecific sites with Tris-buffered solution containing 0.05% (v/v) Tween-20 (TBST) and 5% (w/v) nonfat milk for 1 h at room temperature.
9. Incubate the blocked PVDF membrane with primary antibodies to iPLA$_2$β or iPLA$_2$γ followed by horseradish peroxidase–conjugated secondary antibodies.
10. Detect the regions of antibody binding using enhanced chemiluminescence (Amersham, Arlington Heights, Illinois) after exposure to film (Hyperfilm, Amersham) (Figure 1.1).

Protocol B: Immunofluorescent detection and localization of iPLA$_2$

1. Fix the cells in 35 mm tissue culture dishes and fix with 1% paraformaldehyde for 24 h.
2. Wash the cells twice in PBS followed by a wash in permeabilization buffer containing NaCl 127 mM, KCl 5 mM, NaH$_2$PO$_4$ 1.1 mM, KH$_2$PO$_4$ 0.4 mM, MgCl$_2$ 2 mM, Glucose 5.5 mM, and PIPES 20 mM in 50 mL total volume with additional 5 mL 10% Triton × X100, 12 mL 5% DOC, and 33 mL H$_2$O.
3. Incubate the cells twice in fresh permeabilization buffer for 10 min at room temperature.
4. Block the nonspecific binding sites using buffer containing 800 mg BSA and 0.5 mL fish gelatin in 100 mL of PBS for 1 h at room temperature.

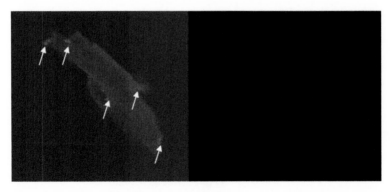

FIGURE 1.2 (See color insert following page 108.) Immunofluorescent localization of iPLA$_2\beta$ in rabbit cardiac myocytes. After appropriate processing (see Protocol B), cells were probed with anti-iPLA$_2\beta$ antibody (1 in 1000 dilution; Cayman Chemical, Ann Arbor, Michigan) and alexa fluor goat anti-chicken 468 secondary antibody (1:1200 dilution, Molecular Probes, Eugene, Oregon). The cells were then mounted in vectashield and examined under a fluorescent microscope. Left panel shows pronounced staining for iPLA$_2\beta$ protein observed at areas of cell–cell junction contact in isolated rabbit cardiac myocytes (white arrows). Negative control with secondary antibody is shown on the right panel.

5. Incubate the cells with primary antibody iPLA$_2\beta$ (1 in 1000 dilution, Cayman Chemical Company, Ann Arbor, Michigan) for 2 h at room temperature followed by three washes with blocking buffer. Incubate the cells with alexa fluor goat anti-chicken 468 secondary antibody (1:1200, Molecular Probes, Eugene, Oregon) for 1 h at room temperature. (It may be necessary to titrate the concentration of primary and secondary antibodies according to the cell type and abundance of protein.)
6. Wash the cells three times in PBS, remove excess liquid, mount cells using vectashield (Vector Chemicals, Burlingame, California), and coverslip and examine under a fluorescent microscope (Figure 1.2).

Protocol C: Measurement of iPLA$_2$ enzyme activity in subcellular fractions using radiolabeled phospholipid substrates

1. Suspend cells or homogenize tissue in ice cold homogenization buffer containing 250 mM sucrose, 10 mM KCl, 10 mM imidazole, 5 mM EDTA, and 2 mM dithiothreitol, pH 7.8.
2. Sonicate suspension on ice for 10 s six times (Fisher Scientific sonic dismembrator, 2.5 power).
3. Prepare subcellular fractions using standard protocols.
4. Determine protein concentration in each fraction.
5. Dilute subcellular factions with homogenization buffer to achieve desired protein concentrations (50 μg cell lysate or 8 μg of membrane protein) in 100 μL volume.

6. Add 100 µL PLA$_2$ assay buffer containing 100 mM Tris, 4 mM EGTA or 1 mM Ca^{2+}, and 10% glycerol, pH 7.0, to each assay tube. Add 100 µL sample to each tube. Make blank tubes for every substrate/buffer condition.

7. Initiate PLA$_2$ reaction by adding 100 µM phospholipid substrate in 5 µL ethanol into each assay tube. Incubate at 37°C for 5 min. We use synthetic (16:0, [^{3}H]18:1) or (16:0, [^{3}H]20:4) PlsCho, PtdCho, or PakCho substrates with a specific activity of 150 dpm/pmol ~68 µCi/µmol (synthesized in the laboratory).

8. Add 100 µL butanol to each reaction tube to stop the reaction. Vortex immediately, add 10 µL arachidonic acid in butanol, vortex again, and centrifuge at 900 × g for 2 min.

9. Spot 25 µL of the butanol layer onto a TLC plate (20 × 20 cm, 250 µM, Analtech, Newark, Delaware), develop in petroleum ether:diethyl ether:acetic acid (70:30:1), and scrape both the fatty acid zone and the origin (unhydrolyzed phospholipid substrate).

10. Count the radioactivity in each zone using a liquid scintillation counter. Use the fatty acid dpm to calculate PLA$_2$ activity and the origin dpm to adjust for 100% recovery.

1.7.1   PAF-AH Assay

PAF-AH catalyzes the hydrolysis of biologically active PAF to inactive lysoPAF. It is classified as a cytosolic iPLA$_2$ isoform that preferentially hydrolyzes phospholipids with short-chain or oxidized fatty acids at the *sn*-2 position. PAF-AH cannot be measured using the PLA$_2$ assay outlined above since it is inactive against long-chain fatty acids at the *sn*-2 position. However, as with the measurement of PLA$_2$ activity, the measurement of PAF should be adjusted with respect to the amount of protein, incubation time, and concentration of substrate to measure optimum activity in the cell or tissue being studied. In our assay system for PAF-AH, we use 2-[acetyl-^{3}H] PAF as a substrate and measure PAF-AH activity in the absence of Ca^{2+}.

Protocol D: Measurement of PAF-AH enzyme activity using a radiolabeled phospholipid substrate

1. Suspend cells or homogenize tissue in ice cold homogenization buffer containing 250 mM sucrose, 10 mM KCl, 10 mM imidazole, 5 mM EDTA, and 2 mM dithiothreitol, pH 7.8.

2. Sonicate suspension six times on ice for 10 s.

3. Determine protein concentration in each sample.

4. Dilute samples with homogenization buffer to achieve desired protein content in 10 µL volume.

5. Mix 10 µL sample aliquots with 40 µL of 0.1 mM [acetyl-^{3}H] PAF (10 mCi/mmol, Perkin Elmer Life Sciences) in polypropylene tubes and incubate for 30 min at 37°C.

6. Stop reaction by adding $50\,\mu L$ acetic acid, followed by 1.5 mL sodium acetate (0.1 M).
7. Pass each reaction mixture through a C_{18} gel cartridge and collect the filtrates directly into scintillation vials. Wash each assay tube with an additional 1.5 mL of sodium acetate, pass through the cartridge, and combine with the original effluent. The radiolabeled PAF substrate will remain bound to the C_{18} column and the amount of radioactive acetate passing through the column represents a measure of the PAF-AH activity in each sample.
8. Measure radioactivity in the column effluent using liquid scintillation counting.

1.8 CONCLUSION

Although the physiologic function of PLA_2 isoforms still largely remains to be elucidated, the ubiquitous nature of these enzymes underscores their importance in various physiologic and pathologic processes. Past research conclusively demonstrates the presence of more than one type of PLA_2 in mammalian cells, each of which can be regulated independently. Further complexity is added by the phenomenon of cross talk between the coexisting PLA_2 isoforms in some cells. The type of stimulus can also determine the isoforms of PLA_2 that can be activated.

Thus, regulation of the intensity and duration of PLA_2 activation is the first critical step in developing specifically targeted pharmacologic inhibitors with a view to controlling their role in multiple disease processes.

LIST OF ABBREVIATIONS

$AACOCF_3$	arachidonyl trifluoromethyl ketone
ARDS	adult respiratory distress syndrome
Asp	aspartate
BEL	bromoenol lactone
$cPLA_2$	cytosolic phospholipase A_2
DSEP	diffusible survival evasion peptide
His	histidine
HUAEC	human umbilical artery endothelial cells
$iPLA_2$	calcium-independent phospholipase A_2
LDL	low-density lipoprotein
MAFP	methyl arachidonyl fluorophosphonate
PAF	platelet-activating factor
PAF-AH	platelet-activating factor acetylhydrolase
PakCho	alkylacyl glycerophosphocholine
PLA_2	phospholipase A_2
PlsCho	plasmenylcholine
PtdCho	phosphatidylcholine
$sPLA_2$	secretory phospholipase A_2

REFERENCES

1. Meyer, M.C. and McHowat, J. (2007) Calcium-independent phospholipase A$_2$-catalyzed plasmalogen hydrolysis in hypoxic human coronary artery endothelial cells. *Am. J. Physiol. Cell. Physiol.* 292:C251–C258.
2. Rastogi, P. and McHowat, J. (2006) Eicosanoids as therapeutic targets in asthma. *Lett. Drug Des. Discov.* 3:668–674.
3. Vinson, S.M. et al. (2005) Neutrophil adherence to bladder microvascular endothelial cells following platelet-activating factor acetylhydrolase inhibition. *J. Pharmacol. Exp. Ther.* 314:1241–1247.
4. Schaloske, R.H. and Dennis, E.A. (2006) The phospholipase A$_2$ superfamily and its group numbering system. *Biochim. Biophys. Acta* 1761:1246–1259.
5. Creer, M.H. and McHowat, J. (2000) Biologic diversity of mammalian phospholipase A$_2$. In: *Recent Research Developments in Lipid Research* (Pandalai, S., Ed.), 4th edn., pp. 13–24, Transworld Research Network, Trivandrum, India.
6. Singer, A.G. et al. (2002) Interfacial kinetics and binding properties of complete set of human and mouse groups I, II, V, X, and XII secreted phospholipases A$_2$. *J. Biol. Chem.* 277:48535–48549.
7. Murakami, M. and Kudo, I. (2001) Diversity and regulatory functions of mammalian secretory phospholipase A$_2$s. *Adv. Immunol.* 77:163–194.
8. Han, S.K. et al. (1999) Roles of Trp31 in high membrane binding and proinflammatory activity of human group V phospholipase A$_2$. *J. Biol. Chem.* 274:11881–11888.
9. Seilhamer, J.J. et al. (1986) Pancreatic phospholipase A$_2$: Isolation of the human gene and cDNAs from porcine pancreas and human lung. *DNA* 5:519–527.
10. Murakami, M., Nakatani, Y., and Kudo, I. (1996) Type II secretory phospholipase A$_2$ associated with cell surfaces via C-terminal heparin-binding lysine residues augments stimulus-initiated delayed prostaglandin generation. *J. Biol. Chem.* 271:30041–30051.
11. Bezzine, S. et al. (2002) On the binding preference of human groups IIA and X phospholipases A$_2$ for membranes with anionic phospholipids. *J. Biol. Chem.* 277:48523–48534.
12. Valentin, E. et al. (1999) Cloning and recombinant expression of a novel mouse-secreted phospholipase A$_2$. *J. Biol. Chem.* 274:19152–19160.
13. Chen, J. et al. (1994) Cloning and recombinant expression of a novel human low molecular weight Ca^{2+}-dependent phospholipase A$_2$. *J. Biol. Chem.* 269:2365–2368.
14. Valentin, E. et al. (2000) Cloning and recombinant expression of human group IIF-secreted phospholipase A$_2$. *Biochem. Biophys. Res. Commun.* 279:223–228.
15. Valentin, E. et al. (2000) Novel human secreted phospholipase A$_2$ with homology to the group III bee venom enzyme. *J. Biol. Chem.* 275:7492–7496.
16. Hanasaki, K. et al. (1999) Purified group X secretory phospholipase A$_2$ induced prominent release of arachidonic acid from human myeloid leukemia cells. *J. Biol. Chem.* 274:34203–34211.
17. Morioka, Y. et al. (2000) Potential role of group X secretory phospholipase A$_2$ in cyclooxygenase-2-dependent PGE$_2$ formation during colon tumorigenesis. *FEBS Lett.* 487:262–266.
18. Gelb, M.H. et al. (2000) Cloning and recombinant expression of a structurally novel human secreted phospholipase A$_2$. *J. Biol. Chem.* 275:39823–39826.
19. Nevalainen, T.J., Haapamaki, M.M., and Gronroos, J.M. (2000) Roles of secretory phospholipases A$_2$ in inflammatory diseases and trauma. *Biochim. Biophys. Acta* 31:83–90.
20. Balboa, M.A. et al. (2003) Localization of group V phospholipase A$_2$ in caveolin-enriched granules in activated P388D1 macrophage-like cells. *J. Biol. Chem.* 278:48059–48065.
21. Ueno, N. et al. (2001) Coupling between cyclooxygenases, terminal prostanoid synthases and phospholipase A$_2$s. *J. Biol. Chem.* 276:34918–34927.

22. MacPhee, M. et al. (1995) The secretory phospholipase A_2 gene is a candidate for the MOM1 locus, a major modifier of *ApcMin*-induced intestinal neoplasia. *Cell* 81:957–966.

23. Kennedy, B.P. et al. (1998) Overexpression of the nonpancreatic secretory group II PLA_2 messenger RNA and protein in colorectal adenomas from familial adenomatous polyposis patients. *Cancer Res.* 58:500–503.

24. Kini, R.M. (2005) Structure–function relationships and mechanism of anticoagulant phospholipase A_2 enzymes from snake venoms. *Toxicon* 45:1147–1161.

25. Murakami, M., Kudo, I., and Inoue, K. (1993) Molecular nature of phospholipases A_2 involved in prostaglandin I_2 synthesis in human umbilical vein endothelial cells: Possible participation of cytosolic and extracellular type II phospholipases A_2. *J. Biol. Chem.* 268:839–844.

26. Mitsuishi, M. et al. (2006) Group V and X secretory phospholipase A_2 prevents adeno-viral infection in mammalian cells. *Biochem. J.* 393:97–106.

27. Cunningham, T.J. et al. (2006) Secreted phospholipase A_2 activity in experimental auto-immune encephalomyelitis and multiple sclerosis. *J. Neuroinflammation* 3:26.

28. Clark, J.D. et al. (1995) Cytosolic phospholipase A_2. *J. Lipid Mediat. Cell Signal.* 12:83–117.

29. Chiba, H. et al. (2004) Cloning of a gene for a novel epithelium-specific cytosolic phos-pholipase A_2, $cPLA_2\Delta$, induced in psoriatic skin. *J. Biol. Chem.* 279:12890–12897.

30. Ohto, T. et al. (2005) Identification of novel cytosolic phospholipase A_2s, murine $cPLA_2$ Δ, ε, and ζ, which form a gene cluster with $cPLA_2$ β. *J. Biol. Chem.* 280:24576–24583.

31. Hanel, A.M., Schuttel, S., and Gelb, M.H. (1993) Processive interfacial catalysis by mammalian 85-kilodalton phospholipase A_2 enzymes on product-containing vesicles: Application to the determination of substrate preferences. *Biochemistry* 32:5949–5958.

32. Reynolds, L.J. et al. (1993) Metal ion and salt effects on the phospholipase A_2, lysophos-pholipase, and transacylase activities of human cytosolic phospholipase A_2. *Biochim. Biophys. Acta* 1167:272–280.

33. Perisic, O. et al. (1998) Crystal structure of a calcium-phospholipid binding domain from cytosolic phospholipase A_2. *J. Biol. Chem.* 273:1596–1604.

34. Hixon, M.S., Ball, A., and Gelb, M.H. (1998) Calcium-dependent and -independent interfacial binding and catalysis of cytosolic group IV phospholipase A_2. *Biochemistry* 37:8516–8526.

35. Veerkamp, J.H., Peeters, R.A., and Maatman, R.G. (1991) Structural and functional features different types of cytoplasmic fatty acid-binding proteins. *Biochim. Biophys. Acta* 1081:1–24.

36. Evans, J.H. et al. (2001) Intracellular calcium signals regulating cytosolic phospholipase A_2 translocation to internal membranes. *J. Biol. Chem.* 276:30150–30160.

37. Schievella, A.R. et al. (1995) Calcium-mediated translocation of cytosolic phos-pholipase A_2 to the nuclear envelope and endoplasmic reticulum. *J. Biol. Chem.* 270:30749–30754.

38. Underwood, K.W. et al. (1998) A novel calcium-independent phospholipase A_2, $cPLA_2\gamma$, that is prenylated and contains homology to $cPLA_2$. *J. Biol. Chem.* 273:21926–21932.

39. Hirabayashi, T. and Shimizu, T. (2000) Localization and regulation of cytosolic phos-pholipase A_2. *Biochim. Biophys. Acta* 1488:124–138.

40. Tay, A. et al. (1994) Isolation of promoter for cytosolic phospholipase A_2 ($cPLA_2$). *Biochim. Biophys. Acta* 1217:345–347.

41. Wick, M.J. et al. (2005) Lung Kruppel-like factor (LKLF) is a transcriptional activator of the cytosolic phospholipase A_2 alpha promoter. *Biochem. J.* 387:239–246.

42. Cowan, M.J. et al. (2004) The role of TFIID, the initiator element and a novel 5′ TFIID binding site in the transcriptional control of the TATA-less human cytosolic phospholi-pase A_2-alpha promoter. *Biochim. Biophys. Acta* 1680:145–157.

43. Pickard, R.T. et al. (1999) Molecular cloning of two new human paralogs of 85-kDa cytosolic phospholipase A$_2$. *J. Biol. Chem.* 274:8823–8831.
44. Uozumi, N. and Shimizu, T. (2002) Roles for cytosolic phospholipase A$_2$ alpha as revealed by gene-targeted mice. *Prostaglandins Other Lipid Mediat.* 68–69:59–69.
45. Qiu, Z.H. et al. (1998) The role of calcium and phosphorylation of cytosolic phospholipase A$_2$ in regulating arachidonic acid release in macrophages. *J. Biol. Chem.* 273:8203–8211.
46. Oozumi, N. et al. (1997) Role of cytosolic phospholipase A$_2$ in allergic response and parturition. *Nature* 390:618–622.
47. Fujishima, H. et al. (1999) Cytosolic phospholipase A$_2$ is essential for both the immediate and the delayed phases of eicosanoid generation in mouse bone marrow-derived mast cells. *Proc. Natl. Acad. Sci. USA* 96:4803–4807.
48. Nagase, T. et al. (2000) Acute lung injury by sepsis and acid aspiration: A key role for cytosolic phospholipase A$_2$. *Nat. Immunol.* 1:42–46.
49. Takaku, K. et al. (2000) Suppression of intestinal polyposis in Apc Δ716 knockout mice by an additional mutation in the cytosolic phospholipase A$_2$ gene. *J. Biol. Chem.* 275:34013–34016.
50. Sugimoto, Y. et al. (1997) Failure of parturition in mice lacking the prostaglandin F receptor. *Science* 277:681–683.
51. Bonventre, J.V. et al. (1997) Reduced fertility and postischaemic brain injury in mice deficient in cytosolic phospholipase A$_2$. *Nature* 390:622–625.
52. Ma, Z. et al. (1999) Human pancreatic islets express mRNA species encoding two distinct catalytically active isoforms of group VI phospholipase A$_2$ (iPLA$_2$) that arise from an exon-skipping mechanism of alternative splicing of the transcript from the *iPLA$_2$* gene on chromosome 22q13.1. *J. Biol. Chem.* 274:9607–9616.
53. Jenkins, C.M. et al. (2001) Identification of the calmodulin-binding domain of recombinant calcium-independent phospholipase A$_2$ implications for structure and function. *J. Biol. Chem.* 276:7129–7135.
54. Mancuso, D.J., Jenkins, C.M., and Gross, R.W. (2000) The genomic organization, complete mRNA sequence, cloning, and expression of a novel human intracellular membrane-associated calcium-independent phospholipase A$_2$. *J. Biol. Chem.* 275:9937–9945.
55. van Tienhoven, M. et al. (2002) Human neuropathy target esterase catalyzes hydrolysis of membrane lipids. *J. Biol. Chem.* 277:20942–20948.
56. Jenkins, C.M. et al. (2004) Identification, cloning, expression, and purification of three novel human calcium-independent phospholipase A$_2$ family members possessing triacylglycerol lipase and acylglycerol transacylase activities. *J. Biol. Chem.* 279:48968–48975.
57. McHowat, J. and Creer, M.J. (2000) Selective plasmalogen substrate utilization by thrombin-stimulated Ca(2+)-independent PLA$_2$ in cardiomyocytes. *Am. J. Physiol. Heart Circ. Physiol.* 278:H1933–H1940.
58. McHowat, J. and Creer, M.J. (1998) Calcium-independent phospholipase A$_2$ in isolated rabbit ventricular myocytes. *Lipids* 33:1203–1212.
59. McHowat, J. et al. (2001) Endothelial cell PAF synthesis following thrombin stimulation utilizes Ca(2+)-independent phospholipase A$_2$. *Biochemistry* 40:14921–14931.
60. Barbour, S.E. and Dennis, E.A. (1993) Antisense inhibition of group II phospholipase A$_2$ expression blocks the production of prostaglandin E$_2$ by P388D1 cells. *J. Biol. Chem.* 268:21875–21882.
61. Atsumi, G. et al. (2000) Distinct roles of two intracellular phospholipase A$_2$s in fatty acid release in the cell death pathway: Proteolytic fragment of type IVA cytosolic phospholipase A$_2$ inhibits stimulus-induced arachidonate release, whereas that of group VI Ca^{2+}-independent phospholipase A$_2$ augments spontaneous fatty acid release. *J. Biol. Chem.* 275:18248–18258.

62. Zhang, L., Peterson, B.L., and Cummings, B.S. (2005) The effect of inhibition of Ca^{2+}-independent phospholipase A_2 on chemotherapeutic-induced death and phospholipid profiles in renal cells. *Biochem. Pharmacol.* 70:1697–1706.

63. McHowat, J. et al. (2001) Clinical concentrations of doxorubicin inhibit activity of myocardial membrane-associated, calcium-independent phospholipase A_2. *Cancer Res.* 61:4024–4029.

64. Cummings, B.S., McHowat, J., and Schnellmann, R.G. (2004) Role of an endoplasmic reticulum Ca^{2+}-independent phospholipase A_2 in cisplatin-induced renal cell apoptosis. *J. Pharmacol. Exp. Ther.* 308:921–928.

65. Seegers, H.C., Gross, R.W., and Boyle, W.A. (2002) Calcium-independent phospholipase A_2-derived arachidonic acid is essential for endothelium-dependent relaxation by acetylcholine. *J. Pharmacol. Exp. Ther.* 302:918–923.

66. Balboa, M.A. and Balsinde, J. (2002) Involvement of calcium-independent phospholipase A_2 in hydrogen peroxide-induced accumulation of free fatty acids in human U937 cells. *J. Biol. Chem.* 277:40384–40389.

67. Ramanadham, S. et al. (2004) Apoptosis of insulin-secreting cells induced by endoplasmic reticulum stress is amplified by overexpression of group VI A calcium-independent phospholipase A_2 ($iPLA_2\beta$) and suppressed by inhibition of $iPLA_2\beta$. *Biochemistry* 43:918–930.

68. Roshak, A.K. et al. (2000) Human calcium-independent phospholipase A_2 mediates lymphocyte proliferation. *J. Biol. Chem.* 275:35692–35698.

69. Mancuso, D.J. et al. (2003) Cardiac ischemia activates calcium-independent phospholipase A_2 beta, precipitating ventricular tachyarrhythmias in transgenic mice: Rescue of the lethal electrophysiologic phenotype by mechanism-based inhibition. *J. Biol. Chem.* 278:22231–22236.

70. Tjoelker, L.W. et al. (1995) Anti-inflammatory properties of a platelet-activating factor acetyl hydrolase. *Nature* 374:549–553.

71. Min, J.H. et al. (2001) Platelet-activating factor acetylhydrolases: Broad substrate specificity and lipoprotein binding does not modulate the catalytic properties of the plasma enzyme. *Biochemistry* 40:4539–4549.

72. Nakajima, K. et al. (1997) Activated mast cells release extracellular type platelet activating factor acetylhydrolase that contributes to autocrine inactivation of platelet-activating factor. *J. Biol. Chem.* 272:19708–19713.

73. Matsuzawa, A. et al. (1997) Protection against oxidative stress-induced cell death by intracellular platelet-activating factor acetylhydrolase II. *J. Biol. Chem.* 272:32315–32320.

74. Karasawa, K., Qiu, X., and Lee, T. (1999) Purification and characterization from rat kidney membranes of a novel platelet-activating factor (PAF)-dependent transacetylase that catalyzes the hydrolysis of PAF, formation of PAF analogs, and C2 ceramide. *J. Biol. Chem.* 274:8655–8861.

75. Hattori, M. et al. (1995) Cloning and expression of a cDNA encoding the beta subunit (30-kDa subunit) of bovine brain platelet-activating factor acetylhydrolase. *J. Biol. Chem.* 270:31345–31352.

76. Adachi, H. et al. (1995) cDNA cloning of human cytosolic platelet-activating factor acetylhydrolase gamma subunit and its mRNA expression in human tissues. *Biochem. Biophys. Res. Commun.* 214:180–187.

77. Schevitz, R.W. et al. (1995) Structure-based design of the first potent and selective inhibitor of human nonpancreatic secretory phospholipase A_2. *Nat. Struct. Biol.* 6:458–465.

78. Rosenthal, M.D. and Franson, R.C. (1989) Oligomers of prostaglandin B1 inhibit arachidonic acid mobilization in human neutrophils and endothelial cells. *Biochim. Biophys. Acta* 1006:278–286.

79. Rastogi, P., Beckett, C.S., and McHowat, J. (2007) Prostaglandin production in human coronary artery endothelial cells is modulated differentially by selective phospholipase A_2 inhibitors. *Prostaglandins Leukot. Essent. Fatty Acids* 76:205–212.

80. Cunningham, T.J. et al. (2004) Systemic treatment of cerebral cortex lesions in rats with a new secreted phospholipase A_2 inhibitor. *J. Neurotrauma* 21:1683–1691.

81. Conde-Frieboes, K. et al. (1996) Activated ketones as inhibitors of intracellular Ca^{2+}-dependent and Ca^{2+}-independent phospholipase A_2. *J. Am. Chem. Soc.* 118:5519–5525.

82. Reynolds, Y.C., Balsinde, J., and Dennis, E.A. (1996) Irreversible inhibition of Ca^{2+}-independent phospholipase A_2 by methyl arachidonyl fluorophosphonate. *Biochim. Biophys. Acta* 1302:55–60.

83. Kell, P.J. et al. (2003) Inhibition of platelet-activating factor (PAF) acetylhydrolase by methyl arachidonyl fluorophosphonate potentiates PAF synthesis in thrombin-stimulated human coronary artery endothelial cells. *J. Pharmacol. Exp. Ther.* 307:1163–1167.

84. Ghomashchi, F. et al. (2001) A pyrrolidine based specific inhibitor of cytosolic phospholipase A_2 alpha blocks arachidonic acid release in a variety of mammalian cells. *Biochim. Biophys. Acta* 1523:160–166.

85. Hazen, S.L. et al. (1991) Suicide inhibition of canine myocardial cytosolic calcium-independent phospholipase A_2: Mechanism-based discrimination between calcium-dependent and -independent phospholipases A_2. *J. Biol. Chem.* 266:7227–7232.

86. Jenkins, C.M. et al. (2002) Identification of calcium-independent phospholipase A_2 ($iPLA_2b$), and not $iPLA_2\gamma$, as the mediator of arginine vasopressin-induced arachidonic acid release in A-10 smooth muscle cells: Enantioselective mechanism-based discrimination of mammalian $iPLA_2$s. *J. Biol. Chem.* 277:32807–32814.

87. Balsinde, J. and Dennis, E.A. (1996) Bromoenol lactone inhibits magnesium dependent phosphatidate phosphohydrolase and blocks triacylglycerol biosynthesis in mouse P388D1 macrophages. *J. Biol. Chem.* 271:31937–31941.

88. Wu, T. et al. (1997) Antisense inhibition of 85-kDa $cPLA_2$ blocks arachidonic acid release from airway epithelial cells. *Am. J. Physiol. Lung Cell. Mol. Physiol.* 273:L331–L338.

89. Su, X. et al. (2004) Small interfering RNA knockdown of calcium-independent phospholipases A_2 β or γ inhibits the hormone-induced differentiation of 3T3-L1 preadipocytes. *J. Biol. Chem.* 279:21740–21748.

90. Yarger, D.E. et al. (2000) A continuous fluorometric assay for phospholipase A_2 activity in brain cytosol. *J. Neurosci. Meth.* 100:127–133.

91. Yang, H.C. et al. (1999) Group-specific assays that distinguish between the four major types of mammalian phospholipase A_2. *Anal. Biochem.* 269:278–288.

2 Methods of Measuring the Activity and Expression of Phospholipases C

Ryan A. Frieler, John A. Watt, and Thad A. Rosenberger

CONTENTS

2.1 INTRODUCTION

Phospholipases C (PLC) impart a fundamental role in the metabolism of membrane phospholipids, and their activity produces lipid intermediates that are involved in cellular signal transduction [1,2]. Activation of this enzyme class is involved in the regulation of many cellular biochemical processes including calcium release, cell cycle control, and gene expression [2,3]. Further, the activation and/or disruption of PLC-mediated signaling occur in tumor malignancy, Alzheimer's disease, the progression of acute myeloid leukemia, as well as many other diseases [4–6]. The numerous roles that PLC have in regulating and maintaining cellular function makes measuring the changes in their expression and activity very important. Further, knowing how these signaling mechanisms are altered during disease processes may aid in understanding disease pathology and help to identify pharmacologic targets of interest. This chapter outlines the different methods for analyzing the PLC found in mammalian tissues with an emphasis on determining their activity and expression.

2.1.1 THE PLC FAMILY

The structural components of the different PLC isoforms are important not only in understanding how the PLC are regulated but also when deciding on the techniques needed to analyze their activity and expression. Based on the substrate specificity demonstrated by the different PLC isoforms, the family is divided into two classes: phosphatidylcholine (PtdCho-) and phosphoinositide (PtdIns-) specific PLC. To date there are a total of 13 isoforms of PtdIns-specific PLC that constitute six different families. When activated by intracellular and/or extracellular signals, PtdIns-specific PLC hydrolyzes phosphatidylinositol-4,5-bisphosphate (PtdIns(4,5)P_2), generating two biologically active intracellular signaling molecules: diacylglycerol and inositol-1,4,5-trisphosphate (Ins(1,4,5)P_3) (Figure 2.1). Similarly, PtdCho-specific PLC hydrolyzes PtdCho, generating diacylglycerol and phosphocholine. The diacylglycerol produced from both reactions can activate protein kinase C (PKC), while the Ins(1,4,5)P_3 can act as a calcium-mobilizing second messenger, inducing the rapid release of intracellular calcium. PtdIns-specific PLC are well studied, and the different isoforms have been identified and cloned. Thorough reviews describing the PtdIns-specific PLC are available [2,7,8].

PtdCho-specific PLC have also been identified and are thought to have similar roles in cellular signaling, although the characterization of this PLC class is not as well defined. In this regard, little structural or genetic characterization is available for those known PtdCho-specific PLC isoforms. For these reasons, the remainder of the discussion and the methods outlined in this chapter will focus primarily on the analysis of PtdIns-specific PLC. Those topics that make explicit reference to the PtdCho-specific PLC will be pointed out when necessary.

2.1.2 THE PTDINS-SPECIFIC PLC FAMILY

Currently, six families of mammalian PtdIns-specific PLC have been identified. These families are characterized based on amino acid sequence similarities and

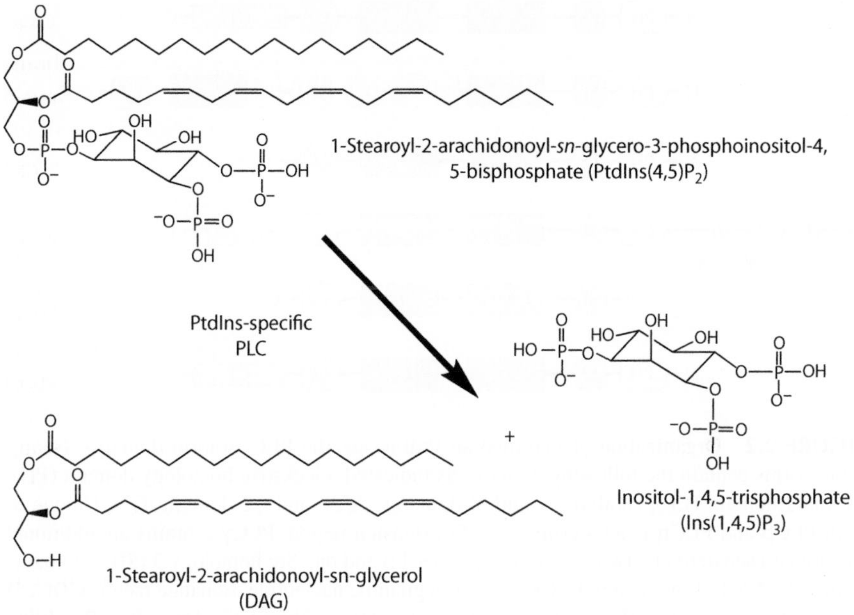

FIGURE 2.1 Generalized schematic of the hydrolysis of phosphatidylinositol-4,5-bisphosphate (PtdIns(4,5)P$_2$), a naturally occurring substrate, by phosphatidylinositol-specific phospholipase C (PtdIns-specific PLC) showing the subsequent hydrolysis products: 1-stearoyl-2-arachidonoyl-*sn*-glycerol (DAG) and inositol-1,4,5-trisphosphate (Ins(1,4,5)P$_3$).

their structural organization (Figure 2.2). Of these families, a total of 13 different isoforms are known: β1–4, γ1–2, δ1, 3, 4, ε1, ζ1, η1–2. It is important to note that a bovine phospholipase C δ2 (PLCδ2) was found to be a homolog of PLCδ4. While reference to this isoform can be found in the literature, there are currently only three accepted delta isoforms [9]. All of the PtdIns-specific PLC isoforms have been purified, their cDNA sequences cloned, and the isoforms have a molecular mass in the range of 70–260 kDa (Table 2.1).

Additional variations in the PtdIns-specific PLC isoforms exist as a result of post-transcriptional splicing mechanisms. For example, two different subtypes of PLCβ1 occur as a result of alternative splicing at the C-terminal region that have different molecular masses (140 and 150 kDa) [10]. Similarly, splice variants for PLCβ1, PLCβ2, PLCβ4, PLCε, PLCδ4, PLCη1, and PLCη2 are identified [10–13] and others likely exist. Proteins that contain domains similar to the PtdIns-specific PLC isoforms but lack PLC-like catalytic activity are also known. These proteins do not have the critical amino residues within the catalytic domain, which render them inactive. These PLC-like isoforms have been named PLC-related catalytically inactive proteins (PRIP), PRIP-1 and PRIP-2. These PLC-related proteins are also referred to as phospholipase C-like (PLC-L) proteins, PLC-L1 and PLC-L2 [14,15]. Because there are so many PtdIns-specific PLC isoforms, it is important to understand that more information is needed to determine which isoforms are responsible for hydrolytic activity following stimulation.

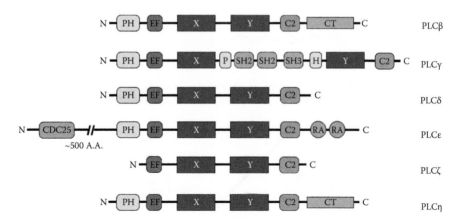

FIGURE 2.2 Organization of mammalian PtdIns-specific PLC isoform domains. Nearly all isoforms contain the following domains as indicated: pleckstrin homology domain (PH), EF-hand domain (EF), catalytic X and Y domains (XY), and C2 domain (C2). Unique to both PLCβ and PLCη is a C-terminus (CT) extension region. PLCγ contains an additional PH domain separated by two Src homology 2 (SH2) and one Src homology 3 (SH3) domains within the XY-linking region. PLCε contains a guanine nucleotide exchange factor (CDC25) and two Ras-association (RA) domains. (Reproduced from Rhee, S.G., *Annu. Rev. Biochem.*, 70, 281, 2001. With permission.)

TABLE 2.1
Classification and Tissue Distribution of PtdIns-Specific-PLC Isoforms

PLC Isoform	Tissue Distribution	Molecular Weight (kDa)	References
PLCβ1	Brain, lung, adrenal gland, kidney	97–152	[10,59]
PLCβ2	Hematopoietic cells		[60]
PLCβ3	Brain, liver, parotid gland, kidney		[32,59]
PLCβ4	Brain, retina, kidney		[59,61–63]
PLCγ1	Brain, liver, skeletal muscle, heart, lung, thymus, kidney, testes	155	[59,64]
PLCγ2	Lung, thymus, spleen, kidney, hematopoietic cells		[59,64,65]
PLCδ1	Brain, heart, lung, testes, kidney	85–90	[59,66]
PLCδ3	Brain, heart, skeletal muscle		[67]
PLCδ4	Brain, skeletal muscle, testes		[68]
PLCε1	Lung, liver, heart, skeletal muscle, small intestine	230–260	[29]
PLCζ1	Sperm	70	[69]
PLCη1	Brain, spinal cord	115–155	[12]
PLCη2	Eye, brain, lung		[11,70]

2.1.3   CHARACTERISTIC FUNCTIONAL UNIT STRUCTURES FOUND IN THE PTDINS-SPECIFIC PLC FAMILY

The functional PtdIns-specific PLC isoforms are modular enzymes that usually contain four distinct domains: pleckstrin homology (PH), EF-Hand, catalytic X and Y, and C2 (Figure 2.2). The PH, EF-Hand, and C2 domains are regulatory in function, whereas the X and Y domain is the catalytic site, where the hydrolysis of the phosphomonoester moiety on the PtdIns(4,5)P$_2$ molecule occurs. These domains impart the characteristic binding and catalytic activity that define this class of enzyme. There is one exception, PLCζ, which lacks a PH domain. Other splice variants exist that can lack entire domain regions. The PLCδ1 was the first PLC isoform crystal structure determined and has revealed many insights concerning the structure–function relationship of this class of phospholipases. These insights include substrate specificity, mechanism of catalysis, and membrane-binding dynamics.

The catalytic X and Y domains found in these enzymes demonstrate high amino acid sequence homology (40%–60%) between the different PtdIns-specific PLC isoforms. The X and Y domains form a modified triose phosphate isomerase (TIM) α/β barrel, which is the active site of PtdIns(4,5)P$_2$ hydrolysis. Within this region, the hydrophobic residues, Trp-555, Leu-320, and Phe-360, surround the active site and form a hydrophobic rim, which allows the enzyme to penetrate the membrane bilayer and bind the inositol phospholipid [16]. Structural studies show that there are several conserved amino acid residues within the catalytic domain that are responsible for substrate preference. The amino acid residues, Lys-438, Arg-549, and Ser-522, interact with the 4-phosphate, and the Lys-440 interacts with the 5-phosphate stabilizing the phosphoinositide substrate [16]. The requirement for phosphoinositides is definite, with the preference being PtdIns(4,5)P$_2$ > PtdIns(4)P > PtdIns due to the tertiary structure of the substrate. Further, the amino acid residues, Glu-341 and Arg-549, prevent phosphoinositides containing a phosphate on the three position of the inositol ring from binding due to steric interference [17]. The active site also contains a calcium-binding region, making calcium a requirement for catalytic activity. The calcium ion interacts with several negatively charged residues within the active site (Glu-390, Glu-341, and Asp-343) as well as with Asn-312 and the 2-hydroxyl group on the inositol ring [16]. Calcium availability is an important requirement for catalytic activity and will be discussed later with regard to PLC assay design.

Hydrolysis of the phosphoinositide substrate is proposed to occur in a double-displacement mechanism. First, a nucleophilic attack by the 2-hydroxyl group of the substrate on the phosphorus bond forms a cyclic inositol phosphate [17]. The second phosphohydrolase step is mediated by the hydrolysis of water, where the imidazole side chain on His-356 deprotonates a water molecule allowing it to make a nucleophilic attack on the cyclic inositol phosphate. Calcium likely has a stabilizing effect at this step as well. Further, the region between the X and Y catalytic domains can range from ~40 to 400 amino acid residues depending on the specific PtdIns-PLC isoform [2]. Although the function of the XY-linking region is not well defined, it likely has an important regulatory role. For example, the XY-linking region of PLCδ1 inhibits catalytic activity when it is bound to calmodulin [18]. Furthermore, a cluster of basic residues in the XY-linking region helps anchor PLCζ to membranes [19].

Nearly all PtdIns-specific PLC contain a pleckstrin homology (PH) domain at the N-terminus with the exception of PLCζ. PLCγ also contains an additional PH domain divided by two Src homology 2 (SH2) and one Src homology 3 (SH3) domains within the XY-linking region [2]. The functional role of the PH domain is to allow PtdIns-specific PLC to associate with the membrane by facilitating binding of the enzyme to phospholipids [20]. The PH domains of the different PtdIns-specific PLC families preferentially bind to specific molecules that can influence the association of the enzyme with the membrane. The amino acid residues of the PH domains between different isoforms of PtdIns-specific PLC are nonconserved, perhaps contributing to the enzymes' binding specificity. The PH domain of PLCγ isoform binds $PtdIns(3,4,5)P_3$, whereas the PH domain of PLCδ isoform binds both $Ins(1,4,5)P_3$ and $PtdIns(4,5)P_2$ [13,21,22]. Some PtdIns-specific PLC PH domains can also activate enzymatic activity. Several isoforms of PLCβ are activated when the PH domain associates with various types of guanine nucleotide binding proteins (G-proteins) [23]. This is an important consideration for PLC assay design because it is ideal to mimic natural membrane binding properties. Therefore, when designing PLC assays, phosphoinositide micelles are used to mimic this environment and to promote enzyme association with the substrate.

The C2 domains bind calcium and interact with phospholipid membranes [24]. Although most C2 domains bind calcium and require it for membrane interactions, it is not always necessary for catalytic activity [25]. The domain consists of ~120 residues with a β-sandwich made up of eight antiparallel β-strands. Structural studies show that the C2 domain contains a calcium-binding region which has three loops (CBR1, 643–653; CBR2, 675–680; CBR3, 706–714) [17]. Electrostatic interactions facilitate protein–membrane interactions and allow the enzyme to associate with the membrane. According to the tether and fix model [24], the PH domain tethers the enzyme to the membrane and the C2 domain fixes it in the proper orientation. The C2 domain of PLCδ1 has a preference to bind phosphatidylserine, and experiments show that it forms a calcium–enzyme–phosphatidylserine ternary complex [17,26]. In contrast, the C2 domain of PLCβ does not interact with the lipid membrane; rather, it acts as an effector of the $G_q\alpha$ subunit [23]. The necessity for calcium binding to C2 domains to illicit an association with the membrane demonstrates the requirement of calcium in PLC assay buffers.

The function of EF-hand domains in PtdIns-specific PLC is not clear but appears to have an essential role since the deletion of this domain in PLCδ1 inactivates the enzyme [27]. The EF-hand domain is a helix-turn-helix motif that contains a calcium-binding region, although many isoforms do not bind calcium at this domain. Evidence shows that calcium binding to the C2 domain enhances the binding of the PH domain to the membrane [28]. The crystal structure of PLCδ1 suggests that its EF-hand domain does not bind calcium; rather, it acts as a flexible region connecting the PH domain to the catalytic domain [24]. Regardless, the EF-hand domain is required for catalytic activity, although the principal role appears to be enhancing the flexibility of the enzyme.

Several of the PtdIns-specific PLC isoforms contain additional regulatory domains. For example, PLCγ contains two SH2 domains and one SH3 domain, and PLCε contains two Ras-association (RA) domains and a CDC25 RasGEF domain [29]. These domains have a role in the regulation and activation of these PLC isoforms. PLCβ and PLCη contain an extension on the carboxy-terminal end which also has a role in activation [11,30].

2.1.4 ACTIVATION AND REGULATION OF PTDINS-SPECIFIC PLC

A significant number of extracellular signaling molecules such as hormones, neurotransmitters, and growth factors elicit cellular responses through the activation of PtdIns-specific PLC [31]. These molecules bind to transmembrane receptors that directly or indirectly activate PLC. There are various modes of activation for the different isoforms of PtdIns-specific PLC. The two most common and well understood modes of activation are by G-proteins and receptor protein tyrosine kinases (RPTK). Of the G-protein super family, heterotrimeric G-protein subunits, $G_q\alpha$ and $G\beta\gamma$, and small GTPases, Ras, Rac, and Rho, have all been identified as activators of PtdIns-specific PLC. The PLCβ, PLCε, and PLCη isoforms are activated by G-proteins, whereas the PLCγ isoforms are predominantly activated by RPTK.

PLCβ isoforms are widely known to be activated by $G_q\alpha$ and $G\beta\gamma$ subunits of G-protein-coupled receptors in response to signaling molecules such as acetylcholine, vasopressin, histamine, TSH, and many others [31]. Four $G_q\alpha$ subunits (αq, $\alpha 11$, $\alpha 14$, $\alpha 16$) activate PLCβ isoforms through interactions with basic residues in the carboxy-terminal extension region as well as the C2 domain [13,23,30]. In contrast, the $G_q\alpha$ subunits do not activate PLCγ, PLCδ, or PLCε.

PLCβ1, β2, and β3, but not β4, isoforms are activated by G-protein $G\beta\gamma$ subunits [32]. The $G\beta\gamma$ subunits activate PLCβ isoforms by interactions with the carboxy-terminal region and by interactions with the PH domain of several PtdIns-specific PLC [23,32]. Some isoforms require both $G_q\alpha$ and $G\beta\gamma$ in order for activation to occur [33]. The PLCβ isoforms are also regulated by Rac, a small GTPase that modulates enzyme activity by interacting with the PH domain [8].

Activation of PLCγ occurs through RPTK. When RPTK are bound by polypeptide signaling molecules such as epidermal growth factor (EGF), platelet-derived growth factor (PDGF), fibroblast growth factor (FGF), and nerve growth factor (NGF) [31], autophosphorylation of specific tyrosine residues occurs. The phosphorylated tyrosine residues then bind the SH2 domain of PLCγ1 causing phosphorylation at Tyr-771, Tyr-783, and Tyr-1254 and activation of enzyme activity [34]. Other modes of activation for PLCγ are by non-receptor protein tyrosine kinases as well as upstream G-proteins.

Activation of PLCδ is not completely understood although it is apparent that calcium has a role. Calcium binds to the catalytic X and Y, EF, and C2 domains and is required for PtdIns(4,5)P_2 binding. Studies show that increased intracellular calcium is adequate to activate PLCδ [35]. Further, PLCβ2 activation results in the inhibition of PLCδ1. PLCε is activated by heterotrimeric G-proteins although most data reveal that small GTPases play a much larger role. PLCε contains a CDC25 RasGEF domain at the N-terminal and two RA domains at the C-terminal of PLCε that are necessary for association with Ras proteins.

The mechanisms that trigger the activation of PLCζ remain unknown. However, PLCζ is active during fertilization and is a sperm factor that activates calcium oscillations necessary for embryo development. PLCη is the most recent PtdIns-specific

PLC isoform identified. Little evidence is available concerning the activation and regulation of this isoform, although PLCη2 is activated by Gβγ subunits from heterotrimeric G-protein-coupled receptors [11,36]. There is a high degree of complex regulatory mechanisms and modes of activation for the different PtdIns-PLC that remain to be determined.

Certain PtdIns-specific PLC isoforms are also involved in nuclear signaling. Nuclear PtdIns-specific PLC isoforms belong to the subfamilies β, γ, δ, and ζ, although PLCβ1 appears to be the predominant isoform [37–39]. PLCβ1 exists as two spliced variants, PLCβ1a and PLCβ1b, both of which are found in the nucleoplasm. The PLCβ1b isoform appears to be restricted to the nucleus [38], while PLCγ and PLCδ isoforms are translocated from the cytoplasm. Various functions of nuclear PtdIns-PLC are known. PLCβ1 is involved in controlling cell proliferation and differentiation by regulating the checkpoints that control progression through the G1 phase of the cell cycle [3]. When nuclear PtdIns-specific PLC is activated, its byproduct, diacylglycerol, causes PKC to translocate to the nucleus and initiate other signaling events. The PLCγ and PLCδ isoforms, although not well understood, are also identified as having a role in cell cycle regulation. Additionally, the importance of nuclear PtdIns-specific PLC in cell cycle control is evident based on studies which demonstrate that the absence of PLCβ1 is involved in the progression of myelodysplastic syndrome to acute myeloid leukemia [6].

2.2 EXPRESSION, DISTRIBUTION, AND LOCALIZATION OF PTDINS-SPECIFIC PLC

The different PtdIns-specific PLC isoforms have been detected in numerous tissues and cell types (Table 2.1) through the use of immunochemical and biochemical techniques. The mRNA and protein expression, distribution, and localization of different PLC isoforms vary and often reflect the functional role of the enzyme. For example, PLCζ is found in sperm and has an important role in fertilization and reproduction. Pathological changes can alter the expression, distribution, and localization of PLC, making a thorough understanding of the techniques used to analyze these changes necessary. Northern and Western blot and reverse transcription polymerase chain reaction (RT-PCR) are used to identify general tissue distribution and expression whereas *in situ* hybridization and immunohistochemical labeling allow for localization to specific cellular phenotypes and examination of localized expression patterns.

2.2.1 RNA ANALYSIS

Similar to methods of analyzing protein expression, mRNA analysis provides a means to examine temporal and spatial patterns of gene expression. It is important to note that mRNA expression does not always reflect protein expression. Posttranscriptional and posttranslational regulations result in differences between cellular mRNA and protein levels and localization. Several commonly utilized methods to analyze mRNA expression include Northern blot analysis, RT-PCR, and *in situ* hybridization.

Northern blot analysis and RT-PCR quantify mRNA expression while *in situ* hybridization identifies the tissue distribution. RT-PCR is less time consuming and provides greater sensitivity than Northern blot analysis, which makes it an excellent technique to examine PtdIns-specific-PLC expression. *In situ* hybridization provides detailed histological and cellular localization of PtdIns-specific PLC mRNA transcripts, which can be quantified, but with less sensitivity than RT-PCR. These methods of mRNA analysis are typically more sensitive than those used for protein analysis allowing for the detection of low-level gene expression. The use of complementary primers or probes also provides a high degree of specificity. Each of these methods has been extensively performed and numerous methods have been developed and optimized. For this reason, methods for RT-PCR and *in situ* hybridization will be briefly discussed and pertinent relevance to PtdIns-specific-PLC will be emphasized.

2.2.2 RT-PCR

The major steps involved in determining mRNA levels by RT-PCR include purification of RNA from crude homogenates, annealing of primers, reverse transcription, amplification, and quantification. The crucial step in achieving success is the development of proper amplification primers. Many different oligonucleotide primers can be generated for PtdIns-specific PLC isoforms based on known cDNA or mRNA sequences available through GenBank (Table 2.2). Considerations for optimal primer sequence design include length, melting temperature, and nucleotide composition. Most commercially available primer design software takes these parameters into account. It is also necessary to consider gene sequence similarities between the different isoforms of PtdIns-specific PLC. In order to design a primer that is specific for a particular PtdIns-specific PLC isoform, it cannot occur in the highly conserved catalytic X and Y domains. In contrast, to study the expression pattern of the entire PtdIns-specific PLC family, it would be necessary to design a primer sequence within a conserved region. Once the appropriate primers have been selected, oligonucleotide primers are synthesized with a DNA synthesizer or are purchased commercially.

Obtaining a pure RNA sample is essential for RT-PCR, and many methods have been developed to purify RNA from tissue extracts. The most commonly used method to purify RNA employs guanidine isothiocyanate, a strong denaturant that also inhibits RNAse activity. This widely used method using guanidine isothiocyanate was first described by Chirgwin et al. [40]. In this method, tissue is homogenized in 4 M guanidine isothiocyanate containing 0.1 M 2-mercaptoethanol and 0.5% *N*-lauroylsarcosine. RNA can then be purified by cesium chloride centrifugation or precipitated with ethanol. A more recent method by Chomczynski and Sacchi [41] combines guanidine isothiocyanate with a phenol chloroform extraction. Tissue is homogenized in a 4 M guanidine isothiocyanate containing 25 mM sodium citrate, 0.1 M 2-mercaptoethanol, and 0.5% *N*-lauroylsarcosine. The homogenate is extracted with phenol and chloroform at a ratio of (5:1, by vol.). The RNA found in the aqueous extract is then precipitated out of solution using ethanol. Modifications of this method are the basis for many commercial RNA purification reagents and kits such as TRIzole.

TABLE 2.2
RT-PCR Primer Sequences Generated from cDNA Sequences for Rat PtdIns-Specific PLC Isoforms

PLC Isoform	Accession Number	RT-PCR Primer Sequences
PLCβ1	M20636	Forward, 5'-TTTTCGGCAGACCGGAAGCGA-3'
		Reverse, 5'-TGCTGTTGGGCTCGTACTTCT-3'
PLCβ2	AJ011035	Forward, 5'-CAAGTGGGACGATGAAACCT-3'
		Reverse, 5'-GGGCTCAGCTGCATTTTAAG-3'
PLCβ3	DQ120508	Forward, 5'-CAACATGGAGGTGGACACAC-3'
		Reverse, 5'-ACGAAACTGAAGCTGCAGGT-3'
PLCβ4	L15556	Forward, 5'-CTGGAAGAGTGAAGGCAAGG-3'
		Reverse, 5'-GGCCAGTTTCATCCAGTGTT-3'
PLCγ1	J03806	Forward, 5'-GTGGATCGTAACCGAGAGGA-3'
		Reverse, 5'-CTCCTCAATCTCTCGCAAGG-3'
PLCγ2	J05155	Forward, 5'-AGGCTTCTTGGACATCATGG-3'
		Reverse, 5'-GAGCTGAAGACCATCTTGCC-3'
PLCδ1	M20637	Forward, 5'-TTCAGTTGGGTCCTTGTACC-3'
		Reverse, 5'-CAATGGAGAAGCATCGATCC-3'
PLCδ3	XM_221004[a]	Forward, 5'-AGTTTCTTAGCTCTGGACCG-3'
		Reverse, 5'-AAGAAGATGTGCTGCGAAGC-3'
PLCδ4	U16655	Forward, 5'-TCTCATTCTCAGTGGAGAGC-3'
		Reverse, 5'-GAACTCTTCCACCAGATAGC-3'
PLCε1	AF323615	Forward, 5'-ATGACGTCCGAAGAAATGGC-3'
		Reverse, 5'-GGTGTTCTTCTGTGACAGAG-3'
PLCζ1	AY885259	Forward, 5'-CAGCGCTGGAATTACTCACA-3'
		Reverse, 5'-ACTGTCCTCGGGGAGAATTT-3'
PLCη1[b]	BC055005	Forward, 5'-CGGAATTCATGGCAGACCTTGAAGTG-3'
		Reverse, 5'-ACTGGTCGACTCAGATCTGTACCAGACAG-3'
PLCη2[b]	BC052329	Forward, 5'-CAAAAGCCAGAAGCCAAGTC-3'
		Reverse, 5'-ATCTCAGGAAGGGTCCCAGT-3'

Source: Rozen, S. and Skaletsky, H.J., Primer3 on the WWW for general users and for biologist programmers, in Krawetz, S. and Misener, S., (eds.), *Bioinformatics Methods and Protocols: Methods in Molecular Biology*, Humana Press, Totowa, NJ, pp. 365–386, 2000. With permission.

Note: Primer sequences were generated using Primer3 software Version 0.4.0.

[a] Predicted primer sequence.

[b] Primer sequences were generated for mouse PtdIns-specific PLC isoforms from cDNA sequences.

Total RNA is generally adequate for RT-PCR and most other RNA techniques; however, methods to purify mRNA are available if necessary. These methods exploit the affinity of poly(A) tails present on most mRNA molecules for oligodeoxythymidylate (oligo(dT)) molecules. Oligo(dT) cellulose columns are easily prepared and RNA preparations are loaded on to the column. The mRNA containing poly(A) tails first bind to the oligo(dT) molecules and are then eluted off after all the contaminants have been washed off the column. Several types of commercially available mRNA purification kits are available and make use of oligo(dT) molecules. Most importantly, when purifying RNA, it is necessary to use RNAse-free reagents and materials. RNAses will readily degrade RNA, therefore tubes and reagents are commonly treated with diethylpyrocarbonate or other RNAse inhibitors.

Following RT-PCR amplification of RNA, the reaction products are analyzed and quantified. Quantification methods are dependent on the particular type of RT-PCR. Real-time RT-PCR and end-point RT-PCR are commonly used methods to quantify RNA. Real-time RT-PCR provides an automated method to quantify reaction products during each amplification cycle. It allows for the analysis of multiple samples with little effort, but costs are considerably higher than end-point RT-PCR. End-point RT-PCR requires the reaction products to be separated by gel electrophoresis and quantified. Three commonly used types of end-point RT-PCR are relative RT-PCR, competitive RT-PCR, and comparative RT-PCR. Relative RT-PCR includes primers for an internal control in order to allow for sample normalization. The internal control primers and the target gene primers are combined in a multiplex RT-PCR reaction, and the data are expressed as a ratio of target gene product to internal control product. Competitive RT-PCR quantifies gene expression by including a synthetic competitor RNA. The synthetic RNA provides an internal concentration curve and is used to measure the target RNA copy number. In comparative RT-PCR, two RNA samples are combined in one reaction mixture and compete for reaction reagents. The final products are separated and the relative amounts are compared.

2.2.3 IN SITU HYBRIDIZATION

In situ hybridization is an expanding technique that allows for the detection of mRNA expression at the cellular and tissue levels. The process is very similar to immunohistochemical labeling with the exception that it uses a labeled nucleic acid probe to identify mRNA transcripts as opposed to antibodies. As with RT-PCR, constructing the probe is the most important step. Thus, when designing these constructs, one must pay particular attention to the probe length, melting temperature, and nucleotide composition. Similarly, avoiding regions that are conserved between the different PtdIns-specific PLC isoforms is necessary to ensure specificity.

The three main types of probes used for *in situ* hybridization include cDNA, cRNA, or synthetic oligonucleotide probes. Synthetic oligonucleotide probes can range from 20 to 50 bases and are most commonly used. They are constructed using 3'-end labeling techniques and can be easily synthesized using a DNA synthesizer. The cDNA probes are also frequently used and are labeled using nick translation to achieve higher sensitivity. The cRNA probes, on the other hand, range from 50 to 200 bases and are synthesized from cDNA templates. The cRNA probes, however, are more susceptible to degradation. Probes are also designed to contain either radioactive or nonradioactive detection labels. Radioactive probes typically contain ^{35}S, ^{33}P, or ^{3}H radioisotopes and use film or emulsion autoradiography. Nonradioactive probes such as fluorescein, biotin, digoxigenin, or alkaline phosphatase have become increasingly popular, although they are less sensitive than radioactive probes.

In situ hybridization occurs in three steps: pretreatment of tissue, probe hybridization, and posthybridization washes. Fixed tissue can be either frozen and sectioned using a cryostat or embedded in paraffin and sectioned using a microtome. The latter requires a deparaffinization step with xylenes to ensure proper hybridization. Pretreatment of tissue is performed by a series of washes in which the sections are treated with pronase and then acetylated. These steps are carried out in a 1.2 M solution of sodium chloride containing 0.12 M trisodium citrate (SSC) buffer. This step is necessary to increase mRNA accessibility and decrease background. The tissue sections are then dehydrated in increasing concentrations of ethanol (70%, 80%, 90%, and 100%) and then allowed to dry. The labeled probes are reconstituted in hybridization buffer appropriate for the type of probe (i.e., oligonucleotide, cRNA, cDNA). Oligonucleotide probes are most commonly used and have probe concentrations of 0.3–3.0 μg/mL. The specific radioactivity of the probes is typically ~10,000 cpm/ μL. The hybridization buffer should cover the tissue sections while being incubated overnight in a humidity chamber at 37°C–40°C. The hybridization time is optimized to achieve the desired detection sensitivity. Following hybridization, tissue sections are washed four to five times for 30 min in SSC buffer at 37°C, hybridized with the radioactive probes, and the sections then placed on films for autoradiographic exposure. The exposed film is developed, and the probes are quantified using optical density and/or grain count. It is important to include radioactive standards in order to quantify net radioactivity by autoradiography.

2.2.4 IMMUNOHISTOCHEMICAL ANALYSIS

Immunohistochemical labeling is a useful method for determining the distribution and localization of proteins with the caveat that this type of qualitative analysis must be performed and interpreted with extreme caution. Antibody specificity is a major concern when analyzing immunohistochemical data because nonspecific binding can lead to erroneous results. There are many factors that can affect antibody specificity and binding, so proper controls and procedure must be closely adhered to. In this regard, there are several types of controls that can be used: negative (omission) control, positive control, Western blot control, and preabsorption control. Preferably, each control should be performed for each experiment, but availability is often a limiting factor. Negative and positive controls are recommended to demonstrate that the primary antibody is in fact labeling the protein of interest. A tissue that is known to contain the antigen of interest should be used as a positive control. One type of negative control omits the primary antibody and uses blocking serum. There should be no appreciable labeling in this type of negative control, and the presence of labeling would suggest cross reactivity of the secondary antibody. Tissues known to lack PLC as well as transgenic PLC knockouts can also serve as negative controls when used appropriately. Ideally, a negative control would not require any procedural changes.

Western blotting of the tissue homogenate is used as a control to assess antibody specificity and to verify reactivity to the protein of interest. In this control, tissue that is being tested is homogenized and the extract separated by SDS-PAGE followed by immunoblot detection. The Western blot should show antibody specificity to the desired protein and its approximate molecular weight. The presence of additional weak bands may be of concern and must be interpreted accordingly [42]. Preabsorption controls involve incubating the antigen with the antibody with the intention that the antibody will bind the antigen and therefore will not bind the protein in the tissue. This type of control reveals that the antibody binding the antigen is causing the labeling but does not provide any information about the specificity of the antibody. Additionally, multiple antibodies created from different epitopes of the same protein can be used as a specificity control. The antibodies should show reactivity to the same protein by Western blot analysis and must also show similar labeling patterns in tissue.

Finally, there are several procedural factors that affect antibody specificity and binding. Fixative and embedding chemicals can alter the penetration, recognition, and specificity of the antibody. If the tissue is stored in fixative for extended periods of time, the penetration of antibody will be decreased due to the abundance of aldehyde cross-linkages between proteins. Fixatives can also cause alterations in tertiary protein structure, which can prevent antibody binding and affect tissue reactivity and antibody recognition. The proper antibody concentration is also important in achieving proper labeling. Like other immunochemical tests, if the antibody concentration is too high, it can lead to nonspecific binding or steric interference. A dilution series should be performed to determine the appropriate antibody concentration required to achieve optimal labeling. Figure 2.3 summarizes the major procedural steps necessary for immunohistochemical labeling and the assessment of antibody specificity.

Our laboratory uses the method outlined below for immunohistochemical labeling of PtdIns-specific PLC. Here, we perform a cardiac perfusion with heparinized saline followed by periodate–lysine–paraformaldehyde perfusion fixation and vibratome sectioning for immunohistochemical analysis of free-floating brain sections. Alternately, other fixatives and sectioning and labeling techniques may provide adequate results. Figure 2.4 demonstrates the localization of PLCβ1 in rat brain tissue sections by immunohistochemical labeling. Figure 2.5 demonstrates the use of Western blot analysis of rat brain tissue as an antibody specificity control for PLCβ1.

2.2.5 Immunohistochemical Methods

Following rehydration in PBS, all sections used for peroxidase immunocytochemistry must first be pretreated with 0.3% H_2O_2 in phosphate-buffered saline (PBS) (pH 7.4) for 30 min. This is followed by incubation for a minimum of 1 h in a 4% solution of the appropriate normal sera (Vector) in PBS (blocking buffer) to reduce endogenous peroxidase activity and nonspecific staining, respectively. All sections are washed repeatedly in PBS before processing and between incubations. For localization of rabbit anti-PtdIns-PLC-immunoreactivity, sections were incubated sequentially in blocking buffer consisting of PBS with 4% normal goat serum for 1 h at room temperature followed by biotinylated goat anti-rabbit IgG (1:500 in PBS/blocking serum, 1 h; Vector) and avidin–biotin complex in PBS for 1 h (Vector ABC *Elite* kit). Binding of the avidin–biotin reagent is then visualized using a 0.05%

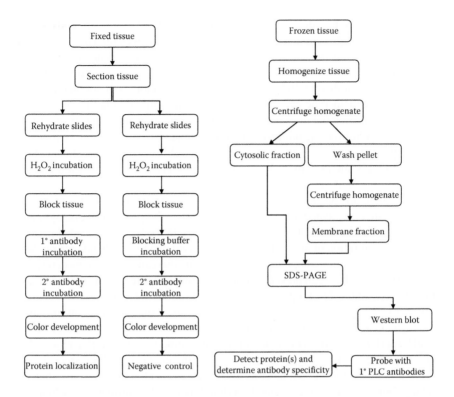

FIGURE 2.3 Flowchart summary of the different steps used for immunohistochemical labeling of PLCβ1 and demonstration of antibody specificity.

solution of diaminobenzidine (Sigma) as chromogen, with the generation of H_2O_2 by the glucose oxidase method [43]. To further assess specificity of the antibody, additional tissue sections are processed as described above with either omission of the primary antibody or using the primary antibody that has been pre-absorbed for 24 h with a 10-fold molar excess of purified antigen when possible. Upon completion of the histochemical preparation, sections are mounted on gelatin-coated microscope slides (3% gelatin solution), dried briefly, and then dehydrated through a series of increasing concentrations of ethanol (70%, 2 × 95%, 2 × 100% for 10 min each), followed by a suitable clearing agent, such as xylene or histoprep (2 × 10 min), and covered with a coverslip fixed in place using Permount (Sigma).

2.3 PHOSPHOLIPASE C ASSAYS

The role of PLC as a signal transduction enzyme gives it the potential to mediate cellular functions in response to aberrant physiological changes. PLC activity can reveal the presence or absence of changes in many regulatory pathways within the cell. Numerous methods to measure PtdIns-specific PLC and PtdCho-specific PLC activity are available and vary mainly in enzyme substrates and methods of

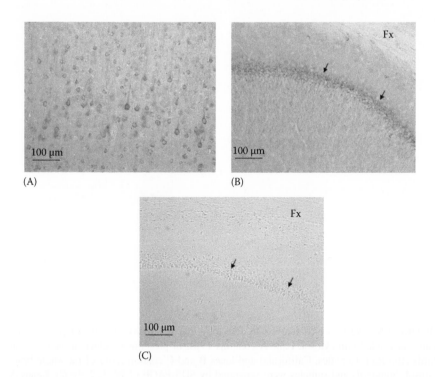

FIGURE 2.4 (See color insert following page 108.) Immunohistochemical localization of PLCβ1 in rat brain. Panel A shows PLCβ1 immunoreactivity in pyramidal neurons located throughout the cortical layers. Panel B shows that neurons within the CA1-3 regions of the hippocampus were also immunoreactive for PLCβ1 (arrows). Scattered neurons were also evident in the hippocampal oriens layer dorsal to CA1-3 regions. Panel C demonstrates that the matching primary omission controls were clear of all immunoreactivity in the CA region (arrows). Abbreviation: Fx, fornix.

detection. Due to enzyme substrate specificities, most assays have been limited to using natural substrates, although, more recently, synthetic analogs have been effectively employed. Detection using radiometric, fluorometric, and chromogenic methods are available.

There are several factors that are important in designing and performing PLC assays. As mentioned above, calcium plays a significant role in the hydrolysis of phosphoinositide substrates and effective concentrations can affect PtdIns-specific PLC activity. Also, PLC activity has a pH dependence, and many isoforms of PtdIns-specific PLC, as well as PtdCho-specific PLC, have different pH optima. The pH affects the ratio of cyclic to acyclic $Ins(1,4,5)P_3$ byproduct release for various isoforms of PtdIns-specific PLC [44]. Finally, detergents are required in order to provide adequate substrate solubilization, and the concentrations can affect PLC activity. Triton X-100, deoxycholate, and cetrimide are detergents that have been used effectively in PLC assays.

Activity can be measured for total PtdIns-specific PLC or for individual PtdIns-specific PLC isoforms that have been purified. Since the specific isoforms

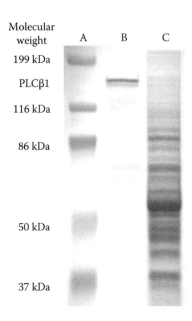

FIGURE 2.5 Western blot analysis to determine the specificity of anti-PLCβ1 antibody using rat whole brain cytosol. Lane A contains 15 μL of prestained molecular weight standards (Bio-Rad, Hercules, California) and lanes B and C contain 25 μg of rat whole brain cytosol. Standards and samples were separated by SDS-PAGE (7.5% T, 2.6% C). Lanes A and B represent standards and protein transferred onto nitrocellulose paper, respectively, and lane C represents brain cytosolic protein stained with Coomassie blue to visualize all cytosolic protein bands. Lanes A and B were probed with anti-PLCβ1 antibody (Santa Cruz Biotechnology, Inc., Santa Cruz, California) at a dilution of 1:1000, followed by visualization using chemiluminescent detection. The single band in Lane B shows the selectivity of the anti-PLCβ1 antibody with regard to total brain cytosolic protein. Parallel samples using rat brain membrane-bound protein extracts were run to rule out nonselective binding of the antibody to membrane-localized proteins.

of PtdIns-specific PLC have been cloned and well characterized, their individual activities have been determined. There are also *in vivo* methods used to measure PtdIns-specific PLC activity in cultured cells. Here, we will focus mainly on methods used to measure total PtdIns-specific PLC activity from primary tissue extracts.

2.3.1 HOMOGENIZATION OF TISSUE FOR PLC ASSAYS

Various methods can be used to prepare samples for PLC assays depending on whether they are from primary tissue or from cultured cells. Here, we describe sample preparation from mammalian tissue for total PLC assays. This method of sample preparation can be used for PtdIns-specific PLC or the PtdCho-specific PLC assay methods. Approximately 1 g of tissue is homogenized in 10 mL of ice-cold homogenizing buffer, 20 mM HEPES (pH 7.4) containing 2 mM EGTA and 5 mM

EDTA and the appropriate protease inhibitors as listed in the reagents. It is important to note that the protease inhibitors should be added to the homogenizing buffer immediately before use, and all steps must be performed at 4°C. The homogenate is centrifuged at $100,000 \times g$ to isolate the cytosolic protein fraction. The cellular pellet is then resuspended in homogenization buffer containing a detergent (Triton X-100, 0.25 mg/mL) and incubated with agitation to dissolve the membrane protein (4°C, 60 min). After centrifugation ($100,000 \times g$), the membrane–protein fraction is collected. The protein concentration is determined using a protein assay, taking into account the presence of Triton X-100 in the buffer, and the cytosolic- and membrane–protein fractions are diluted in homogenizing buffer to a concentration of 0.1–0.5 μg/μL. Protein fractions are kept on ice and assayed immediately or divided into small aliquots and stored at −20°C until use.

2.3.2 RADIOMETRIC PLC ASSAYS

Radiometric assays are the most sensitive and allow for the use of natural substrates; however, these assays are discontinuous and require the separation of reaction products. Radiometric assays are frequently used for determining PtdIns-specific PLC and PtdCho-specific PLC activities. PtdIns-specific PLC assays typically use radiolabeled PtdIns or PtdIns(4,5)P$_2$ substrates, and PtdCho-specific PLC assays use radiolabeled PtdCho. Substrates are labeled at the inositol and choline polar head groups to allow for separation of substrate and reaction byproducts. To measure the specific activity, the radioactivity must be quantified in terms of μmol Ins(1,4,5)P$_3$ or choline byproducts released. Additionally, thin layer chromatography separation allows for quantification of nonhydrolyzed substrate. The following is the method that our laboratory uses for radiometric assay of PtdIns-specific PLC from mammalian tissue homogenates.

2.3.3 ASSAY OF PTDINS-SPECIFIC PLC

Substrate stock (75 μM PtdIns(4,5)P$_2$, 0.05 μCi/mL ^{3}H-PtdIns(4,5)P$_2$) dissolved in chloroform/methanol/water (75:25:1, by vol.) is dried under a steady stream of nitrogen at 45°C. The dry substrate is reconstituted in an equal volume of PtdIns-specific PLC stock solution and then sonicated until the entire dry lipid is in solution (~5 min at 37°C) and micelles of a homogenous size are formed. PtdIns-specific-PLC activity is measured by incubating ~10–20 μg cellular protein with 100 μL of PtdIns-specific PLC assay buffer at 37°C for 10 min. At 10 min, the reaction is stopped by the addition of 0.5 mL of 1 N HCl followed by 0.5 mL of chloroform/methanol (1:1, by vol.). After mixing, the samples are centrifuged to facilitate phase separation and a fraction (400 μL) of the upper aqueous phase is removed for liquid scintillation counting. The concentration of Ins(1,4,5)P$_3$ product formed, typically in units of nmol, can be calculated based on the specific activity of the assay cocktail. The specific activity of the assay cocktail is measured experimentally by determining the amount of lipid phosphorus and the amount of radioactivity in 100 μL of the substrate stock. Lipid phosphorus is measured using a phosphorus assay [45] and the radioactivity is measured by liquid scintillation counting.

2.3.4 Fluorometric PLC Assays

Fluorescent methods to assay PLC have been developed in an effort to minimize the expensive cost and hazards of radiolabeled substrates. The additional benefit of continuous fluorometric assays eliminates the separation of byproducts and allows for better kinetic analysis. PtdIns(4,5)P$_2$ analogs labeled with fluorophores for fluorescent detection have been synthesized and effectively used to measure PtdIns-specific PLC [46]. However, one of the major weaknesses with fluorometric assays for PtdIns-specific PLC is that the specific activity is significantly lower than radiometric data due to steric hindrance of the fluorophore and alterations in the optimal pH requirements. Fluorescently labeled PtdCho has also been used for fluorescent-based PtdCho-specific PLC assays.

Commercial fluorescent PtdCho-specific PLC assay kits are available. The Amplex®Red PtdCho-specific PLC assay kit from Invitrogen (Carlsbad, California) uses N-acetyl-3,7-dihydrophenoxazine (Amplex Red) to measure the activity of PtdCho-specific PLC. After PtdCho-specific PLC hydrolyzes PtdCho to diacylglycerol and phosphocholine, alkaline phosphatase is used to hydrolyze the phosphocholine to choline. Choline oxidase oxidizes choline yielding betaine and hydrogen peroxide. The hydrogen peroxide then reacts with Amplex Red in the presence of horseradish peroxidase and produces fluorescent resorufin. The end product, resorufin, has an excitation maximum and emission maximum at 571 and 585 nm, respectively.

2.3.5 Inhibitors of Phospholipase C Activity

Enzyme inhibitors are very important in elucidating the different functions and regulatory pathways of signaling enzymes. They can also serve a pharmacological purpose. PLC regulation is implicated in numerous diseases, so PLC inhibitors may be potential therapeutic agents. Several inhibitors have been identified for mammalian PtdCho-specific PLC and PtdIns-specific PLC (Table 2.3 and Figure 2.6), some of which produce antiviral and antitumoral effects [47].

TABLE 2.3
Specificity and Inhibitory Concentrations of Reported PLC Inhibitors

Inhibitor	Effective Concentration (µM)	Specificity	References
D609	75–95	PtdCho-specific PLC	[48,50]
U73122	0.5–10	PtdIns-specific PLC	[52]
ET-18-OCH$_3$	0.4–9.6	PtdIns-specific PLC	[53]
Neomycin sulfate	30	PtdIns-specific PLC, PLD	[54]
Aluminum	10–500	PtdIns-specific PLC	[58]
Gro-PIP	380	PtdIns-specific PLC	[55]
Q12713	8.5	PtdIns-specific PLC	[56]
Compound 40/80	13.7	PtdIns-specific PLC, PLA$_2$	[57]

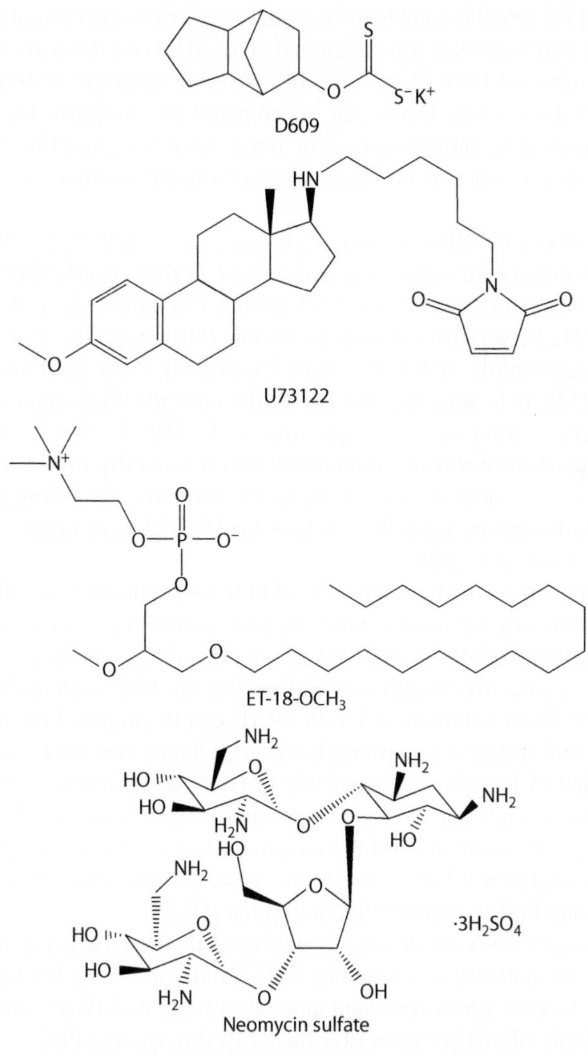

FIGURE 2.6 Common inhibitors of phospholipases C. The potassium xanthate compound, D609, is a selective inhibitor of PtdCho-specific PLC. U73122 and ET-18-OCH$_3$ are selective inhibitors of PtdIns-specific PLC, and the aminoglycoside neomycin sulfate inhibits both PtdIns-specific PLC and phospholipase D.

The compound, tricyclodecan-9-yl xanthogenate (D609), is a competitive inhibitor of PtdCho-specific PLC but not PtdIns-specific PLC [48]. The inhibitory effect and selectivity of D609 has been well established, and it is the most prevalent inhibitor used to study cellular pathways regulated by PtdCho-specific PLC. Although most evidence has shown that D609 is selective for PtdCho-specific PLC, one report shows that D609 can inhibit phorbol ester-stimulated phospholipase D

(PLD) activity [49]. D609 is soluble in H_2O and should be buffered at pH 7.0 because a pH below 6.0 will hydrolyze the compound and a pH above 7.3 will render it inactive. The IC_{50} reported from *in vitro* studies is in the range of 75–95 μM [48,50]. For *in vivo* administration, D609 can be prepared in phosphate-buffered saline and be administered by intraperitoneal or intravenous injection [51]. When using D609, dose–response experiments should be performed for both *in vitro* and *in vivo* studies.

Many inhibitors of PtdIns-specific PLC have been identified, although none have shown specificity for individual isoforms of PtdIns-specific PLC. The compound, 1-(6-((17β-3-methoxyestra-1,3,5(10)-trien-17-yl)amino)hexyl)-1H-pyrrole-2,5-dione (U73122), was first shown to inhibit PtdIns-specific PLC activity in platelets and neutrophils and is the most extensively used inhibitor of PtdIns-specific PLC [52]. It is selective for the PtdIns-specific PLC class of PLC and does not directly inhibit other phospholipases. U73122 is dissolved in organic solvents such as chloroform or ethanol but will not readily dissolve in aqueous solutions. For *in vivo* administration, aqueous solutions containing serum albumin (1–10 mg/mL) can be used. IC_{50} values for U73122 have ranged from 0.5 to 10 μM for *in vitro* studies [52].

Other inhibitors used to study the roles of PtdIns-specific PLC and the signaling pathways in which they are involved include 1-*O*-octadecyl-2-*O*-methyl-sn-glycero-3-phosphorylcholine (ET-18-OCH₃) and neomycin sulfate. The ether–lipid analog ET-18-OCH₃ is a selective inhibitor of PtdIns-specific PLC with an IC_{50} range of 0.4–9.6 μM [53]. Stock solutions of ET-18-OCH₃ can be prepared in chloroform or ethanol. Neomycin sulfate is an aminoglycoside antibiotic that inhibits both PtdIns-specific PLC and PLD activity. It is soluble in H_2O and has an IC_{50} of 30 μM [54]. Additionally, compounds that have not been extensively used in identifying the roles of PtdIns-specific PLC can also inhibit enzyme activity. Aluminum, Q12713, compound 40/80, and glycerol-3-phospho-D-*myo*-inositol-4-phosphate (Gro-PIP) are all capable of causing PtdIns-specific PLC inhibition [55–58].

Gene silencing by RNA interference is another method used to study the functional roles and signaling pathways mediated by PLC. Small interfering RNA (siRNA) can be used to knock down gene expression in order to study the effects. There are commercially available siRNA for most isoforms of PtdIns-specific PLC.

2.4 CONCLUSIONS

A major role in lipid metabolism and signal transduction across cellular membranes is mediated by PLC, and they are involved in regulating many cellular processes. Therefore, methods of PLC analysis are of importance when trying to elucidate the roles that PLC play in both normal and disease states. The structural and regulatory components of PLC as well as several methods to determine the patterns of PLC activity and expression have been described. This information can be applied to many experimental paradigms in order to identify aberrant changes in PLC and pursue potential pharmacological therapies.

ACKNOWLEDGMENTS

The work was supported in part by Grant Number 2P20RR017699-06 from the National Center for Research Resources (NCRR), a component of the National Institutes of Health (NIH).

APPENDIX: COMMONLY USED REAGENTS AND BUFFERS

Heparinized Saline

Reagent	Concentration
NaCl	0.9%
Heparin sodium	1 U/mL

Periodate–Lysine–Paraformaldehyde Fixative

Reagent	Concentration
Paraformaldehyde	2.0%
Sodium phosphate	37.5 mM
Lysine	75 mM
Sodium periodate	10 mM

Homogenizing Buffer Stock Solution

Reagent	Concentration (mM)
HEPES (pH 7.4 at 4°C)	20
EGTA	2
EDTA	5

Add the buffer components and 800 mL of dH_2O (4°C) with stirring. Adjust the pH to 8.0 until all solutes are dissolved and then adjust pH to 7.4 at 4°C. QS to 1 L and store at 4°C indefinitely.

Homogenizing Buffer Working Solution

Add protease inhibitors immediately before use to the homogenizing buffer stock solution to achieve the following final concentrations.

Protease Inhibitor	Concentration
Pepstatin	1.5 μM
Leupeptin	2 μM
Aprotinin	0.2 U/mL
Phenylmethylsulfonyl fluoride	0.5 mM
Dithiothreitol	2 mM
Antipain hydrochloride	10 μg/mL
Sodium fluoride	5 mM
Sodium orthovanadate	1 mM

Homogenizing Buffer + 0.2% Triton X-100 Stock Solution

Reagent	Concentration
HEPES (pH 7.4 at 4°C)	20 mM
EGTA	2 mM
EDTA	5 mM
Triton X-100	0.2%

Homogenizing Buffer + 0.2% Triton X-100 Working Solution

Add protease inhibitors to the homogenizing buffer + 0.2% Triton X-100 stock solution immediately before use to achieve the following final concentrations.

Protease Inhibitor	Concentration
Pepstatin	1.5 μM
Leupeptin	2 μM
Aprotinin	0.2 U/mL
Phenylmethylsulfonyl fluoride	0.5 mM
Dithiothreitol	2 mM
Antipain hydrochloride	10 μg/mL
Sodium fluoride	5 mM
Sodium orthovanadate	1 mM

PtdIns-Specific PLC Stock Solution (pH 5.5)

Reagent	Concentration
Maleic acid	75 mM
$CaCl_2$	1.5 mM
KCl	150 mM
Lithium chloride	15 mM
Triton X-100	0.25 mg/mL

Dissolve the buffer components in 800 mL of dH_2O with stirring. Bring the temperature to 37°C, and adjust the pH to 5.5. QS to 1 L and store at 4°C

PtdIns-Specific PLC Stock Solution (pH 7.4)

Reagent	Concentration
HEPES (pH 7.4 at 37°C)	30 mM
$CaCl_2$	1.5 mM
KCl	150 mM
Lithium chloride	15 mM
Triton X-100	0.5 mg/mL

Dissolve the buffer components in 800 mL of dH_2O with stirring. Bring the temperature to 37°C, and then adjust the pH to 7.4. QS to 1 L and store at 4°C.

Substrate Stock

Prepare the substrate stock with the following concentrations of $PtdIns(4,5)P_2$ and [inositol-2-^{3}H(N)]$PtdIns(4,5)P_2$ in chloroform/methanol/water (75:25:1, by vol.).

Substrate	Concentration
$PtdIns(4,5)P_2$	75 μM
[inositol-2-^{3}H(N)]$PtdIns(4,5)P_2$	0.05 μCi/mL

PtdIns-Specific PLC Assay Buffer (pH 7.4)

Dry down 10 mL of substrate stock under a stream of N_2 and bring up in an equal volume of PtdIns-specific PLC stock solution. Sonicate for 5 min at 37°C.

LIST OF ABBREVIATIONS

D609	tricyclodecan-9-yl xanthogenate
ET-18-OCH$_3$	1-*O*-octadecyl-2-*O*-methyl-*sn*-glycero-3-phosphoryl-choline
Gro-PIP	glycerol-3-phospho-D-*myo*-inositol-4-phosphate
Ins(1,4,5)P$_3$	inositol-1,4,5-trisphosphate
PH	pleckstrin homology
PKC	protein kinase C
PLC	Phospholipases C
PtdCho	phosphatidylcholine
PtdCho-specific-PLC	phosphatidylcholine-specific PLC
PtdIns	phosphatidylinositol
PtdIns(4,5)P$_2$	phosphatidylinositol-4,5-bisphosphate
PtdIns-specific-PLC	phosphoinositide-specific PLC
RA	Ras-association
RPTK	receptor protein tyrosine kinases
RT-PCR	reverse-transcription polymerase chain reaction
SH2	Src homology 2
SH3	Src homology 3
TIM	triose phosphate isomerase
U73122	1-(6-((17β-3-Methoxyestra-1,3,5(10)-trien-17-yl)amino) hexyl)-1H-pyrrole-2,5-dione

REFERENCES

1. Exton, J.H. (1994) Phosphatidylcholine breakdown and signal transduction. *Biochim. Biophys. Acta* 1212:26–42.
2. Rhee, S.G. (2001) Regulation of phosphoinositide-specific phospholipase C. *Annu. Rev. Biochem.* 70:281–312.

3. Faenza, I. et al. (2000) A role for nuclear phospholipase Cβ 1 in cell cycle control. *J. Biol. Chem.* 275:30520–30524.

4. Bertagnolo, V. et al. (2007) Phospholipase C-β 2 promotes mitosis and migration of human breast cancer-derived cells. *Carcinogenesis* 28:1638–1645.

5. Shimohama, S. et al. (1991) Aberrant accumulation of phospholipase C-δ in Alzheimer brains. *Am. J. Pathol.* 139:737–742.

6. Lo Vasco, V.R. et al. (2004) Inositide-specific phospholipase Cβ 1 gene deletion in the progression of myelodysplastic syndrome to acute myeloid leukemia. *Leukemia* 18:1122–1126.

7. Rebecchi, M.J. and Pentyala, S.N. (2000) Structure, function, and control of phospho-inositide-specific phospholipase C. *Physiol. Rev.* 80:1291–1335.

8. Harden, T.K. and Sondek, J. (2006) Regulation of phospholipase C isozymes by ras superfamily GTPases. *Annu. Rev. Pharmacol. Toxicol.* 46:355–379.

9. Irino, Y. et al. (2004) Phospholipase C δ-type consists of three isozymes: Bovine PLCδ2 is a homologue of human/mouse PLCδ4. *Biochem. Biophys. Res. Commun.* 320:537–543.

10. Bahk, Y.Y. et al. (1994) Two forms of phospholipase C-β 1 generated by alternative splicing. *J. Biol. Chem.* 269:8240–8245.

11. Zhou, Y. et al. (2005) Molecular cloning and characterization of PLC-eta2. *Biochem. J.* 391:667–676.

12. Hwang, J.I. et al. (2005) Molecular cloning and characterization of a novel phospholi-pase C, PLC-eta. *Biochem. J.* 389:181–186.

13. Bae, Y.S. et al. (1998) Activation of phospholipase C-γ by phosphatidylinositol 3,4,5-trisphosphate. *J. Biol. Chem.* 273:4465–4469.

14. Harada, K. et al. (2005) Role of PRIP-1, a novel Ins(1,4,5)P3 binding protein, in Ins(1,4,5)P3-mediated Ca2+signaling. *J. Cell Physiol.* 202:422–433.

15. Uji, A. et al. (2002) Molecules interacting with PRIP-2, a novel Ins(1,4,5)P3 binding protein type 2: Comparison with PRIP-1. *Life Sci.* 72:443–453.

16. Ellis, M.V. et al. (1998) Catalytic domain of phosphoinositide-specific phospholipase C (PLC). Mutational analysis of residues within the active site and hydrophobic ridge of plcdelta1. *J. Biol. Chem.* 273:11650–11659.

17. Essen, L.O. et al. (1997) Structural mapping of the catalytic mechanism for a mamma-lian phosphoinositide-specific phospholipase C. *Biochemistry* 36:1704–1718.

18. Sidhu, R.S. et al. (2005) Regulation of phospholipase C-δ1 through direct interactions with the small GTPase Ral and calmodulin. *J. Biol. Chem.* 280:21933–21941.

19. Nomikos, M. et al. (2007) Binding of phosphoinositide-specific phospholipase C-zeta (PLC-zeta) to phospholipid membranes: Potential role of an unstructured cluster of basic residues. *J. Biol. Chem.* 282:16644–16653.

20. Lemmon, M.A. and Ferguson, K.M. (2000) Signal-dependent membrane targeting by pleckstrin homology (PH) domains. *Biochem. J.* 350(Pt 1):1–18.

21. Garcia, P. et al. (1995) The pleckstrin homology domain of phospholipase C-delta 1 binds with high affinity to phosphatidylinositol 4,5-bisphosphate in bilayer membranes. *Biochemistry* 34:16228–16234.

22. Ferguson, K.M. et al. (1995) Structure of the high affinity complex of inositol trisphos-phate with a phospholipase C pleckstrin homology domain. *Cell* 83:1037–1046.

23. Wang, T. et al. (1999) Selective interaction of the C2 domains of phospholipase C-β1 and -β2 with activated Gαq subunits: An alternative function for C2-signaling modules. *Proc. Natl. Acad. Sci. USA* 96:7843–7846.

24. Essen, L.O. et al. (1996) Crystal structure of a mammalian phosphoinositide-specific phospholipase C δ. *Nature* 380:595–602.

25. Grobler, J.A. and Hurley, J.H. (1998) Catalysis by phospholipase C δ1 requires that Ca^{2+} bind to the catalytic domain, but not the C2 domain. *Biochemistry* 37:5020–5028.
26. Lomasney, J.W. et al. (1999) Activation of phospholipase C δ1 through C2 domain by a Ca(2+)-enzyme-phosphatidylserine ternary complex. *J. Biol. Chem.* 274:21995–22001.
27. Nakashima, S. et al. (1995) Deletion and site-directed mutagenesis of EF-hand domain of phospholipase C-δ 1: Effects on its activity. *Biochem. Biophys. Res. Commun.* 211:365–369.
28. Yamamoto, T. et al. (1999) Involvement of EF hand motifs in the Ca(2+)-dependent binding of the pleckstrin homology domain to phosphoinositides. *Eur. J. Biochem.* 265:481–490.
29. Kelley, G.G. et al. (2001) Phospholipase C(epsilon): A novel Ras effector. *EMBO J.* 20:743–754.
30. Kim, C.G. et al. (1996) The role of carboxyl-terminal basic amino acids in Gqα-dependent activation, particulate association, and nuclear localization of phospholipase C-β1. *J. Biol. Chem.* 271:21187–21192.
31. Noh, D.Y. et al. (1995) Phosphoinositide-specific phospholipase C and mitogenic signaling. *Biochim. Biophys. Acta* 1242:99–113.
32. Jhon, D.Y. et al. (1993) Cloning, sequencing, purification, and Gq-dependent activation of phospholipase C-β 3. *J. Biol. Chem.* 268:6654–6661.
33. Offermanns, S. et al. (1997) Defective platelet activation in G α(q)-deficient mice. *Nature* 389:183–186.
34. Kim, J.W. et al. (1990) Tyrosine residues in bovine phospholipase C-γ phosphorylated by the epidermal growth factor receptor *in vitro*. *J. Biol. Chem.* 265:3940–3943.
35. Kim, Y.H. et al. (1999) Phospholipase C-δ1 is activated by capacitative calcium entry that follows phospholipase C-β activation upon bradykinin stimulation. *J. Biol. Chem.* 274:26127–26134.
36. Zhou, Y. et al. (2008) Activation of human phospholipase C-η2 by Gβγ. *Biochemistry* 47:4410–4417.
37. Manzoli, L. et al. (2005) Nuclear phospholipase C: Involvement in signal transduction. *Prog. Lipid Res.* 44:185–206.
38. Bahk, Y.Y. et al. (1998) Localization of two forms of phospholipase C-β1, a and b, in C6Bu-1 cells. *Biochim. Biophys. Acta* 1389:76–80.
39. Larman, M.G. et al. (2004) Cell cycle-dependent Ca2+ oscillations in mouse embryos are regulated by nuclear targeting of PLCzeta. *J. Cell Sci.* 117:2513–2521.
40. Chirgwin, J.M. et al. (1979) Isolation of biologically active ribonucleic acid from sources enriched in ribonuclease. *Biochemistry* 18:5294–5299.
41. Chomczynski, P. and Sacchi, N. (1987) Single-step method of RNA isolation by acid guanidinium thiocyanate–phenol–chloroform extraction. *Anal. Biochem.* 162:156–159.
42. Holmseth, S. et al. (2006) Specificity controls for immunocytochemistry. *Anat. Embryol. (Berl.)* 211:257–266.
43. Itoh, K. et al. (1979) Application of coupled oxidation reaction to electron microscopic demonstration of horseradish peroxidase: Cobalt-glucose oxidase method. *Brain Res.* 175:341–346.
44. Kim, J.W. et al. (1989) Cyclic and noncyclic inositol phosphates are formed at different ratios by phospholipase C isozymes. *Biochem. Biophys. Res. Commun.* 163:177–182.
45. Rouser, G. et al. (1966) Quantitative analysis of phospholipids by thin-layer chromatography and phosphorus analysis of spots. *Lipids* 1:85–86.
46. Hendrickson, H.S. (1994) Fluorescence-based assays of lipases, phospholipases, and other lipolytic enzymes. *Anal. Biochem.* 219:1–8.

47. Muller-Decker, K. (1989) Interruption of TPA-induced signals by an antiviral and anti-tumoral xanthate compound: Inhibition of a phospholipase C-type reaction. *Biochem. Biophys. Res. Commun.* 162:198–205.

48. Amtmann, E. (1996) The antiviral, antitumoural xanthate D609 is a competitive inhibitor of phosphatidylcholine-specific phospholipase C. *Drugs Exp. Clin. Res.* 22:287–294.

49. Kiss, Z. and Tomono, M. (1995) Compound D609 inhibits phorbol ester-stimulated phospholipase D activity and phospholipase C-mediated phosphatidylethanolamine hydrolysis. *Biochim. Biophys. Acta* 1259:105–108.

50. Schutze, S. et al. (1992) TNF activates NF-kappa B by phosphatidylcholine-specific phospholipase C-induced "acidic" sphingomyelin breakdown. *Cell* 71:765–776.

51. Joshi, G. et al. (2005) In vivo protection of synaptosomes from oxidative stress mediated by Fe^{2+}/H_2O_2 or 2,2-azobis-(2-amidinopropane) dihydrochloride by the glutathione mimetic tricyclodecan-9-yl-xanthogenate. *Free Radic. Biol. Med.* 38:1023–1031.

52. Bleasdale, J.E. et al. (1990) Selective inhibition of receptor-coupled phospholipase C-dependent processes in human platelets and polymorphonuclear neutrophils. *J. Pharmacol. Exp. Ther.* 255:756–768.

53. Powis, G. et al. (1992) Selective inhibition of phosphatidylinositol phospholipase C by cytotoxic ether lipid analogues. *Cancer Res.* 52:2835–2840.

54. Lipsky, J.J. and Lietman, P.S. (1982) Aminoglycoside inhibition of a renal phosphatidylinositol phospholipase C. *J. Pharmacol. Exp. Ther.* 220:287–292.

55. Cruz-Rivera, M. et al. (1990) Glycerol-3-phospho-D-myo-inositol 4-phosphate (Gro-PIP) is an inhibitor of phosphoinositide-specific phospholipase C. *Biochim. Biophys. Acta* 1042:113–118.

56. Ogawara, H. et al. (1993) An inhibitor of inositol-phospholipid-specific phospholipase C. *Biochim. Biophys. Acta* 1175:289–292.

57. Bronner, C. et al. (1987) Compound 48/80 is a potent inhibitor of phospholipase C and a dual modulator of phospholipase A2 from human platelet. *Biochim. Biophys. Acta* 920:301–305.

58. Nostrandt, A.C. et al. (1996) Inhibition of rat brain phosphatidylinositol-specific phospholipase C by aluminum: Regional differences, interactions with aluminum salts, and mechanisms. *Toxicol. Appl. Pharmacol.* 136:118–125.

59. Cha, S.H. et al. (1998) Distributional patterns of phospholipase C isozymes in rat kidney. *Nephron* 80:314–323.

60. Park, D. et al. (1992) Cloning, sequencing, expression, and Gq-independent activation of phospholipase C-β 2. *J. Biol. Chem.* 267:16048–16055.

61. Kim, M.J. et al. (1998) A cytosolic, Gαq- and βγ-insensitive splice variant of phospholipase C-β4. *J. Biol. Chem.* 273:3618–3624.

62. Min, D.S. et al. (1993) Purification of a novel phospholipase C isozyme from bovine cerebellum. *J. Biol. Chem.* 268:12207–12212.

63. Min, D.S. et al. (1993) A G-protein-coupled 130 kDa phospholipase C isozyme, PLC-β 4, from the particulate fraction of bovine cerebellum. *FEBS Lett.* 331:38–42.

64. Homma, Y. et al. (1989) Tissue- and cell type-specific expression of mRNAs for four types of inositol phospholipid-specific phospholipase C. *Biochem. Biophys. Res. Commun.* 164:406–412.

65. Mizuguchi, M. et al. (1991) Phospholipase C isozymes in neurons and glial cells in culture: An immunocytochemical and immunochemical study. *Brain Res.* 548:35–40.

66. Lee, W.K. et al. (1999) Molecular cloning and expression analysis of a mouse phospholipase C-δ1. *Biochem. Biophys. Res. Commun.* 261:393–399.

67. Lin, F.G. et al. (2001) Downregulation of phospholipase C δ3 by cAMP and calcium. *Biochem. Biophys. Res. Commun.* 286:274–280.
68. Lee, S.B. and Rhee, S.G. (1996) Molecular cloning, splice variants, expression, and purification of phospholipase C-δ 4. *J. Biol. Chem.* 271:25–31.
69. Saunders, C.M. et al. (2002) PLC zeta: A sperm-specific trigger of Ca(2+) oscillations in eggs and embryo development. *Development* 129:3533–3544.
70. Nakahara, M. et al. (2005) A novel phospholipase C, PLC(eta)2, is a neuron-specific isozyme. *J. Biol. Chem.* 280:29128–29134.

3 Methods for Measuring the Activity and Expression of Phospholipase D

Wenjuan Su and Michael A. Frohman

CONTENTS

3.1 PHOSPHOLIPASE D: A HISTORICAL PERSPECTIVE

3.1.1 IDENTIFICATION OF PLD ACTIVITY

Phospholipase D (PLD) was defined in 1947 by Hanahan and Chaikoff in carrot extracts as a phospholipid-specific phosphodiesterase that hydrolyzed phosphatidylcholine (PtdCho) to generate phosphatidic acid (PtdOH) and choline [1,2]. PLD activity was thought to be absent in animal tissues, in retrospect, due to the abundant levels of inhibitors that are found in crude extracts, until Saito and Kanfer demonstrated 30 years later that rat brain preparations could release choline and ethanolamine from PtdCho and phosphatidylethanolamine (PtdEtn), respectively [3]. PLD is now known to be present in all organisms [4].

More generally, PLD is now viewed as a transphosphatidylase (Figure 3.1) that can use water for hydrolysis or can use glycerol or short-chain primary alcohols, such as ethanol or 1-butanol, to generate phosphatidylalcohol products [5–7]. Phosphatidylalcohols are not normally found in biological membranes, are relatively inert with respect to the functions mediated by PtdOH, and are relatively metabolically stable in comparison to PtdOH [8]. Hence, eliciting phosphatidylbutanol formation by exposing cells to low levels of 1-butanol has become widely used as a convenient means to record and assay PLD activity [9], as discussed in Section 3.4. Moreover, addition of high levels of alcohol has been used as a means to divert PLD away from producing PtdOH, and thus as a way to inhibit PLD-mediated pathways [10,11], with the caveat that the amounts of alcohol required to fully block PLD production of PtdOH can also inhibit levels of phosphoinositides [11] or blunt signaling pathways, for example, insulin receptor signaling [10] or p42/44 mitogen-activated protein kinase activation [12,13].

PLD activity is found in most cellular compartments in mammalian cells, including the plasma membrane, cytosol, endoplasmic reticulum (ER), Golgi, and nucleus [14–18]. This derives in part from the unique localization of the individual isoforms,

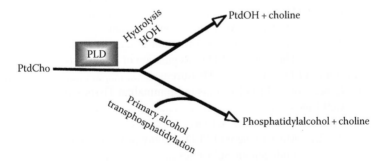

FIGURE 3.1 Overview of PLD action. PLD hydrolyzes PtdCho to produce PtdOH and choline, and can perform transphosphatidylation reactions using primary alcohols to generate phosphatidylalcohols.

and, in part, because the isoforms cycle through multiple subcellular compartments during signaling events [19–22]. In addition, a highly divergent PLD family member is found on the surface of mitochondria that hydrolyzes cardiolipin to yield PtdOH [23], and another divergent PLD family member with unknown activity is found in the lumen of the ER [24].

Activation of classical PLD in mammalian cells is triggered by a wide range of stimuli that signal through G-protein-coupled receptors and receptor tyrosine kinases [25–29]. Although there are reports that PLD can, in some cases, interact directly with the receptors [30] or G-proteins [31], in general, the activation mechanism is thought to be downstream of protein kinase C (PKC) activation and mobilization of small GTPases of the Rho and ARF families [7,32–34]. However, exactly how PLD activation by these cytosolic factors is coordinated by receptor stimulation is complex, and it remains unclear despite extensive efforts by many groups. Detailed regulatory mechanisms of PLD activity is discussed in Section 3.2.

Once activated, PLD-generated PtdOH has several important roles in cell function, including helping to regulate membrane vesicle trafficking during regulated exocytosis, endocytosis, and phagocytosis [35–37]. PLD-generated PtdOH may also promote proliferation, survival, and motility for metastatic cancer cells, as discussed in Section 3.3.

3.1.2 CLONING AND CHARACTERIZATION OF MAMMALIAN PLD GENES

The first PLD gene identified was in castor bean [38]. The publication of this sequence led to the realization that a gene in yeast that had been well characterized as being required for sporulation was in fact a PLD [39], providing immediate identification of and insight into a critical cell biological process mediated by PLD generation of PtdOH. Cloning and analysis in parallel of the first mammalian gene, denoted PLD1 [40,41], revealed that it encoded for a phosphatidylinositol-4,5-bisphosphate $(PtdIns(4,5)P_2)$-dependent, ARF-, Rho-, and PKC-stimulated activities similar to the one that had been defined previously using partially purified biochemical preparations [32,33]. A second isoform, denoted PLD2, was subsequently identified, which exhibited a high basal activity that was also $PtdIns(4,5)P_2$-dependent but otherwise unresponsive to the well-studied cytosolic factors that stimulated PLD1 [19]. Taken together, the plant, yeast, and human PLD genes comprised a new gene family that shared in common a pair of half-catalytic sites denoted "HKD"s for the three most highly conserved residues [40], and a pleckstrin homology (PH) domain, a phox (PX) domain, and a $PtdIns(4,5)P_2$-binding site later shown to be important for how the enzymes associate with specific subcellular membrane compartments [20,21].

3.1.3 THE PLD SUPERFAMILY

Members of the PLD superfamily are defined by the presence of an HKD catalytic site (formally known as $HxK(x)_4D(x)_6GSxN$). Although the classic definition of the family involves hydrolysis of PtdCho to generate PtdOH and choline, some of the family members possess quite divergent activities [4], including ones that use cardiolipin [23] or other phospholipids as substrates, or use the phosphatidyltransferase capacity to synthesize new lipids (cardiolipin synthase and phosphatidylserine synthase [5,42,43]). The superfamily also includes endonucleases (Nuc) that use the phosphodiesterase activity to cleave the backbone of DNA [44], and pox virus envelope proteins and their mammalian counterparts that are required for virion formation through an unknown biochemical mechanism [5], and the protein Tdp1, which resolves stalled topoisomerase–DNA complexes involving covalent links between the protein and the DNA by again using the phosphodiesterase activity to sever them [45].

3.1.4 PLD PRIMARY STRUCTURE

The PLD enzymes have a characteristic modular structure that includes common catalytic domains and regulatory sequences [7,34,46]. Figure 3.2 shows the domain organization of PLD1 and highlights defined sites of interaction with regulatory proteins and lipids.

The defining feature of PLD is the HKD catalytic domain [40]. Each HKD domain forms half of a catalytic site [5,47,48]. The PH domain mediates the association of PLD1 with lipid rafts, which facilitates transit of the enzyme to endosomes after its translocation to the plasma membrane upon agonist stimulation [20]. PH domains can bind PtdIns(4,5)P_2, and the PLD1 PH domain has been reported to exhibit binding to phospholipids; however, as described below, the stimulation of PLD by PtdIns(4,5)P_2 is mediated by a different region of the protein, and deletion of the PH domain does not impair its intrinsic enzymatic activity [49,50]. PX domains are found in many signaling molecules and mediate protein–protein interactions or binding to phosphatidylinositol phosphates [51]. The PLD PX domain has been reported to bind to phosphatidylinositol-3,4,5-triphosphate (PtdIns(3,4,5)P_3) as determined by surface plasmon resonance [52]. In other assays, affinity for PtdIns(5) P [20] and binding to the epidermal growth factor (EGF) receptor [30] have also been observed. The PX domain may play a role to facilitate internalization of PLD1 from the plasma membrane [20].

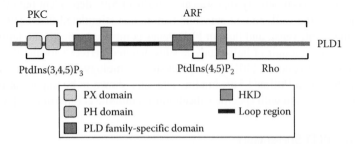

FIGURE 3.2 Schematic of motifs and domains found within mammalian PLD1. HKD domains are essential for catalysis. The PX and PH domains are lipid-binding domains important for the regulation of PLD localization. The loop region may play a negative regulatory role for PLD activity. The PtdIns(4,5)P_2-binding domain is important for PLD localization and activity. Regions where phospholipid and protein effectors are known to interact are highlighted.

PLD1 and PLD2 activation are highly dependent on PtdIns(4,5)P$_2$. The activation requires a sequence of conserved basic and aromatic amino acids known as a polybasic motif, which mediates the interaction of this and other proteins with negatively charged lipids [53,54]. The affinity for PtdIns(4,5)P$_2$ may additionally facilitate the translocation of PLD to the plasma membrane surface [20].

PLD1, *Drosophila* PLD, and *Caenorhabditis elegans* PLD have a conserved loop region that is not found in PLD2 or yeast PLD. Deletion of this region from PLD1 increases its basal activity threefold, indicating that the loop may possess a negative regulatory element that contributes to the low basal activity of PLD1 in comparison with PLD2 and yeast PLD [50].

The N-terminus of PLD is poorly conserved. Removing the N-terminus from PLD1 does not affect its basal activity but does prevent it from responding to PKC [55]. The PLD1 N-terminus may also interact with Gβγ [31]. In contrast, removing the N-terminus of PLD2 eliminates its high basal activity and renders it responsive to ARF GTPases but not Rho proteins. Unlike the N-terminus, the C-terminus of PLD is relatively well conserved among mammals. Deletion or modification of the C-terminus abolishes its activity [50,56], which has led to the suggestion that it stabilizes the active site conformation [56].

3.2   REGULATION OF PLD ACTIVITY

PLD is activated secondary to G-protein-coupled receptor and receptor tyrosine kinase stimulation, although the connection to PLD appears to be indirect, subsequent to the mobilization and activation of cytosolic factors. PLD activity is tightly regulated by a large number of factors including small GTPases, PKC, phosphoinositides, and posttranslational modification and inhibitory factors. The regulation is complex, and many of these factors act in concert to regulate PLD activity in either synergistic or antagonistic ways.

3.2.1   SMALL GTPASES

Several members of the small GTPase family are implicated in the activation of PLD, most prominently, members of the Rho and ARF subfamilies [57–62]. The Rho family members interact directly with PLD1 [40,41,63,64], while this has been harder to demonstrate for ARF family members [33,50,65–67]. Nonetheless, agonist stimulation, such as fMet-Leu-Phe (fMLP) stimulation of HL-60 cells, triggers ARF translocation to membranes, resulting in colocalization with PLD1 and its activation [65,68]. PLD activation by Rho may also involve indirect mechanisms. For example, since the inhibition of PLD by Rho inactivation can be largely rescued by the addition of PtdIns(4,5)P$_2$, it has been suggested that the primary effect of Rho on PLD is the regulation of the production of PLD's required cofactor, PtdIns(4,5)P$_2$, by stimulating the enzyme that generates it, PtdIns(4)P-5 kinase [69–71]. Rho may also affect PLD activity via its effector, Rho kinase, since overexpression of Rho kinase markedly increases M3 muscarinic receptor–mediated PLD activation [71].

3.2.2 PROTEIN KINASE C

PKC is a well-known PLD1 activator: overexpression of PKCα and PKCβ1 increases PLD responses to endothelin, thrombin, and platelet-derived growth factor (PDGF) [72–74]. *In vitro* studies also demonstrated that PLD1 interacts with and can be activated by PKCα and PKCβ1 and -2 in a phorbol 12-myristate 13-acetate (PMA)-enhanced manner [50,75,76]. Regulation of PLD activity by PKC may also act through indirect mechanisms involving phosphorylation or activation of PLCγ (reviewed in [34]), although it should be noted that PKC stimulation of PLD1 activity occurs *in vitro* in the absence of ATP; hence, it takes place through a phosphorylation-independent mechanism [77].

3.2.3 PHOSPHOINOSITIDES

PLD dependence on $PtdIns(4,5)P_2$ was first demonstrated by Brown et al. in HL-60 cells [32,33] and confirmed by a subsequent study of the cloned PLD1 and PLD2 isoforms [19,40,67]. $PtdIns(3,4,5)P_3$ also activates PLD [41], although the significance of this activation is not clear, since agonist-induced PLD activity can be observed in the absence of measurable PI3-kinase activity [41,78]. Alternately, $PtdIns(3,4,5)P_3$ may facilitate PLD stimulation indirectly by promoting ARF or Rho activation [78,79].

3.2.4 POSTTRANSLATIONAL MODIFICATION AND LIPIDATION

A variety of kinases have been implicated in PLD regulation including serine/threonine (Ser/Thr) kinase, AMP-dependent protein kinase, and receptor or nonreceptor tyrosine kinases [34]. Phosphorylation is proposed to act both directly and indirectly to regulate PLD. PLD activity is observed in response to EGF, PDGF, and insulin [80]. Phosphorylation and PLD activity can be inhibited by the protein tyrosine kinase inhibitors, genestein and herbmycin A; downregulation of PKC; or the PKC inhibitors, Ro 31-8220 and calphostin C [63]. Signaling-dependent tyrosine phosphorylation of PLD2 has also been reported [81,82]. However, the ultimate significance of PLD phosphorylation in the context of its function remains unclear.

Palmitoylation also occurs on PLD1 and appears to be important for localization and regulation of the enzyme [83]. Sugars et al. identified the palmitoylated residues as cysteine 240 and 241 within the PH domain [84], and double mutation of these residues resulted in the loss of 50%–80% of the activity of the wild-type protein *in vivo*. The mutations also led to a change in localization; whereas the wild-type protein localizes in the perinuclear region in a punctate distribution, the mutant protein localizes to the plasma membrane [84]. The mutant protein exhibits reduced Ser and Thr phosphorylation and weakened membrane association [85]. It has been proposed that the palmitoylation may help to facilitate the movement of PLD1 into lipid rafts on the plasma membrane and, from there, recycling to the perinuclear region [20].

3.3   FUNCTIONS OF PLD

3.3.1   OVERVIEW

PLD and PtdOH have been implicated in a broad range of cellular processes and diseases, including vesicular trafficking and endocytosis, secretion and diabetes, cytoskeleton reorganization and cell migration, phagocytosis, the oxidative respiratory burst and inflammation, and cell survival and oncogenesis [36]. In addition to acting as an intracellular messenger, PLD-generated PtdOH can also undergo conversion to other bioactive lipids.

3.3.2   PLD-GENERATED PTDOH IN LIPID METABOLISM

PtdOH can be converted into other lipids that are important in cellular signaling and lipid metabolism. For example, PtdOH dephosphorylation by phosphatidic acid phosphohydrolases (PAP) is a significant route for the generation of diacylglycerol (DAG) in stimulated cells [86,87]. PtdOH can also be deacylated by phospholipase A to form the cell surface receptor active compound lysophosphatidic acid (lyso-PtdOH), and PtdOH may be an important source of arachidonic acid for the synthesis of prostaglandins and leukotrienes [88–90].

3.3.3   VESICULAR TRAFFICKING, SECRETION, AND ENDOCYTOSIS

ARF stimulation of PLD1 is proposed to be important in the regulation of vesicle transport between the ER and Golgi [16,91,92] since this process can be blocked by primary alcohol and rescued by the provision of exogenous PtdOH [16,91,93]. PLD2, in contrast, has been implicated in the mechanism underlying receptor internalization by endocytosis for the EGF receptor, the μ-opioid receptor MOR1, and the angiotensin II receptor [22,94–97]. PLD1 and PLD2 have both been implicated to have a role in macrophage phagocytosis [37,98].

Roles for PLD1 in vesicular trafficking are best established for secretion (exocytosis). PLD1 activity regulates exocytosis in neuroendocrine cells [35,99,100], mast cells [101], pancreatic β-cells [102], neutrophils [103], and adipocytes [104], potentially by affecting the step at which the transport vesicles fuse with the plasma membrane [35,100,104,105]. In this context, PtdOH may act as a fusogenic lipid to induce negative membrane curvature and bilayer instability that promotes membrane fusion. In mast cells and neutrophils, PLD activity has also been implicated in degranulation and oxidative respiratory burst, which are critical for host defense.

3.3.4   CYTOSKELETON ORGANIZATION

PLD activity can be stimulated by $PtdIns(4,5)P_2$, ARF, and Rho GTPases, which are well-defined regulators of membrane transport and actin-reorganization processes. In turn, PtdOH activates type I PIP kinase, the enzyme responsible for generating $PtdIns(4,5)P_2$ [11,106–108]. This has been shown to occur dynamically in membrane ruffling, in cooperation with ARF6 [109]. PtdOH formation, especially by PLD1, can also induce stress fiber formation in certain cell types [78,110,111]. For example, in myogenic differentiation induced by arginine vasopressin, actin fiber formation is mediated by Rho GTPases and PLD1 and involves PLD-induced PtdIns(4,5)P$_2$ production along the actin fibers [112]. Thus, it is now proposed that the reciprocal stimulation of PLD and PIP kinase forms a rapid feed-forward stimulation loop that triggers localized and explosive production of PtdOH and PtdIns(4,5)P$_2$, which may then direct reorganization of the actin cytoskeleton.

3.3.5 PLD AND CANCER

3.3.5.1 PLD as an Alternate Survival Signal

PLD activity is elevated in many types of human cancer, including breast, colon, gastric, and kidney cancers [113]. PLD2 point mutations and deletions are found in breast cancer [114]. Several lines of evidence indicate that increased PLD activity can provide an alternate survival signal for cancer cells [34]. Mammalian target of rapamycin (mTOR), a Ser/Thr kinase that acts downstream of PI 3-kinase/Akt survival signaling, binds PtdOH at its rapamycin-interaction site and is activated by PtdOH [113,115,116]. Rapamycin is an mTOR inhibitor, and breast cancer cell lines with high levels of PLD exhibit rapamycin resistance [117], whereas inhibiting PLD activity using siRNA triggers apoptosis. The PtdIns 3-kinase inhibitor, L294002, does not block the survival signal conferred by elevated PLD activity [118], indicating that PLD/mTOR functions via a pathway parallel to the PtdIns 3-kinase/Akt survival signaling pathway.

3.3.5.2 PLD in Cell Transformation

Several studies have implicated PLD in cooperating with tyrosine kinases to promote cell transformation [119]. Activated v-Src (but not c-Src) transforms fibroblasts in culture and increases PLD activity [119–121]. However, overexpression of both c-Src and PLD1 or PLD2 also transforms rat fibroblasts [122], suggesting that the endogenous machinery suffices in the setting of unregulated signaling. PLD activity is also implicated in cell transformation by the oncogenes, v-Ras and v-Raf [119]. Recent work has revealed an unexpected role of PLD2 in Ras activation in response to EGF stimulation [123], in which the PLD2-generated PtdOH acts upstream of Ras by recruiting its immediate activator, Sos, to translocate to the plasma membrane, which is a key step in Ras activation leading to cell transformation.

3.3.5.3 The Role of PLD in Regulation of Metastasis

Metastasis is a key issue in cancer therapy since it causes 90% of the deaths from solid tumors [124]. Metastasis is a complicated multistep process in which cell motility/migration is essential. Cytoskeleton networks play a crucial part in cell motility and have been studied extensively in tumor invasion. PLD stimulates cell protrusion in v-Src transformed cells and to be required for EGF-induced membrane ruffling [109,125]. PLD activity is also implicated in tumor invasion: MDA-MB-231 human breast cancer cells with high levels of PLD activity migrate and invade in matrigel whereas MCF-7 cells with relatively low PLD activity do not. Moreover, the ability of the MDA-MB-231 cells to migrate and invade matrigel is dependent on PLD and mTOR [126]. These data suggest that PLD contributes to cell migration and invasiveness, and therefore inhibiting PLD activity could be a potential target in metastatic cancer therapy.

3.4 EXPRESSION OF PLD AND ACTIVITY MEASUREMENT

3.4.1 Expression of PLD Isoforms in Mammalian Tissues and Cell Lines

PLD1 and PLD2 mRNA are expressed in a wide variety of cell and tissue types [67,127–130], with particularly high levels of PLD1 found in secretory cells, and high levels of PLD2 in lung, kidney, and the heart and brain [67,127,128,130,131]. Cell-type specific expression is seen during embryogenesis. For example, PLD1 is found in ventricular cells of the spinal cord and brain whereas PLD2 is expressed in the hippocampus [130], while in adult brains PLD1 is expressed primarily in neurons and PLD2 in astrocytes [132]. However, the significance of these expression patterns is not known. Due to the absence of good antibodies, it is not clear whether the mRNA levels correspond well to protein levels, and further studies are needed to yield a clear picture of PLD expression in mammalian tissues.

Although numerous groups have published western analysis and immunofluorescence staining of overexpressed and endogenous PLD1 and PLD2, several issues should be noted by investigators interested in these methodologies. First, since PLD is a membrane-bound protein, a special sample buffer (a urea-containing sample buffer) is required to avoid having the protein become trapped at the edge of the stacking gel [41]. Second, PLD2 generally migrates as a single band, but PLD1 migrates as 2–4 bands due to variable phosphorylation [133]. As well, of concern for both western blotting and immunofluorescence staining, many commercial and privately generated antisera recognize many nonspecific bands, so all experiments to examine endogenous proteins should be validated using RNAi. Finally, since overexpression of PLD1 and PLD2 may cause altered localization [134,135], care should be taken in interpreting localization of the overexpressed proteins.

Many commonly used cell lines express both isoforms of PLD, although levels of expression vary markedly between cell types. Cos-7 cells express very low levels of both isoforms and have therefore been used widely as a model system to study the functions of PLD using overexpression approaches [19,76]. Many cell types of hematopoietic origin also express only low levels of PLD mRNA. HL-60 cells express PLD1, but little or undetectable PLD2 [76,127,130]. Both isoforms are expressed in a number of cancer cell lines such as PC-3 (prostate cancer), DLD (colon carcinoma), MDA-MB-231 (breast carcinoma), MCF-7 (breast cancer), and HT1080 (fibrosarcoma) [136]. Recently, the expression level of PLD2 was evaluated in human colorectal carcinoma samples by real-time PCR, and it was found that the expression level varied from tumor to tumor, in correlation with tumor size and survival of the patients [137]. These findings support the hypothesis that PLD may play an important role in tumorigenesis as discussed in Section 3.3.

3.4.2 MEASUREMENT OF PLD ACTIVITY

Methods used for the determination of PLD activity can be divided into two categories: assays *in vitro* using exogenously provided substrates and assays employing endogenous phospholipid substrates to measure PLD activity in intact or broken cells [9]. Both PtdOH and choline, the PLD reaction products, can be isolated and quantitated as indicators of PLD activity. Details of assay procedures and how to choose an appropriate method are discussed in the following sections.

3.4.2.1 Measurement of PLD Activity *in Vivo* Using Endogenous Substrates

In order to determine PLD activity in living cells, the endogenous PLD products (PtdOH, choline, or phosphatidylalcohol) have to be labeled by incorporation of a suitable tracer that is usually radioactive (Table 3.1). The radiolabeled PLD products are then analyzed by a variety of chromatography procedures and quantitated by liquid scintillation counting.

3.4.2.1.1 Transphosphatidylation

PtdOH does not readily serve as a good measure of PLD activity because it is rapidly converted to DAG. However, addition of primary alcohols to cells results in PLD-catalyzed formation of phosphatidylalcohols, which are relatively stable and can be easily resolved from other labeled species present in cellular phospholipid extracts by thin layer chromatography (TLC).

Cellular PtdCho is labeled using [^{3}H]-palmitic, [^{3}H]-oleic, or [^{3}H]-arachidonic acids for 12 h [138]. Following labeling, 0.1%–0.5% (v/v) primary alcohol, usually ethanol or 1-butanol, is added to the cells 5–10 min before agonist treatment. The reaction is terminated using organic solvents to extract the labeled phospholipids and the organic phase recovered and dried by vacuum. Phosphatidylalcohol is analyzed by TLC on silica gel plates using a variety of different solvent systems. A mixture of 2,2,4-trimethylpentane:ethyl acetate:acetic acid:water (50:110:20:100) is a particularly effective solvent system [139]. The plate can be exposed on film and/or the radioactive material recovered from the plate by scraping and quantified by liquid scintillation counting. For specific procedure details, see [9].

Although this method is the most widely used method for measuring PLD activity, there are several limitations. First, phosphatidylalcohols are not necessarily completely stable. Second, in some cases, alcohols can inhibit agonist stimulation. For example, butanol inhibits insulin-induced PLD activation [10]. If these issues pose problems for the specific experiments intended, the following assays represent alternative approaches.

3.4.2.1.2 Headgroup Release

This procedure measures the formation of [^{3}H]-choline by the PLD-catalyzed hydrolysis reaction. To label PtdCho, the cells are incubated with [methyl-^{3}H]-choline chloride for 1–2 days. After agonist stimulation, the reaction is terminated by addition of ice-cold methanol and the aqueous phase recovered [140]. The major radiolabeled lipid species in this phase consist of choline, glycerophosphocholine, and

TABLE 3.1
Summary of PLD Assay Procedures

Assay Type	Source of Substrate	Product Analysis	Advantages	Drawbacks
In vivo				
Transphosphatidylation	Labeling of endogenous PtdCho with fatty acids	TLC analysis	Highly sensitive transphosphatidylation traps metabolically labile PtdOH products as stable phosphatidyl alcohol	PLD activities with selectivity for PtdCho substrates with different acyl chain compositions may exist
Headgroup release	Labeling of endogenous PtdCho with [³H]-choline	Chromatographic resolution of choline and phosphocholine	Rapid analysis of large numbers of samples	Choline and phosphocholine are rapidly interconverted
In vitro				
Headgroup release	Methyl-[³H]PtdCho	Release of water-soluble choline	Simple, rapid, and straightforward	Product analysis required to resolve choline from phospho-glycerophosphocholine
Transphosphatidylation	Radiolabeled or fluorescent PtdCho (acyl chain or diester phosphate group)	TLC	Highly sensitive	Quantitation of radiolabeled products require scraping from TLC plate
Coupled assays	Unlabeled PtdCho	Coupled detection of choline using choline kinase choline oxidase	Sensitive, nonradioactive suitable for high throughput assays	Some procedures require sample preparation and purified enzymes to avoid competing activities
Conductimetric	Unlabeled PtdCho	Measurement of pH-dependent conductance change	Real-time, noninvasive	Requires highly purified components
Membrane surface charge	Unlabeled PtdCho	Measurement of membrane charge change using metachromic dye	Real-time, noninvasive	Requires highly purified components, restricted substrate phospholipid composition

phosphocholine, which can be resolved by TLC, column chromatography, or HPLC. Cation-exchange chromatography using small gravity-fed 1 mL resin columns is very effective to quickly process a large number of samples. Glycerophosphocholine is collected in the eluant because it does not bind to the resin. Phosphocholine is eluted by water and choline by 1 M HCl, which can then be quantitated by liquid scintillation counting [138,139].

3.4.2.2 Measurement of PLD Activity *in Vitro*

PLD activity *in vitro* can be determined for purified enzyme and cell and tissue extracts using a headgroup release assay, transphosphatidylation, coupled assays, and conductimetric and spectrophotometric assays. The differences between these assays lie in the choice of substrate and methods used to detect and quantitate the products.

3.4.2.2.1 *Headgroup Release Assay*

Similar to the assay *in vivo* mentioned above, this procedure measures the formation of [3H]-choline by PLD-mediated hydrolysis of [3H]-labeled PtdCho [32]. A lipid mixture consisting of PtdEtn, [3H]-labeled PtdCho, PtdIns(4,5)P$_2$, and PtdSer (16: 1.4:1:2) is used to present the substrate to the enzyme [19,40,41]. The phospholipid mixture is dried under a stream of N$_2$ and resuspended by sonication to form liposomes, which are then incubated with the PLD source in an assay buffer that has approximately physiological ionic strength and composition. After a 30 min reaction period, the [3H]-choline product is separated from unreacted [3H]-PtdCho by the precipitation of unreacted lipid substrate with BSA and TCA. The radiolabeled choline is then quantitated by liquid scintillation counting. The major shortcoming of this method is that activated phospholipase B (PLB) and phospholipase C (PLC), if present in the PLD source, can generate glycerophosphocholine or phosphocholine, respectively, which are also water-soluble. Choline, phosphocholine, and glycerophosphocholine can be separated by cation-exchange chromatography using either HPLC or gravity-fed columns as described above for the headgroup release assay *in vivo* [4,138,139]. In general, however, PLB and PLC do not generate a strong enough signal to prevent using the simple version of the assay from being used.

3.4.2.2.2 *Transphosphatidylation Assay*

Again, this assay is based on the property of PLD-catalyzed transphosphatidylation. Similar to the assay *in vivo*, the substrate PtdCho has to be labeled by either radioactive or fluorescent tracers. PtdCho can be labeled in the diester phosphate with [32P] or with [3H] in one of the acyl chains. Alternatively, a commercially available fluorescent derivative of PtdCho (BODIPY-PtdCho) has been shown to be an effective substrate for PLD1 and PLD2. However, BODIPY-PtdCho is a 1-alkyl-2-acyl PtdCho analog, and several pieces of evidence have indicated that PLD activity in mammalian cells distinguishes substrates with this type of chain linkage from substrates having diacyl chains linked to their glycerol backbone. Since [3H]-acylPtdCho is commercially available from several sources, it is commonly used in this procedure. Alternatively, [32P]-PtdCho can be isolated from a lipid mixture of [32P]-PO$_4$-labeled tissue or cell lines if elevated sensitivity is needed. The radiolabeled or fluorescent

PtdCho are mixed with carrier lipids, PtdEtn and PtdIns(4,5)P$_2$, and dispersed by sonication into liposomes. In the incubation, the substrates are combined with the enzyme source and a primary alcohol added to measure PLD-catalyzed transphosphatidylation. Generally, 1%–5% of primary alcohol is used to maximize the transphosphatidylation efficiency. The reaction is terminated by ice-cold methanol and the organic phase recovered. Unreacted PtdCho can be separated from the PtdOH or phosphatidylalcohol by TLC using the solvent system described previously.

3.4.2.2.3   Coupled Assays

A number of different PLD assays have been reported that use enzymatic conversion of choline by choline kinase or choline oxidase to generate radiolabeled or spectrophotometrically detectable products.

The *radioenzymatic assay for choline* employs commercially available yeast choline and [γ-^{32}P]-ATP to generate [^{32}P]-phosphocholine, which can be isolated by anion-exchange chromatography. This assay is highly sensitive and has been used to measure agonist-stimulated increases in choline in several cell types [141,142].

The *enzyme-coupled spectrophotometric assay* couples PLD-catalyzed formation of choline to the oxidation of NADH using choline kinase, pyruvate kinase, and lactate dehydrogenase, which are commercially available. Formation of NAD$^+$ is monitored by a decrease in absorbance of the reaction mixture at 340 nm [143]. This method is especially useful for detailed kinetic analyses because enzymatic activity can be continuously monitored.

Choline-oxidative-catalyzed formation of hydrogen peroxide can be measured by an amperometric peroxidase electrode or by coupling to the peroxidase-catalyzed formation of a dye complex that can be detected by absorbance at 500 nM. The sensitivity limit of this assay is 100 nmol. The major drawback of this method is interference from contaminating activities, which limits the use of this assay to purified PLD preparations.

3.4.2.2.4   Conductimetric/Charge-Dependent Assays

Hydrolysis of PtdCho by PLD generates two independently mobile ionic species: PtdOH, which increases the net negative charge of the lipid/aqueous interface and cationic choline that increases the solution conductance. Both of these phenomena have been used to design assay procedures for the determination of PLD activity. Measurement of solution electrical conductance is a simple method to detect the hydrolytic reactions catalyzed by phospholipases like PLD. The procedure measures the change in cell conductance under constant voltage conditions, which allows PLD activity to be determined in real time [144]. However, this procedure is limited to systems where the source of enzyme is free from contaminating activities. Spectrophotometric assays have also been used to monitor the change of membrane surface charge caused by PLD activity, based on metachromic dyes that produce a change in the absorbance spectrum when they interact with negatively charged groups located on part of a polymeric chain. The dye safrazine is useful to measure PLD activity. The dye binds to negatively charged phospholipids such as PtdOH, which leads to a decrease in absorbance at 520 nm with isobestic points at 475 and 560 nm (whereas binding to PtdCho does not change the absorbance of the dye). The absorbance change in A560/520 caused by the conversion of PtdCho to PtdOH can thus be used as a measure of PLD activity.

In general, substrate presentation and the source of enzyme are the key factors to be considered in selecting the optimal PLD assay. Other factors to be considered include whether it is necessary to monitor the activity in real time and the pros and cons of using methods that employ radioactive labeling. For assay of PLD activities *in vitro*, the headgroup-release or transphosphatidylation assays using radiolabeled substrates are efficient and simple methods. For real-time measurement, the enzyme-coupled, conductance measurement or membrane-charge-dependent assays are the best approaches. To assess PLD stimulation by extracellular stimuli in whole cells, the method of choice is the transphosphatidylation assay *in vivo*.

LIST OF ABBREVIATIONS

DAG	diacylglycerol
EGF	epidermal growth factor
ER	endoplasmic reticulum
fMLP	fMet-Leu-Phe
HKD	$HxK(x)_4D(x)_6GSxN$
HPLC	high performance liquid chromatography
lysoPtdOH	lysophosphatidic acid
MOR	μ-opioid receptor
mTOR	mammalian target of rapamycin
NADPH	nicotinamide adenine dinucleotide phosphate-oxidase
PAP	phosphatidic acid phosphohydrolases
PDGF	platelet-derived growth factor
PH	plekstrin homology
PKC	protein kinase C
PLB	phospholipase B
PLC	phospholipase C
PLD	phospholipase D
PtdCho	phosphatidylcholine
PtdEtn	phosphatidylethanolamine
$PtdIns(4,5)P_2$	phosphatidylinositol-4,5-bisphosphate
$PtdIns(3,4,5)P_3$	phosphatidylinositol-3,4,5-triphosphate
PtdOH	phosphatidic acid
PX	phox
TLC	thin layer chromatography

REFERENCES

1. Hanahan DJaC, I.L. (1947) The phosphorus-containing lipides of the carrot. *J. Biol. Chem.* 168:233–240.
2. Hanahan DJaC, I.L. (1948) On the nature of the phosphorus-containing lipides of cabbage leaves and their relation to a phospholipid-splitting enzyme contained in these leaves. *J. Biol. Chem.* 172:191–198.

3. Saito, M. and Kanfer, J. (1975) Phosphatidohydrolase activity in a solubilized preparation from rat brain particulate fraction. *Arch. Biochem. Biophys.* 169:318–323.

4. Liscovitch, M. et al. (1985) High-performance liquid chromatography of water-soluble choline metabolites. *Anal. Biochem.* 151:182–187.

5. Sung, T.C. et al. (1997) Mutagenesis of phospholipase D defines a superfamily including a trans-Golgi viral protein required for poxvirus pathogenicity. *EMBO J.* 16:4519–4530.

6. Yang, S.F., Freer, S., and Benson, A.A. (1967) Transphosphatidylation by phospholipase D. *J. Biol. Chem.* 242:477–484.

7. Jenkins, G.M. and Frohman, M.A. (2005) Phospholipase D: A lipid centric review. *Cell. Mol. Life Sci.* 62:2305–2316.

8. Wakelam, M.J., Hodgkin, M., and Martin, A. (1995) The measurement of phospholipase D-linked signaling in cells. *Methods Mol. Biol.* 41:271–278.

9. Morris, A.J., Frohman, M.A., and Engebrecht, J. (1997) Measurement of phospholipase D activity. *Anal. Biochem.* 252:1–9.

10. Emoto, M. et al. (2000) A role for phospholipase D in GLUT4 glucose transporter translocation. *J. Biol. Chem.* 275:7144–7151.

11. Skippen, A. et al. (2002) Mechanism of ADP ribosylation factor-stimulated phosphatidylinositol 4,5-bisphosphate synthesis in HL60 cells. *J. Biol. Chem.* 277:5823–5831.

12. Pannequin, J. et al. (2007) Phosphatidylethanol accumulation promotes intestinal hyperplasia by inducing ZONAB-mediated cell density increase in response to chronic ethanol exposure. *Mol. Cancer Res.* 5:1147–1157.

13. Aroor, A.R. et al. (2002) Phosphotidylethanol mimics ethanol modulation of p42/44 mitogen-activated protein kinase signalling in hepatocytes. *Alcohol Alcohol.* 37:534–539.

14. Balboa, M.A. and Insel, P.A. (1995) Nuclear phospholipase D in Madin-Darby canine kidney cells. Guanosine 5′-O-(thiotriphosphate)-stimulated activation is mediated by RhoA and is downstream of protein kinase C. *J. Biol. Chem.* 270:29843–29847.

15. Ktistakis, N.T. et al. (1995) Phospholipase D is present on Golgi-enriched membranes and its activation by ADP ribosylation factor is sensitive to brefeldin A. *Proc. Natl. Acad. Sci. USA* 92:4952–4956.

16. Ktistakis, N.T. et al. (1996) Evidence that phospholipase D mediates ADP ribosylation factor-dependent formation of Golgi coated vesicles. *J. Cell Biol.* 134:295–306.

17. Provost, J.J. et al. (1996) Tissue-specific distribution and subcellular distribution of phospholipase D in rat: Evidence for distinct RhoA- and ADP-ribosylation factor (ARF)-regulated isoenzymes. *Biochem. J.* 319(Pt 1):285–291.

18. Whatmore, J. et al. (1996) ADP-ribosylation factor 1-regulated phospholipase D activity is localized at the plasma membrane and intracellular organelles in HL60 cells. *Biochem. J.* 320(Pt 3):785–794.

19. Colley, W.C. et al. (1997) Phospholipase D2, a distinct phospholipase D isoform with novel regulatory properties that provokes cytoskeletal reorganization. *Curr. Biol.* 7:191–201.

20. Du, G. et al. (2003) Regulation of phospholipase D1 subcellular cycling through coordination of multiple membrane association motifs. *J. Cell Biol.* 162:305–315.

21. Sciorra, V.A. et al. (2002) Dual role for phosphoinositides in regulation of yeast and mammalian phospholipase D enzymes. *J. Cell Biol.* 159:1039–1049.

22. Du, G. et al. (2004) Phospholipase D2 localizes to the plasma membrane and regulates angiotensin II receptor endocytosis. *Mol. Biol. Cell* 15:1024–1030.

23. Choi, S.Y. et al. (2006) A common lipid links Mfn-mediated mitochondrial fusion and SNARE-regulated exocytosis. *Nat. Cell Biol.* 8:1255–1262.

24. Munck, A. et al. (2005) Hu-K4 is a ubiquitously expressed type 2 transmembrane protein associated with the endoplasmic reticulum. *FEBS J.* 272:1718–1726.

25. Besterman, J.M., Duronio, V., and Cuatrecasas, P. (1986) Rapid formation of diacyl-glycerol from phosphatidylcholine: A pathway for generation of a second messenger. *Proc. Natl. Acad. Sci. USA* 83:6785–6789.

26. Daniel, L.W., Waite, M., and Wykle, R.L. (1986) A novel mechanism of diglyceride formation. 12-*O*-tetradecanoylphorbol-13-acetate stimulates the cyclic breakdown and resynthesis of phosphatidylcholine. *J. Biol. Chem.* 261:9128–9132.

27. Boeckino, S.B. et al. (1987) Phosphatidate accumulation in hormone-treated hepato-cytes via a phospholipase D mechanism. *J. Biol. Chem.* 262:15309–15315.

28. Pai, J.K. et al. (1988) Activation of phospholipase D by chemotactic peptide in HL-60 granulocytes. *Biochem. Biophys. Res. Commun.* 150:355–364.

29. Oude Weernink, P.A. et al. (2007) Dynamic phospholipid signaling by G protein-coupled receptors. *Biochim. Biophys. Acta* 1768:888–900.

30. Lee, C.S. et al. (2006) The phox homology domain of phospholipase D activates dynamin GTPase activity and accelerates EGFR endocytosis. *Nat. Cell Biol.* 8:477–484.

31. Preininger, A.M. et al. (2006) Direct modulation of phospholipase D activity by Gβγ. *Mol. Pharmacol.* 70:311–318.

32. Brown, H.A. et al. (1993) ADP-ribosylation factor, a small GTP-dependent regulatory protein, stimulates phospholipase D activity. *Cell* 75:1137–1144.

33. Brown, H.A. et al. (1995) Partial purification and characterization of ARF-sensitive phospholipase D from porcine brain. *J. Biol. Chem.* 270:14935–14943.

34. McDermott, M., Wakelam, M.J., and Morris, A.J. (2004) Phospholipase D. *Biochem. Cell Biol.* 82:225–253.

35. Vitale, N. et al. (2001) Phospholipase D1: A key factor for the exocytotic machinery in neuroendocrine cells. *EMBO J.* 20:2424–2434.

36. Huang, P. and Frohman, M.A. (2007) The potential for phospholipase D as a new thera-peutic target. *Expert Opin. Ther. Targets* 11:707–716.

37. Iyer, S.S. et al. (2004) Phospholipases D1 and D2 coordinately regulate macrophage phagocytosis. *J. Immunol.* 173:2615–2623.

38. Wang, X., Dyer, J.H., and Zheng, L. (1993) Purification and immunological analysis of phospholipase D from castor bean endosperm. *Arch. Biochem. Biophys.* 306:486–494.

39. Rose, K. et al. (1995) Phospholipase D signaling is essential for meiosis. *Proc. Natl. Acad. Sci. USA* 92:12151–12155.

40. Hammond, S.M. et al. (1995) Human ADP-ribosylation factor-activated phosphati-dylcholine-specific phospholipase D defines a new and highly conserved gene family. *J. Biol. Chem.* 270:29640–29643.

41. Hammond, S.M. et al. (1997) Characterization of two alternately spliced forms of phos-pholipase D1. Activation of the purified enzymes by phosphatidylinositol 4,5-bisphos-phate, ADP-ribosylation factor, and Rho family monomeric GTP-binding proteins and protein kinase C-α. *J. Biol. Chem.* 272:3860–3868.

42. Koonin, E.V. (1996) A duplicated catalytic motif in a new superfamily of phosphohy-drolases and phospholipid synthases that includes poxvirus envelope proteins. *Trends Biochem. Sci.* 21:242–243.

43. Ponting, C.P. and Kerr, I.D. (1996) A novel family of phospholipase D homologues that includes phospholipid synthases and putative endonucleases: Identification of dupli-cated repeats and potential active site residues. *Protein Sci.* 5:914–922.

44. Zhao, Y. et al. (1997) Expression, characterization, and crystallization of a member of the novel phospholipase D family of phosphodiesterases. *Protein Sci.* 6:2655–2658.

45. Interthal, H., Pouliot, J.J., and Champoux, J.J. (2001) The tyrosyl-DNA phosphodi-esterase Tdp1 is a member of the phospholipase D superfamily. *Proc. Natl. Acad. Sci. USA* 98:12009–12014.

46. Liscovitch, M. et al. (2000) Phospholipase D: Molecular and cell biology of a novel gene family. *Biochem. J.* 345:401–415.

47. Gottlin, E.B. et al. (1998) Catalytic mechanism of the phospholipase D superfamily proceeds via a covalent phosphohistidine intermediate. *Proc. Natl. Acad. Sci. USA* 95:9202–9207.
48. Stuckey, J.A. and Dixon, J.E. (1999) Crystal structure of a phospholipase D family member. *Nat. Struct. Biol.* 6:278–284.
49. Sung, T.C. et al. (1999) Molecular analysis of mammalian phospholipase D2. *J. Biol. Chem.* 274:494–502.
50. Sung, T.C. et al. (1999) Structural analysis of human phospholipase D1. *J. Biol. Chem.* 274:3659–3666.
51. Xu, Y. et al. (2001) The phox homology (PX) domain, a new player in phosphoinositide signalling. *Biochem. J.* 60:513–530.
52. Stahelin, R.V. et al. (2004) Mechanism of membrane binding of the phospholipase D1 PX domain. *J. Biol. Chem.* 279:54918–54926.
53. Morris, A.J. (2007) Regulation of phospholipase D activity, membrane targeting and intracellular trafficking by phosphoinositides. *Biochem. Soc. Symp.* 74:247–257.
54. Stace, C.L. and Ktistakis, N.T. (2006) Phosphatidic acid- and phosphatidylserine-binding proteins. *Biochim. Biophys. Acta* 1761:913–926.
55. Zhang, Y. et al. (1999) Loss of receptor regulation by a phospholipase D1 mutant unresponsive to protein kinase C. *EMBO J.* 18:6339–6348.
56. Liu, M.Y., Gutowski, S., and Sternweis, P.C. (2001) The C terminus of mammalian phospholipase D is required for catalytic activity. *J. Biol. Chem.* 276:5556–5562.
57. Olson, S.C., Bowman, E.P., and Lambeth, J.D. (1991) Phospholipase D activation in a cell-free system from human neutrophils by phorbol 12-myristate 13-acetate and guanosine 5′-O-(3-thiotriphosphate). Activation is calcium dependent and requires protein factors in both the plasma membrane and cytosol. *J. Biol. Chem.* 266:17236–17242.
58. Ohguchi, K. et al. (1995) Activation of membrane-bound phospholipase D by protein kinase C in HL60 cells: Synergistic action of a small GTP-binding protein RhoA. *Biochem. Biophys. Res. Commun.* 211:306–311.
59. Kuribara, H. et al. (1995) Synergistic activation of rat brain phospholipase D by ADP-ribosylation factor and RhoA p21, and its inhibition by *Clostridium botulinum* C3 exoenzyme. *J. Biol. Chem.* 270:25667–25671.
60. Singer, W.D. et al. (1995) Resolved phospholipase D activity is modulated by cytosolic factors other than ARF. *J. Biol. Chem.* 270:14944–14950.
61. Walker, S.J. et al. (2000) Activation of phospholipase D1 by Cdc42 requires the Rho insert region. *J. Biol. Chem.* 275:15665–15668.
62. Kim, J.H. et al. (1998) Activation of phospholipase D1 by direct interaction with ADP-ribosylation factor 1 and RalA. *FEBS Lett.* 430:231–235.
63. Min, D.S., Kim, E.G., and Exton, J.H. (1998) Involvement of tyrosine phosphorylation and protein kinase C in the activation of phospholipase D by H_2O_2 in Swiss 3T3 fibroblasts. *J. Biol. Chem.* 273:29986–29994.
64. Hodgkin, M.N. et al. (1999) Characterization of the regulation of phospholipase D activity in the detergent-insoluble fraction of HL60 cells by protein kinase C and small G-proteins. *Biochem. J.* 339(Pt 1):87–93.
65. Martin, A. et al. (1996) Activation of phospholipase D and phosphatidylinositol 4-phosphate 5-kinase in HL60 membranes is mediated by endogenous ARF but not Rho. *J. Biol. Chem.* 271:17397–17403.
66. Caumont, A.S. et al. (1998) Regulated exocytosis in chromaffin cells. Translocation of ARF6 stimulates a plasma membrane-associated phospholipase D. *J. Biol. Chem.* 273:1373–1379.
67. Lopez, I., Arnold, R.S., and Lambeth, J.D. (1998) Cloning and initial characterization of a human phospholipase D2 (hPLD2). ADP-ribosylation factor regulates hPLD2. *J. Biol. Chem.* 273:12846–12852.

68. Houle, M.G. et al. (1995) ADP-ribosylation factor translocation correlates with potentiation of GTPγ S-stimulated phospholipase D activity in membrane fractions of HL-60 cells. *J. Biol. Chem.* 270:22795–22800.
69. Schmidt, M. et al. (1996) Inhibition of receptor signaling to phospholipase D by *Clostridium difficile* toxin B. Role of Rho proteins. *J. Biol. Chem.* 271:2422–2426.
70. Schmidt, M. et al. (1996) Restoration of *Clostridium difficile* toxin-B-inhibited phospholipase D by phosphatidylinositol 4,5-bisphosphate. *Eur. J. Biochem.* 240:707–712.
71. Schmidt, M. et al. (1999) A role for rho-kinase in rho-controlled phospholipase D stimulation by the m3 muscarinic acetylcholine receptor. *J. Biol. Chem.* 274:14648–14654.
72. Pai, J.K., Dobek, E.A., and Bishop, W.R. (1991) Endothelin-1 activates phospholipase D and thymidine incorporation in fibroblasts overexpressing protein kinase C beta 1. *Cell Regul.* 2:897–903.
73. Pachter, J.A. et al. (1992) Differential regulation of phosphoinositide and phosphatidylcholine hydrolysis by protein kinase C-β 1 overexpression. Effects on stimulation by α-thrombin, guanosine 5′-*O*-(thiotriphosphate), and calcium. *J. Biol. Chem.* 267:9826–9830.
74. Eldar, H. et al. (1993) Up-regulation of phospholipase D activity induced by overexpression of protein kinase C-α. Studies in intact Swiss/3T3 cells and in detergent-solubilized membranes *in vitro. J. Biol. Chem.* 268:12560–12564.
75. Lee, T.G. et al. (1997) Phorbol myristate acetate-dependent association of protein kinase Cα with phospholipase D1 in intact cells. *Biochim. Biophys. Acta* 1347:199–204.
76. Park, S.K., Min, D.S., and Exton, J.H. (1998) Definition of the protein kinase C interaction site of phospholipase D. *Biochem. Biophys. Res. Commun.* 244:364–367.
77. Singer, W.D. et al. (1996) Regulation of phospholipase D by protein kinase C is synergistic with ADP-ribosylation factor and independent of protein kinase activity. *J. Biol. Chem.* 271:4504–4510.
78. Cross, M.J. et al. (1996) Stimulation of actin stress fibre formation mediated by activation of phospholipase D. *Curr. Biol.* 6:588–597.
79. Chardin, P. et al. (1996) A human exchange factor for ARF contains Sec7- and pleckstrin-homology domains. *Nature* 384:481–484.
80. Voss, M. et al. (1999) Phospholipase D stimulation by receptor tyrosine kinases mediated by protein kinase C and a Ras/Ral signaling cascade. *J. Biol. Chem.* 274:34691–34698.
81. Choi, W.S. et al. (2004) Activation of RBL-2H3 mast cells is dependent on tyrosine phosphorylation of phospholipase D2 by Fyn and Fgr. *Mol. Cell Biol.* 24:6980–6992.
82. Slaaby, R. et al. (1998) PLD2 complexes with the EGF receptor and undergoes tyrosine phosphorylation at a single site upon agonist stimulation. *J. Biol. Chem.* 273:33722–33727.
83. Manifava, M., Sugars, J., and Ktistakis, N.T. (1999) Modification of catalytically active phospholipase D1 with fatty acid *in vivo. J. Biol. Chem.* 274:1072–1077.
84. Sugars, J.M. et al. (1999) Fatty acylation of phospholipase D1 on cysteine residues 240 and 241 determines localization on intracellular membranes. *J. Biol. Chem.* 274:30023–30027.
85. Xie, Z., Ho, W.T., and Exton, J.H. (2001) Requirements and effects of palmitoylation of rat PLD1. *J. Biol. Chem.* 276:9383–9391.
86. Sciorra, V.A. and Morris, A.J. (1999) Sequential actions of phospholipase D and phosphatidic acid phosphohydrolase 2b generate diglyceride in mammalian cells. *Mol. Biol. Cell* 10:3863–3876.
87. Brindley, D.N. and Waggoner, D.W. (1996) Phosphatidate phosphohydrolase and signal transduction. *Chem. Phys. Lipids* 80:45–57.
88. Exton, J.H. (1994) Phosphatidylcholine breakdown and signal transduction. *Biochim. Biophys. Acta* 1212:26–42.

89. Spiegel, S., Foster, D., and Kolesnick, R. (1996) Signal transduction through lipid second messengers. *Curr. Opin. Cell Biol.* 8:159–167.
90. Morris, A.J., Engebrecht, J., and Frohman, M.A. (1996) Structure and regulation of phospholipase D. *Trends Pharmacol. Sci.* 17:182–185.
91. Bi, K., Roth, M.G., and Ktistakis, N.T. (1997) Phosphatidic acid formation by phospholipase D is required for transport from the endoplasmic reticulum to the Golgi complex. *Curr. Biol.* 7:301–307.
92. Pathre, P. et al. (2003) Activation of phospholipase D by the small GTPase Sar1p is required to support COPII assembly and ER export. *EMBO J.* 22:4059–4069.
93. Jones, D., Morgan, C., and Cockcroft, S. (1999) Phospholipase D and membrane traffic. Potential roles in regulated exocytosis, membrane delivery and vesicle budding. *Biochim. Biophys. Acta* 1439:229–244.
94. Shen, Y., Xu, L., and Foster, D.A. (2001) Role for phospholipase D in receptor-mediated endocytosis. *Mol. Cell Biol.* 21:595–602.
95. Koch, T. et al. (2003) ADP-ribosylation factor-dependent phospholipase D2 activation is required for agonist-induced μ-opioid receptor endocytosis. *J. Biol. Chem.* 278:9979–9985.
96. Koch, T. et al. (2004) Phospholipase D2 modulates agonist-induced μ-opioid receptor desensitization and resensitization. *J. Neurochem.* 88:680–688.
97. Bhattacharya, M. et al. (2004) Ral and phospholipase D2-dependent pathway for constitutive metabotropic glutamate receptor endocytosis. *J. Neurosci.* 24:8752–8761.
98. Corrotte, M. et al. (2006) Dynamics and function of phospholipase D and phosphatidic acid during phagocytosis. *Traffic* 7:365–377.
99. Humeau, Y. et al. (2001) A role for phospholipase D1 in neurotransmitter release. *Proc. Natl. Acad. Sci. USA* 98:15300–15305.
100. Vitale, N., Chasserot-Golaz, S., and Bader, M.F. (2002) Regulated secretion in chromaffin cells: An essential role for ARF6-regulated phospholipase D in the late stages of exocytosis. *Ann. NY Acad. Sci.* 971:193–200.
101. Peng, Z. and Beaven, M.A. (2005) An essential role for phospholipase D in the activation of protein kinase C and degranulation in mast cells. *J. Immunol.* 174:5201–5208.
102. Hughes, W.E. et al. (2004) Phospholipase D1 regulates secretagogue-stimulated insulin release in pancreatic β-cells. *J. Biol. Chem.* 279:27534–27541.
103. Stutchfield, J. and Cockcroft, S. (1993) Correlation between secretion and phospholipase D activation in differentiated HL60 cells. *Biochem. J.* 293(Pt 3):649–655.
104. Huang, P. et al. (2005) Insulin-stimulated plasma membrane fusion of Glut4 glucose transporter-containing vesicles is regulated by phospholipase D1. *Mol. Biol. Cell* 16:2614–2623.
105. Vicogne, J. et al. (2006) Asymmetric phospholipid distribution drives *in vitro* reconstituted SNARE-dependent membrane fusion. *Proc. Natl. Acad. Sci. USA* 103:14761–14766.
106. Moritz, A. et al. (1992) Phosphatidic acid is a specific activator of phosphatidylinositol-4-phosphate kinase. *J. Biol. Chem.* 267:7207–7210.
107. Jenkins, G.H., Fisette, P.L., and Anderson, R.A. (1994) Type I phosphatidylinositol 4-phosphate 5-kinase isoforms are specifically stimulated by phosphatidic acid. *J. Biol. Chem.* 269:11547–11554.
108. Jones, D.R., Sanjuan, M.A., and Merida, I. (2000) Type I α phosphatidylinositol 4-phosphate 5-kinase is a putative target for increased intracellular phosphatidic acid. *FEBS Lett.* 476:160–165.
109. Honda, A. et al. (1999) Phosphatidylinositol 4-phosphate 5-kinase α is a downstream effector of the small G protein ARF6 in membrane ruffle formation. *Cell* 99:521–532.
110. Kam, Y. and Exton, J.H. (2001) Phospholipase D activity is required for actin stress fiber formation in fibroblasts. *Mol. Cell Biol.* 21:4055–4066.

111. Porcelli, A.M. et al. (2002) Phospholipase D stimulation is required for sphingosine-1-phosphate activation of actin stress fibre assembly in human airway epithelial cells. *Cell Signal.* 14:75–81.

112. Komati, H. et al. (2005) Phospholipase D is involved in myogenic differentiation through remodeling of actin cytoskeleton. *Mol. Biol. Cell* 16:1232–1244.

113. Foster, D.A. (2007) Regulation of mTOR by phosphatidic acid? *Cancer Res.* 67:1–4.

114. Wood, L.D. et al. (2007) The genomic landscapes of human breast and colorectal cancers. *Science* 318:1108–1113.

115. Yang, Q. and Guan, K.L. (2007) Expanding mTOR signaling. *Cell Res.* 17:666–681.

116. Fang, Y. et al. (2001) Phosphatidic acid-mediated mitogenic activation of mTOR signaling. *Science* 294:1942–1945.

117. Chen, Y., Zheng, Y., and Foster, D.A. (2003) Phospholipase D confers rapamycin resistance in human breast cancer cells. *Oncogene* 22:3937–3942.

118. Chen, Y., Rodrik, V., and Foster, D.A. (2005) Alternative phospholipase D/mTOR survival signal in human breast cancer cells. *Oncogene* 24:672–679.

119. Foster, D.A. and Xu, L. (2003) Phospholipase D in cell proliferation and cancer. *Mol. Cancer Res.* 1:789–800.

120. Song, J.G., Pfeffer, L.M., and Foster, D.A. (1991) v-Src increases diacylglycerol levels via a type D phospholipase-mediated hydrolysis of phosphatidylcholine. *Mol. Cell Biol.* 11:4903–4908.

121. Lu, Z. et al. (1997) Tumor promotion by depleting cells of protein kinase C δ. *Mol. Cell Biol.* 17:3418–3428.

122. Joseph, T. et al. (2001) Transformation of cells overexpressing a tyrosine kinase by phospholipase D1 and D2. *Biochem. Biophys. Res. Commun.* 289:1019–1024.

123. Zhao, C. et al. (2007) Phospholipase D2-generated phosphatidic acid couples EGFR stimulation to Ras activation by Sos. *Nat. Cell Biol.* 9:706–712.

124. Gupta, G.P. and Massague, J. (2006) Cancer metastasis: Building a framework. *Cell* 127:679–695.

125. Shen, Y., Zheng, Y., and Foster, D.A. (2002) Phospholipase D2 stimulates cell protrusion in v-Src-transformed cells. *Biochem. Biophys. Res. Commun.* 293:201–206.

126. Zheng, Y. et al. (2006) Phospholipase D couples survival and migration signals in stress response of human cancer cells. *J. Biol. Chem.* 281:15862–15868.

127. Saqib, K.M. and Wakelam, M.J. (1997) Differential expression of human phospholipase D genes. *Biochem. Soc. Trans.* 25:S586.

128. Steed, P.M. et al. (1998) Characterization of human PLD2 and the analysis of PLD isoform splice variants. *FASEB J.* 12:1309–1317.

129. Katayama, K. et al. (1998) Cloning, differential regulation and tissue distribution of alternatively spliced isoforms of ADP-ribosylation-factor-dependent phospholipase D from rat liver. *Biochem. J.* 329(Pt 3):647–652.

130. Colley, W.C. et al. (1997) Cloning and expression analysis of murine phospholipase D1. *Biochem. J.* 326(Pt 3):745–753.

131. Kodaki, T. and Yamashita, S. (1997) Cloning, expression, and characterization of a novel phospholipase D complementary DNA from rat brain. *J. Biol. Chem.* 272:11408–11413.

132. Zhang, Y. et al. (2004) Increased expression of two phospholipase D isoforms during experimentally induced hippocampal mossy fiber outgrowth. *Glia* 46:74–83.

133. Ganley, I.G. et al. (2001) Interaction of phospholipase D1 with a casein-kinase-2-like serine kinase. *Biochem. J.* 354:369–378.

134. Freyberg, Z., Bourgoin, S., and Shields, D. (2002) Phospholipase D2 is localized to the rims of the Golgi apparatus in mammalian cells. *Mol. Biol. Cell* 13:3930–3942.

135. Freyberg, Z. et al. (2001) Intracellular localization of phospholipase D1 in mammalian cells. *Mol. Biol. Cell* 12:943–955.

136. Meier, K.E. et al. (1999) Expression of phospholipase D isoforms in mammalian cells. *Biochim. Biophys. Acta* 1439:199–213.

137. Saito, M. et al. (2007) Expression of phospholipase D2 in human colorectal carcinoma. *Oncol. Rep*. 18:1329–1334.

138. Wakelam, M.J.O., Hodgkin, M., and Martin, A. (1995). Methods in molecular biology, in *Signal Transduction Protocols*, Humana Press, Totowa, NJ, 1995, pp. 271–278.

139. Cook, S.J. and Wakelam, M.J.O. (1989) Analysis of the water-soluble products of phosphatidylcholine breakdown by ion-exchange chromatography. Bombesin and TPA (12-*O*-tetradecanoylphorbol 13-acetate) stimulate choline generation in Swiss 3T3 cells by a common mechanism. *Biochem. J*. 263:581–587.

140. Bligh, E.G. and Dyer, W.J. (1959) A rapid method of total lipid extraction and purification. *Can. J. Biochem. Physiol*. 37:911–917.

141. Muma, N.A. and Rowell, P.P. (1985) A sensitive and specific radioenzymatic assay for the simultaneous determination of choline and phosphatidylcholine. *J. Neurosci. Methods* 12:249–257.

142. Murray, J.J. et al. (1990) Isolation and enzymic assay of choline and phosphocholine present in cell extracts with picomole sensitivity. *Biochem. J*. 270:63–68.

143. Carman, G.M. et al. (1981) A spectrophotometric method for the assay of phospholipase D activity. *Anal. Biochem*. 110:73–76.

144. Mezna, M. and Lawrence, A.J. (1994) Conductimetric assays for the hydrolase and transferase activities of phospholipase D enzymes. *Anal. Biochem*. 218:370–376.

4 Methods for Measuring the Activity and Expression of Diacylglycerol Kinase

Kaoru Goto, Yasukazu Hozumi,
Tomoyuki Nakano, Alberto M. Martelli, and
Hisatake Kondo

CONTENTS

4.1   INTRODUCTION

Diacylglycerol kinase (DGK) phosphorylates diacylglycerol (DAG) to produce phosphatidic acid (PtdOH). DAG serves not only as a major intermediate product in the synthesis of several kinds of lipids but also as a bioactive molecule [1,2]. Therefore, DGK is thought to participate in signal transduction by modulating the levels of DAG in a variety of cellular responses to extracellular stimuli.

One of the best known functional roles of DGK is in the regulation of protein kinase C (PKC), for which DAG acts as an allosteric activator and whose activity plays a central role in the control of proliferation and differentiation of many different cell types [3–6]. Recent studies show that DAG also activates other proteins including Ras GRP [7], the chimaerins [8], Unc-13 [9], protein kinase D [10], and some mammalian homologues of transient receptor potential proteins [11]. In addition, DAG may modulate the activity of several proteins in the Rho and Ras families, thus potentially affecting other cellular functions such as cytoskeletal reorganization, cellular growth, and carcinogenesis [12,13]. These studies suggest that DAG is more widely implicated in cellular events than previously thought; as such, regulation of its levels are important.

In addition to DAG, PtdOH, the product of DGK activity, may also serve as a potent lipid second messenger and has been reported to regulate a growing list of signaling proteins including PKC-ζ [14], phosphatidylinositol-4-phosphate (PIP) 5-kinase [15], and phospholipase C (PLC)-γl [16]. PtdOH also directly interacts with the mammalian target of rapamycin, which governs cell growth and proliferation by mediating the mitogen- and nutrient-dependent signal transduction that regulates messenger RNA translation in a variety of cells [17]. Taken together, DGK is thought to be one of the key enzymes closely involved in the lipid-mediated cellular signaling events, because it is directly involved in the control of these two lipid messenger molecules.

Molecular and cellular studies have revealed that DGK comprises a family of isozymes and each has a unique character in terms of expression and localization, regulatory mechanism, binding partner, and subcellular localization. This chapter focuses on the characteristics of this enzyme family, including structural, biochemical, morphological, and pathophysiological points of view. We also include historical perspective for the enzyme and its surroundings.

4.2   A HISTORICAL PERSPECTIVE

4.2.1   PHOSPHATIDYLINOSITOL TURNOVER AND DIACYLGLYCEROL

Phospholipids are one of the major components of biological systems and contribute to the compartmentalization of cellular and intracellular environments. Among those, phosphoinositide (PtdIns), a minor component (approximately 4%–6%) of the total membrane lipid, is not just a structural block of the membrane, but serves as a functional molecule. The story about "phospholipid effect" got started a long time ago [18]. In a series of experiments, when pigeon pancreas was treated with acetylcholine or carbamyl choline, a roughly 10-fold overall increase in the incorporation of ^{32}P-labeled phosphate into phospholipids was seen, which correlated with enhanced secretion of amylase. Stimulation of label incorporation varied considerably between phospholipids, being 15-fold for PtdIns, threefold for PtdOH, and much lower for phosphatidylcholine (PtdCho) and phosphatidylethanolamine (PtdEtn) [19–21]. Those findings launched the study on the significance of preferential metabolism of PtdIns lipids. As early as 1962, the PtdIns lipids were shown to turn over rapidly in the brain, and it was suggested that they may play an important physiological role [22].

We now know that receptor stimulation of the plasma membrane phosphoinositide cycle leads to the production of two second messengers, inositol-1,4,5-trisphosphate (IP$_3$) and sn-1,2-diacylglycerol (DAG), which initially came out from two observations. First, agonists that stimulated inositol phospholipid hydrolysis also induced intracellular mobilization of an established second messenger, Ca^{2+} [23]. Measurements in different cell types revealed that the increase in IP$_3$ either preceded or coincided with the onset of the Ca^{2+} signal [24]. Second, PLC-dependent cleavage of phosphatidylinositol-4,5-bisphosphate (PtdIns(4,5)P$_2$) also produced another product, namely DAG. While IP$_3$ is free to diffuse throughout the cytosol, DAG remains within the lipid bilayer and must act there before metabolization. A protein kinase activity from the brain was found to be activated reversibly by lipid-soluble membrane components in the presence of Ca^{2+} [25–27]. Further research revealed that a neutral lipid, DAG, drastically enhanced the reaction velocity while simultaneously reducing the Ca^{2+} requirement of this kinase activity [25,26,28] which became known as PKC.

These results firmly established DAG as an intracellular second messenger through the activation of PKC, which was additionally supported by the observation that a tumor-promoting phorbol ester (tetradecanoylphorbol 13-acetate), a synthetic analogue of DAG, activated PKC in the presence of phospholipids. Phorbol ester induces a wide variety of cellular responses in vitro including altered cell proliferation, differentiation, intracellular communication, and tumor promotion [29,30]. Furthermore, DAG produced in response to growth factors stimulates cell proliferation, and specific oncogenes may alter DAG levels [31]. DAG levels in K-ras-transformed rat kidney cells grown at 34°C and 38°C were increased by 168% and 138%, respectively, in comparison to normal rat kidney (NRK) cells [32] as well as H-, K-, and N-ras-transformed NIH3T3 cells [33]. These findings confirmed the notion that altered levels of DAG second messenger play an important role in cellular transformation.

4.2.2 Discovery of Diacylglycerol Kinase Activity

The enzymatic activity of conversion of DAG to PtdOH was first described in 1959 by Hokin and Hokin [34], who performed an experiment in which DAG prepared from cabbage was phosphorylated by ATP in the presence of brain microsomes and deoxycholate extracts of microsomes. The enzyme which catalyzes the formation of PtdOH via this pathway was termed "diglyceride kinase." Since then it has been revealed that the enzyme is widely distributed in animal tissues [35–42]. However, the enzyme showed a multimodal intracellular distribution in the brain, and the activity is associated with membranes [34,38] and cytoskeleton [43] in addition to the soluble fraction [38], suggesting the multiplicity of DGK in animal organs, tissues, and cells. Because of the difficulties associated with its assay and purification [44,45], it was not until 1983 that its purification to homogeneity was reported [46].

4.2.3 SUBSTRATE SPECIFICITY TOWARD ARACHIDONOYL DIACYLGLYCEROL

Several groups report that DGK activity is found in the cytosol or microsomes of the brain and liver and that they found little or no selectivity among different DAG species. However, Glomset and colleagues found in Swiss 3T3 cells that the membrane-bound and cytosolic DGK utilized DAG in octylglucoside mixed micelles and were activated by phosphatidylserine (PtdSer) although they differed in substrate selectivity [47]. The membrane-bound DGK selectively phosphorylated DAG with arachidonic acid at the sn-2 position. Furthermore, the most important point was that the substrate preference of this enzyme toward the arachidonoyl-DAG was greatly dependent on the assay system [47–49]. Comparison of different assay systems using different detergents, i.e., the octylglucoside mixed micelle assay, deoxycholate assay, and Triton X-100 assay, demonstrates that the substrate selectivity of the arachidonoyl DGK was most evident in the octylglucoside assay. The activity was strongly decreased in the deoxycholate assay as compared with that in the octylglucoside assay. Furthermore, the enzyme exhibited no substrate preference when assayed in the presence of deoxycholate. The activity of arachidonoyl DGK in the Triton X-100 assay was half of that in the octylglucoside assay. No substrate preference was observed for the nonspecific DGKs in any of the assay systems. Therefore, the inability to detect substrate specificity in the previous studies could be ascribed to the characteristics of the arachidonoyl DGK, including its thermal lability and the detergents used in the assay system. Glomset had purified the arachidonoyl DGK from bovine testis to apparent homogeneity and estimated the molecular mass as 58,000 [48], although isolation of cDNA clone for that enzyme was done using PCR with degenerate primers for the amino acid sequences conserved among the catalytic domains of DGK isozymes known at the time [50].

With regard to the functional implications of substrate specificity toward the arachidonoyl-DAG, one possibility is that the arachidonoyl-specific DGKε may attenuate specifically the signal of arachidonoyl-DAG derived mostly from the breakdown of PtdIns, although the other isozymes are also capable of catalyzing arachidonoyl-DAG nonspecifically [51]. The other possibility is that if the DGKε works specifically on the arachidonate-containing species of DAG, then multiple cycles would progressively enrich PtdIns with arachidonate [52,53]. In either case, it is clear that DGKε may reflect an activity of the cellular PtdIns cycle.

4.3   DIACYLGLYCEROL

4.3.1   PROPERTIES OF DIACYLGLYCEROL

To evaluate the role of DAG in signal transduction, we need to consider three major parameters including (1) the kinetics of DAG production, (2) the molecular species of DAG, and (3) the subcellular compartment of DAG formation.

First, DAG production occurs in a biphasic manner. For example, in IIC9 fibroblasts, high concentrations of α-thrombin (14 nM) induce a biphasic increase in DAG mass. The early phase peaks at 15 s and a later phase peaks at 5 min after the stimulation [54]. DAG levels remain elevated for at least 4 h as long as catalytically active α-thrombin is present. A number of other systems also show similar biphasic response for DAG production [55]. It should be noted that the growth factors, epidermal growth factor (EGF) and platelet-derived growth factor (PDGF), in contrast to α-thrombin, induce monophasic increase in DAG levels at all mitogenic concentrations in IIC9 cells [54,56], while PDGF-induced DAG production is biphasic in NIH3T3 cells [57]. These differences may be ascribed to the ability of the agonists to stimulate multiple, cell type-specific DAG-generating pathways.

Second, DAG constitutes at least 50 structurally distinct molecular species, whose fatty-acyl groups can be polyunsaturated, diunsaturated, monounsaturated or saturated [2,58]. In general, fatty acids are attached to the sn-2-glycerol carbon via an ester linkage, while the fatty acid at sn-1 may be linked via ester, ether, or alkenyl ether [59]. Combination of a variety of acyl chains at sn-1 and sn-2 positions contributes to diversity and distinct profiles of DAG species. In this regard, it should be noted that DAG derived from PtdIns is largely composed of polyunsaturated acyl chains, i.e., 1-stearoyl-2-arachidonoyl species, whereas DAG originating from phosphatidylcholine (PtdCho) contains monounsaturated and saturated chains [51]. It is very hard to predict the extent to which a particular DAG species contributes to the signaling, although, with regard to PKC activation, there is some preference for polyunsaturated DAG species, while saturated DAG are generally poor activators; diunsaturated DAG are more active; and polyunsaturated DAG, such as 1-stearoyl-2-arachidonoyl DAG, are the most potent [60,61].

Third, in addition to molecular species, it is becoming more important to consider the microenvironment in which the molecular species is found, i.e., the subcellular compartment in which the DAG are located. It was previously assumed that DAG formation by agonist-induced PtdIns hydrolysis occurred at the plasma membrane. However, quantification of induced DAG mass revealed that not all of the induced DAG could be present at the plasma membrane. In α-thrombin-stimulated fibroblasts [54,56,62], the total amount of induced DAG represents nearly 1% of the total mass of cellular lipid. If all of this DAG formation occurred at the plasma membrane, it would result in devastating changes in the physical properties of the membrane structure [63]. The small early phase of DAG production (possibly derived from PtdIns) occurs in the plasma membrane while the larger later phase (possibly derived from PtdCho) is produced in internal membranes [64]. In this regard, it was reported that isolated rat liver nuclear envelop retains the ability to synthesize in vitro PtdOH, PtdIns4P, and PtdIns(4,5)P$_2$ [65]. In addition, a robust, but transient, increase in nuclear DAG is evident in quiescent IIC9 cells in response to α-thrombin [66,67]. These observations suggest that DAG production in response to stimuli must occur not only at the plasma membrane but also at other sites including the nucleus.

4.3.2 METABOLISM OF DIACYLGLYCEROL

As described above, DAG is produced transiently in cells and metabolized in response to stimuli at variable rates. There are four types of reactions that remove DAG second messenger signals including (1) phosphorylation by DGK to PtdOH, (2) lipolytic breakdown to yield monoacylglycerols and free fatty acids, (3) conversion to either PtdCho or triacylglycerol (TAG), and (4) formation of bisphosphatidic acid. In mammalian cells, it is difficult to define the contribution of DGK to the regulation of steady-state levels of DAG because DAG is a common precursor for glycerolipid synthesis and can be metabolized potentially by a number of enzymes such as DAG lipase (see Chapter 5). It is thought that DGK acts at the early phase of PtdIns signaling and that the contribution of DGK to metabolic processing of cellular DAG is quantitatively minor when compared to other enzymes. However, it becomes clear that the cellular DAG pool is spatially and functionally segregated and that DGK is critically involved in attenuating DAG produced in a signaling complex containing DGK or at certain restricted intracellular sites such as the plasma membrane, internal membranes, and nucleus [68].

4.4 MOLECULAR CLONING

4.4.1 80 kDa DIACYLGLYCEROL KINASE

An 80 kDa soluble DSG was purified to homogeneity from pig brain by Kanoh et al. [46]. This group purified another DGK from porcine thymus [69], which was then partially sequenced to synthesize oligonucleotide probes to isolate cDNA clones. Molecular cloning of 80 kDa DGK was the first to be reported among DAG-utilizing enzymes and revealed that the primary structure encodes 734 amino acids (including the initial methionine), contains two cysteine-rich zinc finger–like sequences, two Ca^{2+}-binding EF-hand motifs, and an ATP-binding site that is commonly found in protein kinases [70]. Interestingly, the enzyme was found to be expressed most abundantly in the thymus, then the spleen, and less abundantly in the brain.

4.4.2 MAMMALIAN DIACYLGLYCEROL KINASE FAMILY

Following this report, human [71] and rat [72] homologues of this enzyme were isolated. Morphological investigation in rat brain revealed that mRNA for 80 kDa DGK is expressed in glial cells, but not in neurons. In general, neurons are considered the leading player for the signal propagation in the brain, while glial cells are considered supportive. The unexpected localization of 80 kDa DGK in supportive cells in the brain, together with the previous report suggesting other immunologically distinct DGK [73], motivated researchers to isolate other forms of DGK. Intensive screening in the brain led to the isolation of another DGK having a molecular mass of 90 kDa

[74]. Intriguingly, this novel DGK is expressed in neurons only in a limited region of the brain, suggesting that the "neuronal type" is not a single entity but is composed of DGK that might localize differentially to other neurons. Amino acid sequences from two distinct DGK made it possible to predict conserved sequences of the enzyme and to synthesize degenerate primers for "PCR cloning." In addition, information from EST program and genome projects of several species, including human and mouse, further accelerated the identification of more DGK by looking for sequence homology. Those days were just a period of "Great Voyage" seeking new isozymes. In 1996, as many as six cloning papers were published from several species including human, rat, and hamster, which is called the "Gold rush" year for DGK.

To date, 10 mammalian DGK isozymes have been identified [12,75–82]. Based on the structural motifs, they are classified into five groups (Figure 4.1). Class I comprises the α [70,71], β [74], γ [83,84]; class II, the δ [85], η [86], and κ [87]; class III, the ε [50]; class IV, the ζ [88,89] and ι [90]; and class V, the θ [91]. In addition, additional splice variants have been reported in many of the isozymes, which may be due to alternative splicing; DGKβ [92], -γ [84], -δ [93], -η [94], -ι [95], and -ζ [96].

All of the mammalian DGK share a conserved catalytic domain in the C-terminal region and two cysteine-rich Zn^{2+}-finger motifs (three for DGKθ) similar to the C1A and C1B motifs of PKC, but lack certain consensus residues present in phorbol ester-binding proteins. However, class II isozymes have bipartite catalytic domains. It is thought that the cysteine-rich Zn^{2+}-finger motifs bind DAG and present it to the catalytic domain, though this has never been demonstrated in a conclusive manner.

DGK isozymes can be distinguished by the presence of additional domains that conceivably confer to each isozyme specific functions in biological processes,

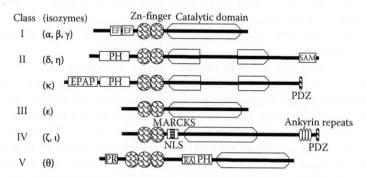

FIGURE 4.1   Schematic representation of the DGK family in mammals. To date, 10 isozymes have been identified, which are classified into 5 groups based on their structures. Zn^{2+}-finger motifs are included in all of the isozymes. Note that class II isozymes have bipartite catalytic domain. Major motifs or domains are shown. EF, Ca^{2+}-binding EF hand; Zn-finger, cysteine-rich, Zn^{2+}-finger motif; PH, pleckstrin-homology domain; SAM, sterile α motif; EPAP, 33 tandem repeats of Glu-Pro-Ala-Pro; PDZ, PSD-95/discs large/zona occludens-1 domain; MARCKS, sequence homologous to myristoylated alanine-rich C-kinase substrate phosphorylation site domain; NLS, nuclear localization signal; Ankyrin repeats; PR, proline- and glycine-rich domain; RA, Ras-association domain.

sensitivity to different regulatory mechanisms, and a differential intracellular local-ization. Indeed, these motifs or domains are likely to play a role in lipid–protein and protein–protein interactions in various signaling pathways. Class I DGK have Ca^{2+}-binding EF domains in their N-terminal half so that they are active in the pres-ence of Ca^{2+}. Class II isozymes have a pleckstrin homology (PH)-like domain at their N-terminal portion, but no specific function has been ascribed to this domain, even though it could be involved in interaction with membranes. Among this class, DGKδ and -η contain at the C-terminus a sterile α motif (SAM) domain [85,94], which is shown to localize DGKδ to the endoplasmic reticulum (ER) where it regu-lates ER-to-Golgi traffic [97]. On the other hand, DGKκ contains a unique motif, 33 tandem repeats of Glu-Pro-Ala-Pro (EPAP repeats), at the N-terminus [87]. DGKε that belongs to class III is the only isozyme which has no domains with obvious regulatory functions, but shows substrate preference toward arachidonoyl-DAG at the sn-2 position [50]. Class IV isozymes display four C-terminal ankyrin repeats, a PDZ (postsynaptic density protein-95/Discs large/zona occludens-1) domain, and nuclear localization signal (NLS) which overlaps with a region homologous to the phosphorylation site of the PKC substrate "myristoylated alanine-rich C-kinase sub-strate (MARCKS)" [88,89]. Class V DGKθ has a region with a weak homology to a PH domain located in the middle of its sequence. This domain overlaps with a Ras-associating domain [91].

4.4.3 DIACYLGLYCEROL KINASES IN OTHER SPECIES

The mammalian system shows diversities of molecular structures and biological actions of DAG and PtdOH, together with their various subcellular signaling sites, which may be the reason why so many isozymes have been diversified for the DGK family in the course of evolution. On the other hand, only one or a few isozymes have been reported in *Escherichia coli* [98], *Dictyostelium discoideum* [99], *Drosophila melanogaster* [100–102], and *Arabidopsis thaliana* [103,104]. No DGK gene has been isolated in yeast.

Bacteria contain only one DGK enzyme, which is encoded by the *dgkA* gene. It is predicted to be a 13 kDa membrane protein consisting of three transmembrane α-helical segments, an amphipathic helix, and an α-helix and exhibits the activ-ity of phosphorylating other lipids such as ceramide [32,98]. *Drosophila rdgA* gene encodes dmDGK2, which contains a similar structure to class IV isozymes and shows 49% identity to human DGKι [90,100]. Intriguingly, mutations in this gene result in early onset retinal degeneration and blindness [105,106], probably because of uncontrolled Ca^{2+} influx in *rdgA* photoreceptors via two classes of light-sensitive channels: a highly Ca^{2+}-permeable channel (TRP) and a nonselective cation chan-nel (TRPL) [107]. In *Caenorhabditis elegans*, DGK1 (ceDGK1), which shows 38% identity with human DGKθ, may be involved in serotonin inhibition of synaptic transmission. Mutations of ceDGK1 result in abnormality of animal locomotion and egg-laying behavior [9,108], where signaling via presynaptic DAG-binding protein, UNC13, is impaired. These simpler organism models could provide useful systems to investigate functional implications of DGK in mammals.

4.5   CELL BIOLOGY AND PATHOPHYSIOLOGICAL IMPLICATIONS

4.5.1   SUBCELLULAR LOCALIZATION AND TRANSLOCATION

It is of particular interest that most, if not all, of the isozymes show abundant expression in the central nervous system. Furthermore, they are expressed differentially in various regions of the brain [109]. These comparative data suggest that specific types of neurons express specific sets of the isozymes and that more than one isozyme is coexpressed in a single neuron in most of the brain regions. Protein purification experiments had also showed that a single cell can contain at least three different DGK isozymes [110,111]. But a fundamental question remains open: why do so many isozymes exist in a single neuron?

Analyses of the subcellular locations of several of the DGK isozymes provided evidence that some are mainly cytosolic, some are mainly associated with membranes, while some are localized to the nucleus. These distinct properties of subcellular localization have been directly compared by the experiments using transfected cultured cells under the same conditions, which reveal that the activity of DGKα is mostly detected in the soluble fraction, while those of DGKβ and DGKε are mostly in the particulate fraction, and those of DGKγ and DGKζ are observed in both of the fractions [112]. Considering that DGK contain no apparent membrane-spanning domains, these data suggest that DGK are peripheral membrane proteins that are recruited to the membrane, depending on the characteristic properties of reversible or irreversible membrane association under physiological conditions. In addition, DGK isozymes are targeted to differential subcellular sites in cDNA-transfected COS7 cells under the same experimental conditions; DGKβ colocalizes with actin filaments, DGKγ with the Golgi complex, and DGKε with the ER [112]. It remains unexplored whether the localization in cultured cell lines is a true indicator for other cell types and native cells *in situ*. Such doubts arise because each cell type has its unique regulatory mechanism for the subcellular elements, such as organelles and cytoskeletons, which are composed of a variety of molecules, in order to accommodate the cell to various environments. Immunocytochemical study using specific antibodies needs to be performed.

Another point that should be noted is that the subcellular localization may represent translocation, i.e., shuttling between the cytosol and the particulate structures and/or between the organelles. In the case of DGK, the substrate DAG is membrane-bound and therefore DGK must bind to the membrane, at least transiently, in order to function. In this regard, DGK translocates from the cytosol to membranes in homogenates of either rat brain or rat liver, depending on the content of membrane-bound DAG, although the isozyme(s) involved remained unknown at that time [113]. This suggests that DGK may be partially, if not completely, activated by its own substrate DAG and that DGK itself might be one of the targets of a second messenger DAG. More specifically, DGKα translocates from the cytoplasm to the plasma membrane upon T cell activation, where subsequent phosphorylation of DAG causes a rapid release of the protein back to the cytoplasm [114], showing a reversible translocation in this model. Under pathological conditions, transient ischemia promotes DGKζ, an isozyme that normally localizes to the nucleus in neurons, to translocate from the nucleus to the perikaryal cytoplasm in pyramidal neurons of the hippocampal CA1 at a very early phase of ischemic insult and it never relocates back to the nucleus during the time course of reperfusion [115] (see Section 4.5.4.1.1). Delayed neuronal death occurs in hippocampal CA1 neurons after 48–72 h reperfusion in this transient forebrain ischemia [116]. Rapid and irreversible disappearance of DGKζ from the nucleus of ischemic neurons might be involved in this neuronal death although the details remain unclear. Taken together, it is conceivable that a variety of factors may influence the subcellular localization of DGK under pathophysiological conditions.

4.5.2   Binding Assay Using Unilamellar Lipid Vesicles

Enzyme assays using detergents are simple and can provide useful information on biochemical characteristics of the enzyme, but they are also unphysiological. Assays with unilamellar lipid vesicles, which simulate the lipid bilayer of biological membranes, are employed to characterize both interfacial binding and catalytic activation [117,118]. In these studies, a unilamellar lipid vesicle assay system is used to identify lipids and medium components that influence the ability of the most abundant soluble DGK in Swiss 3T3 cells, a cytosolic, Ca^{2+}-independent enzyme, to bind to membranes. Stable, 100 nm diameter vesicles are used because they are large enough to have a curvature similar to that of planar bilayers in cells, but small enough to have relatively uniform size distribution. The results provided evidence that (1) anionic phosphoglycerides are positive effectors of binding, PtdCho is a negative effector, and PtdEtn and unesterified cholesterol are essentially neutral diluents, (2) DAG and some of its structural analogues are also important positive effectors of enzyme binding, and (3) addition of ATP to the medium increases their effects. Together, those studies suggest that the enzyme may bind to vesicles via at least two types of binding sites: one type that requires anionic phospholipids and is enhanced by Mg^{2+} but inhibited by PtdCho, and another that requires DAG and is enhanced by ATP.

The previous report that showed DAG-dependent DGK translocation [113] may be explained by the observations described above. The point is that DGK may contain DAG-binding sites in addition to the enzyme's substrate binding site [118]. All of the mammalian DGK share cysteine-rich zinc-finger motifs similar to the C1A and C1B motifs of PKC but lacking certain consensus residues present in phorbol ester-binding proteins [12,75,81]. It is plausible that Zn^{2+}-finger motifs may promote the binding of the enzyme to vesicles or membranes. It is also conceivable that the presence of DAG in the vesicles may alter the physical properties of the vesicle surface. Physical effects of DAG on enzyme–vesicle interactions appear to have been observed in other systems [119,120].

4.5.3   Metabolism of Diacylglycerol in the Nucleus

Many reports suggest the existence of phospholipid biosynthesis in the nucleus [76], although it was first described that isolated rat liver nuclear envelope retains the ability to synthesize *in vitro* PtdOH, Ptd4P, and Ptd(4,5)P$_2$ [65]. In envelope-deprived nuclei from murine erythroleukemia cells, synthesis of Ptd4P as well as Ptd(4,5)P$_2$ *in vitro* occurred, showing that inositol lipid metabolism also occurs in the internal domains of the nucleus [121]. The nuclear Ptd(4,5)P$_2$ synthesis was more pronounced if the cells were differentiated along the erythroid pathway by dimethyl sulfoxide (DMSO). These studies promoted further investigation on the functional aspects of nuclear phosphoinositide cycle.

With regard to the regulatory mechanism, in insulin-like growth factor-I (IGF-I)-stimulated Swiss 3T3 cells, a rapid decrease in the mass of PtdIns(4,5)P$_2$ occurs within the nuclei, with a concomitant increase in nuclear DAG and translocation of PKC to the nuclear region, while bombesin stimulation causes similar inositol lipid changes at the plasma membrane, but has no effect on nuclear inositide levels and causes a translocation of PKC to postnuclear membrane [122]. This suggests that nuclear PtdIns signaling system is regulated differentially from the well-known plasma membrane system and these two phosphoinositide cycles are indeed uncoupled.

There are two distinct subnuclear pools of DAG [123]. One is highly polyunsaturated and is likely derived from the hydrolysis of PtdIns through the action of PtdIns-specific PLC. The other pool is highly diunsaturated and monounsaturated species (more than 90% of the total nuclear DAG) and is derived from various pathways, one of which may be through the hydrolysis of PtdCho. Interestingly, two separate pools of DAG are independently regulated in the nucleus. In MEL cells, the differentiation along the erythroid pathway by DMSO leads to a decrease in the levels of diunsaturated and monounsaturated DAG when assayed using intact nuclei. On the other hand, during progression through the cell cycle, no changes are seen in the levels of monounsaturated and diunsaturated DAG. In contrast, however, there is a threefold increase in the nuclear PtdOH labeling 2 h after release of cells from a nocodazole-induced block at G$_2$/M phase. The increase in PtfOH is completely inhibited by PtdIns-PLC inhibitors, and more than 70% of the PtdOH is highly polyunsaturated species. Therefore, it is suggested that the nuclear DAG pools are not always accessible to the nuclear DGK, and DAG with different acyl chains may be metabolically compartmentalized in the nucleus.

Two different DAG generation pathways are also reported in the human promyelocytic cell line, HL-60 [124]. When these cells are induced to proliferate with IGF-I, PtdIns(4,5)P$_2$-mediated DAG is generated in the nucleus, and a selective translocation of PKCβII occurs. On the other hand, when induced to differentiate toward a granulocyte-like phenotype by a variety of chemicals, including DMSO, there is a rise in the nuclear DAG mass, which is derived from PtdCho by the action of phospholipase D (PLD) and a nuclear translocation of PKCα [125]. These data clearly show that diverging DAG pathways are operated after differentiation, during cell-cycle progression, and upon receptor stimulation in the nucleus of various cell types, although effectors may vary depending on cascades. Considering the diversity of DAG, such as source and acyl chain composition, DAG pathway might be an efficient and useful machinery to control a variety of cellular events.

As for the metabolism of nuclear DAG, a study was performed to find out how the nuclear DAG signal is switched off [126]. When quiescent Swiss 3T3 cells are stimulated with a mitogenic concentration of IGF-I, there is a rapid and sustained increase in the nuclear DAG levels up to 60 min. After the beginning of the treatment, nuclear DGK activity is unchanged up to 30 min of IGF-I stimulation, and then it starts to rise, reaching the maximum at 90 min. These results show that an inverse relationship exists in the nucleus of IGF-I-stimulated Swiss 3T3 cells between the levels of

DAG and DGK activity. Treatment of cells with DGK inhibitors (R59022, 5 μM or R59949, 1 μM, from Calbiochem, La Jolla, California) blocks the IGF-I-dependent rise in nuclear DGK activity and maintains elevated intranuclear levels of DAG. The data indicate that DGK is a key player in regulating stimulated cell growth as DGK inhibitors potentiate the mitogenic effect of IGF-I.

4.5.4 ANIMAL STUDIES

4.5.4.1 Brain

4.5.4.1.1 Cerebral Ischemia and Infarction

Cerebral ischemia is a major research area in medicine and has long been investigated extensively using various animal models, because cessation of blood supply for only a short period, caused by heart attack or arterial occlusion, may lead to serious brain damage due to neuronal death. Previous studies showed that the most prominent changes in the lipids in ischemic brain is a rapid decrease in the levels of polyphosphoinositides and a parallel accumulation of DAG [127], suggesting an involvement of DGK.

Transient forebrain ischemic model was used to investigate whether DGK is involved in this process [115]. In this model, DGKζ translocates from the nucleus to the perikaryal cytoplasm in pyramidal neurons of the hippocampal CA1 region at a very early phase of ischemic insult and remains in the cytoplasm with decreasing intensity during the time course of reperfusion. This phenomenon is observed specifically in hippocampal CA1 neurons, but not in neurons of the other hippocampal areas on cerebral cortex under transient ischemic conditions. Rapid disappearance of DGKζ from the nucleus of ischemic neurons suggests decreased metabolism of DAG in the nucleus.

Delayed neuronal death is known to occur in hippocampal CA1 neurons after 48–72h reperfusion in the transient forebrain ischemia, but not in neurons of other areas of the brain [116]. From the previous study on cerebral ischemia [115], it might be hypothesized that transient ischemia induces the removal of DGKζ from the nucleus, which leads to the reduced nuclear DGK activity. This may result in sustained increase in DAG levels or decrease in PtdOH levels in the nucleus, which might trigger signals in the nuclear process of delayed neuronal death in hippocampal neurons.

Hypoxia is associated with ischemic brain injury and it has two types of effects on cellular metabolism and gene expression in mammalian systems [128]. The first type constitutes rapid and reversible effects on cell signaling, contractility, ion flux, and redox state, which are critical for cellular function. The second type constitutes slower transcriptional responses that are largely dependent on the hypoxia-inducible factors (HIF), whose target genes enable long-term cellular survival and adaptation to hypoxic conditions [129]. Among the immediate effects of hypoxia is a rapid inhibition of mRNA translation. Translation, especially at the initiation step, is highly regulated and sensitive to various kinds of cellular stresses that are known to regulate translation initiation factors and complexes, such as eukaryotic initiation factor (eIF) 4F and eIF2-GTP-Met-tRNA ternary complexes [130]. Rapid disappearance of DGKζ in the hippocampal neurons in the early phase of ischemia may be related to the early response to hypoxia described above, although it remains to be elucidated whether this phenomenon is supportive or destructive for ischemic neurons.

Cerebral infarction is caused by extended durations of ischemia and leads to neuronal death. The molecular mechanisms responsible for neuronal damage are incompletely understood; however, it is widely accepted that in the early phase of ischemia, glutamate plays a predominant role in the pathogenesis of ischemic brain injury [131,132]. This excitatory amino acid leads to a massive influx of Ca^{2+} that activates a variety of catabolic processes, which subsequently lead to cell death. Following the early phase of neuronal death after ischemia, the late inflammatory process causes the activation of glial and inflammatory cells and this response may play a critical role in the development of brain damage [133]. Blood-derived neutrophils and monocytes additionally infiltrate the area of cerebral infarction, where the blood–brain barrier is breached. During a rat focal middle cerebral artery occlusion (MCAO), a focal ischemic insult lasting 90 min produces an infarct whose volume is maximum 3 days later [134]. This rapid destruction of tissue is consistent with a process dominated by ischemic necrosis [135]. It is reported that after 3–7 days of 1 h MCAO model, the infarct is covered by round cells that is immunoreactive for ED-1, a marker for activated phagocytes/macrophages [136].

In this respect, DGKζ is also shown to be involved in both the early necrotic process and the late inflammatory process of cerebral infarction [137]. Hypoperfusion of blood flow greatly attenuates DGKζ in cortical neurons of the afflicted area immediately after 90 min arterial occlusion, while neuronal marker proteins such as MAP-2 still remain intact. Because cytoskeletal proteins, such as MAP-2, are generally decreased and may participate in the initial phase of neuronal dysfunction, disappearance of DGKζ-immunoreactivity in ischemic cortical neurons may be quite an early event preceding ischemia-induced neuronal degeneration. This phenomenon is not seen on DGKι, which remains unchanged in the cytoplasm of cortical neurons in this early phase of ischemia. Together with the results in transient cerebral ischemia, DGKζ may play a major role in the nuclear events of ischemic neurons at an early stage. It should be noted that hippocampal CA1 neurons show different response of DGKζ after transient ischemia as described above. In transient ischemic model, DGKζ is translocated from the nucleus to the cytoplasm in CA1 neurons and never relocates to the nucleus throughout reperfusion. However, in the infarction model it seems that DGKζ never comes back to the cytoplasm of the afflicted neocortical neurons. Although models used between those studies are not exactly same, different responses of DGKζ may be ascribed to different vulnerability of distinct neurons to hypoxic insult.

Furthermore, DGKζ appears in nonneuronal cells in the late inflammatory phase of infarction [137]. Ischemic necrosis is followed by the activation of glial and inflammatory cells, and this response may play a critical role in the development of brain damage. It is revealed that DGKζ is detected in activated phagocytes/macrophages, suggesting a possible involvement in phagocytic process. In addition, DGKζ is also shown to be detected in reconstructing endothelial cells. It is reported that hypoxia induces upregulations of vascular endothelial growth factor receptor, HIF-1, and HIP-2 in this MCAO model [138] and that PtdOH, which is produced by DGK, is shown to activate HIF-1 [139,140]. DGKζ-immunoreactivity detected in the reconstructed endothelial cells suggests that this isozyme may play a role in the cascade involved with HIF-1.

4.5.4.1.2 Energy Balance in the Hypothalamus

Dietary fat metabolism plays a critical role in the development of obesity. In rodents, the concentration of fat in the diet, but not protein or carbohydrate, is shown to be positively correlated with the amount of body fat mass and that free access to a high-fat diet causes obesity and hyperinsulinemia [141]. Dietary fat also affects plasma levels of leptin, a hormone that exerts a key function in regulating food intake and body weight [142]. Serum leptin levels are strongly, positively correlated with body fat mass [143], and leptin controls energy balance through its long form receptor (Ob-Rb) on neurons in the hypothalamus [144].

A search for genes that are functionally linked to both dietary fat and Ob-Rb in the hypothalamus has revealed that DGKζ interacts, via its ankyrin repeats, with Ob-Rb in yeast two-hybrid system [145]. It is also shown that a high-fat diet stimulates DGKζ expression in the hypothalamus. Moreover, hypothalamic DGKζ expression is found to be reduced in obese animals and strongly, inversely related to both body fat mass and serum leptin level. These results support an idea that reduced activity of DGKζ may contribute to the accumulation of body fat, suggesting that the activity of DGKζ is specifically regulated by the interaction of its ankyrin repeats with a leptin-stimulated Ob-Rb. Considering that DGKζ mRNA is higher in the wild-type lean rats and inbred mice, elevated DGKζ mRNA in the mutant, morbidly obese ob-/ob- or db-/db- mice may be interpreted to mean that these animals regard themselves as "lean" [142] and consequently upregulate the gene expression of DGKζ. This is supported by the evidence that Ob-Rb mutation may cause a reduction of DGK activity in obese Zucker rats, which have a mutant Ob-Rb [146] and exhibit elevated DAG levels and PKC activity [147,148]. Importantly, DGKζ is the only isozyme detected in this experiment, and other DGK isozymes are not affected by fat consumption and do not interact with Ob-Rb. Together, the enzymatic activity of DGKζ may be activated in response to a high-fat diet, and DGKζ may participate in the control of body fat accumulation in the leptin signaling pathway. Nuclear localization of DGKζ suggests a possible involvement of this molecule on the nuclear events such as transcriptional control of fat metabolism–related molecules.

4.5.4.1.3 Emotion

DGKβ shows a unique distribution pattern in the rat brain. Its mRNA is detected in the caudate-putamen, accumbens nucleus, olfactory tubercle, olfactory nucleus, and frontal cortex [74]. These are well-known regions that receive dopaminergic input, suggesting possible involvement of this isozyme in dopaminergic transmission. The diverse physiological actions of dopamine are mediated by at least five distinct G-protein-coupled receptor subtypes [149]. Two D1-like receptor subtypes (D1 and D5) couple to the G-protein Gs and activate adenylyl cyclase. The other receptor subtypes belong to the D2-like subfamily (D2, D3, and D4) and are prototypic of G-protein-coupled receptors that inhibit adenylyl cyclase and activate K^+ channels.

Human DGKβ is expressed as a complex series of mRNA via alternative splicing and differential polyadenylation signals [92]. Hence, the human DGKβ locus can potentially generate up to 16 different isoforms. This suggests a degree of diversification of human DGKβ activity, which may involve the modulation of isoform expression, enzymatic activity, and subcellular localization. This is important because human DGKβ EST is annotated in GenBank as being differentially expressed in bipolar disorder patients and corresponds to the 3′ region of human DGKβ transcripts encoding the isoforms lacking the C-terminal region (designated SV3′ DGKβ). It is shown that, despite similar enzymatic activity, the wild type and SV3′ DGKβ may display differences in their subcellular localization, suggesting differential substrate accessibility and subsequently altered signal transduction. The dopamine system is one of the major systems involved in the control of emotional and cognitive behavior. The altered signaling of DGKβ in the caudate-putamen could possibly affect the dopaminergic transmission, although further information is unavailable at the moment.

4.5.4.2 Heart

4.5.4.2.1 Cardiac Hypertrophy

Cardiac hypertrophy is a well-known major risk factor for the development of heart failure and death [150]. The signaling molecules involved in the progression of cardiac hypertrophy include heterotrimeric G-protein-coupled receptor agonists such as endothelin-1 (ET-1) and angiotensin II [151]. Previous studies in human heart failure and animal models of heart failure, including genetically engineered mice, clearly demonstrated that activation of PKC plays a pivotal role in these conditions [152–155], suggesting that DGK may be intimately involved in cardiac dysfunctions.

Pressure-overloaded cardiac hypertrophy model of rats is generated by ascending aortic banding and is assessed in terms of the changes in the mRNA levels of DGK isozymes during the initiation and progression of the disease [156]. At 28 days after surgery, the expression of DGKε mRNA is significantly decreased in the left ventricular (LV) myocardium of the aortic-banded rats, while DGKζ mRNA remains unchanged. However, DGKζ protein is found to translocate from the particulate to the cytosolic compartment in these rats at this stage. Concomitantly, the DAG content is significantly increased 1.7-fold in the hypertrophied LV myocardium compared with that from sham-operated rats. PKCδ protein is also increased in the particulate fraction of the hypertrophied LV tissues. It is well known that the process of PKC activation includes its translocation from the cytosol to the particulate fraction and subsequent binding to DAG [157]. The increase in DAG content may be ascribed to the decreased mRNA expression for DGKε and/or the translocation of DGKζ to the cytosolic compartment. However, no significant difference in the fatty acid profiles of DAG, including arachidonate, in the LV myocardium between the two groups supports the idea that the increased DAG may be due to the decreased DGKζ activity in the particulate fraction because of its translocation, but the expression of arachidonoyl-specific DGKε might also have some effects. Decreased expression of DGKε suggests downregulation of phosphoinositide signaling cascade under stress conditions, which might be related to cardiac dysfunctions.

Cardiomyocyte hypertrophy has been studied by using a recombinant adenovirus-encoding rat DGKζ (Ad-DGKζ) [158]. G-protein-coupled receptor agonists are shown to develop cardiac hypertrophy and the authors found that ET-1 stimulates DGKζ mRNA expression in rat neonatal cardiomyocytes, leading them to examine the functional role of DGKζ in this model. In cultured cardiomyocytes,

ET-1-induced translocation of PKCε is blocked by Ad- DGKζ. Ad- DGKζ also inhibits ET-1-induced activation of ERK and activator protein-1 (AP-1) DNA-binding activity. Furthermore, in cardiomyocytes transfected with Ad- DGKζ, ET-1 fails to cause gene induction of atrial natriuretic factor, increases in [³H]-leucine uptake, and increases in cellular surface area. From these data, it is concluded that DGKζ blocks ET-1-induced activation of the PKCε-ERK-AP1 signaling pathway, atrial natriuretic factor gene induction, and the resulting cardiomyocyte hypertrophy. These observations seem consistent with those of pressure-overload hypertrophy model induced by aortic banding, and collectively suggest that DGKζ plays a key role in cardiac hypertrophy through the regulation of different isoforms of PKC.

This is supported by studies in transgenic mice with cardiac-specific overexpression of DGKζ (DGKζ-TG), where continuous administration of angiotensin II and phenylephrine cause PKC translocation, gene induction of atrial natriuretic factor, and subsequent cardiac hypertrophy in wild-type mice. However, in DGKζ-TG mice, neither translocation of PKC nor upregulation of atrial natriuretic factor gene expression is observed after angiotensin II and phenylephrine infusion [159]. Furthermore, phenylephrine-induced increases in myocardial DAG levels are completely blocked in the hearts of these mice, suggesting that DGKζ controls cellular levels of DAG, which regulates downstream PKC activity.

In cardiac hypertrophy model created by thoracic transverse aortic constriction, increases in heart weight and interventricular thickness, dilatation of the left ventricular cavity, and decreases in left ventricular systolic function are attenuated in DGKζ-TG mice [160]. In addition, cardiac fibrosis and gene induction of type I and III collagens are blocked in these mice. In DGKζ-TG mice, translocation of PKCα leading to its activation is inhibited in the aortic constriction, which might be a possible mechanism for the preservation of left ventricular function, attenuation of fibrosis, and blockade of profibrotic gene induction. This idea is supported by a study, *in vitro*, showing that DGKζ associates with PKCα and inhibits its activity [161,162]. These studies clearly demonstrate that DGKζ negatively regulates hypertrophic signaling cascade, suppresses cardiac hypertrophy and fibrosis, and prevents impaired left ventricular systolic function caused by pressure overload at the animal level.

4.5.4.2.2 Myocardial Infarction

Myocardial infarction (MI) is caused by prolonged cessation of coronary blood flow. In the chronic stage, MI induces LV remodeling including myocyte necrosis, thinning of the infarcted myocardium, dilatation of the ventricular cavity, eccentric myocardial hypertrophy, and interstitial fibrosis [163]. Early changes in the LV architecture after MI are compensatory phenomena of the heart to adapt to the consequences of functional losses in myocardial function and initially preserve cardiac performance. However, the infarcted heart progressively dilates, and cardiac fibrosis is formed in the noninfarcted area when these processes develop once after a large MI.

In the infarcted heart, the expression of DGKζ is enhanced in the peripheral zone of the necrotic area and at the border zone 3 and 7 days after MI [164]. The enhanced DGKζ expression in these areas appears to be attributed to granulocytes and macrophages that infiltrate to remove the residual debris. In quantitative analysis of DGKζ, expression returns to normal in the infarcted area by postoperative day

21. In contrast, DGKε mRNA level in the infarcted area is greatly reduced compared with that in sham-operated LV. Unexpectedly, the expression level of DGKε is also reduced significantly in the viable LV compared with that of sham-operated rats throughout postoperative period, while DGKζ expression remains unchanged in the viable myocardium. These data suggest that DGKε is downregulated not only in the infarcted area but also in the viable myocardium during the myocardial remodeling process. The decreased expression of DGKε suggests downregulation of phosphoinositide signaling cascade in the myocardium, which may be linked to cardiac remodeling.

The direct effect of DGKζ on LV remodeling after MI has been examined using transgenic mice with cardiac-specific overexpression of DGKζ (DGKζ-TG) described previously [165]. Similar to pressure-overload model, DGKζ exerts beneficial effects on the infarcted heart as LV chamber dilatation, reduction of LV systolic function, and increases in LV weight and lung weight are all attenuated in DGKζ-TG mice. In the noninfarcted area, fibrosis fraction and upregulation of profibrotic genes, such as transforming growth factor-β1 and collagen type I and III, are blocked in these mice, increasing the survival rate at 4 weeks as MI is higher in DGKζ-TG mice than in wild-type mice. These results strongly suggest that DGKζ suppresses LV structural remodeling and fibrosis, which improves survival after MI.

Taken together, DGKζ plays a critical role in cardiac function, especially under pathological conditions, by preventing cardiac hypertrophy and LV remodeling, although other isozymes such as DGKε could exert some effects on these conditions.

4.5.4.3 Knockout Mice

The physiological functions of DGK isozymes have been investigated in mice with targeted disruption of DGK, including DGKα [166], DGKδ [167], DGKε [168], DGKζ [169], and DGKι [170].

DGKα and DGKζ are implicated in T cell receptor signaling, especially T cell anergy. In anergic T cells, DAG metabolism is altered as is activation of Ras and Erk [171,172]. This is supported by the studies demonstrating that transcription of DGK genes is increased in anergic cells and that both DGKα and DGKζ are transcriptionally downregulated in activated T cells [166]. Overexpression of DGKα results in the functional and biochemical characteristics of the anergic state [166,173]. Knockout study confirms that DGKα-deficient T cells have more DAG-dependent TCR signaling and that anergy induction is impaired in DGKα-deficient mice [166]. It is interesting to note that DGKζ -null T cells are also hyperresponsive to TCR stimulation, as manifested by enhanced activation of the Ras-ERK cascade, expression of activation markers, proliferation *ex vivo* and *in vivo*, and antiviral immune responses [169]. These T cells also have enhanced proliferation and IL-2 production after anergy-producing stimulation [166]. All these studies clearly indicate that DGKα and DGKζ function as physiological negative regulators of TCR signaling. However, a functional difference between these two DGK isozymes remains unclear, although it is speculated that the activity and subcellular localization might be differentially regulated in T cells.

DGKε-deficient mice have provided an intriguing tool to examine how arachidonoyl-DAG metabolism is involved in epilepsy [168], which frequently develops in previously normal nervous tissue, secondary to trauma, tumor, or stroke [174]. The pathogenesis of temporal lobe epilepsy is still not fully understood, although it is shown that metabotropic glutamate receptor (mGluR) agonists potentiate the depolarization of basolateral amygdala neurons from brain slices of kindled rats [175] and that signal transduction mediated by mGluR is correlated with an increase in DAG production by hydrolysis of PtdIns(4,5)P$_2$ [176]. In addition, DAG is involved in the potentiation of excitatory glutamatergic neurotransmission and promotes an efficient and sustained glutamate-mediated signaling in postsynaptic neurons [177].

In this respect, DGKε-deficient mice are more resistant to electroconvulsive shock (ECS), displaying shorter tonic seizures and faster recovery than DGKε +/+ mice. This behavioral response is paralleled by lower degradation of brain PtdIns(4,5)P$_2$ after ECS. Synaptic transmission and plasticity were also examined in dentate granular cells of DGKε-deficient mice. There are no abnormalities in basic membrane properties, including resting membrane potential, input resistance, and action potential generation in hippocampal slices from the knockout mice. However, the potentiation of EPSP amplitude by high-frequency stimulation was significantly reduced in cells from DGKε −/− mice compared with that in DGKε +/+ mice. Although resting levels of PtdIns4P and PtdIns(4,5)P$_2$ are similar in the cerebral cortex from DGKε −/− and DGKε +/+ mice, some changes are detected, i.e., arachidonoyl-PtdIns(4,5) P$_2$ levels are reduced in DGKε −/− mice under resting conditions. DGKε −/− mice also show a significant decrease of stearoyl- and arachidonoyl-PtdIns4P after ECS, which is not observed in DGKε +/+ mice.

What accounts for these different changes in the levels of inositol lipids under resting conditions and after ECS in DGKε −/− and DGKε +/+ mice? One plausible explanation follows: First, neuronal polyphosphoinositides are maintained by synthesis *de novo* via PtdOH, whereas the DAG-DGKε pathway contributes to their resynthesis after synaptic activity–induced PtdIns(4,5)P$_2$ hydrolysis. Therefore, despite the deficiency in DAG-DGKε pathway, other DGK and/or *de novo* synthesis pathway may partly compensate to generate arachidonoyl-PtdOH that is channeled to inositol lipids. Second, because synaptic activity, ECS in this case, induces degradation and resynthesis of inositol lipids, the levels of PtdIns4P and PtdIns(4,5)P$_2$ reflect the balance of these two pathways. In DGKε +/+ mice, degradation of PtdIns(4,5)P$_2$ occurs at a faster rate than its replenishment from PtdIns4P by PtdIns4P 4-phosphate 5-kinase while PtdIns4P is being replenished through the DAG-PtdOH-phosphatidylinositol pathway. This results in a decrease of PtdIns(4,5) P$_2$ and no detectable changes in PtdIns4P. On the other hand, in DGKε −/− mice, the decrease in PtdIns4P may indicate its active phosphorylation to PtdIns(4,5)P$_2$ and its slower replenishment from the DAG-DGKε −/− pathway. Because type I PtdIns4P 4-phosphate 5-kinase ε is highly expressed in the brain and greatly stimulated by PtdOH [178,179], deficiency of DGKε will likely not force resynthesis of PtdIns(4,5) P$_2$. These results suggest that PtdIns4P degradation by PLC and/or its dephosphorylation to PtdIns should also be considered under DGKε-deficient conditions.

DGKδ-deficient mice are born with open eyelids and die shortly after birth because of respiratory difficulty [167]. It was found that DGKδ deficiency reduces EGF receptor (EGFR) protein expression and activity, which may be due to increased threonine phosphorylation of EGFR by PKC. This is supported by an experiment *in vitro* showing that EGFR expression is significantly reduced in DGKδ-knockdown SCC-9 cells, a squamous cell cancer line. In addition, DGKδ deficiency also affects other targets downstream of PKC, such as MARCKS and keratin 6, suggesting that aberrant phosphorylation of multiple PKC targets likely contributes to the phenotype of the mutant mice, including the open eyes at birth and respiratory distress. Knockout study clearly indicates that DGKδ has an important role in lung and keratinocyte functions.

DGKι binds and regulates the activity of RasGRP3 and predominantly reduces Rap1 activation [170], which may lead to increased active Rap in these mice that would interfere with Ras signaling [12]. When transgenic mice carrying v-Ha-Ras protooncogene, which make them prone to develop skin cancer, are crossed with wild-type or DGKι knockout mice, the DGKι-deficient mice developed fewer tumors in response to a phorbol ester or after wounding, suggesting that deleting DGKι in mice reduces tumor formation in a Ras-dependent manner. It should be pointed out that DGKι is predominantly expressed in the brain and retina under physiological conditions [90,95]. Predominant expression of this isozyme in postmitotic neurons suggests that this isozyme may also be implicated in other mechanisms in neurons.

4.6 ASSAYS FOR DIACYLGLYCEROL KINASE

4.6.1 Measurement of Diacylglycerol Kinase Activity

As noted, some difficulties were associated with the assay of the enzyme [44,45]. This is because of the property of DGK, an interfacial enzyme acting on the membrane lipid, which must bind to the membrane, at least transiently, in order to function. In addition, as DGK show distinct patterns of distribution in cells, each isozyme may have characteristic affinity to the membranes depending on lipid and medium components. Therefore, one assay is not always appropriate for all of the isozymes, which is exemplified by Section 4.2.3.

There are three major assay systems reported to measure the enzymatic activity for DGK: (1) the deoxycholate assay, (2) the octylglucoside/PtdSer mixed micelle assay, and (3) the Triton X-100 assay, each of which is named after the detergent employed for lipid solubilization. DGK activity is estimated as the amount of ^{32}P incorporated from $[\gamma\text{-}^{32}P]ATP$ into PtdOH.

The deoxycholate assay was initially introduced by Kanoh and Ohno [180]. The reaction mixture (final volume $100\,\mu L$) contains $50\,mM$ Tris–HCl (pH 7.4), $50\,mM$ NaCl, $1\,mM$ deoxycholate, $1\,mM$ EGTA, $1\,mM$ DTT, $20\,mM$ NaF, $10\,mM$ MgCl$_2$, $2\,mM$ $[\gamma\text{-}^{32}P]ATP$ ($10,000$–$20,000\,cpm/nmol$), $0.5\,mM$ DAG, and enzyme sample (usually $10\,\mu L$) in a final volume of $100\,\mu L$ [46,181]. Chloroform solution of DAG is evaporated to dryness under N_2 and sonicated in the Tris–HCl buffer containing deoxycholate, NaF, and DTT. The sonication is conducted at 0°C at maximum intensity using a cuphorn (Branson Sonifier) until the turbidity becomes visibly constant. The DAG suspension should be prepared freshly for each experiment. The reaction is started by adding ATP and enzyme sample and run for 2–3 min at 30°C and then stopped by adding $50\,\mu L$ of concentrated HCl to the tubes, followed by $1.5\,mL$ of water. The reaction products are extracted with $1.0\,mL$ of 1-butanol and the butanol extracts are washed once with 1 mL dH$_2$O saturated with butanol. The radioactivity of the washed butanol extracts is measured in toluence/Triton X-100 scintillation fluid.

The octylglucoside/PtdSer mixed micelle assay is a modification of the mixed micellar assay developed for the membrane-bound DGK from *E. coli* [98]. Final conditions for this system (50 μL) are 50 mM MOPS (pH 7.2), 100 mM NaCl, 75 mM octylglucoside, 3.5 mM PtdSer, 1 mM EGTA, 1 mM DTT, 20 mM NaF, 0.5 mM [γ-^{32}P]ATP (10,000–20,000 cpm/nmol), 2 mM DAG, and enzyme sample (usually 10 μL), which is based on the previous studies [48,49,182] with some modifications. Similar to the deoxycholate assay, lipids are dried under N$_2$ and sonicated in the buffer containing octylglucoside. It is essential that the lipids are solubilized completely. The solution should be completely clear, and there should be no remaining lipid film at the bottom of the tube. The reaction is started by adding ATP and enzyme sample and continues for 3 min at 30°C. After addition of 100 μL of 1 M HCl and carrier PtdOH (25 μg), the lipids are extracted with 250 μL of chloroform/methanol (1:1, v/v). The solvent phase is separated by a brief centrifugation, and the lower phase is washed once with 100 μL of methanol/0.1 M HCl (1:1, v/v) and dried under N$_2$. Lipids are dissolved in a small volume of chloroform/methanol/water (75:25:2, v/v) and spotted on silica gel thin layer plates (Merck). Ascending chromatography is carried out with the upper phase of ethyl acetate/2,2,4-trimethylpentane (isooctane)/acetic acid/water (9:5:2:1, v/v). The band of PtdOH is detected by iodine vapor or autoradiography and is scraped and counted for radioactivity.

The Triton X-100 assay is essentially performed similar to the octylglucoside assay described above, but octylglucoside is substituted with 1% Triton X-100 [49].

4.6.2 ACTIVITY OF ARACHIDONOYL DIACYLGLYCEROL KINASE

The octylglucoside/PtdSer mixed micelle assay is best suited for the detection of enzymatic activity toward arachidonoyl-DAG. As noted in Section 4.2.3, the substrate selectivity of the arachidonoyl DGK was most evident in the octylglucoside assay while the activity was strongly decreased in the deoxycholate assay as compared with that in the octylglucoside assay [49,183]. Furthermore, the enzyme exhibited no substrate preference in the deoxycholate assay. At present, arachidonoyl DGK has been cloned from human cDNA library, which is known as DGKε [50]. The activity of arachidonoyl DGK in the Triton X-100 assay was half of that in the octylglucoside assay. A quick screen for the presence of arachidonoyl-DGK can be performed by incubating a tissue extract simultaneously with 18:0/20:4 DAG and 10:0/10:0 DSG as substrates, and measuring the respective amounts of long chain [^{32}P]PtdOH containing 20:4 *n*−6 and [^{32}P]PtdOH containing 10:0 are formed [184]. The long chain [^{32}P]PtdOH and [^{32}P]PtdOH formed during the assay are separated with the thin-layer chromatography system described above. PtdOH migrates below long chain PtdOH.

The activity of the arachidonoyl-DGK varies with the mole fraction of DAG in the assay mixture rather than the absolute DAG concentration. The mole fraction is defined as the ratio of the DAG concentration to the concentration of octylglucoside in micelles, the latter concentration being equal to the difference between the total octylglucoside concentration and the concentration of octylglucoside monomers (critical micelle concentration). The concentration of octylglucoside monomers varies with ionic strength and the composition of the micelles.

The enzyme uses DAG containing arachidonic acid in the *sn*-2 position 5–10 times more efficiently than other naturally occurring DAG and 4 times efficiently than *sn*-1,2-dioleoylglycerol [184]. A synthetic DAG containing arachidonic acid in the *sn*-1 position is used 10 times less efficiently, showing that the arachidonic acid must be in the *sn*-2 position.

Variations in some of the assay conditions result in variations in the arachidonoyl-DGK activity. It seems that the most important factor is the nature of the detergent used in the assay. In addition, the enzyme activity is decreased in the absence of glycerol, at both low and high ionic strength, and in the absence of PtdSer.

4.7   CONCLUDING REMARKS

About half a century has passed since the enzymatic activity of DGK was first discovered in the brain microsomes. Initial studies based on enzymology revealed the DGK activities with distinct properties of the regulatory mechanism in various tissues and cells, suggesting that DGK consists of a family of isozymes. This prediction has been confirmed by biochemical purification followed by molecular cloning of the isozymes. Morphological investigation of the mRNA and protein for each molecule discloses heterogeneity of the expression and localization of the DGK family in organs and cells, which raises a new question: why are so many isozymes necessary and why are two or more isozymes coexpressed in a single cell?

Thanks to recent advances in technology, we have gained large amounts of information on this enzyme, although we sometimes wonder how much these data are relevant to physiological and/or pathological phenomena at the organism level, which makes us anxious that we might get lost in a deluge of information. Now, there has come an exciting era. Genome projects provide the entire genome information on human, monkey, mouse, and other lower vertebrates, and also some invertebrates. Together with valuable tools of knockout mice for DGK, we could try to understand the functional roles of this enzyme in relation to human health and diseases as well as from an evolutional point of view.

ACKNOWLEDGMENTS

The work was supported by Grants-in-Aid and the twenty-first century Center of Excellence Program from the Ministry of Education, Science, Culture, and Sports of Japan (K.G.), and Italian MUR PRIN 2005 (A.M.M.).

LIST OF ABBREVIATIONS

DAG	diacylglycerol
DGK	diacylglycerol kinase
ECS	electroconvulsive shock

EGF   epidermal growth factor
IGF1   insulin-like growth factor 1
IP$_3$ inositol-tris-phosphate
MARCKS   myristoylated alanine-rich C-kinase substrate
MCAO   middle cerebral artery occlusion
MI   myocardial infarction
NLS   nuclear localization signal
PDGF   platelet-derived growth factor
PDZ   postsynaptic density protein-95/Discs large/Zona occludens-1
PH   pleckstrin homology
PLC   phospholipase C
PtdEtn   phosphatidylethanolamine
PtdIns(4)P   phosphatidylinositol-4-phosphate
PtdIns(4,5)P$_2$ phosphatidylinositol-4,-5-bis-phosphate
PtdCho   phosphatidylcholine
PtdOH   phosphatidic acid
PtdSer   phosphatidylserine
SAM   sterile α motif
TAG   triacylglycerol

REFERENCES

1. English, D., Cui, Y., and Siddiqui, R.A. (1996) Messenger functions of phosphatidic acid. *Chem. Phys. Lipids* 80:117–132.
2. Wakelam, M.J. (1998) Diacylglycerol—When is it an intracellular messenger? *Biochim. Biophys. Acta* 1436:117–126.
3. Ron, D. and Kazanietz, M.G. (1999) New insights into the regulation of protein kinase C and novel phorbol ester receptors. *FASEB J.* 13:1658–1676.
4. Nishizuka, Y. (1992) Intracellular signaling by hydrolysis of phospholipids and activation of protein kinase C. *Science* 258:607–614.
5. Newton, A.C. (1995) Protein kinase C: Structure, function, and regulation. *J. Biol. Chem.* 270:28495–28498.
6. Hurley, J.H. et al. (1997) Taxonomy and function of C1 protein kinase C homology domains. *Protein Sci.* 6:477–480.
7. Ebinu, J.O. et al. (1998) RasGRP, a Ras guanyl nucleotide-releasing protein with calcium- and diacylglycerol-binding motifs. *Science* 280:1082–1086.
8. Caloca, M.J. et al. (1999) Beta2-chimaerin is a novel target for diacylglycerol: Binding properties and changes in subcellular localization mediated by ligand binding to its C1 domain. *Proc. Natl. Acad. Sci. USA* 96:11854–11859.
9. Nurrish, S., Segalat, L., and Kaplan, J.M. (1999) Serotonin inhibition of synaptic transmission: Galpha(0) decreases the abundance of UNC-13 at release sites. *Neuron* 24:231–242.
10. Baron, C.L. and Malhotra, V. (2002) Role of diacylglycerol in PKD recruitment to the TGN and protein transport to the plasma membrane. *Science* 295:325–328.
11. Hofmann, T. et al. (1999) Direct activation of human TRPC6 and TRPC3 channels by diacylglycerol. *Nature* 397:259–263.
12. Topham, M.K. (2006) Signaling roles of diacylglycerol kinases. *J. Cell. Biochem.* 97:474–484.

13. Takai, Y., Sasaki, T., and Matozaki, T. (2001) Small GTP-binding proteins. *Physiol. Rev.* 81:153–208.

14. Limatola, C. et al. (1994) Phosphatidic acid activation of protein kinase C-zeta over-expressed in COS cells: Comparison with other protein kinase C isotypes and other acidic lipids. *Biochem. J.* 304:1001–1008.

15. Jenkins, G.H., Fisette, P.L., and Anderson, R.A. (1994) Type I phosphatidylinositol 4-phosphate 5-kinase isoforms are specifically stimulated by phosphatidic acid. *J. Biol. Chem.* 269:11547–11554.

16. Jones, G.A. and Carpenter, G. (1993) The regulation of phospholipase C-gamma 1 by phosphatidic acid. Assessment of kinetic parameters. *J. Biol. Chem.* 268:20845–20850.

17. Fang, Y. et al. (2001) Phosphatidic acid-mediated mitogenic activation of mTOR signaling. *Science* 294:1942–1945.

18. Hokin, L.E. and Hokin, M.R. (1956) Metabolism of phospholipids *in vitro*. *Can. J. Biochem. Physiol.* 34:349–360.

19. Hokin, L.E. (1985) Receptors and phosphoinositide-generated second messengers. *Annu. Rev. Biochem.* 54:205–235.

20. Hokin, M.R. and Hokin, L.E. (1954) Effects of acetylcholine on phospholipids in the pancreas. *J. Biol. Chem.* 209:549–558.

21. Hokin, M.R. and Hokin, L.E. (1953) Enzyme secretion and the incorporation of P32 into phospholipids of pancreas slices. *J. Biol. Chem.* 203:967–977.

22. Brockerhoff, H. and Ballou, C.E. (1962) Phosphate incorporation in brain phospho-inositides. *J. Biol. Chem.* 237:49–52.

23. Michell, R.H. (1975) Inositol phospholipids and cell surface receptor function. *Biochim. Biophys. Acta* 415:81–147.

24. Berridge, M.J. (1987) Inositol trisphosphate and diacylglycerol: Two interacting second messengers. *Annu. Rev. Biochem.* 56:159–193.

25. Takai, Y. et al. (1979) Unsaturated diacylglycerol as a possible messenger for the activation of calcium-activated, phospholipid-dependent protein kinase system. *Biochem. Biophys. Res. Commun.* 91:1218–1224.

26. Takai, Y. et al. (1979) Calcium-dependent activation of a multifunctional protein kinase by membrane phospholipids. *J. Biol. Chem.* 254:3692–3695.

27. Kaibuchi, K., Takai, Y., and Nishizuka, Y. (1981) Cooperative roles of various membrane phospholipids in the activation of calcium-activated, phospholipid-dependent protein kinase. *J. Biol. Chem.* 256:7146–7149.

28. Kishimoto, A. et al. (1980) Activation of calcium and phospholipid-dependent protein kinase by diacylglycerol, its possible relation to phosphatidylinositol turnover. *J. Biol. Chem.* 255:2273–2276.

29. Nishizuka, Y. (1984) The role of protein kinase C in cell surface signal transduction and tumour promotion. *Nature* 308:693–698.

30. Blumberg, P.M. (1980) *In vitro* studies on the mode of action of the phorbol esters, potent tumor promoters: Part 1. *Crit. Rev. Toxicol.* 8:153–197.

31. Bell, R.M. (1986) Protein kinase C activation by diacylglycerol second messengers. *Cell* 45:631–632.

32. Preiss, J. et al. (1986) Quantitative measurement of *sn*-1,2-diacylglycerols present in platelets, hepatocytes, and ras- and sis-transformed normal rat kidney cells. *J. Biol. Chem.* 261:8597–8600.

33. Fleischman, L.F., Chahwala, S.B., and Cantley, L. (1986) Ras-transformed cells: Altered levels of phosphatidylinositol-4,5-bisphosphate and catabolites. *Science* 231:407–410.

34. Hokin, M.R. and Hokin, L.E. (1959) The synthesis of phosphatidic acid from diglyceride and adenosine triphosphate in extracts of brain microsomes. *J. Biol. Chem.* 234:1381–1386.

35. Farese, R.V., Sabir, M.A., and Larson, R.E. (1981) Effects of adrenocorticotropin and cycloheximide on adrenal diglyceride kinase. *Biochemistry* 20:6047–6051.
36. Billah, M.M., Lapetina, E.G., and Cuatrecasas, P. (1979) Phosphatidylinositol-specific phospholipase-C of platelets: Association with 1,2-diacyglycerol-kinase and inhibition by cyclic-AMP. *Biochem. Biophys. Res. Commun.* 90:92–98.
37. Daleo, G.R., Piras, M.M., and Piras, R. (1976) Diglyceride kinase activity of microtubules. Characterization and comparison with the protein kinase and ATPase activities associated with vinblastine-isolated tubulin of chick embryonic muscles. *Eur. J. Biochem.* 68:339–346.
38. Call, F.L. II and Rubert, M. (1973) Diglyceride kinase in human platelets. *J. Lipid Res.* 14:466–474.
39. Lapetina, E.G. and Hawthorne, J.N. (1971) The diglyceride kinase of rat cerebral cortex. *Biochem. J.* 122:171–179.
40. Prottey, C. and Hawthorne, J.N. (1967) The biosynthesis of phosphatidic acid and phosphatidylinositol in mammalian pancreas. *Biochem. J.* 105:379–392.
41. Sastry, P.S. and Hokin, L.E. (1966) Studies on the role of phospholipids in phagocytosis. *J. Biol. Chem.* 241:3354–3361.
42. Hokin, L.E. and Hokin, M.R. (1963) Diglyceride kinase and other path ways for phosphatidic acid synthesis in the erythrocyte membrane. *Biochim. Biophys. Acta* 67:470–484.
43. Daleo, G.R., Piras, M.M., and Piras, R. (1974) The presence of phospholipids and diglyceride kinase activity in microtubules from different tissues. *Biochem. Biophys. Res. Commun.* 61:1043–1050.
44. Walsh, J.P. and Bell, R.M. (1986) *sn*-1,2-Diacylglycerol kinase of *Escherichia coli*. Structural and kinetic analysis of the lipid cofactor dependence. *J. Biol. Chem.* 261:15062–15069.
45. Walsh, J.P. and Bell, R.M. (1986) *sn*-1,2-Diacylglycerol kinase of *Escherichia coli*. Mixed micellar analysis of the phospholipid cofactor requirement and divalent cation dependence. *J. Biol. Chem.* 261:6239–6247.
46. Kanoh, H., Kondoh, H., and Ono, T. (1983) Diacylglycerol kinase from pig brain. Purification and phospholipid dependencies. *J. Biol. Chem.* 258:1767–1774.
47. MacDonald, M.L. et al. (1988) A membrane-bound diacylglycerol kinase that selectively phosphorylates arachidonoyl-diacylglycerol. Distinction from cytosolic diacylglycerol kinase and comparison with the membrane-bound enzyme from *Escherichia coli*. *J. Biol. Chem.* 263:1584–1592.
48. Walsh, J.P. et al. (1994) Arachidonoyl-diacylglycerol kinase from bovine testis. Purification and properties. *J. Biol. Chem.* 269:21155–21164.
49. Lemaitre, R.N. et al. (1990) Distribution of distinct arachidonoyl-specific and nonspecific isoenzymes of diacylglycerol kinase in baboon (*Papio cynocephalus*) tissues. *Biochem. J.* 266:291–299.
50. Tang, W. et al. (1996) Molecular cloning of a novel human diacylglycerol kinase highly selective for arachidonate-containing substrates. *J. Biol. Chem.* 271:10237–10241.
51. Holub, B.J. and Kuksis, A. (1978) Metabolism of molecular species of diacylglycerophospholipids. *Adv. Lipid Res.* 16:1–125.
52. Glomset, J.A. (1996) *Advances in Lipobiology*, Cross, R.W. ed., pp. 61–100. JAI Press, Greenwich, CT.
53. Prescott, S.M. and Majerus, P.W. (1981) The fatty acid composition of phosphatidylinositol from thrombin-stimulated human platelets. *J. Biol. Chem.* 256:579–582.
54. Wright, T.M. et al. (1988) Kinetic analysis of 1,2-diacylglycerol mass levels in cultured fibroblasts. Comparison of stimulation by alpha-thrombin and epidermal growth factor. *J. Biol. Chem.* 263:9374–9380.
55. Exton, J.H. (1990) Signaling through phosphatidylcholine breakdown. *J. Biol. Chem.* 265:1–4.

56. Pessin, M.S., Baldassare, J.J., and Raben, D.M. (1990) Molecular species analysis of mitogen-stimulated 1,2-diglycerides in fibroblasts. Comparison of alpha-thrombin, epidermal growth factor, and platelet-derived growth factor. *J. Biol. Chem.* 265:7959–7966.

57. Fukami, K. and Takenawa, T. (1989) Quantitative changes in polyphosphoinositides 1,2-diacylglycerol and inositol 1,4,5-trisphosphate by platelet-derived growth factor and prostaglandin F2 alpha. *J. Biol. Chem.* 264:14985–14989.

58. Hodgkin, M.N. et al. (1998) Diacylglycerols and phosphatidates: Which molecular species are intracellular messengers? *Trends Biochem. Sci.* 23:200–204.

59. Cook, H.W. (1991) *Biochemistry of Lipids, Lipoproteins and Membranes*, D.E.a.V. Vance, J., ed., pp. 141–169. Benjamin/Cummings, Menlo Park, CA.

60. Schachter, J.B., Lester, D.S., and Alkon, D.L. (1996) Synergistic activation of protein kinase C by arachidonic acid and diacylglycerols *in vitro*: Generation of a stable membrane-bound, cofactor-independent state of protein kinase C activity. *Biochim. Biophys. Acta* 1291:167–176.

61. Marignani, P.A., Epand, R.M., and Sebaldt, R.J. (1996) Acyl chain dependence of diacylglycerol activation of protein kinase C activity *in vitro*. *Biochem. Biophys. Res. Commun.* 225:469–473.

62. Pessin, M.S. and Raben, D.M. (1989) Molecular species analysis of 1,2-diglycerides stimulated by alpha-thrombin in cultured fibroblasts. *J. Biol. Chem.* 264:8729–8738.

63. Siegel, D.P. et al. (1989) Physiological levels of diacylglycerols in phospholipid membranes induce membrane fusion and stabilize inverted phases. *Biochemistry* 28:3703–3709.

64. Martin, T.F., Hsieh, K.P., and Porter, B.W. (1990) The sustained second phase of hormone-stimulated diacylglycerol accumulation does not activate protein kinase C in GH3 cells. *J. Biol. Chem.* 265:7623–7631.

65. Smith, C.D. and Wells, W.W. (1983) Phosphorylation of rat liver nuclear envelopes. II. Characterization of *in vitro* lipid phosphorylation. *J. Biol. Chem.* 258:9368–9373.

66. Leach, K.L. et al. (1992) Alpha-thrombin stimulates nuclear diglyceride levels and differential nuclear localization of protein kinase C isozymes in IIC9 cells. *J. Biol. Chem.* 267:21816–21822.

67. Jarpe, M.B., Leach, K.L., and Raben, D.M. (1994) Alpha-thrombin-induced nuclear *sn*-1,2-diacylglycerols are derived from phosphatidylcholine hydrolysis in cultured fibroblasts. *Biochemistry* 33:526–534.

68. Kanoh, H., Yamada, K., and Sakane, F. (2002) Diacylglycerol kinases: Emerging downstream regulators in cell signaling systems. *J. Biochem.* (*Tokyo*) 131:629–633.

69. Sakane, F., Yamada, K., and Kanoh, H. (1989) Different effects of sphingosine, R59022 and anionic amphiphiles on two diacylglycerol kinase isozymes purified from porcine thymus cytosol. *FEBS Lett.* 255:409–413.

70. Sakane, F. et al. (1990) Porcine diacylglycerol kinase sequence has zinc finger and E-F hand motifs. *Nature* 344:345–348.

71. Schaap, D. et al. (1990) Purification, cDNA-cloning and expression of human diacylglycerol kinase. *FEBS Lett.* 275:151–158.

72. Goto, K. et al. (1992) Gene cloning, sequence, expression and *in situ* localization of 80 kDa diacylglycerol kinase specific to oligodendrocyte of rat brain. *Brain Res. Mol. Brain Res.* 16:75–87.

73. Yamada, K. and Kanoh, H. (1988) Occurrence of immunoreactive 80 kDa and non-immunoreactive diacylglycerol kinases in different pig tissues. *Biochem. J.* 255:601–608.

74. Goto, K. and Kondo, H. (1993) Molecular cloning and expression of a 90-kDa diacylglycerol kinase that predominantly localizes in neurons. *Proc. Natl. Acad. Sci. USA* 90:7598–7602.

75. Sakane, F. et al. (2007) Diacylglycerol kinases: Why so many of them? *Biochim. Biophys. Acta* 1771:793–806.
76. Evangelisti, C. et al. (2007) Nuclear diacylglycerol kinases: Emerging downstream regulators in cell signaling networks. *Histol. Histopathol.* 22:573–579.
77. Goto, K., Hozumi, Y., and Kondo, H. (2006) Diacylglycerol, phosphatidic acid, and the converting enzyme, diacylglycerol kinase, in the nucleus. *Biochim. Biophys. Acta* 1761:535–541.
78. Luo, B. et al. (2004) Diacylglycerol kinases. *Cell. Signal.* 16:983–989.
79. Goto, K. and Kondo, H. (2004) Functional implications of the diacylglycerol kinase family. *Adv. Enzyme Regul.* 44:187–199.
80. Martelli, A.M. et al. (2002) Diacylglycerol kinases in nuclear lipid-dependent signal transduction pathways. *Cell Mol. Life Sci.* 59:1129–1137.
81. van Blitterswijk, W.J. and Houssa, B. (2000) Properties and functions of diacylglycerol kinases. *Cell. Signal.* 12:595–605.
82. Sakane, F. and Kanoh, H. (1997) Molecules in focus: Diacylglycerol kinase. *Int. J. Biochem. Cell Biol.* 29:1139–1143.
83. Goto, K., Funayama, M., and Kondo, H. (1994) Cloning and expression of a cytoskeleton-associated diacylglycerol kinase that is dominantly expressed in cerebellum. *Proc. Natl. Acad. Sci. USA* 91:13042–13046.
84. Kai, M. et al. (1994) Molecular cloning of a diacylglycerol kinase isozyme predominantly expressed in human retina with a truncated and inactive enzyme expression in most other human cells. *J. Biol. Chem.* 269:18492–18498.
85. Sakane, F. et al. (1996) Molecular cloning of a novel diacylglycerol kinase isozyme with a pleckstrin homology domain and a C-terminal tail similar to those of the EPH family of protein-tyrosine kinases. *J. Biol. Chem.* 271:8394–8401.
86. Klauck, T.M. et al. (1996) Cloning and characterization of a glucocorticoid-induced diacylglycerol kinase. *J. Biol. Chem.* 271:19781–19788.
87. Imai, S. et al. (2005) Identification and characterization of a novel human type II diacylglycerol kinase, DGK kappa. *J. Biol. Chem.* 280:39870–39881.
88. Goto, K. and Kondo, H. (1996) A 104-kDa diacylglycerol kinase containing ankyrin-like repeats localizes in the cell nucleus. *Proc. Natl. Acad. Sci. USA* 93:11196–11201.
89. Bunting, M. et al. (1996) Molecular cloning and characterization of a novel human diacylglycerol kinase zeta. *J. Biol. Chem.* 271:10230–10236.
90. Ding, L. et al. (1998) The cloning and characterization of a novel human diacylglycerol kinase, DGKiota. *J. Biol. Chem.* 273:32746–32752.
91. Houssa, B. et al. (1997) Cloning of a novel human diacylglycerol kinase (DGKtheta) containing three cysteine-rich domains, a proline-rich region, and a pleckstrin homology domain with an overlapping Ras-associating domain. *J. Biol. Chem.* 272:10422–10428.
92. Caricasole, A. et al. (2002) Molecular cloning and characterization of the human diacylglycerol kinase beta (DGKbeta) gene: Alternative splicing generates DGKbeta isotypes with different properties. *J. Biol. Chem.* 277:4790–4796.
93. Sakane, F. et al. (2002) Alternative splicing of the human diacylglycerol kinase delta gene generates two isoforms differing in their expression patterns and in regulatory functions. *J. Biol. Chem.* 277:43519–43526.
94. Murakami, T. et al. (2003) Identification and characterization of two splice variants of human diacylglycerol kinase eta. *J. Biol. Chem.* 278:34364–34372.
95. Ito, T. et al. (2004) Cloning and characterization of diacylglycerol kinase iota splice variants in rat brain. *J. Biol. Chem.* 279:23317–23326.
96. Ding, L. et al. (1997) Alternative splicing of the human diacylglycerol kinase zeta gene in muscle. *Proc. Natl. Acad. Sci. USA* 94:5519–5524.

97. Nagaya, H. et al. (2002) Diacylglycerol kinase delta suppresses ER-to-Golgi traffic via its SAM and PH domains. *Mol. Biol. Cell* 13:302–316.
98. Loomis, C.R., Walsh, J.P., and Bell, R.M. (1985) sn-1,2-Diacylglycerol kinase of *Escherichia coli*. Purification, reconstitution, and partial amino- and carboxyl-terminal analysis. *J. Biol. Chem.* 260:4091–4097.
99. De La Roche, M.A. et al. (2002) *Dictyostelium discoideum* has a single diacylglycerol kinase gene with similarity to mammalian theta isoforms. *Biochem. J.* 368:809–815.
100. Masai, I. et al. (1993) Drosophila retinal degeneration A gene encodes an eye-specific diacylglycerol kinase with cysteine-rich zinc-finger motifs and ankyrin repeats. *Proc. Natl. Acad. Sci. USA* 90:11157–11161.
101. Harden, N. et al. (1993) A Drosophila gene encoding a protein with similarity to diacylglycerol kinase is expressed in specific neurons. *Biochem. J.* 289:439–444.
102. Masai, I. et al. (1992) Molecular cloning of a Drosophila diacylglycerol kinase gene that is expressed in the nervous system and muscle. *Proc. Natl. Acad. Sci. USA* 89:6030–6034.
103. Gomez-Merino, F.C. et al. (2004) AtDGK2, a novel diacylglycerol kinase from *Arabidopsis thaliana*, phosphorylates 1-stearoyl-2-arachidonoyl-sn-glycerol and 1,2-dioleoyl-sn-glycerol and exhibits cold-inducible gene expression. *J. Biol. Chem.* 279:8230–8241.
104. Katagiri, T., Mizoguchi, T., and Shinozaki, K. (1996) Molecular cloning of a cDNA encoding diacylglycerol kinase (DGK) in *Arabidopsis thaliana*. *Plant Mol. Biol.* 30:647–653.
105. Hotta, Y. and Benzer, S. (1970) Genetic dissection of the Drosophila nervous system by means of mosaics. *Proc. Natl. Acad. Sci. USA* 67:1156–1163.
106. Harris, W.A. and Stark, W. S. (1977) Hereditary retinal degeneration in *Drosophila melanogaster*. A mutant defect associated with the phototransduction process. *J. Gen. Physiol.* 69:261–291.
107. Raghu, P. et al. (2000) Constitutive activity of the light-sensitive channels TRP and TRPL in the Drosophila diacylglycerol kinase mutant, rdgA. *Neuron* 26:169–179.
108. Miller, K.G., Emerson, M.D., and Rand, J.B. (1999) Goalpha and diacylglycerol kinase negatively regulate the Gqalpha pathway in *C. elegans*. *Neuron* 24:323–333.
109. Goto, K. and Kondo, H. (1999) Diacylglycerol kinase in the central nervous system—Molecular heterogeneity and gene expression. *Chem. Phys. Lipids* 98:109–117.
110. Stathopoulos, V.M. et al. (1990) Identification of two cytosolic diacylglycerol kinase isoforms in rat brain, and in NIH-3T3 and ras-transformed fibroblasts. *Biochem. J.* 272:569–575.
111. Yada, Y. et al. (1990) Purification and characterization of cytosolic diacylglycerol kinases of human platelets. *J. Biol. Chem.* 265:19237–19243.
112. Kobayashi, N. et al. (2007) Differential subcellular targeting and activity-dependent subcellular localization of diacylglycerol kinase isozymes in transfected cells. *Eur. J. Cell. Biol.* 86:433–444.
113. Besterman, J.M. et al. (1986) Diacylglycerol-induced translocation of diacylglycerol kinase: Use of affinity-purified enzyme in a reconstitution system. *Proc. Natl. Acad. Sci. USA* 83:9378–9382.
114. Sanjuan, M.A. et al. (2001) Role of diacylglycerol kinase alpha in the attenuation of receptor signaling. *J. Cell Biol.* 153:207–220.
115. Ali, H. et al. (2004) Selective translocation of diacylglycerol kinase zeta in hippocampal neurons under transient forebrain ischemia. *Neurosci. Lett.* 372:190–195.
116. Kirino, T. (1982) Delayed neuronal death in the gerbil hippocampus following ischemia. *Brain Res.* 239:57–69.
117. Thomas, W.E. and Glomset, J.A. (1999) Affinity purification and catalytic properties of a soluble, Ca^{2+}-independent, diacylglycerol kinase. *Biochemistry* 38:3320–3326.

118. Thomas, W.E. and Glomset, J.A. (1999) Multiple factors influence the binding of a soluble, Ca^{2+}-independent, diacylglycerol kinase to unilamellar phosphoglyceride vesicles. *Biochemistry* 38:3310–3319.
119. Bell, J.D. et al. (1996) Relationships between bilayer structure and phospholipase A2 activity: Interactions among temperature, diacylglycerol, lysolecithin, palmitic acid, and dipalmitoylphosphatidylcholine. *Biochemistry* 35:4945–4955.
120. Dawson, R.M. et al. (1984) Long-chain unsaturated diacylglycerols cause a perturbation in the structure of phospholipid bilayers rendering them susceptible to phospholipase attack. *Biochem. Biophys. Res. Commun.* 125:836–842.
121. Cocco, L. et al. (1987) Synthesis of polyphosphoinositides in nuclei of Friend cells. Evidence for polyphosphoinositide metabolism inside the nucleus which changes with cell differentiation. *Biochem. J.* 248:765–770.
122. Divecha, N., Banfic, H., and Irvine, R.F. (1991) The polyphosphoinositide cycle exists in the nuclei of Swiss 3T3 cells under the control of a receptor (for IGF-I) in the plasma membrane, and stimulation of the cycle increases nuclear diacylglycerol and apparently induces translocation of protein kinase C to the nucleus. *EMBO J.* 10:3207–3214.
123. D'Santos, C.S. et al. (1999) Nuclei contain two differentially regulated pools of diacylglycerol. *Curr. Biol.* 9:437–440.
124. Neri, L.M. et al. (2002) Proliferating or differentiating stimuli act on different lipid-dependent signaling pathways in nuclei of human leukemia cells. *Mol. Biol. Cell.* 13:947–964.
125. Martelli, A.M. et al. (1999) Multiple biological responses activated by nuclear protein kinase C. *J. Cell Biochem.* 74:499–521.
126. Martelli, A.M. et al. (2000) Enhanced nuclear diacylglycerol kinase activity in response to a mitogenic stimulation of quiescent Swiss 3T3 cells with insulin-like growth factor I. *Cancer Res.* 60:815–821.
127. Kunievsky, B., Bazan, N.G., and Yavin, E. (1992) Generation of arachidonic acid and diacylglycerol second messengers from polyphosphoinositides in ischemic fetal brain. *J. Neurochem.* 59:1812–1819.
128. Arsham, A.M., Howell, J.J., and Simon, M.C. (2003) A novel hypoxia-inducible factor-independent hypoxic response regulating mammalian target of rapamycin and its targets. *J. Biol. Chem.* 278:29655–29660.
129. Hudson, C.C. et al. (2002) Regulation of hypoxia-inducible factor 1alpha expression and function by the mammalian target of rapamycin. *Mol. Cell Biol.* 22:7004–7014.
130. DeGracia, D.J. et al. (2002) Molecular pathways of protein synthesis inhibition during brain reperfusion: Implications for neuronal survival or death. *J. Cereb. Blood Flow Metab.* 22:127–141.
131. Ikegaya, Y. et al. (2001) Rapid and reversible changes in dendrite morphology and synaptic efficacy following NMDA receptor activation: Implication for a cellular defense against excitotoxicity. *J. Cell Sci.* 114:4083–4093.
132. Choi, D.W. (1990) Cerebral hypoxia: Some new approaches and unanswered questions. *J. Neurosci.* 10:2493–2501.
133. Lehrmann, E. et al. (1998) Microglia and macrophages are major sources of locally produced transforming growth factor-beta1 after transient middle cerebral artery occlusion in rats. *Glia* 24:437–448.
134. Sun, Y. et al. (2003) VEGF-induced neuroprotection, neurogenesis, and angiogenesis after focal cerebral ischemia. *J. Clin. Invest.* 111:1843–1851.
135. Wei, L. et al. (2004) Necrosis, apoptosis and hybrid death in the cortex and thalamus after barrel cortex ischemia in rats. *Brain Res.* 1022:54–61.
136. Kato, H. et al. (2000) Expression of microglial response factor-1 in microglia and macrophages following cerebral ischemia in the rat. *Brain Res.* 882:206–211.

137. Nakano, T. et al. (2006) Diacylglycerol kinase zeta is involved in the process of cerebral infarction. *Eur. J. Neurosci.* 23:1427–1435.
138. Marti, H.J. et al. (2000) Hypoxia-induced vascular endothelial growth factor expression precedes neovascularization after cerebral ischemia. *Am. J. Pathol.* 156:965–976.
139. Temes, E. et al. (2004) Role of diacylglycerol induced by hypoxia in the regulation of HIF-1alpha activity. *Biochem. Biophys. Res. Commun.* 315:44–50.
140. Aragones, J. et al. (2001) Evidence for the involvement of diacylglycerol kinase in the activation of hypoxia-inducible transcription factor 1 by low oxygen tension. *J. Biol. Chem.* 276:10548–10555.
141. Bray, G.A. and Popkin, B.M. (1998) Dietary fat intake does affect obesity! *Am. J. Clin. Nutr.* 68:1157–1173.
142. Friedman, J.M. and Halaas, J.L. (1998) Leptin and the regulation of body weight in mammals. *Nature* 395:763–770.
143. Frederich, R.C. et al. (1995) Leptin levels reflect body lipid content in mice: Evidence for diet-induced resistance to leptin action. *Nat. Med.* 1:1311–1314.
144. Spanswick, D. et al. (1997) Leptin inhibits hypothalamic neurons by activation of ATP-sensitive potassium channels. *Nature* 390:521–525.
145. Liu, Z., Chang, G.Q., and Leibowitz, S.F. (2001) Diacylglycerol kinase zeta in hypothalamus interacts with long form leptin receptor. Relation to dietary fat and body weight regulation. *J. Biol. Chem.* 276:5900–5907.
146. Phillips, M.S. et al. (1996) Leptin receptor missense mutation in the fatty Zucker rat. *Nat. Genet.* 13:18–19.
147. Avignon, A. et al. (1996) Chronic activation of protein kinase C in soleus muscles and other tissues of insulin-resistant type II diabetic Goto-Kakizaki (GK), obese/aged, and obese/Zucker rats. A mechanism for inhibiting glycogen synthesis. *Diabetes* 45:1396–1404.
148. Considine, R.V. et al. (1995) Protein kinase C is increased in the liver of humans and rats with non-insulin-dependent diabetes mellitus: An alteration not due to hyperglycemia. *J. Clin. Invest.* 95:2938–2944.
149. Missale, C. et al. (1998) Dopamine receptors: From structure to function. *Physiol. Rev.* 78:189–225.
150. Levy, D. et al. (1990) Prognostic implications of echocardiographically determined left ventricular mass in the Framingham Heart Study. *N. Engl. J. Med.* 322:1561–1566.
151. Morgan, H.E. and Baker, K.M. (1991) Cardiac hypertrophy. Mechanical, neural, and endocrine dependence. *Circulation* 83:13–25.
152. Bowling, N. et al. (1999) Increased protein kinase C activity and expression of Ca^{2+}-sensitive isoforms in the failing human heart. *Circulation* 99:384–391.
153. Takeishi, Y. et al. (1998) *In vivo* phosphorylation of cardiac troponin I by protein kinase Cbeta2 decreases cardiomyocyte calcium responsiveness and contractility in transgenic mouse hearts. *J. Clin. Invest.* 102:72–78.
154. Takeishi, Y. et al. (1999) Responses of cardiac protein kinase C isoforms to distinct pathological stimuli are differentially regulated. *Circ. Res.* 85:264–271.
155. Takeishi, Y. et al. (2000) Transgenic overexpression of constitutively active protein kinase C epsilon causes concentric cardiac hypertrophy. *Circ. Res.* 86:1218–1223.
156. Yahagi, H. et al. (2005) Differential regulation of diacylglycerol kinase isozymes in cardiac hypertrophy. *Biochem. Biophys. Res. Commun.* 332:101–108.
157. Newton, A.C. (1997) Regulation of protein kinase C. *Curr. Opin. Cell Biol.* 9:161–167.
158. Takahashi, H. et al. (2005) Adenovirus-mediated overexpression of diacylglycerol kinase-zeta inhibits endothelin-1-induced cardiomyocyte hypertrophy. *Circulation* 111:1510–1516.
159. Arimoto, T. et al. (2006) Cardiac-specific overexpression of diacylglycerol kinase zeta prevents Gq protein-coupled receptor agonist-induced cardiac hypertrophy in transgenic mice. *Circulation* 113:60–66.

160. Harada, M. et al. (2007) Diacylglycerol kinase zeta attenuates pressure overload-induced cardiac hypertrophy. *Circ. J.* 71:276–282.

161. Luo, B., Prescott, S.M., and Topham, M.K. (2003) Protein kinase C alpha phosphorylates and negatively regulates diacylglycerol kinase zeta. *J. Biol. Chem.* 278:39542–39547.

162. Luo, B., Prescott, S.M., and Topham, M.K. (2003) Association of diacylglycerol kinase zeta with protein kinase C alpha: Spatial regulation of diacylglycerol signaling. *J. Cell Biol.* 160:929–937.

163. Pfeffer, M.A. (1995) Left ventricular remodeling after acute myocardial infarction. *Annu. Rev. Med.* 46:455–466.

164. Takeda, M. et al. (2001) Gene expression and *in situ* localization of diacylglycerol kinase isozymes in normal and infarcted rat hearts: Effects of captopril treatment. *Circ. Res.* 89:265–272.

165. Niizeki, T. et al. (2007) Cardiac-specific overexpression of diacylglycerol kinase zeta attenuates left ventricular remodeling and improves survival after myocardial infarction. *Am. J. Physiol. Heart Circ. Physiol.* 292:H1105–H1112.

166. Olenchock, B.A. et al. (2006) Disruption of diacylglycerol metabolism impairs the induction of T cell anergy. *Nat. Immunol.* 7:1174–1181.

167. Crotty, T. et al. (2006) Diacylglycerol kinase delta regulates protein kinase C and epidermal growth factor receptor signaling. *Proc. Natl. Acad. Sci. USA* 103:15485–15490.

168. Rodriguez de Turco, E.B. et al. (2001) Diacylglycerol kinase epsilon regulates seizure susceptibility and long-term potentiation through arachidonoyl-inositol lipid signaling. *Proc. Natl. Acad. Sci. USA* 98:4740–4745.

169. Zhong, X.P. et al. (2003) Enhanced T cell responses due to diacylglycerol kinase zeta deficiency. *Nat. Immunol.* 4:882–890.

170. Regier, D.S. et al. (2005) Diacylglycerol kinase iota regulates Ras guanyl-releasing protein 3 and inhibits Rap1 signaling. *Proc. Natl. Acad. Sci. USA* 102:7595–7600.

171. Li, W. et al. (1996) Blocked signal transduction to the ERK and JNK protein kinases in anergic CD4+ T cells. *Science* 271:1272–1276.

172. Fields, P.E., Gajewski, T.F., and Fitch, F.W. (1996) Blocked Ras activation in anergic CD4+ T cells. *Science* 271:1276–1278.

173. Zha, Y. et al. (2006) T cell anergy is reversed by active Ras and is regulated by diacylglycerol kinase-alpha. *Nat. Immunol.* 7:1166–1173.

174. Simonato, M. (1993) A pathogenetic hypothesis of temporal lobe epilepsy. *Pharmacol. Res.* 27:217–225.

175. Keele, N.B. et al. (2000) Epileptogenesis up-regulates metabotropic glutamate receptor activation of sodium-calcium exchange current in the amygdala. *J. Neurophysiol.* 83:2458–2462.

176. Conn, P.J. and Pin, J.P. (1997) Pharmacology and functions of metabotropic glutamate receptors. *Annu. Rev. Pharmacol. Toxicol.* 37:205–237.

177. Bazan, N.G., Rodriguez de Turco, E.B., and Allan, G. (1995) Mediators of injury in neurotrauma: Intracellular signal transduction and gene expression. *J. Neurotrauma* 12:791–814.

178. Anderson, R.A. et al. (1999) Phosphatidylinositol phosphate kinases, a multifaceted family of signaling enzymes. *J. Biol. Chem.* 274:9907–9910.

179. Ishihara, H. et al. (1998) Type I phosphatidylinositol-4-phosphate 5-kinases. Cloning of the third isoform and deletion/substitution analysis of members of this novel lipid kinase family. *J. Biol. Chem.* 273:8741–8748.

180. Kanoh, H. and Ohno, K. (1981) Partial purification and properties of diacylglycerol kinase from rat liver cytosol. *Arch. Biochem. Biophys.* 209:266–275.

181. Kanoh, H., Sakane, F., and Yamada, K. (1992) Diacylglycerol kinase isozymes from brain and lymphoid tissues. *Methods Enzymol.* 209:162–172.

182. Sakane, F. et al. (1991) Porcine 80-kDa diacylglycerol kinase is a calcium-binding and calcium/phospholipid-dependent enzyme and undergoes calcium-dependent translocation. *J. Biol. Chem.* 266:7096–7100.
183. MacDonald, M.L. et al. (1988) Regulation of diacylglycerol kinase reaction in Swiss 3T3 cells. Increased phosphorylation of endogenous diacylglycerol and decreased phosphorylation of didecanoylglycerol in response to platelet-derived growth factor. *J. Biol. Chem.* 263:1575–1583.
184. Lemaitre, R.N. and Glomset, J.A. (1992) Arachidonoyl-specific diacylglycerol kinase. *Methods Enzymol.* 209:173–182.

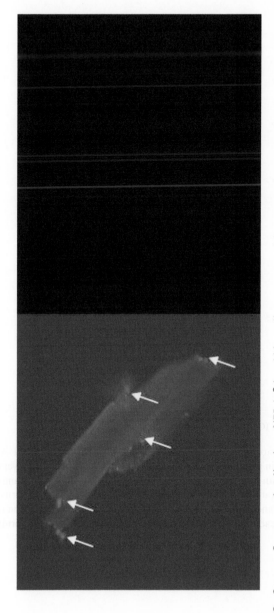

FIGURE 1.2 Immunofluorescent localization of iPLA$_2$β in rabbit cardiac myocytes. After appropriate processing (see Protocol B), cells were probed with anti-iPLA$_2$β antibody (1 in 1000 dilution; Cayman Chemical, Ann Arbor, Michigan) and alexa fluor goat anti-chicken 468 secondary antibody (1:1200 dilution, Molecular Probes, Eugene, Oregon). The cells were then mounted in vectashield and examined under a fluorescent microscope. Left panel shows pronounced staining for iPLA$_2$β protein observed at areas of cell–cell junction contact in isolated rabbit cardiac myocytes (white arrows). Negative control with secondary antibody is shown on the right panel.

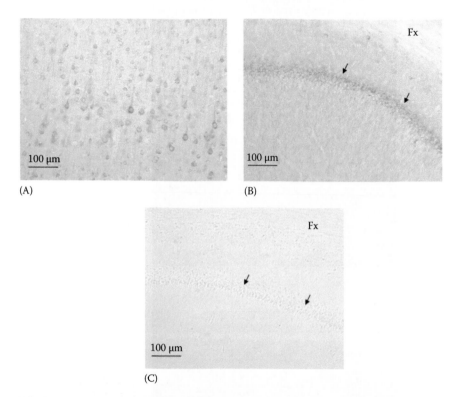

(A)

(B)

(C)

FIGURE 2.4 Immunohistochemical localization of PLCβ1 in rat brain. Panel A shows PLCβ1 immunoreactivity in pyramidal neurons located throughout the cortical layers. Panel B shows that neurons within the CA1-3 regions of the hippocampus were also immunoreactive for PLCβ1 (arrows). Scattered neurons were also evident in the hippocampal oriens layer dorsal to CA1-3 regions. Panel C demonstrates that the matching primary omission controls were clear of all immunoreactivity in the CA region (arrows). Abbreviation: Fx, fornix.

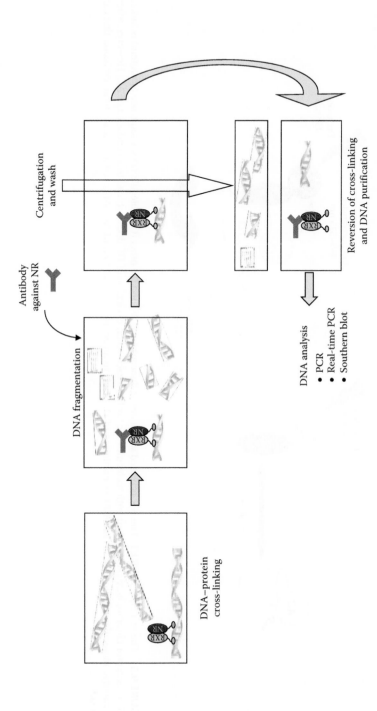

FIGURE 9.4 The principle of ChIP. After cell treatment, cross-linking of protein–DNA is first done with formaldehyde. Cells are lysed and DNA is fragmented by restriction enzyme cut or sonication in 200–1000 bp fragment length. DNA-bound NR is then targeted by a specific antibody. The complex is precipitated and free fraction is washed. Cross-linking is then reversed and DNA is purified. Purified DNA can be analyzed by PCR, real-time PCR, or Southern blot (NR, nuclear receptor; PCR, polymerase chain reaction).

Centrifugation and wash

Antibody against NR

DNA fragmentation

DNA–protein cross-linking

DNA analysis
• PCR
• Real-time PCR
• Southern blot

Reversion of cross-linking and DNA purification

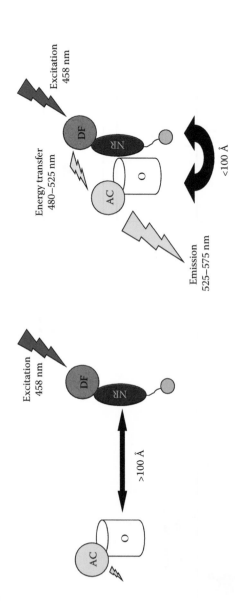

FIGURE 9.5 The principle of FRET. The NR is linked with a donor fluorophore (DF) and the other (O) molecule (DNA sequence or protein) is linked with an acceptor chromophore (AC). When both molecules are far from each other, excitation of DF does not lead to an emission of AC. However, when the two dye molecules are close (less than 100 Å), excitation of DF leads to an energy transfer toward AC. AC excitation leads to an emission that can be read.

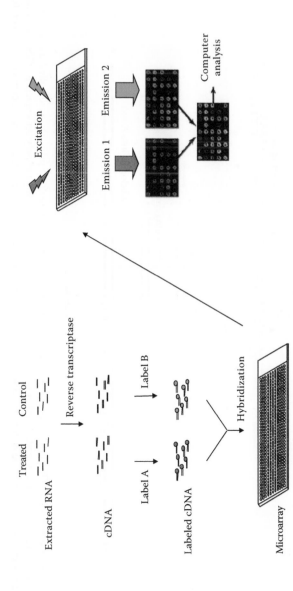

FIGURE 9.6 The principle of microarray. cDNA microarray scheme. Total RNA from control and treated cells or animal tissues is first reverse transcribed and then fluorescently labeled with a different dye. Cye3 or Cye5 is usually used. Both samples are mixed and hybridized to the microarray under stringent conditions. The chip is afterward excited by laser. Emissions are scanned and computer analyzed. Images obtained are merged and ratio between control coloration and treated sample coloration is used to indicate the change in expression level. (Adapted from Duggan, D.J. et al., *Nat. Genet.* 21, 10, 1999.)

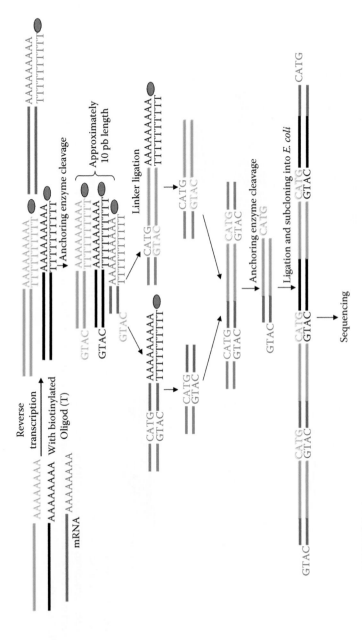

FIGURE 9.7 The principle of SAGE. mRNA is reverse-transcribed using biotinylated oligo(dT) primers. A so-called anchoring enzyme cleaves cDNA approximately 10bp before the 3' end. A linker is added to cleaved cDNA. After removing the poly(A) sequence, cDNA are linked together. A second cleavage by the anchoring enzyme is done to allow the ligation of multiple cDNA fragments. Ligations are cloned into *E. coli* bacteria and can be sequenced.

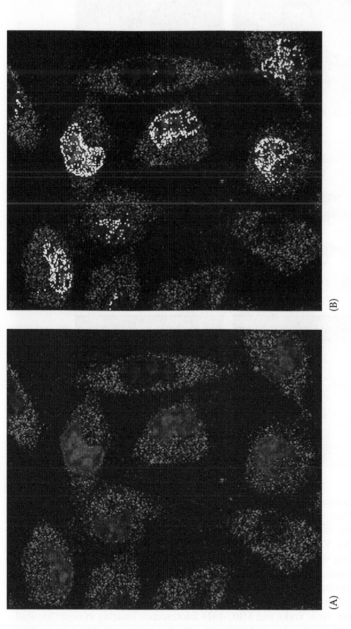

(A)

(B)

FIGURE 10.6 Fatty acid nuclear localization in cultured L-cells. L-cells were seeded onto chambered cover glass, culture media was replaced with PBS containing 1.25 μM of SYTO59 nucleic acid stain, and cells were incubated at room temperature for 30 min. Background images were obtained, followed by the addition of 100 nM BODIPY C-16, and images were acquired 30 min postincubation on 0.3 μm confocal slices with a X63 Plan-Fluor oil immersion objective, N.A.1.45, an Axiovert 135 microscope (Zeiss, Carl Zeiss Inc., Thornwood, New York), and MRC-1024 fluorescence imaging system (Bio-Rad, Hercules, California). Syto59 and BODIPY-C16 probes were excited at 488/568 lines with a krypton-argon laser (Coherent, Sunnyvale, California), and emission was recorded by a photomultiplier after passing through a 522/D35 or 680/32 emission filter, respectively. (A) Merged images of BODIPY C16 (green) and SYTO59 (red) fluorescence. (B) Colocalized pixels representing BODIPY C-16 nuclear localization are shown in white.

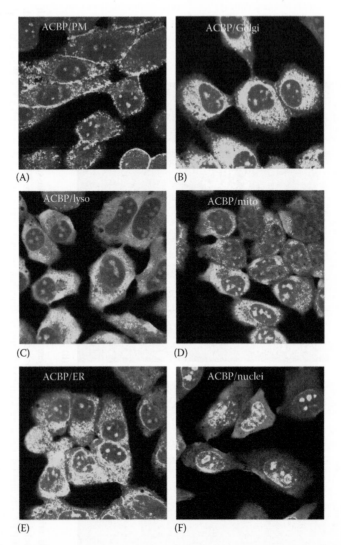

FIGURE 10.7 Intracellular distribution of endogenous ACBP in fixed rat hepatoma cells by indirect immunofluorescence confocal microscopy. T-7 rat hepatoma cells were seeded onto chambered slides, the culture media was removed, and cells were fixed with 4% glutaraldehyde at 37°C for 30 min. Residual aldehyde was quenched with ammonium chloride, and nonspecific binding was blocked by incubation in 5% FBS in Hank's solution. Primary antibodies raised against ACBP were produced locally in rabbit or rat, while all other primary and secondary antibodies were commercially purchased. Images were acquired with an X63 Plan-Fluor oil immersion objective, N.A.1.45, an Axiovert 135 microscope (Zeiss, Carl Zeiss, Inc., Thornwood, New York), and MRC-1024 fluorescence imaging system (Bio-Rad, Hercules, California). Colocalization of ACBP with: (A) cholera toxin B (PM caveolae/lipid raft marker), (B) wheat germ agglutinin (Golgi marker), (C) cathepsin D (lysosomal marker), (D) mitochondrial heat sock protein 70 (mitochondrial marker), (E) concanavalin A (ER marker), (F) Hoechst 33342 (nuclear marker). Yellow pixels represent colocalized pixels.

5 Methods for Measuring Endocannabinoid Production and Expression and Activity of Enzymes Involved in the Endocannabinoid System

Tiziana Bisogno, Luciano De Petrocellis,
and Vincenzo Di Marzo

CONTENTS

5.1 INTRODUCTION

The finding, in the early 1990s, of specific G-protein-coupled receptors for the psychoactive component of *Cannabis sativa*, (−)-Δ^9-tetrahydrocannabinol (THC) [1], led to the discovery of a whole endogenous signaling system now known as the endocannabinoid system (ECS). The ECS consists of the aforementioned cannabinoid receptors, the endocannabinoids (EC), and the proteins responsible for their metabolism. Mammalian tissues contain at least two types of cannabinoid receptors, CB_1 and CB_2 [2], and several types of endogenous compounds that activate these receptors have been identified (Figure 5.1). The two most studied EC are anandamide (AEA) [3] and 2-arachidonoyl-glycerol (2-AG) [4,5]. These molecules are biosynthesized via the cleavage of their membrane lipid precursors, *N*-arachidonoyl-phosphatidylethanolamine (*N*-ArPE) and diacylglycerols (DAG). The EC are inactivated by intracellular hydrolyzing enzymes, fatty acid amide hydrolase (FAAH), and monoacylglycerol lipases (MAGL) ([6] for review). In this chapter, we review some of the methodologies necessary to study the ECS. In particular, we focus our attention on the analytical and enzymatic procedures that allow for the study of the regulation of EC levels and have helped in establishing the physiological role of the ECS and its involvement in several pathological conditions.

5.1.1 BIOSYNTHESIS OF ENDOCANNABINOIDS

AEA and 2-AG are not stored in resting cells but, unlike other mediators, are synthesized "on demand." This concept suggests that they are synthesized when and where necessary and then immediately released from the cell, via an as-yet-unidentified mechanism, to act in an autocrine or paracrine manner at cannabinoid receptors [6]. The biosynthetic processes are activated by the elevation of the intracellular Ca^{2+} that follows either Ca^{2+} entry (such as after neuronal depolarization) or mobilization from intracellular stores (such as following stimulation of metabotropic receptors coupled to $G_{q/11}$ proteins), or both [7–13]. In fact, in the brain, AEA and, particularly,

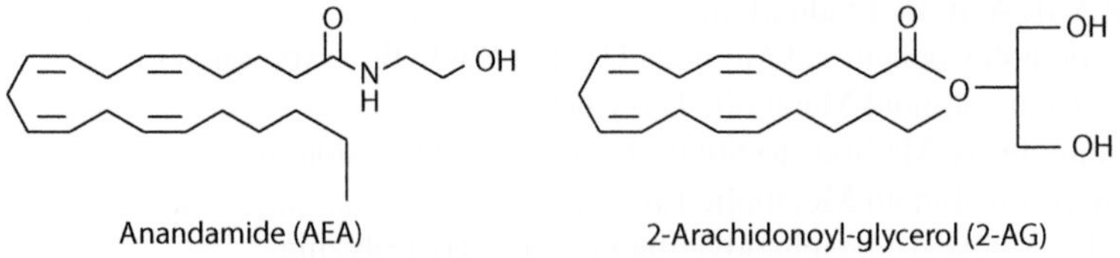

Anandamide (AEA)   2-Arachidonoyl-glycerol (2-AG)

FIGURE 5.1   Chemical structures of the two best studied EC, anandamide (AEA) and 2-arachidonoyl glycerol (2-AG).

2-AG are biosynthesized via the action of Ca^{2+}-sensitive enzymes and then released from the postsynaptic neuron to activate presynaptic cannabinoid receptors.

5.1.1.1 Biosynthesis of AEA

AEA is formed by a phospholipid-dependent pathway (Figure 5.2) consisting of the enzymatic hydrolysis of the corresponding membrane phosphoglyceride precursor N-ArPE [14], catalyzed by an N-acyl-phosphatidylethanolamine-selective phospholipase D (NAPE-PLD) [15–19] that was only recently cloned [20]. NAPE-PLD is quite different from other phospholipase D (PLD) enzymes in that it exhibits no selectivity for a particular fatty acid moiety at the sn-1, sn-2, and N positions of

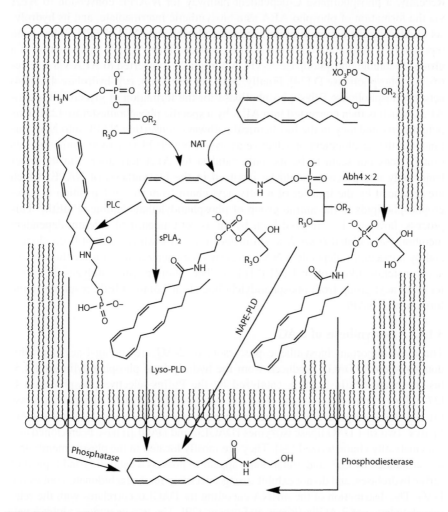

FIGURE 5.2 Major biosynthetic pathways for AEA. Abbreviations: Abh4, α/β-hydrolase 4; lyso-PLD, lyso-phospholipase D; sPLA$_2$, soluble phospholipase A$_2$, PLC, phopholipase C; NAPE-PLD, phospholipase D selective for N-acyl-phosphatidylethanolamines; NAT, *trans*-N-acyltransferase.

N-acyl-phosphatidylethanolamine (NAPE). Therefore, this enzyme is also responsible for the formation of other *N*-acyl-ethanolamines (NAE). The amino acid sequence showed that this enzyme has no homology with other known PLD enzymes but is classified as a member of the zinc metallohydrolase family of the β-lactamase fold [20]. A close precursor–product relationship exists between *N*-ArPE and AEA [21] and evidence has been reported that NAPE-PLD, when overexpressed in cells, upregulates the levels of NAE [22]. Although NAPE are now well recognized as NAE precursors, alternative pathways for *N*-ArPE conversion to AEA have been proposed that could explain why several tissues and organs of NAPE-PLD-deficient mice contain lower levels of saturated NAE, but not of AEA, than wild-type mice. Recently, a phospholipase C-dependent pathway for *N*-ArPE conversion to AEA via the formation of phospho-AEA as a biosynthetic intermediate, and its hydrolysis to AEA by the action of protein tyrosine phosphatase N22 (PTPN22) has been reported [23]. Additionally, 2-lyso-NAPE produced form *N*-ArPE by the action of a group IB soluble phospholipase A_2 (PLA_2) might be converted to AEA via a selective lysophospholipase D [24]. Finally, in the mouse brain, α/β-hydrolase 4 (Abh4) acts as phospholipase B/lysophospholipase for the formation of glycerol-phospho-AEA, which is then converted into AEA by a specific phosphodiesterase [25]. This potential redundancy in the biochemical pathways converting NAPE into NAE has hindered the development of selective assays of NAPE-PLD activity, although this protein does not seem to be the only catalyst for AEA formation. On the other hand, only one process should be involved in the biosynthesis of NAPE, which are produced by the transfer of a fatty acyl chain from the *sn*-1 position of glycerophospholipids to the amino group of phosphatidylethanolamine (PtdEth). This process, however, is catalyzed both by an as-yet-unidentified calcium-dependent membrane-associated *trans*-*N*-acyltransferase (NAT), which has not been cloned, and by a calcium-independent NAT recently characterized as a rat lecithin-retinol acyltransferase-like protein 1 (RLP-1) [26]. The latter enzyme catalyzes the transfer of an acyl group from phosphatidylcholine (PtdCho) to PtdEth, resulting in the formation of NAPE.

5.1.1.2 Biosynthesis of 2-AG

The most important biosynthetic precursors of 2-AG are the *sn*-1-acyl-2-arachidonoylglycerols (DAG), produced from the hydrolysis of phosphatidylinositol-4,5-bis-phosphate (PtdIns(4,5)P_2), catalyzed by the PtdIns-selective phospholipase C [27–30], or of phosphatidic acid (PtdOH), catalyzed by a PtdOH phosphohydrolase [31–33]. DAG (Figure 5.3) are then converted into 2-AG by *sn*-1 selective-DAG lipases. Two *sn*-1 DAG lipase isozymes (DAGLα and DAGLβ) have been cloned and enzymatically characterized [34]. They are mostly located in the plasma membrane, are stimulated by Ca^{2+} and glutathione, appear to possess a catalytic triad typical of serine hydrolases, and do not exhibit strong selectivity for 2-arachidonate-containing DAG. The distribution of the mRNA encoding for DAGLα correlates with the relative abundance of 2-AG in tissues and organs [29]. The two enzymes exhibit a pattern of expression that fits with the proposed role of 2-AG as a mediator of neurite growth, during brain development [35], or as retrograde signal-mediating depolarization-induced suppression of neurotransmission and heterosynaptic plasticity in the

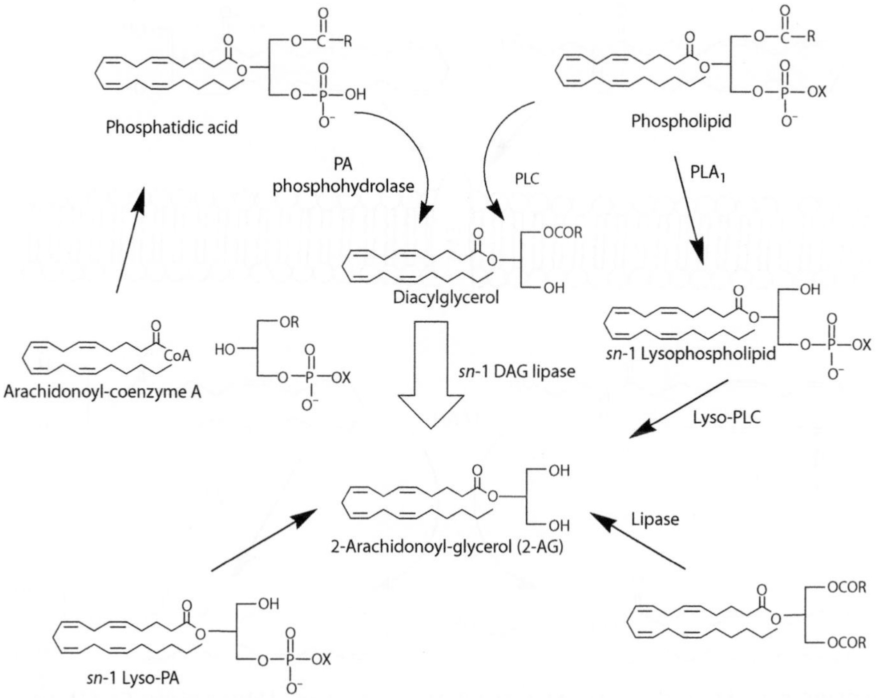

FIGURE 5.3 Biosynthetic pathways for 2-AG. Abbreviations: PA, phosphatidic acid; PLC, phospholipase C; PLA$_1$, phospholipase A$_1$; DAG, diacylglycerol.

adult brain [36]. Indeed, DAGLα is localized to the postsynaptic dendritic spines establishing synapses with CB$_1$-expressing axons [37,38]. Recently, a biosynthetic pathway responsible for 2-AG formation that does not require the sequential activation of PLCβ and DAGL was suggested [13,39]. The hippocampal levels of AEA, but not of 2-AG, are reduced in transgenic mice lacking G$_{αq}$/G$_{α11}$ proteins, indicating that the basal biosynthesis of AEA in this brain area is under the tonic control of receptors coupled to these G-proteins whereas 2-AG formation is not [12]. On the other hand, in G$_{αq}$/G$_{α11}$ null mice, the levels of AEA, but not 2-AG, are still stimulated by kainic acid as much as in wild-type mice. These data suggest that the stimulus-induced biosynthesis of 2-AG, but not of AEA, does require G$_{αq}$/G$_{α11}$ in order to occur [12].

5.1.2 ENDOCANNABINOID DEGRADATION

The inactivation pathways of AEA and 2-AG involve a two-step mechanism (Figure 5.4). Both EC are transported across the membrane and are inactivated by intracellular hydrolyzing enzymes. It is not known whether EC are taken up by cells via a plasma membrane transporter (the putative endocannabinoid membrane transporter [EMT]) or through a more complicated protein-mediated process. Furthermore, it is unclear if the same mechanism is responsible for the release from cells of *de novo* biosynthesized EC, necessary for these compounds to interact with cannabinoid receptors. The EMT has not been isolated or cloned, but indirect evidence suggesting its

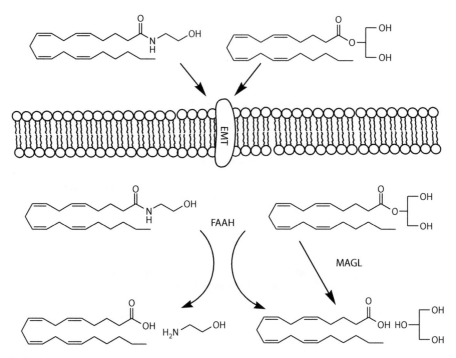

FIGURE 5.4 Mechanisms for AEA and 2-AG inactivation. Abbreviations: FAAH, fatty acid amide hydrolase; MAGL, monoacylglycerol lipase; EMT, endocannabinoid membrane transporter.

existence was recently discussed in comprehensive reviews [40,41]. In brief, there are data to support the existence of specific protein(s), different from FAAH, facilitating AEA transport across the membrane [42–47]. This evidence opposes the hypothesis that FAAH is the only enzyme mediating AEA cellular uptake [48].

Once inside the cell, the EC are degraded through different mechanisms, depending on their chemical nature. The enzyme that is capable of catalyzing the hydrolysis of AEA and 2-AG to arachidonic acid (ARA) and ethanolamine or glycerol was purified and cloned from rat liver microsomes [49]. FAAH also recognizes other fatty acid primary amides as substrate, such as the N-acyltaurines [50] and N-acyl amino acids [51,52]. FAAH is the oldest and the best characterized enzyme involved in the degradation of EC. The cloning, structural and kinetic properties, distribution in the body, and crystal structure have been described [53]. FAAH is unusual among the "amidase signature" (AS) enzymes since it is an integral membrane enzyme. It is mainly expressed in microsomal membranes and has an alkaline optimal pH. Site-directed mutagenesis and crystallographic x-ray studies showed that, unlike most serine hydrolases, the FAAH catalytic mechanism involves a Ser-Ser-Lys triad [53]. The promoter region on the FAAH gene is targeted by progesterone and leptin, which usually upregulate the enzyme, and by estrogens and glucocorticoids, which downregulate its expression [54–57]. FAAH is abundantly expressed throughout the central nervous system, and FAAH-positive neurons in the brain are found in proximity to nerve terminals that contain CB_1 cannabinoid receptors. This evidence supports a role of this enzyme in EC deactivation. The physiopathological relevance of FAAH is becoming more and more evident, thanks to studies utilizing its chemical or genetic inactivation. In particular, despite its atypical catalytic mechanism, FAAH is inhibited by most classical serine hydrolase inhibitors including trifluoromethyl ketones, fluorophosphonate, carbamates, and α-ketoheterocycle compounds. FAAH inhibitors that are efficacious *in vivo* [58–60] have provided evidence that the tonic control of FAAH may have therapeutic value for the treatment of several disorders [61–63]. However, the specific physiopathological role of FAAH is currently being investigated in transgenic mice lacking the enzyme, i.e., the "FAAH-knockout" mice, and FAAH gene polymorphisms associated with human pathological conditions are also studied [64–66]. Recently, the functional proteomic discovery of a second membrane-associated AS enzyme in humans, which displays FAAH activity, was reported [67]. This enzyme, named FAAH-2, exhibits only 20% homology with FAAH and is expressed in several mammalian species but not in rodents.

Although FAAH can catalyze 2-AG hydrolysis [68], 2-AG levels, unlike those of AEA, are not increased in FAAH "knockout" mice [69]. This observation is in agreement with previously reported evidence of the existence of other enzymes catalyzing 2-AG inactivation that are different from FAAH [70,71]. Indeed, a MAGL inactive on AEA and with high homology with other human and mouse MAGL was cloned from the rat [72–74]. The observed complementary localization in the rat brain for MAGL and FAAH, i.e., presynaptic and postsynaptic, respectively, suggests different roles for the two EC in the CNS [75]. RNA interference–mediated silencing of MAGL expression greatly enhances 2-AG, but not AEA, accumulation in HeLa cells, suggesting a primary role for MAGL lipase in the degradation of 2-AG [76]. The same authors also carried out immunodepletion experiments, suggesting that MAGL accounts for only 50% of the total 2-AG-hydrolyzing activity in soluble fractions of rat brain. Furthermore, evidence of a second MAGL activity that controls 2-AG levels in intact microglial cells has been provided [77]. The novel MAGL lipase expressed mainly in mitochondria and nuclei is pharmacologically distinguishable from the cloned MAGL and FAAH.

5.1.3 Physiopathological Function of the Endocannabinoid System

Considerable progress in understanding the functions of the ECS has also been made by measuring the tissue, blood, and cerebrospinal fluid concentrations of EC in either physiological or pathological conditions. The physiological functions of the EC and their corresponding potential pathological implications were recently discussed in comprehensive reviews [2,6,63,78–80]. The cannabinoid receptor antagonists, or drugs capable of manipulating selectively the levels of the EC, have then been used to confirm the existence of cause–effect relationships between alterations of EC levels and the "homeostatic" attempt to counteract certain pathological conditions. In particular, disorders in which the ECS is upregulated only in the tissues participating in the disease were reported. These disorders can be potentially treated with inhibitors of EC degradation ("indirect agonists") or with cannabinoid receptor agonists. Furthermore, EC have been shown to also participate in the symptoms or progress of diseases, and, in this case, treatment with antagonists or inhibitors of synthesis might be effective [6,63]. In many diseases, the control of EC levels occurs through various mechanisms that reflect the availability, expression, activity, and regulation of EC biosynthetic and inactivating enzymes [7,81–83]. In this context, a multidisciplinary approach involving the organic and analytical chemists together with pharmacologists and enzymologists led to the development of specific and sensitive methodologies for the quantitative analysis of the various components of the ECS. Hereafter, we review the methodological and technical approaches for the quantification of EC levels in biological systems, of the catalytic activity of their metabolic enzymes, as well as of their mRNA and protein expression.

5.2 EXTRACTION, PURIFICATION, AND QUANTIFICATION OF ENDOCANNABINOIDS

Because EC and other NAE are involved in the regulation of different physiopathological processes, it was important to develop methods to accurately detect and quantify these molecules in cells, tissues, and biological fluids. Established methods for the quantification of EC in tissues have used primarily solvent-based extraction methods, followed by sample cleanup procedures, and analysis by gas chromatography-mass spectrometry (GC-MS) or liquid chromatography-mass spectrometry (LC-MS). A rapid postmortem increase of EC levels that may cause an artifactual overestimation of their endogenous levels was reported [84,85]. To minimize postmortem changes, cells and tissues have to be frozen, possibly in liquid nitrogen, immediately after preparation, and then kept frozen at −80°C until extraction. Owing to the lipid nature of EC, which are polyunsaturated fatty acid derivatives, they are easily oxidized and hydrolyzed and bind to plastic and glass. Therefore, these properties must be considered during tissue homogenization and extraction procedures (Figure 5.5). In particular, several organic solvents such as chloroform/methanol (2:1 by vol.), diethyl ether, ethyl acetate/hexane (9:1 by vol.) and ethyl acetate, often used to extract ARA and other fatty acids from tissues [17,85–87], were compared in their ability to extract EC from cell cultures or tissues [88–91]. The recovery of EC obtained using these organic solvents are sufficient to allow for a reliable quantification of compounds after partial or full purification of the lipid extract. Indeed, silica-gel open-bed chromatography and/or thin-layer chromatography (TLC), eluted with increasing concentrations of methanol in chloroform, have been employed for sample fractionation. The introduction of high-performance liquid chromatography (HPLC) methods mainly by using reverse phase, RP-C18 columns significantly improves the cleanup procedure. Several analytical HPLC methods using spectrophotometric detection are described with the aim of developing procedures with common applicability that do not require the use of complex and costly instrumentation. The development of fluorogenic benzodioxazole derivatization of AEA and other NAE that can be used with HPLC coupled to a fluorescence detector was reported. The fluorimetric detection of the derivatives was made at 560 nm with excitation at 450 nm and the detection limits for AEA was 20 fmol on column [92]. A fluorescence-based method that has been applied to study AEA biosynthesis involves AEA derivatization with 1-anthroyl cyanide, followed by HPLC-fluorescence analysis [93]. The dansyl ester

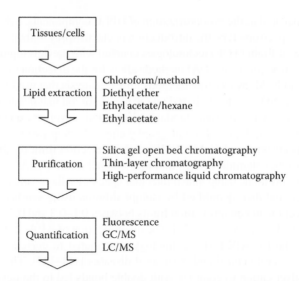

FIGURE 5.5 Schematic description of an analytic procedure for the quantification of EC.

of AEA, obtained by the treatment of the compound with an excess of dansyl chloride and dimethylaminopyridine in acetone, followed by detection at 255 nm, was also used to measure AEA in media from cultured hepatocytes [94].

Considerable progress in the understanding of the physiological and pathological roles of EC was achieved after the development of quantitative measurement techniques based on mass spectrometric methods. Electron-impact (EI) gas chromatography-mass spectrometry (GC/MS) methods based on isotope dilution, used for the identification and quantification of AEA, 2-AG, and other NAE in biological matrices, have been described. NAE were treated with Bis(trimethylsilyl)trifluoroacetamide (BSTFA) or N-methyl-N-(trimethylsilyl)trifluoroacetamide (MSTFA) at room temperature, for 2 h, to obtain the trimethylsilylether (TMS) derivatives or overnight with acetic anhydride in pyridine (1:4) in an atmosphere of nitrogen to synthesize the acetoxy derivatives. These derivatives allow for the identification of NAE using the selected-ion monitoring (SIM) detection mode [95–109]. Pentafluorobenzyl ester derivatives of AEA and their quantification by gas chromatography-negative ion chemical ionization-mass spectrometry (GC-NICI-MS) are also described [85]. These pentafluorobenzoyl-pentafluoropropionate-derivatives (PFBz-PFP-derivatives) also allowed the quantification of AEA and N-ArPE in various brain regions and in the spinal cord, testis, and spleen of rats [110]. Protocols for EC analysis via GC/MS of the native analytes in rat brain, human neuroblastoma CHP100, and lymphoma U937 cells are also available. Compounds are detected in the SIM mode measuring their dehydrated forms. The detection limit of this method is 20 ± 10 pmol, and no traces of 1(3)-AG were found [111].

The low sensitivity of GC-MS methods requires relatively large amounts of tissue or biological fluids to observe measurable EC levels and, therefore, lengthy sample preparation times as well as a derivatization step [112]. These problems led to the development of a higher sensitivity liquid chromatography-mass spectrometric techniques. In particular, the miniaturization of HPLC components, hybridization of multistage mass spectrometers, the introduction of electrospray ionization (ESI), and MS-MS or time-of-flight (TOF) technologies coupled with the development of information-dependent acquisition (IDA) methods allow for the detection of low amounts of these compounds. Many existing methods for EC analysis that use LC-MS (single quadrupole) in the SIM mode are based on the precursor ion mass/charge ratio (m/z) compared with synthetic internal standards. Koga and coworkers quantified AEA and its analogues using liquid chromatography-atmospheric pressure chemical ionization-mass spectrometry (LC-APCI-MS) with a detection limit of 200 fmol [89]. AEA, palmitoylethanolamide (PEA), and oleylethanolamide (OEA) were detected in rat plasma in SIM mode using quasimolecular species as diagnostic ions produced by an ESI probe and then quantified by isotope dilution using synthetic deuterium standards. Detection and quantification limits between 0.1–0.3 and 0.5–1.2 pmol per sample, respectively, are reported [113]. The quantification of AEA, 2-AG, and other NAE by using the LC-APCI-MS technology contributed to correlate changes of EC levels with several central and peripheral diseases [114–127]. The well-known ability of the silver cation to complex with double bonds led to the development of more sensitive LC/MS methods to quantify EC. The presence of four double bonds in the arachidonate moiety of AEA and 2-AG provides coordination sites for Ag^+, and the easy vaporization of the adducts allows their identification as $[M+Ag]^+$ entity using ESI [90]. This method significantly reduces the detection limits, which were 13 and 14 fmol on column for 2-AG and AEA, respectively, and its applicability to other neutral lipids is also possible. LC-MS-MS methods, where accurate masses are reported (<10 ppm) and the full fragmentation pattern and HPLC retention time of endogenous material is compared with that of synthetic material, led to the identification of other EC and new, related bioactive compounds [51,128]. Huang and coworkers, using solid phase extraction and (LC)-MS/MS on the ion trap and hybrid quadrupole/time-of-flight (QTOF) instruments, reported the identification of N-arachidonoyl-dopamine (NADA), an endogenous ligand of cannabinoid CB_1 and transient receptor potential vanilloid-1 channels, in nervous tissues, including the striatum, hippocampus, cerebellum, and dorsal root ganglia [128]. While the QTOF performs well for the identification of novel compounds, multiple-reaction monitoring (MRM) scans on triple quadrupole MS coupled with HPLC is the preferred method for quantifying trace levels of endogenous compounds. However, MRM experiments cannot replace product ion scans because only one or a few major fragment ions are used in these experiments while a full range mass spectrum is required for the identification of new compounds [129]. More recently, a sensitive and specific liquid chromatography-tandem mass spectrometry method for the simultaneous identification and quantitative profiling of AEA, 2-AG and six other related compounds in brain tissue has been described [91]. The authors used an extraction solvent consisting of ethyl acetate:hexane (9:1 by vol.) to ensure recovery of AEA, 2-AG, noladin ether, arachidonoyl-glycine, OEA, PEA, virodhamine, and 2-linoleoyl-glycerol (2-LG) from rat brain. LC methods were optimized to guarantee resolution among all analytes. A C8 HyPurity Advance column and a gradient elution, with a mobile phase of acetonitrile, formic acid, and ammonium acetate, at pH 3.6, offer the best analyte resolution and peak shape. A detection limit of 25 fmol on column for all analytes except 2-AG, noladin ether, and 2-LG were reported, and AEA, 2-AG, 2-LG, OEA, and PEA, but not virodhamine, noladin ether, and arachidonyl glycine, were detected in different brain regions [91]. An LC-ESI-MS-MS method for simultaneous quantification of AEA, PEA, and OEA was also recently reported. AEA and OEA are identified as $[M + Ag]^+$ ions as previously reported for AEA [90], while the mass spectrum of PEA, which does not have any double bond, was obtained as $[M + 1]^+$ [130].

5.3 QUANTIFICATION OF NAPE

It is important to quantify not only AEA but also its biosynthetic precursor, N-ArPE. Methods for NAPE quantification are useful in determining NAT activity in homogenates. NAPE, including N-ArPE, are extracted from cells and tissues using the same experimental procedures described above to extract AEA and other NAE and then purified by using open-bed chromatography. Fractions eluted with chloroform/methanol 1:1 by vol. (containing NAPE) [88] can be quantified by spectroscopic and mass spectrometric analyses or digested with PLD [110,131,132]. NAPE-containing fractions were reported to be digested with a yield approaching 100% by using PLD from *Streptomyces chromofuscus* (10 units/mL) for 2 h at 37°C in a two-phase mixture of ethyl ether and 50 mM Tris–HCl, pH 7.4. After the digestion, the ether phase is dried down and the amount of AEA released from the reaction of N-ArPE quantified as described above [21,99,133,134]. Phosphorus-31 nuclear magnetic resonance (^{31}P-NMR) spectroscopy is used to identify and quantify NAPE in extracts of animal tissues [132]. Lipids are extracted using a modification of the procedure used by Folch et al. [135], and the extracts are thoroughly washed with potassium ethylenediaminetetraacetic acid (K-EDTA, 0.1 M). The lower organic phases are isolated and evaporated to dryness under a stream of nitrogen. The lipids are then redissolved in $CDCl_3$–CH_3OH–H_2O 100.0:29.9:5.2 (v/v/v) for NMR analysis. Each NMR allowed for the identification of NAPE subclasses such as 1,2-diacyl-*sn*-glycero-3-phospho-(N-acyl)-ethanolamine ($NAPE_{DIACYL}$) and 1-(1′-alkenyl)-2-acyl-*sn*-glycero-3-phospho-(N-acyl)-ethanolamine ($NAPE_{PLAS}$). The development of a fast assay for NAPE quantification based on anionic phospholipid precipitation in the presence of zirconium [136] is also reported. The method is based on evidence that tetravalent zirconium forms insoluble salts with phosphates under strongly acidic solutions [137]. However, the applicability of this method to NAPE precipitation required a slight modification consisting of the addition of NaOH. This is necessary to obtain a pH over the pK_a of NAPE and a phosphate carrier, NaH_2PO_4, to improve the yield of zirconium-phosphate precipitation. The specificity of the zirconium assay toward LysoNAPE and GP-NAE as substrates was also tested. When NAPE and lysoNAPE were coincubated, 100% of NAPE was precipitated by zirconium together with more than 95% lysoNAPE.

5.4 NAPE-PLD AND NAT ACTIVITY EVALUATION

Studies on the existence of NAE and on their biosynthetic pathways were performed long before anyone understood their true physiological relevance [15,138–140].

Different types of phospholipase, hydrolase, and *trans*-acylase enzymes are suggested to be involved in AEA formation but only one of these, the NAPE-PLD, has been cloned and characterized. Evaluation of the enzymatic activities responsible for AEA and other NAE formation has been performed using several different experimental procedures. The main methods developed so far were focused toward measuring the amount of AEA produced from *N*-ArPE hydrolysis via NAPE-PLD-like catalytic activity. Either NAPE-like chromatographic fractions or radiolabeled NAPE can be used as substrates.

Methods assessing the conversion of *N*-[³H]ArPE into [³H]AEA, through the separation of the latter compound by TLC on silica gel plate or by HPLC are also reported [14,20,133]. Fezza and coworkers developed a sensitive method to measure NAPE-PLD activity based on reverse phase-HPLC (RP-HPLC) and in-line scintillation counting (Figure 5.6a). An easy procedure to homogenize tissues following the incubation of whole organs in the presence of a mixture of [³H]-*N*-ArPE and *N*-ArPE in a buffer containing 0.1% Triton X-100 was reported. The amount of [³H]-AEA released from [³H]-*N*-ArPE hydrolysis was used as a measure of specific NAPE-PLD activity. In this way, the resolution and reproducibility of HPLC, performed by using a C18 column with a mobile phase of methanol/water/acetic acid, were combined with the sensitivity afforded by the use of radioisotopes. The reported recovery of [³H]-AEA is approximately 90% after RP-HPLC separation, which is higher than that after TLC (30% or even less). Furthermore, the same authors optimized the method by correcting for the NAPE-PLD obtained by measuring the formation of [³H]-AEA by adding the amount of [³H]-ARA produced from [³H]-AEA hydrolysis. NAPE, AEA, and ARA can also be identified by UV detection at 204 nm and then quantified by using a scintillation counter if an in-line scintillation counter is not available [141].

The development of a specific and sensitive method to assess the unambiguously precursor–product relationship between *N*-ArPE and AEA is becoming more and more complicated since new alternative biosynthetic pathways responsible for *N*-ArPE conversion to AEA have been reported. In this new biosynthetic scenario, it appears evident that NAPE-PLD activity might be overestimated, when measured by quantifying the formation of the AEA product, in particular, when [³H]-*N*-ArPE labeled on the *N*-acyl moiety is used as substrate. For these reasons, sensitive methods for NAT activity evaluation that account for NAPE formation independently from its hydrolysis are needed. Evidence for a NAT enzymatic activity was reported in mouse neuroblastoma cells N18TG2 and J774 macrophages prelabeled overnight with either [¹⁴C]ethanolamine or [³H]arachidonic acid [133]. Lipids were extracted and then loaded onto Pasteur pipettes containing silica gel eluted with increasing concentration of methanol in chloroform in order to fractionate AEA- and NAPE-like materials. The fraction eluted with methanol/chloroform 6:4 (by vol.), which contains the NAPE, was purified on analytical silica-coated plastic plates using CHCl₃:MeOH:NH₃ (85:15:1 by vol.) as the developing system. The band (Rf. 0.4) comigrating with synthetic standard of *N*-palmitoyl-phosphatidylethanolamine was scraped off the plate and extracted from the silica gel. The metabolites were then subjected to digestion with *S. chromofuscus* PLD, yielding a mixture of lipids which all co-chromatographed with

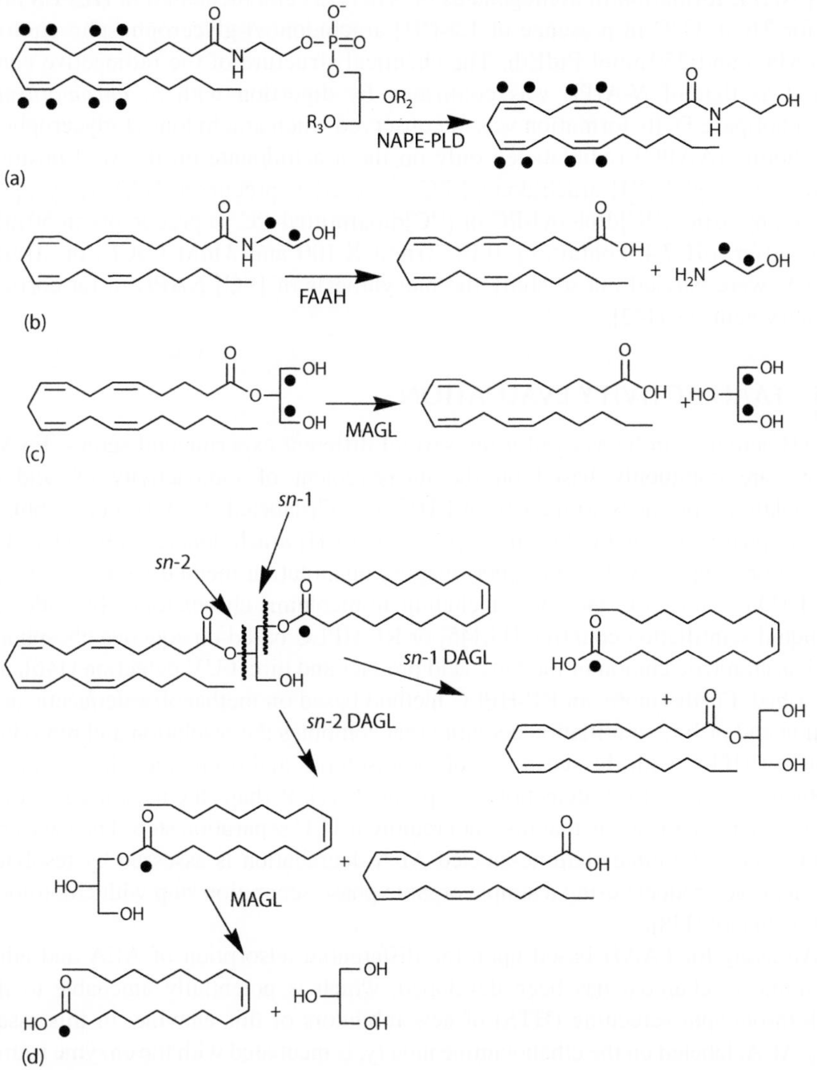

FIGURE 5.6 Schematic representation of the enzymatic assays described in this chapter. Labeled substrates (full circle identifies the radioisotopic labeling) have been selected among those used in most current assays. Radiolabeled AEA (a) is quantified by scintillation counting after RP-HPLC purification as a measure of NAPE-PLD activity. Radiolabeled ethanolamine (b) and glycerol (c) are quantified by using a β-counter after an organic phase separation step with chloroform and methanol. (d) [14C]-oleic acid is quantified, after TLC purification, as the product of both sn-1-selective-DAGL acting on 1-[14C]-oleoyl-2-arachidonoyl-glycerol and MAGL acting on 1-[14C]-oleoyl-glycerol produced by sn-2-DAGL. Abbreviations: FAAH, fatty acid amide hydrolase; MAGL, monoacylglycerol lipase; NAPE-PLD, phospholipase D selective for N-acyl-phosphatidylethanolamines; DAGL, diacylglycerol lipase.

synthetic NAE in both TLC and HPLC analyses. The same authors [27] reported [³H]-NAPE formation in homogenates of N18TG2 cells incubated in HEPES buffer for 2h at 37°C in presence of 1,2-[³H]-arachidonoyl-glycerophospho-choline (ARAPC) and 250nmol PtdEth. The chemical structure of the radioactive compound as that of *N*-ArPE was confirmed by digestion with *S. chromofuscus* phospholipase D. Its formation was not observed when arachidonoyl-glycerophosphocholine (AAPC) radiolabeled only on the arachidonate on the *sn*-2 position or *sn*-1-stearoyl-2-[³H]-arachidonoyl-PC was used as precursor. NAT assays, performed by using [¹⁴C]dioleoyl-PC or [¹⁴C]dipalmitoyl-PC as precursors in 50 mM Tris buffer, pH 7.4, containing 0.1% Triton X-100 and 3 mM $CaCl_2$, or 10 mM EGTA, were carried out to study the biosynthesis of [¹⁴C]-NAPE in rat cortical primary neurons [142].

5.5 FAAH ACTIVITY EVALUATION

FAAH activity can be assayed using several different experimental setups. FAAH assays are commonly based on the measurement of radioactivity of acid or ethanolamine products formed from [³H]- or [¹⁴C]-labeled AEA (Figure 5.6b). A normal-phase TLC method for the separation of [³H]-arachidonate [143] and a TLC separation coupled with a phosphor-imaging quantitation method were also developed [144]. Non-TLC methods, including minicolumn chromatography followed by liquid scintillation counting [14,145] or RP-HPLC based on aqueous phosphoric acid–acetonitrile elution of the fatty acid product and direct UV detection [146], are described. Furthermore, an RP-HPLC method based on methanol/water/acetic acid elution and in-line scintillation counting that combines the resolution and reproducibility of HPLC with the sensitivity of radioisotopes and overcomes the sensitivity problems of a direct UV detection, is reported [147]. Perhaps the most used method is a radioenzymatic assay that does not require a TLC separation step. The measurement of [1,2-¹⁴C]-ethanolamine-labeled AEA degradation is assessed by resolving the cleavage products using a simple organic phase separation step with chloroform and methanol [148].

An assay for FAAH based upon the differential adsorption of AEA and ethanolamine to charcoal has been developed, which is potentially amenable to the high-throughput screening (HTS) of new inhibitors of this enzyme. In this assay, [³H]-AEA, labeled on the ethanolamine moiety, is incubated with the enzyme source (membranes from T84 cells) for 60 min at room temperature in microtiter plates and then transferred into multiscreen filter plates containing charcoal. The filter plates are assembled over plates containing scintillant and centrifuged, and the plates are counted in a microplate scintillation counter. The substrate remains bound to the charcoal whereas the [³H]ethanolamine product formed is transferred to the scintillation microplates [149]. A slight modification of the above experimental procedure showed that the assay gave a better extraction efficiency when acidic rather than alkaline charcoal solutions were used to stop the reaction, and a more favorable sample/blank ratio was also obtained [150].

Several alternative approaches to the use of radiolabeled substrates are reported that assess FAAH activity. One of the first assays was reported by Patterson et al.

[151] using oleamide as substrate. In particular, by using an ion-selective ammonia electrode (ATI/Orion), the potency of potential inhibitors of oleamide hydrolysis was evaluated by directly measuring ammonia formation as the product of the reaction. However, this assay is limited to large reaction volumes by the size of the electrode. One of the first fluorescence displacement assay was derived by the observation that the hydrolysis of AEA and oleamide releases fatty acids that can bind to liver fatty acid–binding protein (FABP) and displace the fluorescent fatty acid probe, 11-(5-dimethylaminonaphthalenesulfonyl)-undecanoic acid (DAUDA) [152]. This assay requires the isolation and delipidation of FABP from rat liver or recombinant sources. In addition, the assay must be calibrated to normalize for the binding of AEA and oleamide to this protein. An assay that uses various p-nitroanilides (pNA) bearing acyl chains of 6–20 carbons in length as substrates for FAAH has been described [153]. The use of pNA substrates allows for the precise monitoring of enzymatic hydrolysis rates by following the increase in UV absorbance at 382 nm due to the release of p-nitroaniline. While this substrate provides an opportunity to study FAAH kinetics and substrate specificity, the sensitivity of the colorimetric signal is limited and susceptible to interference from colored compounds. A novel spectrophotometric method for FAAH activity based on an enzyme-coupled assay of oleamide hydrolysis is reported [154]. In particular, to measure the rate of ammonia released from oleamide hydrolysis, the authors developed a dual-enzyme assay containing FAAH and L-glutamate dehydrogenase (GDH), since GDH catalyzes the condensation of ammonia and 2-oxoglutarate to L-glutamate, using NADH as a coenzyme, which has a large molar extinction at 340 nm. This assay uses a reduction in NADH through this coupled reaction, and oleamide hydrolysis is directly proportional to the decrease of absorbance as NADH is oxidized to NAD^+.

Routine methods to identify FAAH inhibitors through the HTS screening of large libraries of chemical compounds have started to appear in the literature. Different fluorescent assays were designed to be simple, specific, and HTS compatible. Arachidonyl 7-amino, 4-methyl coumarin amide (AAMCA) was synthesized as a stable and specific FAAH substrate. The hydrolysis of nonfluorescent AAMCA to produce ARA and a highly fluorescent 7-amino, 4-methyl coumarin (AMC λ_{ex} 355 nm, λ_{em} 460 nm) can be readily detected using a fluorometer [155]. Recently, an HTS assay for the identification of FAAH inhibitors using a novel substrate, decanoyl 7-amino-4-methyl coumarin (D-AMC), was developed [156]. This fluorescent assay was validated using a panel of known FAAH inhibitors and purified recombinant human FAAH. This assay was adapted to a 384-well assay format to screen a large library of compounds (>600,000 compounds) and hence identify new FAAH inhibitors.

5.6 DAGL ACTIVITY EVALUATION

Before its identification as an EC, 2-AG was known as an intermediate in phosphoglyceride and diglyceride/triglyceride metabolism. In fact, 2-AG is one of the degradation products of DAG through the action of DAG lipase and therefore a product of the termination of protein kinase C/DAG-mediated intracellular signaling [157,158].

As a substrate for MAGL, 2-AG generates ARA, the precursor of a plethora of bioactive intracellular and extracellular mediators in cells where ARA release is not catalyzed by G-protein-coupled PLA$_2$ enzymes. Monoacylglycerol and diacylglycerol lipases are assayed using as substrate *rac*-1-*S*-decanoyl-1-mercapto-2,3-propanediol and *rac*-1, 2-*S,O*-didecanoyl-1-mercapto-2,3-propanediol, respectively [159]. The hydrolysis of both compounds produces a free thiol group that reacts with specific thiol capture reagents and can be monitored continuously on a spectrophotometer. With the use of 5,5′-dithiobis(2-nitrobenzoic acid (DTNB) and 4,4′dithiodipyridine (DTP), the complete pH range necessary for the assay of lipases can be covered [160,161].

Evidence of DAG lipase activities responsible for the formation of 2-AG acting as an EC was obtained in intact cells in an "indirect way." In particular, by labeling overnight cells with [³H-]-ARA in serum-free medium, the release of 2-AG was measured after stimulation with the Ca^{2+}-ionophore ionomycin [28,31,162]. Although this method was useful to show the involvement of 1-acyl-2-arachidonoylglycerol as direct 2-AG biosynthetic precursors, it required a heavy clean up procedure to purify [³H]-2-AG from the lipidic extract that contains free [³H]-ARA and other glycolipids and phospholipids labeled with [³H]-ARA. The DAG needed to specifically identify the DAGL activities involved in the biosynthesis of 2-AG are not available commercially. However, *sn*-1-stearoyl-2-[³H]-arachidonoylglycerol and *sn*-1-arachidonoyl-2-[³H]-arachidonoylglycerol suitable for this purpose can be prepared by the digestion of *sn*-1-stearoyl-2-[³H]arachidonoylphosphatidylcholine and *sn*-1-arachidonoyl-2-[³H]arachidonoylphosphatidylcholine, respectively, by incubation with *Clostridium perfringens* phospholipase C in Tris–HCl, pH 7.4, for 2 h at 37°C, followed by lipid extraction and purification by open-bed silica chromatography. Later on, commercially available synthetic 1-acyl-2-[¹⁴C]-arachidonoylglycerol was used to evaluate the presence in cell homogenates of an enzymatic DAGL activity responsible for 2-AG formation. The limitation of this method is the possible wrong evaluation of the DAGL activity present in the cells. In particular, 2-[¹⁴C]-AG, the product of the reaction of DAGL on DAG labeled on *sn*-2 position, is further hydrolyzed to [¹⁴C]-ARA and glycerol by MAGL enzymes. These might then become the driving force of the reaction by subtracting 2-[¹⁴C]-AG to the equilibrium. Therefore, 1-[¹⁴C]acyl-2-arachidonoyl-glycerols were synthesized in order to monitor DAGL activity through the production of [¹⁴C]-fatty acids, which are not substrates for other lipases. A useful synthetic procedure that allows the production of DAG selectively labeled on *sn*-1 or *sn*-2 position is reported (Figure 5.7). In brief, the compounds are obtained from the R (−)-solketal esterified with either unlabeled or ¹⁴C-labeled oleic acid using *N*′-(3-dimethylaminopropyl)-*N*-ethylcarbodiimide hydrochloride/4-dimethylaminopyridin (EDC/DMAP) and deprotecting the acetonide with hydrochloride/methanol. The primary alcoholic group is protected selectively with triisopropylsilyl chloride whereas the free secondary alcohol is esterified with various fatty acids, either unlabeled or [¹⁴C]-labeled. Finally, the *sn*-1,2-diacylglycerol with two different acyl groups in positions *sn*-1 and 2 was obtained by removing selectively the silyl group with tetrabutylammonium fluoride/acetic acid. This synthetic procedure was successfully applied to obtain all the different DAG used to enzymatically characterize the first specific *sn*-1 DAGL,

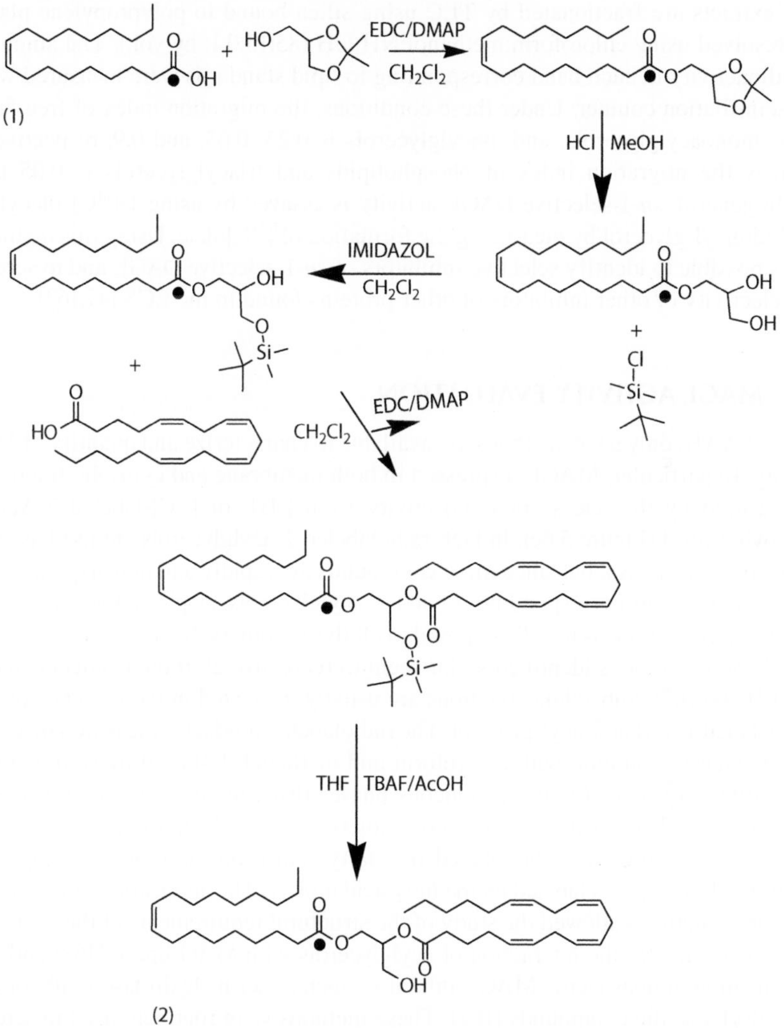

FIGURE 5.7 Synthesis of 1-[^{14}C]-oleoyl-2-arachidonoyl-glycerol. [^{14}C]-Oleic acid (1) is converted to 1-[^{14}C]-oleoyl-2-arachidonoyl-glycerol (2) in five steps as detailed; full circle identifies the radioisotopic labeling. Abbreviations: DMAP, 4-dimethylaminopyridin; EDC, N'-(3-dimethylaminopropyl)-N-ethylcarbodiimide hydrochloride; TBAF, tetrabutylammonium fluoride 1.0 M solution in tetrahydrofuran.

DAGLα, and DAGLβ, which are responsible for the last step of the formation of EC 2-AG [34]. In particular, 10,000×g membrane fractions from cells in which recombinant human DAGLα or β cDNA was overexpressed are obtained by sequential centrifugation of cell homogenates. The 10,000×g membrane fractions are incubated at pH 7.0 at 37°C for 18 min in the presence of different radiolabeled substrates (Figure 5.6d). After the incubation, lipids are extracted three times with 2 vol. chloroform/methanol 2:1 (by vol.), and the extracts lyophilized under vacuum. The lipid extracts are fractionated by TLC using silica bound to polypropylene plates and resolved using chloroform:methanol:NH$_4$OH (85:15:0.1, by vol.). The amounts of radioactivity in each band corresponding to lipid standards were measured with a β-scintillation counter. Under these conditions, the migration index of free fatty acids, monoacylglycerols, and diacylglycerols is 0.25, 0.65, and 0.9, respectively, whereas the migration index of phospholipids and triacylglycerols is 0.05 and 1.0. In general, sn-1-selective DAGL activity is assayed by using 1-[^{14}C]-oleoyl-2-arachidonoyl-glycerol by measuring the formation of [^{14}C]oleic. Using this method, it was possible to identify selective inhibitors of sn-1-selective DAGL and to screen the selectivity of other inhibitors of other proteins found in the ECS [47,163].

5.7   MAGL ACTIVITY EVALUATION

Unlike FAAH, only a few methods are available to characterize and quantify MAGL activity. In particular, MAGL, expressed in both membrane and cytosolic fractions, is measured by the release of radioactivity from [^{3}H]- or [^{14}C]-labeled 2-AG or 2-oleoylglycerol (Figure 5.6c). In fact, radiolabeled 2-acylglycerols are used as substrates in these assays, but since these compounds are rapidly and nonenzymatically converted into the corresponding 1- and 3-acylglycerols in physiological buffers [164,165], the enzyme is usually exposed to all these isomers. Interestingly, however, most MAGL activities identified so far seem to recognize all three isomers equally well [74,166,167]. Subcellular fractions are usually incubated in the presence of the radiolabeled 1-acyl or 2-acyl-glycerol. The radiolabeled product is measured by using organic phase separation with chloroform and methanol, followed by counting the radioactivity associated with the aqueous phases that contain radioactive glycerol, if the substrate is labeled on the glycerol moiety. Conversely, normal-phase TLC is used for the separation of the labeled free fatty acid from the mono-acyl-glycerol substrate, if the latter is labeled on the fatty acid moiety. The application of the aforementioned methods allowed the study of the structural requirements of the acyl side chain necessary for the interaction of acylglycerols with MAG lipase [168] and the identification of nonspecific MAGL inhibitors such as serine hydrolase inhibitors or sulfhydryl-specific compounds [167]. These methods were routinely used to screen the potential inhibitors of other proteins of the ECS [24,163,169,170] and enabled the recent identification of disulfide derivatives, such us tetraethylthiuram disulfide, as potent inhibitors of MAGL [171].

Degradation of 2-AG was studied in a slightly more complicated way in rat cerebellar membranes prepared under conditions that were previously used to monitor CB$_1$ receptor-dependent G-protein activation [172]. Membranes were incubated with 50 µM 2-AG, acetonitrile was added to stop the enzymatic reaction, and the pH of the samples was simultaneously decreased to 3.0 with phosphoric acid (added together with acetonitrile) to stabilize 2-AG against a possible postincubation chemical acyl migration to form the 1(3)-isomers. The specific 2-AG-degrading enzyme activity was determined based on the formation of ARA, quantified by using HPLC, during the 90 min incubation period. The relative concentrations of 2-AG and ARA were determined by comparing the corresponding peak areas. By using this procedure, the sensitivity of MAGL to sulfhydrilic reagent was confirmed and *N*-arachidonoylmaleimide identified as an MAGL inhibitor, which suggested that the presence of a sulfhydryl group is essential for substrate recognition [165,173,174].

5.8 IMMUNOHISTOCHEMICAL METHODS USED TO STUDY THE EXPRESSION OF ENDOCANNABINOID METABOLIC ENZYMES

Significant progress has been made toward the understanding of the localization and distribution of EC metabolic enzymes, thanks to the contribution of several techniques such as immunohistochemistry, Western blotting, and real-time RT-PCR. A commercially available NAPE-PLD rabbit polyclonal antibody raised against human NAPE-PLD 378–390 amino acid sequence (LKHGESRYLNNDD) is used for immunohistochemistry on formalin-fixed paraffin-embedded sections [22]. Alternatively, a polyclonal NAPE-PLD purified antibody raised against recombinant mouse NAPE-PLD, developed in rabbits, can also be used [20,175]. Rat FAAH 561–579 amino acid sequence (CLRFMREVEQLMTPQKQPS) [176] has been used in rabbits to raise a polyclonal antibody that displays cross-reactivity to human, rat, murine, and ferret [177]. A polyclonal antibody raised in rabbits against the human MAGL 1–14 amino acid sequence, and cross-reacting with human, murine, rat, and bovine MAGL, is available through Cayman Chemicals (Ann Arbor, Michigan). No commercially available antibodies exist to date for DAGL, although various research groups have developed anti-DAGLα polyclonal antibodies (Patrick Doherty, King's College London, London, United Kingdom; Kenneth Mackie University of Washington, Seattle, Washington), including an antibody recognizing the 42 C-terminal amino acids (1003–1044) of this enzyme [37]. Immunohistochemistry, immunofluorescence, and immunoelectron microscopy techniques contributed to the discovery that enzymes for 2-AG biosynthesis and degradation are complementarily distributed in cerebellar neurons [34,37,75,178]. In the cerebellum, DAGLα was predominantly expressed in Purkinje cells, where it localizes on the dendritic surface and occasionally on the somatic surface, whereas MAGL is localized in molecular layer neurons, which make synapses with dendritic spines of Purkinje's cells. DAGLα immunoreactivity (ir) was also observed in the hippocampal pyramidal cells and was distributed in the spine head and neck, or both, thus indicating that DAGLα is essentially targeted to postsynaptic spines in cerebellar and hippocampal neurons, although its fine distribution within and around spines is differently regulated between the two types of neurons. MAGL-ir localizes to axon terminals of granule cells, CA3 pyramidal cells, and some interneurons. The expression of cannabinoid receptors and EC biosynthesizing enzymes (NAPE-PLD and DAGL-α) was also a subject of our recent study conducted in pancreatic and adipose tissue sections [179].

FAAH distribution has been intensively studied in the mammalian brain using immunohistochemical techniques, by using an immunopurified, polyclonal antibody to the C-terminal region of FAAH [180–185]. Furthermore, the localization of FAAH in rat and goldfish retina has also been reported [185,186].

5.9 QUANTITATIVE METHODS TO STUDY THE mRNA EXPRESSION OF ENDOCANNABINOID METABOLIC ENZYMES

The reverse transcription polymerase chain reaction (RT-PCR) is the method of choice for any experiments requiring sensitive, specific, and reproducible detection and quantification of low-abundance mRNA. Five methods are in common use for the quantification of transcription: Northern blotting [187] and *in situ* hybridization [187], RNAse protection assays [188,189], RT-PCR [190], and cDNA arrays [191]. *In situ* hybridization is the only one that allows localization of transcripts to specific cells within a tissue but the main limitation is its low sensitivity [192]. RT-PCR is an *in vitro* method used to amplify defined sequences of RNA, and is the most sensitive of the quantification methods. This very high sensitivity, however, reduces the specificity of the reaction, by increasing the possible weight of contaminations and the detection of false positives. Therefore, caution and appropriate controls are needed when using RT-PCR.

The first step in an RT-PCR assay is the reverse transcription of the RNA template into cDNA, involving an RNA-dependent DNA polymerase (reverse transcriptase) generating a stable cDNA pool, followed by its exponential amplification in a PCR reaction. It is necessary to treat the RNA sample with RNAse-free DNAse, and primers should be designed to bind to separate exons in order to avoid false positive results arising from genomic DNA contamination. Primer (and probe) selection is based on estimated T_m, the desire for small amplicon size and the location of the sequence. There are several alternative software tools available for probe selection. The application of fluorescence techniques to RT-PCR, together with suitable instrumentation capable of combining amplification, detection, and quantification led to the monitoring of PCR reactions in real-time during the PCR, suitable to quantify nucleic acids. The simplest method uses a fluorescent dye (SYBR® Green) that binds specifically to double-stranded DNA [193]. The unbound dye exhibits little fluorescence in solution, but increases the fluorescence signal as PCR proceeds. RT-PCR product verification is obtained by plotting fluorescence as a function of temperature to generate a melting curve of the amplicon, whose T_m depends on its nucleotide composition. The specificity of this method is determined entirely by the primers used, because any double-stranded DNA will generate fluorescence. Other methods (molecular beacons, hybridization probes, and the TaqMan® assay) are based on the hybridization of fluorescence-labeled probes to the correct amplicon, obviating the need for post-PCR analysis to confirm identity of the amplicon. Fluorescence values are recorded during every cycle and represent the basis to determine the amount of product amplified.

Quantification of mRNA transcription can be either relative or absolute. RT-PCR-specific errors in the quantification of mRNA transcripts are minimized and corrected for sample-to-sample variation by amplifying, simultaneously with the target, a cellular RNA that serves as an internal reference against which other RNA values can be normalized [194]. The ideal internal standard should be expressed at a constant level among different tissues of an organism, at all stages of development, and should be unaffected by the experimental treatment and should also be expressed at roughly the same level as the RNA under study. The mRNA specifying the housekeeping genes, glyceraldehyde-3-phosphate-dehydrogenase (GAPDH), β-actin, ribosomal RNAs (rRNA), cyclophilin, and hypoxanthine phosphoribosyl transferase, are used [195]. In our laboratory, we relate all copy numbers to total RNA concentration, which is measured using RiboGreen® RNA quantitation (Molecular Probes) or by performing total RNA electrophoretic separation on microfabricated chips and subsequent fluorescence detection via laser on an Agilent 2100 Bioanalyzer. The latter method has the advantage to determine the integrity of the starting RNA, which is a critical step in gene expression analysis. A software algorithm is capable of assessing RNA quality standardization and RNA integrity assessment, facilitating the comparison of samples. Either method can detect as little as 5 ng/mL RNA.

5.9.1 N-ACYL-PHOSPHATIDYLETHANOLAMINE-HYDROLYZING PHOSPHOLIPASE D

Homo sapiens: Genomic context: chromosome: 7; Location: 7q22.1; Homo sapiens N-acyl-phosphatidylethanolamine-hydrolyzing phospholipase D (NAPE-PLD), mRNA Acc. Nr. NM_001122838 (variant 1), Acc. Nr. NM_198990 (variant 2) (Table 5.1) [20].

NAPE-PLD expression in bone marrow samples was assayed by quantitative real time PCR; RNA was extracted using Qiagen RNAeasy kit (Qiagen Sciences, Valencia, California) or TRIzol®, cDNA synthesized using SuperScript Strand Synthesis System for RT-PCR (Invitrogen Corp., Carlsbad, California). Primer Express software (PE Applied Biosystem, Carlsbad, California) was used to design primers and probes for TaqMan hybridization. GAPDH expression was measured in each sample to ascertain cDNA quality and to establish a reference standard for making comparisons [196]. In other experiments, total RNA was also extracted from subcutaneous adipose tissue by TRIzol reagent and underwent a second extraction step by means of the Perfect RNA Mini kit (Eppendorf, AG, Hamburg, Germany) and treated with DNA-free kit (Ambion, Inc., Austin, Texas) to digest genomic DNA. First strand cDNA was synthesized using SuperScript II reverse transcriptase and random hexamer primers. Pure adipocyte cultures, obtained from Cell Application, Inc. (San Diego, California), were used as control [197].

Mus musculus: Genomic context: chromosome: 5; Location: 5 A3; Mus musculus cDNA sequence AB112350, mRNA, Acc. Nr. NM_178728 (Table 5.2) [20].

Mouse uteri were processed for mRNA analysis and RT reaction was performed using SuperScript II RT kit (Invitrogen). Real-time quantitative PCR was performed using SYBR Green Supermix (BioRad, Hercules, California), while mouse GAPDH served as the reference gene for analyzing the relative abundance of NAPE-PLD [198]. In another study, total RNA from mouse adipocyte cell line 3T3F442A was extracted using RNA purification kit. RNA was further treated with RNAse-free DNAse I (Ambion DNA-free™ kit); RNA was evaluated by RiboGreen-RNA quantitation kit and reverse transcribed with SuperScript® III in the presence of hexanucleotide mixture. Real-time cDNA quantitation was performed with iQ™ SYBR Green Supermix. Optimized primers were designed by "Beacon Designer®" software (Biosoft International) and synthesized by MWG-Biotech AG (Germany). Relative expression analysis was corrected for PCR efficiency and normalized with respect to reference genes, β-actin and hypoxanthine phosphoribosyltransferase [123].

TABLE 5.1

List of Useful Resources for the Genes of Interest in *Homo sapiens*

Abbreviated Gene Name	Chrom Location	Ref Seq. ID	Translated Protein	Applied Biosystem TaqMan	Amplicon Length	Exon Boundary
NAPE-PLD	7q22.1	NM_001122838.1 (variant 1)	NP_001116310.1	Hs 00419593_m1	79	3–4
		NM_198990.4 (variant 2)	NP_945341.3	Hs 00419594_m1	100	4–5
MAGL	3q21.3	NM_007283.5 (variant 1)	NP_009214.1	Hs00996009_g1	64	1–2
		NM_001003794.1 (variant 2)	NP_001003794.1	Hs00200752_m1	66	2–3
				Hs00996003_m1	61	3–4
				Hs00996004_m1	64	4–5
				Hs00996005_m1	69	5–6
				Hs00996006_m1	90	6–7
				Hs00996007_g1	70	7–8
DAGL-α	11q12.2	NM_006133.1	NP_006124.1	Hs01023072_g1	74	2–3
				Hs01023061_m1	67	3–4
				Hs01023062_g1	61	4–5
				Hs00391374_m1	55	5–6
				Hs00391375_m1	57	6–7
				Hs00391376_m1	109	7–8
				Hs01023054_m1	79	8–9
				Hs01023073_m1	92	9–10
				Hs01023064_g1	62	10–11
				Hs01023065_g1	83	11–12
				Hs01023066_g1	112	12–13
				Hs00391382_m1	79	13–14
				Hs01023067_g1	81	14–15

Gene	Location	mRNA	Protein	Probe ID	Value	Exon boundary
FAAH	1p35-p34	NM_001441.2 (variant 1)	NP_001432.2			
		NM_174912. (variant 2)	NP_777572.1			
				Hs00391384_g1	59	15–16
				Hs01023068_m1	81	16–17
				Hs00391386_m1	83	17–18
				Hs00391387_g1	63	18–19
				Hs01023069_m1	59	19–20
				Hs01038674_m1	77	1–2
				Hs00155015_m1	58	2–3
				Hs01038673_g1	71	3–4
				Hs01038675_g1	69	4–5
				Hs01038676_g1	69	5–6
				Hs01038677_g1	95	6–7
				Hs01038678_m1	87	7–8
				Hs01038659_g1	88	8–9
				Hs01038660_m1	122	9–10
				Hs01038661_m1	62	10–11
				Hs01038662_m1	59	11–12
				Hs01038663_g1	71	12–13
				Hs01038664_m1	74	13–14
				Hs01038665_g1	87	14–15
				Hs00415899_m1	114	1–2
				Hs00398731_m1	105	3–4
				Hs00398732_m1	106	4–5
				Hs01006043_m1	87	5–6
				Hs01006044_m1	148	6–7
				Hs01006045_m1	84	7–8
				Hs01006046_m1	75	8–9
				Hs01006038_m1	103	9–10
				Hs01006039_m1	86	10–11

TABLE 5.2
List of Useful Resources for the Genes of Interest in *Mus musculus*

Abbreviated Gene Name	Chrom Location	Ref Seq. ID	Translated Protein	Applied Biosystem TaqMan	Amplicon Length	Exon Boundary
NAPE-PLD	5 A3	NM_178728.5	NP_848843.1	Mm00724596_m1	85	4–5
				Mm01167825_m1	66	—
				Mm01167826_m1	152	—
MAGL	6 D1	NM_011844.3	NP_035974.1	Mm01183763_m1	65	4–5
DAGL-α	19 A	NM_198114.2	NP_932782.2	Mm00813830_m1	69	8–9
				Mm01701557_m1	122	9–10
				Mm01701558_g1	82	10–11
				Mm01701554_m1	83	11–12
FAAH	4 D1	NM_010173.3	NP_034303.2	Mm00515683_m1	76	1–2
				Mm00515684_m1	62	2–3
				Mm01191805_m1	78	4–5
				Mm01191806_m1	74	5–6
				Mm01191807_m1	84	6–7
				Mm01191808_m1	105	7–8
				Mm01191809_m1	88	8–9
				Mm00515691_m1	60	9–10
				Mm00515692_m1	91	10–11
				Mm01191801_m1	59	11–12
				Mm00515694_m1	70	12–13

Rattus norvegicus: *Genomic context: chromosome: 4; Location: 4q11; Rattus norvegicus N-acyl-phosphatidylethanolamine-hydrolyzing phospholipase D (NAPE-PLD), mRNA, Acc. Nr. NM_199381* (Table 5.3) [20].

Total RNA was extracted from various brain regions or whole brain with TRIzol and reverse transcription performed using hexa-deoxyribonucleotide mixture as primers and Moloney murine leukemia virus-reverse transcriptase. Real-time quantitative PCR analysis was performed with primers and TaqMan probes based on the sequences from GenBank using Primer Express software (Applied Biosystems), TaqMan Rodent GAPDH Control Reagents were used as an endogenous control [199]. In another study, total RNA from rat brain, prefrontal cortex, dorsal striatum, and cerebellum was obtained using TRIzol (Invitrogen) and first-strand synthesized using random hexamer primer and M-MuLV reverse transcriptase. Resulting cDNA was used as the template for real-time quantitative PCR performed with SYBR green, endogenous control: β-glucoronidase [200].

5.9.2 Monoglyceride Lipase

Homo sapiens: Genomic context: chromosome: 3; Location: 3q21.3; Homo sapiens monoglyceride lipase (MAGL), transcript variant 1, mRNA Acc. Nr. NM_007283 and transcript variant 2, mRNA Acc. Nr. NM_001003794 [72,201].

Quantitative PCR was performed to study the expression of monoglyceride lipase in medullary breast cancer cell lines, ductal breast cancer and normal breast epithelia [202], and subcutaneous adipose tissue, brain and spleen RNA [142].

Mus musculus: Genomic context: chromosome: 6; Location: 6 D1; Mus musculus monoglyceride lipase (MAGL), Acc. Nr. NM_011844 [72].

Real-time cDNA quantitation was performed on RNA from mouse adipocyte cell line 3T3F442A reverse transcribed with SuperScript III in the presence of hexanucleotide mixture with iQ SYBR Green Supermix (BioRad). Relative expression analysis was corrected for PCR efficiency and normalized with respect to reference genes, β-actin and hypoxanthine phosphoribosyltransferase [123].

Rattus norvegicus: Genomic context: chromosome: 4; Location: 4q34; Rattus norvegicus monoglyceride lipase (MAGL), mRNA Acc. Nr. NM_138502 [74].

Total RNA was isolated from Wistar rat brain and human cell line HeLa cells with RNAqueous, retrotranscribed with oligo(dT) primer using SuperScript II and real-time PCR performed with primers/TaqMan probe designed with Primer Express software (Applied Biosystems), using GAPDH as external standard [76]; real-time quantitative PCR from cDNA from prefrontal cortex, dorsal striatum, and cerebellum rat brain (control: β-glucoronidase) was performed as previously described [200].

5.9.3 Diacylglycerol Lipase α

Homo sapiens: Genomic context: Chromosome: 11; Location: 11q12.2; Homo sapiens diacylglycerol lipase α (DAGLA), mRNA. Acc. Nr. NM_006133 [34].

TaqMan RT-PCR was used to measure the relative level of human diacylglycerol lipase α (DAGLα) transcripts in various male and female human tissues. RT-PCR was performed with primers based on sequences encoded by exon 19 to normalize

TABLE 5.3

List of Useful Resources for the Genes of Interest in *Rattus norvegicus*

Abbreviated Gene Name	Chrom Location	Ref Seq. ID	Translated Protein	Applied Biosystem TaqMan	Amplicon Length	Exon Boundary
NAPE-PLD	4q11	NM_199381.1	NP_955413.1	Rn001786260_m1	69	1–2
				Rn001786261_m1	64	2–3
				Rn001786262_m1	71	3–4
MAGL	4q34	NM_138502.2	NP_612511.1	Rn00593297_m1	78	2–3
				Rn00684475_m1	63	4–5
				Rn00449277_m1	66	5–6
				Rn01425259_m1	113	6–7
				Rn01425260_m1	65	7–8
DAGL-α	1q43	NM_001005886.1	NP_001005886.1.1	Rn01498063_g1	65	2–3
				Rn01498064_m1	64	3–4
				Rn01498065_m1	60	5–6
				Rn01498066_m1	62	6–7
				Rn01498067_g1	62	7–8
				Rn01454299_g1	69	12–13
				Rn01454300_g1	79	13–14
				Rn01454301_g1	87	14–15
				Rn01454302_g1	61	15–16
				Rn01023068_m1	81	16–17
				Rn01454303_m1	61	17–18
				Rn01454304_m1	67	19–20
FAAH	5q36	NM_024132.3	NP_077046.1	Rn00577086_m1	63	2–3

data and to correct for variations in RNA and/or cDNA quality and quantity; parallel TaqMan assays were run for two housekeeping genes: GAPDH and β-actin [34]; TaqMan quantitative real-time PCR was also performed on subcutaneous adipose tissue as previously described [197].

Mus musculus: Genomic context: chromosome: 19; Location: 19 A; Mus musculus diacylglycerol lipase α (DAGL-α), mRNA. Acc. Nr. NM_198114 [34].

TaqMan RT-PCR was used to measure the relative level of mouse DAGLα transcripts in various tissues. RT-PCR was performed as previously described [34] and real-time cDNA quantitation was performed on RNA from mouse adipocyte cell line 3T3F442A as previously described [123]. In another set of experiments, total RNA was extracted with TRIzol from Neuro-2a cells, first-strand complementary DNA synthesized using SuperScript II, and reverse transcribed using oligo(dT) primers. TaqMan real-time quantitative PCR was conducted with primer/probe sets designed with Primer Express software (Applied Biosystems). DAGLα mRNA levels were normalized using GADPH as an internal standard [203].

5.9.4 FATTY ACID AMIDE HYDROLASE

Homo sapiens: Genomic context: chromosome: 1; Location: 1p35-p34; Homo sapiens FAAH, mRNA. Acc. Nr. NM_001441 [204]; Homo sapiens fatty acid amide hydrolase 2 (FAAH2), mRNA. Acc. Nr. NM_174912 [67].

Total RNA was extracted from first trimester human placenta of differing gestational age using TRIzol; the extracted RNA was treated with RNase-free DNase I and reverse transcribed with oligo(dT) primer using SuperScript II (Invitrogen). TaqMan probes and primers for FAAH were obtained from Applied Biosystems, the relative expression of FAAH mRNA normalized to the amount of 18S rRNA [205]. In another series of studies, total RNA from preadipocytes, isolated adipocytes, or adipose tissue biopsies was isolated by the Qiagen Rneasy and measured with RNA 6000 Chip and 2100 Bioanalyzer (Agilent), and gene expression measured for real-time PCR and normalized for human GAPDH and 18S rRNA [206]. The same methodologies were applied to process RNA from visceral and subcutaneous adipose tissue from Caucasian men and women who underwent open abdominal surgery for gastric banding, cholecystectomy, appendectomy, abdominal injuries, or explorative laparatomy [206]. Human FAAH mRNA gene expression was measured by real-time PCR relative to 18S rRNA [207]; TaqMan quantitative real-time PCR was also performed on subcutaneous adipose tissue as previously described [197]. Real-time RT-PCR assays were used also to measure relative FAAH mRNA concentrations in the whole blood of 30 healthy donors and 8 sepsis patients to ascertain whether downregulation takes place during sepsis, indicating that mRNA expression of FAAH in human whole blood correlates with the degree of sepsis, and may be an interesting biomarker for predicting the onset of septic shock [208].

Mus musculus: Genomic context: chromosome: 4; Location: 4 56.5 cM; Mus musculus FAAH, mRNA. Acc. Nr. NM_010173 [204].

Real-time cDNA quantitation RNA from mouse adipocytes was performed as previously described and relative expression analysis corrected for PCR efficiency and normalized with respect to reference genes, β-actin and hypoxanthine phosphoribosyltransferase [123].

Rattus norvegicus: Genomic context: chromosome: 5; Location: 5q36; Rattus norvegicus FAAH, mRNA. Acc. Nr. NM_024132 [49].

RNA was isolated from serum-starved activated rat HSC cells and rat hepatocytes using TRIzol (Invitrogen), after DNase treatment and RNA was reverse-transcribed using random hexamer primers. Real-time PCR was performed and probes and primers for rat FAAH and 18S rRNA (for normalization) were designed by ABI [209]. In another study, real-time quantitative PCR from cDNAs from prefrontal cortex, dorsal striatum, and cerebellum of rat brain (control: β-glucoronidase) was performed as previously described [200].

5.10 CONCLUSIONS

After the discovery of the cannabinoid receptors, their endogenous agonists and the proteins involved in their synthesis and inactivation, significant progress has been made toward the understanding of their role in vital functions as well as during pathological conditions. This progress has led to the identification of new drugs now on the market. For example, Acomplia®, a CB_1 receptor antagonist/inverse agonist that has been used in several countries for the treatment of obesity and related metabolic disorders, and Sativex®, a *Cannabis* extract used in Canada against neuropathic pain associated with multiple sclerosis, can be considered milestones in the 40 years of intensive research on plant and endogenous cannabinoids. The contribution of chemists, biologists, pharmacologists, and enzymologists, supported by the development and subsequent improvement of analytical methods, allowed for the current understanding of EC regulation and function. Further efforts are now necessary to more fully understand this signaling system and to develop more specific and sensitive methodologies.

LIST OF ABBREVIATIONS

AAMCA	arachidonyl 7-amino, 4-methyl coumarin amide
AAPC	arachidonoyl-glycerophosphocholine
AMC	7-amino, 4-methyl coumarin
APCI	atmospheric pressure chemical ionization
ARA	arachidonic acid
Abh4	α/β-hydrolase 4
AEA	anandamide
2-AG	2-arachidonoyl-glycerol
AS	amidase signature
BSTFA	bis(trimethylsilyl)trifluoroacetamide
CB_1	cannabinoid receptor type 1
CB_2	cannabinoid receptor type 2
DAG	diacylglycerols
DAGLα	diacylglycerol lipase α
D-AMC	decanoyl 7-amino-4-methyl coumarin

DAUDA	(11-(5-dimethylaminonaphthalenesulfonyl)-undecanoic acid)
EC	endocannabinoids
ECS	endocannabinoid system
EDC/DMAP	N'-(3-dimethylaminopropyl)-N-ethylcarbodiimide hydrochloride/4-dimethylaminopyridin
EI	electron-impact
EMT	endocannabinoid membrane transporter
ESI	electrospray ionization
FAAH	fatty acid amide hydrolase
FABP	fatty acid–binding protein
GAPDH	glyceraldehyde-3-phosphate-dehydrogenase
GC-MS	gas chromatography-mass spectrometry
GC-NICI-MS	gas chromatography-negative ion chemical ionization-mass spectrometry
GDH	L-glutamate dehydrogenase
HPLC	high-performance liquid chromatography
HTS	high-throughput screening
IDA	information-dependent acquisition
LC-APCI-MS	liquid chromatography-atmospheric pressure chemical ionization-mass spectrometry
LC-MS	liquid chromatography-mass spectrometry
2-LG	2-linoleoyl-glycerol
MAGL	monoacylglycerol lipases
MRM	multiple-reaction monitoring
MSTFA	N-methyl-N-(trimethylsilyl)trifluoroacetamide
NADA	N-arachidonoyl-dopamine
NAE	N-acyl-ethanolamines
NAPE	N-acyl-phosphatidylethanolamine
N-ArPE	N-arachidonoyl-phosphatidylethanolamine
NAPE$_{DIACYL}$	1,2-diacyl-sn-glycero-3-phospho-(N-acyl)-ethanolamine
NAPE$_{PLAS}$	1-(1'-alkenyl)-2-acyl-sn-glycero-3-phospho-(N-acyl)-ethanolamine
NAPE-PLD	N-acyl-phosphatidylethanolamine-selective phospholipase D
NAT	$trans$-N-acyltransferase
OEA	oleylethanolamide
PEA	palmitoylethanolamide
PFBz-PFP	pentafluorobenzoyl-pentafluoropropionate
PLA$_2$	phospholipase A$_2$
PLD	phospholipase D
^{31}P-NMR	phosphorus-31 nuclear magnetic resonance
PtdCho	phosphatidylcholine
PtdEth	phosphatidylethanolamine
PtdIns(4,5)P$_2$	phosphatidylinositol-4,5-bis-phosphate
PtdOH	phosphatidic acid
PTPN22	protein tyrosine phosphatase N22
Q/TOF	quadrupole/time-of-flight
RLP-1	rat lecithin-retinol acyltransferase-like protein 1

RT-PCR   reverse transcription polymerase chain reaction
SIM   selected-ion monitoring
THC $(-)$-Δ^9-tetrahydrocannabinol
TLC   thin-layer chromatography
TMS   trimethylsilylether
TOF   time-of-flight

REFERENCES

1. Gaoni, Y. and Mechoulam, R. (1971) The isolation and structure of Δ-1-tetra-hydrocannabinol and other neutral cannabinoids from hashish. *J. Am. Chem. Soc.* 93:217–224.
2. Howlett, A.C. et al. (2004) Cannabinoid physiology and pharmacology: 30 years of progress. *Neuropharmacology* 47(Suppl 1):345–358.
3. Devane, W.A. et al. (1992) Isolation and structure of a brain constituent that binds to the cannabinoid receptor. *Science* 258:1946–1949.
4. Mechoulam, R. et al. (1995) Identification of an endogenous 2-monoglyceride, present in canine gut, that binds to cannabinoid receptors. *Biochem. Pharmacol.* 50:83–90.
5. Sugiura, T. et al. (1995) 2-Arachidonoylglycerol: A possible endogenous cannabinoid receptor ligand in brain. *Biochem. Biophys. Res. Commun.* 215:89–97.
6. Di Marzo, V. and Petrosino, S. (2007) Endocannabinoids and the regulation of their levels in health and disease. *Curr. Opin. Lipidol.* 18:129–140.
7. Witting, A. et al. (2004) P2X7 receptors control 2-arachidonoylglycerol production by microglial cells. *Proc. Natl. Acad. Sci. USA* 101:3214–3219.
8. van der Stelt, M. et al. (2005) Anandamide acts as an intracellular messenger amplifying Ca^{2+} influx via TRPV1 channels. *EMBO J.* 24:3026–3037.
9. Jung, K.M. et al. (2005) Stimulation of endocannabinoid formation in brain slice cultures through activation of group I metabotropic glutamate receptors. *Mol. Pharmacol.* 68:1196–1202.
10. Hashimotodani, Y. et al. (2005) Phospholipase Cbeta serves as a coincidence detector through its Ca^{2+} dependency for triggering retrograde endocannabinoid signal. *Neuron* 45:257–268.
11. Maejima, T. et al. (2005) Synaptically driven endocannabinoid release requires Ca^{2+}-assisted metabotropic glutamate receptor subtype 1 to phospholipase Cβ4 signaling cascade in the cerebellum. *J. Neurosci.* 25:6826–6835.
12. Wettschureck, N. et al. (2006) Forebrain-specific inactivation of Gq/G11 family G proteins results in age-dependent epilepsy and impaired endocannabinoid formation. *Mol. Cell. Biol.* 26:5888–5894.
13. Edwards, D.A., Kim, J., and Alger, B.E. (2006) Multiple mechanisms of endocannabinoid response initiation in hippocampus. *J. Neurophysiol.* 95:67–75.
14. Di Marzo, V. et al. (1994) Formation and inactivation of endogenous cannabinoid anandamide in central neurons. *Nature* 372:686–691.
15. Schmid, P.C. et al. (1983) Metabolism of *N*-acylethanolamine phospholipids by a mammalian phosphodiesterase of the phospholipase D type. *J. Biol. Chem.* 258:9302–9306.
16. Schmid, H.H.O., Schmid, P.C., and Natarajan, V. (1990) *N*-Acylated glycerophospholipids and their derivatives. *Prog. Lipid Res.* 29:1–43.
17. Schmid, H.H., Schmid, P.C., and Natarajan, V. (1996) The *N*-acylation-phosphodiesterase pathway and cell signalling. *Chem. Phys. Lipids* 80:133–142.
18. Hansen, H.S. et al. (1998) Formation of *N*-acyl-phosphatidylethanolamines and *N*-acetylethanolamines: Proposed role in neurotoxicity. *Biochem. Pharmacol.* 55:719–725.

19. Schmid, H.H. and Berdyshev, E.V. (2002) Cannabinoid receptor-inactive *N*-acylethanolamines and other fatty acid amides: Metabolism and function. *Prostaglandins Leukot. Essent. Fatty Acids* 66:363–376.

20. Okamoto, Y. et al. (2004) Molecular characterization of a phospholipase D generating anandamide and its congeners. *J. Biol. Chem.* 279:5298–5305.

21. Bisogno, T. et al. (1999) Brain regional distribution of endocannabinoids: Implications for their biosynthesis and biological function. *Biochem. Biophys. Res. Commun.* 256:377–380.

22. Okamoto, Y. et al. (2005) Mammalian cells stably overexpressing *N*-acylphosphatidylethanolamine-hydrolysing phospholipase D exhibit significantly decreased levels of *N*-acylphosphatidylethanolamines. *Biochem. J.* 389:241–247.

23. Liu, J. et al. (2006) A biosynthetic pathway for anandamide. *Proc. Natl. Acad. Sci. USA* 103:13345–13350.

24. Sun, Y.X. et al. (2004) Biosynthesis of anandamide and *N*-palmitoylethanolamine by sequential actions of phospholipase A2 and lysophospholipase D. *Biochem. J.* 380:749–756.

25. Simon, G.M. and Cravatt, B.F. (2006) Endocannabinoid biosynthesis proceeding through glycerophospho-*N*-acyl ethanolamine and a role for alpha/beta-hydrolase 4 in this pathway. *J. Biol. Chem.* 281:26465–26472.

26. Jin, X.H. (2007) Discovery and characterization of a Ca^{2+}-independent phosphatidylethanolamine *N*-acyltransferase generating the anandamide precursor and its congeners. *J. Biol. Chem.* 282:3614–3623.

27. Di Marzo, V. et al. (1996) Potential biosynthetic connections between the two cannabimimetic eicosanoids, anandamide and 2-arachidonoyl-glycerol, in mouse neuroblastoma cells. *Biochem. Biophys. Res. Commun.* 227:281–288.

28. Stella, N., Schweitze, P., and Piomelli, D. (1997) A second endogenous cannabinoid that modulates long-term potentiation. *Nature* 388:773–778.

29. Kondo, S. et al. (1998) 2-arachidonoylglycerol, an endogenous cannabinoid receptor agonist: Identification as one of the major species of monoacylglycerols in various rat tissues, and evidence for its generation through Ca^{2+}-dependent and -independent mechanisms. *FEBS Lett.* 429:152–156.

30. Berdyshev, E.V. et al. (2001) Activation of PAF receptors results in enhanced synthesis of 2-arachidonoylglycerol (2-AG) in immune cells. *FASEB J.* 15:2171–2178.

31. Bisogno, T. et al. (1999) Phosphatidic acid as the biosynthetic precursor of the endocannabinoid 2-arachidonoylglycerol in intact mouse neuroblastoma cells stimulated with ionomycin. *J. Neurochem.* 72:2113–2119.

32. Carrier, E.J. et al. (2004) Cultured rat microglial cells synthesize the endocannabinoid 2-arachidonylglycerol, which increases proliferation via a CB2 receptor-dependent mechanism. *Mol. Pharmacol.* 65:999–1007.

33. Oka, S. et al. (2005) Evidence for the involvement of the cannabinoid CB2 receptor and its endogenous ligand 2-arachidonoylglycerol in 12-*O*-tetradecanoylphorbol-13-acetate-induced acute inflammation in mouse ear. *J. Biol. Chem.* 280:18488–18497.

34. Bisogno, T. et al. (2003) Cloning of the first *sn* 1-DAG lipases points to the spatial and temporal regulation of endocannabinoid signalling in the brain. *J. Cell Biol.* 163:463–468.

35. Williams, E.J., Walsh, F.S., and Doherty, P. (2003) The FGF receptor uses the endocannabinoid signaling system to couple to an axonal growth response. *J. Cell Biol.* 160:481–486.

36. Chevaleyre, V. and Castillo, P.E. (2003) Heterosynaptic LTD of hippocampal GABAergic synapses: A novel role of endocannabinoids in regulating excitability. *Neuron* 38:461–472.

37. Yoshida, T. et al. (2006) Localization of diacylglycerol lipase-α around postsynaptic spine suggests close proximity between production site of an endocannabinoid, 2-arachidonoyl-glycerol, and presynaptic cannabinoid Cβ1 receptor. *J. Neurosci.* 26:4740–4751.

38. Katona, I. et al. (2006) Molecular composition of the endocannabinoid system at glutamatergic synapses. *J. Neurosci.* 26:5628–5637.

39. Brenowitz, S.D., Best, A.R., and Regehr, W.G. (2006) Sustained elevation of dendritic calcium evokes widespread endocannabinoid release and suppression of synapses onto cerebellar Purkinje cells. *J. Neurosci.* 26:6841–6850.

40. McFarland, M.J. and Barker, E.L. (2004) Anandamide transport. *Pharmacol. Ther.* 104:117–135.

41. De Petrocellis, L., Cascio, M.G., and Di Marzo, V. (2004) The endocannabinoid system: A general view and latest additions. *Br. J. Pharmacol.* 141:765–774.

42. Ligresti, A. et al. (2004) Further evidence for the existence of a specific process for the membrane transport of anandamide. *Biochem. J.* 380:265–272.

43. Fegley, D. et al. (2004) Anandamide transport is independent of fatty-acid amide hydrolase activity and is blocked by the hydrolysis-resistant inhibitor AM1172. *Proc. Natl. Acad. Sci. USA* 101:8756–8761.

44. Oddi, S. et al. (2005) Confocal microscopy and biochemical analysis reveal spatial and functional separation between anandamide uptake and hydrolysis in human keratinocytes. *Cell. Mol. Life Sci.* 62:386–395.

45. Moore, S.A. et al. (2005) Identification of a high-affinity binding site involved in the transport of endocannabinoids. *Proc. Natl. Acad. Sci. USA* 102:17852–17857.

46. Dickason-Chesterfield, A.K. et al. (2006) Pharmacological characterization of endocannabinoid transport and fatty acid amide hydrolase inhibitors. *Cell. Mol. Neurobiol.* 26:407–423.

47. Ortar, G. et al. (2008) Carbamoyl tetrazoles as inhibitors of endocannabinoid inactivation: A critical revisitation. *Eur. J. Med. Chem.* 43:62–72.

48. Kaczocha, M. et al. (2006) Anandamide uptake is consistent with rate-limited diffusion and is regulated by the degree of its hydrolysis by fatty acid amide hydrolase. *J. Biol. Chem.* 281:9066–9075.

49. Cravatt, B.F. et al. (1996) Molecular characterization of an enzyme that degrades neuromodulatory fatty-acid amides. *Nature* 384:83–87.

50. Saghatelian, A. et al. (2006) A FAAH-regulated class of *N*-acyl taurines that activates TRP ion channels. *Biochemistry* 45:9007–9015.

51. Huang, S.M. et al. (2001) Identification of a new class of molecules, the arachidonyl amino acids, and characterization of one member that inhibits pain. *J. Biol. Chem.* 276:42639–42644.

52. Cascio, M.G. et al. (2004) A structure-activity relationship study on *N*-arachidonoyl-amino acids as possible endogenous inhibitors of fatty acid amide hydrolase. *Biochem. Biophys. Res. Commun.* 314:192–196.

53. McKinney, M.K. and Cravatt, B.F. (2005) Structure and function of fatty acid amide hydrolase. *Annu. Rev. Biochem.* 74:411–432.

54. Puffenbarger, R.A. et al. (2001) Characterization of the 5′-sequence of the mouse fatty acid amide hydrolase. *Neurosci. Lett.* 314:21–24.

55. Waleh, N.S. et al. (2002) Transcriptional regulation of the mouse fatty acid amide hydrolase gene. *Gene* 291:203–210.

56. Maccarone, M. et al. (2003) Leptin activates the anandamide hydrolase promoter in human T lymphocytes through STAT3. *J. Biol. Chem.* 278:13318–13324.

57. Maccarone, M. et al. (2003) Progesterone activates fatty acid amide hydrolase (FAAH) promoter in human T lymphocytes through the transcription factor Ikaros: Evidence for a synergistic effect of leptin. *J. Biol. Chem.* 278:32726–32732.

58. Kathuria, S. et al. (2003). Modulation of anxiety through blockade of anandamide hydrolysis. *Nat. Med.* 9:76–81.

59. Lichtman, A.H. et al. (2004) Reversible inhibitors of fatty acid amide hydrolase that promote analgesia: Evidence for an unprecedented combination of potency and selectivity. *J. Pharmacol. Exp. Ther.* 311:441–448.

60. Maione, S. et al. (2007) Analgesic actions of *N*-arachidonoyl-serotonin, a fatty acid amide hydrolase inhibitor with antagonistic activity at vanilloid TRPV1 receptors. *Br. J. Pharmacol.* 150:766–781.

61. Maccarrone, M. (2006) Fatty acid amide hydrolase: A potential target for next generation therapeutics. *Curr. Pharm. Des.* 12:759–772.

62. Piomelli, D. et al. (2006) Pharmacological profile of the selective FAAH inhibitor KDS-4103 (URB597). *CNS Drug Rev.* 12:21–38.

63. Di Marzo, V. et al. (2007) Endocannabinoids and related compounds: Walking back and forth between plant natural products and animal physiology. *Chem. Biol.* 7:741–756.

64. Flanagan, J.M. et al. (2006) The fatty acid amide hydrolase 385 A/A (P129T) variant: Haplotype analysis of an ancient missense mutation and validation of risk for drug addiction. *Hum. Genet.* 120:581–588.

65. Sipe, J.C. et al. (2005) Overweight and obesity associated with a missense polymorphism in fatty acid amide hydrolase (FAAH). *Int. J. Obes.* 29:755–759.

66. Morita, Y. et al. (2005) A nonsynonymous polymorphism in the human fatty acid amide hydrolase gene did not associate with either methamphetamine dependence or schizophrenia. *Neurosci. Lett.* 376:182–187.

67. Wei, B.Q. et al. (2006) A second fatty acid amide hydrolase with variable distribution among placental mammals. *J. Biol. Chem.* 281:36569–36578.

68. Di Marzo, V. and Deutsch, D.G. (1998) Biochemistry of the endogenous ligands of cannabinoid receptors. *Neurobiol. Dis.* 5:386–404.

69. Lichtman, A.H. et al. (2002) Pharmacological activity of fatty acid amides is regulated, but not mediated, by fatty acid amide hydrolase *in vivo. J. Pharmacol. Exp. Ther.* 302:73–79.

70. Di Marzo, V. et al. (1999) Biosynthesis and inactivation of the endocannabinoid 2-arachidonoylglycerol in circulating and tumoral macrophages. *Eur. J. Biochem.* 264:258–267.

71. Goparaju, S.K. et al. (1999) Enzymes of porcine brain hydrolyzing 2-arachidonoylglycerol, an endogenous ligand of cannabinoid receptors. *Biochem. Pharmacol.* 57:417–423.

72. Karlsson, M. et al. (2001) Exon-intron organization and chromosomal localization of the mouse monoglyceride lipase gene. *Gene* 272:11–18.

73. Ho, S.Y., Delgado, L., and Storch, J. (2002) Monoacylglycerol metabolism in human intestinal Caco-2 cells: Evidence for metabolic compartmentation and hydrolysis. *J. Biol. Chem.* 277:1816–1823.

74. Dinh, T.P. et al. (2002) Brain monoglyceride lipase participating in endocannabinoid inactivation. *Proc. Natl. Acad. Sci. USA* 99:10819–10824. Erratum in: *Proc. Natl. Acad. Sci. USA* 99:13961.

75. Gulyas, A.I. et al. (2004) Segregation of two endocannabinoid-hydrolyzing enzymes into pre- and postsynaptic compartments in the rat hippocampus, cerebellum and amygdala. *Eur. J. Neurosci.* 20:441–458.

76. Dinh, T.P., Kathuria, S., and Piomelli, D. (2004) RNA interference suggests a primary role for monoacylglycerol lipase in the degradation of the endocannabinoid 2-arachidonoylglycerol. *Mol. Pharmacol.* 66:1260–1264.

77. Muccioli, G.G. et al. (2007) Identification of a novel endocannabinoid-hydrolyzing enzyme expressed by microglial cells. *J. Neurosci.* 11:2883–2889.

78. Pertwee, R.G. (2006) The pharmacology of cannabinoid receptors and their ligands: An overview. *Int. J. Obes.* 30(Suppl 1):S13–S18; Cannabinoid pharmacology: The first 66 years. *Br. J. Pharmacol.* 147(Suppl 1):S163–S171.

79. Di Marzo, V. (2006) A brief history of cannabinoid and endocannabinoid pharmacology as inspired by the work of British scientists. *Trends Pharmacol. Sci.* 27:134–140.

80. Klein, T.W. (2005) Cannabinoid-based drugs as anti-inflammatory therapeutics. *Nat. Rev. Immunol.* 5:400–411.

81. Maccarrone, M. (2005) The endocannabinoid system in the brain: From biology to therapy. *Curr. Drug Targets CNS Neurol. Disord.* 4:613.

82. Valenti, M. et al. (2004) Differential diurnal variations of anandamide and 2-arachidonoyl-glycerol levels in rat brain. *Cell. Mol. Life Sci.* 61:945–950.

83. Di Marzo, V. et al. (2001) Leptin-regulated endocannabinoids are involved in maintaining food intake. *Nature* 410:822–825.

84. Sugiura, T., Yoshinaga, N., and Waku, K. (2001) Rapid generation of 2-arachidonoyl-glycerol, an endogenous cannabinoid receptor ligand, in rat brain after decapitation. *Neurosci. Lett.* 297:175–178.

85. Kempe, K. et al. (1996) Isotope dilution mass spectrometric measurements indicate that arachidonylethanolamide, the proposed endogenous ligand of the cannabinoid receptor, accumulates in rat brain tissue post mortem but is contained at low levels in or is absent from fresh tissue. *J. Biol. Chem.* 271:17287–17295.

86. Schmid, P.C. et al. (1995) Occurrence and postmortem generation of anandamide and other long-chain *N*-acylethanolamines in mammalian brain. *FEBS Lett.* 375:117–120.

87. Cadas, H., di Tomaso, E., and Piomelli, D. (1997) Occurrence and biosynthesis of endogenous cannabinoid precursor, *N*-arachidonoyl phosphatidylethanolamine, in rat brain. *J. Neurosci.* 17:1226–1242.

88. Fontana, A. et al. (1995) Analysis of anandamide, an endogenous cannabinoid substance, and of other natural *N*-acylethanolamines. *Prostaglandins Leukot. Essent. Fatty Acids* 53:301–308.

89. Koga, D. et al. (1997) Liquid chromatographic-atmospheric pressure chemical ionization mass spectrometric determination of anandamide and its analogs in rat brain and peripheral tissues. *J. Chromatogr. B Biomed. Sci. Appl.* 690:7–13.

90. Kingsley, P.J. and Marnett, L.J. (2003) Analysis of endocannabinoids by Ag+ coordination tandem mass spectrometry. *Anal. Biochem.* 314:8–15.

91. Richardson, D. et al. (2007) Quantitative profiling of endocannabinoids and related compounds in rat brain using liquid chromatography-tandem electrospray ionization mass spectrometry. *Anal. Biochem.* 360:216–226.

92. Koga, D. et al. (1995) High-performance liquid chromatography and fluorometric detection of arachidonylethanolamide (anandamide) and its analogues, derivatized with 4-(*N*-chloroformylmethyl-*N*-methyl)amino-7-*N,N*-dimethylaminosulphonyl-2,1,3-benzoxadiazole (DBD-COCl). *Biomed. Chromatogr.* 9:56–57.

93. Sugiura, T. et al. (1996) Transacylase-mediated and phosphodiesterase-mediated synthesis of *N*-arachidonoylethanolamine, an endogenous cannabinoid-receptor ligand, in rat brain microsomes. Comparison with synthesis from free arachidonic acid and ethanolamine. *Eur. J. Biochem.* 240:53–62.

94. Yagen, B. and Burstein, S. (2000) Novel and sensitive method for the detection of anandamide by the use of its dansyl derivative. *J. Chromatogr. B Biomed. Sci. Appl.* 740:93–99.

95. Giuffrida, A. and Pomelli, D. (1998) Isotope dilution GC/MS determination of anandamide and other fatty acylethanolamides in rat blood plasma. *FEBS Lett.* 422:373–376.

96. Bisogno, T. et al. (1997) Occurrence and metabolism of anandamide and related acyl-ethanolamides in ovaries of the sea urchin *Paracentrotus lividus*. *Biochim. Biophys. Acta* 1345:338–348.

97. Bisogno, T. et al. (1997) Biosynthesis, uptake, and degradation of anandamide and palmitoylethanolamide in leukocytes. *J. Biol. Chem.* 272:3315–3323.
98. Bisogno, T. et al. (1997) The sleep inducing factor oleamide is produced by mouse neuroblastoma cells. *Biochem. Biophys. Res. Commun.* 239:473–479.
99. Bisogno, T. et al. (1999) Biosynthesis and inactivation of *N*-arachidonoylethanolamine (anandamide) and *N*-docosahexaenoylethanolamine in bovine retina. *Arch. Biochem. Biophys.* 370:300–307.
100. Di Marzo, V. et al. (2000) Enhanced levels of endogenous cannabinoids in the globus pallidus are associated with a reduction in movement in an animal model of Parkinson's disease. *FASEB J.* 10:1432–1438.
101. Gonzales, S. et al. (1999) Identification of endocannabinoids and cannabinoid Cβ(1) receptor mRNA in the pituitary gland. *Neuroendocrinology* 70:137–145.
102. Di Marzo, V. et al. (2000) Enhancement of anandamide formation in the limbic forebrain and reduction of endocannabinoid contents in the striatum of Δ9-tetrahydrocannabinol-tolerant rats. *J. Neurochem.* 74:1627–1635.
103. Matias, I. et al. (2001) Evidence for an endocannabinoid system in the central nervous system of the leech *Hirudo medicinalis. Brain Res. Mol. Brain Res.* 87:145–159.
104. Pinto, L. et al. (2002) Endocannabinoids as physiological regulators of colonic propulsion in mice. *Gastroenterology* 123:227–234.
105. Izzo, A.A. et al. (2003) An endogenous cannabinoid tone attenuates cholera toxin-induced fluid accumulation in mice. *Gastroenterology* 125:765–774.
106. Sepe, N. et al. (1998) Bioactive long chain *N*-acylethanolamines in five species of edible bivalve molluscs. Possible implications for mollusc physiology and sea food industry. *Biochim. Biophys. Acta* 1389:101–111.
107. Izzo, A.A. et al. (2001) Cannabinoid CB1-receptor mediated regulation of gastrointestinal motility in mice in a model of intestinal inflammation. *Br. J. Pharmacol.* 134:563–570.
108. Baker, D. et al. (2001) Endocannabinoids control spasticity in a multiple sclerosis model. *FASEB J.* 15:300–302.
109. Lastres-Becker, I. et al. (2001) Changes in endocannabinoid transmission in the basal ganglia in a rat model of Huntington's disease. *Neuroreport* 12:2125–2129.
110. Yang, H.Y. et al. (1999) GC/MS analysis of anandamide and quantification of *N*-arachidonoylphosphatidylethanolamides in various brain regions, spinal cord, testis, and spleen of the rat. *J. Neurochem.* 72:1959–1968.
111. Maccarrone, M. et al. (2001) Gas chromatography-mass spectrometry analysis of endogenous cannabinoids in healthy and tumoral human brain and human cells in culture. *J. Neurochem.* 76:594–601.
112. Hardison, S., Weintraub, S.T., and Guiffrida, A. (2006) Quantification of endocannabinoids in rat biological samples by GC/MS: Technical and theoretical considerations. *Prostaglandins Other Lipid Mediat.* 81:106–112.
113. Giuffrida, A., Rodriguez de Fonseca, F., and Piomelli, D. (2000) Quantification of bioactive acylethanolamides in rat plasma by electrospray mass spectrometry. *Anal. Biochem.* 1:87–93.
114. De Marchi, N. et al. (2003) Endocannabinoid signalling in the blood of patients with schizophrenia. *Lipids Health Dis.* 2:5.
115. Ligresti, A. et al. (2003) Possible endocannabinoid control of colorectal cancer growth. *Gastroenterology* 125:677–687.
116. Bifulco, M. et al. (2004) A new strategy to block tumor growth by inhibiting endocannabinoid inactivation. *FASEB J.* 18:1606–1608.
117. van der Stelt, M. et al. (2005) A role for endocannabinoids in the generation of parkinsonism and levodopa-induced dyskinesia in MPTP-lesioned non-human primate models of Parkinson's disease. *FASEB J.* 19:1140–1142.

118. van der Stelt, M. et al. (2006) Endocannabinoids and β-amyloid-induced neurotoxicity *in vivo*: Effect of pharmacological elevation of endocannabinoid levels. *Cell. Mol. Life Sci*. 63:1410–1424.

119. D'Argenio, G. et al. (2006) Up-regulation of anandamide levels as an endogenous mechanism and a pharmacological strategy to limit colon inflammation. *FASEB J*. 20:568–570.

120. Melis, M. et al. (2006) Protective activation of the endocannabinoid system during ischemia in dopamine neurons. *Neurobiol. Dis*. 24:15–27.

121. Mestre, J. et al. (2005) Pharmacological modulation of the endocannabinoid system in a viral model of multiple sclerosis. *J. Neurochem*. 6:1327–1339.

122. Bisogno, T. et al. (2007) Symptom-related changes of endocannabinoid and palmitoyle-thanolamide levels in brain areas of R6/2 mice, a transgenic model of Huntington's disease. *Neurochem. Int*. 52:307–313.

123. Matias, I. et al. (2006) Regulation, function, and dysregulation of endocannabinoids in models of adipose and beta-pancreatic cells and in obesity and hyperglycemia. *J. Clin. Endocrinol. Metab*. 91:3171–3180.

124. Matias, I. et al. (2006) Changes in endocannabinoid and palmitoylethanolamide levels in eye tissues of patients with diabetic retinopathy and age-related macular degeneration. *Prostaglandins Leukot. Essent. Fatty Acids* 75:413–418.

125. Cote, M. et al. (2007) Circulating endocannabinoid levels, abdominal adiposity and related cardiometabolic risk factors in obese men. *Int. J. Obes*. 31:692–699.

126. Gonthier, M.P. et al. (2007) Identification of endocannabinoids and related compounds in human fat cells. *Obesity* 15:837–845.

127. Karsak, M. et al. (2007) Attenuation of allergic contact dermatitis through the endocannabinoid system. *Science* 5830:1494–1497.

128. Huang, S.M. et al. (2002) An endogenous capsaicin-like substance with high potency at recombinant and native vanilloid VR1 receptors. *Proc. Natl. Acad. Sci. USA* 99:8400–8405.

129. Tan, B. et al. (2006) Targeted lipidomics: Discovery of new fatty acyl amides. *AAPS J*. 8:E461–E465.

130. Schreiber, D. et al. (2007) Determination of anandamide and other fatty acyl ethanolamides in human serum by electrospray tandem mass spectrometry. *Anal. Biochem*. 2:162–168.

131. Moesgaard, B., Jaroszewski, J.W., and Hansen, H.S. (1999) Accumulation of *N*-acyl-ethanolamine phospholipids in rat brains during post-decapitative ischemia: A ^{31}P NMR study. *J. Lipid Res*. 40:515–521.

132. Moesgaard, B. et al. (2000) Age dependent accumulation of *N*-acyl-ethanolamine phospholipids in ischemic rat brain. A ^{31}P NMR and enzyme activity study. *J. Lipid Res*. 41:985–990.

133. Di Marzo, V. et al. (1996) Biosynthesis of anandamide and related acylethanolamides in mouse J774 macrophages and N18 neuroblastoma cells. *Biochem. J*. 316:977–984.

134. Berrendero, F. et al. (1999) Analysis of cannabinoid receptor binding and mRNA expression and endogenous cannabinoid contents in the developing rat brain during late gestation and early postnatal period. *Synapse* 33:181–191.

135. Folch, J. et al. (1957) A simple method for the isolation and purification of total lipides from animal tissues. *J. Biol. Chem*. 226:497–509.

136. Peterson, G., Chapman, K.D., and Hansen, H.S. (2000) A rapid phospholipase D assay using zirconium precipitation of anionic substrate phospholipids: Application to *N*-acylethanolamine formation *in vitro*. *J. Lipid Res*. 41:1532–1538.

137. Burns, D.T., Townshend, A., and Carter, A.H. (1981) *Inorganic Reaction Chemistry, Vol. 2. Reactions of the Elements and Their Compounds*. Ellis Horwood, New York, pp. 501–507.

138. Epps, D.E. et al. (1979) N-Acylethanolamine accumulation in infarcted myocardium. *Biochem. Biophys. Res. Commun.* 90:628–633.

139. Epps, D.E. et al. (1980) Accumulation of N-acylethanolamine glycerophospholipids in infarcted myocardium. *Biochim. Biophys. Acta* 618:420–430.

140. Natarajan, V. et al. (1981) On the biosynthesis and metabolism of N-acylethanolamine phospholipids in infarcted dog heart. *Biochim. Biophys. Acta* 664:445–448.

141. Fezza, F. et al. (2005) Radiochromatographic assay of N-acyl-phosphatidylethanolamine-specific phospholipase D activity. *Anal. Biochem.* 339:113–120.

142. Cadas, H. et al. (1996) Biosynthesis of an endogenous cannabinoid precursor in neurons and its control by calcium and cAMP. *J. Neurosci.* 12:3934–3942.

143. Deutsch, D.G. and Chin, S.A. (1993) Enzymatic synthesis and degradation of anandamide, a cannabinoid receptor agonist. *Biochem. Pharmacol.* 46:791–796.

144. Devane, W.A. and Axelrod, J. (1994) Enzymatic synthesis of anandamide, an endogenous ligand for the cannabinoid receptor, by brain membranes. *Proc. Natl. Acad. Sci. USA* 91:6698–6701.

145. Desarnaud, F. et al. (1995) Anandamide amidohydrolase activity in rat brain microsomes. Identification and partial characterization. *J. Biol. Chem.* 11:6030–6035.

146. Lang, W. et al. (1996) High-performance liquid chromatographic determination of anandamide amidase activity in rat brain microsomes. *Anal. Biochem.* 238:40–45.

147. Maccarrone, M., Bari, M., and Finazzi Agrò, A. (1999) A sensitive and specific radiochromatographic assay of fatty acid amide hydrolase activity. *Anal. Biochem.* 267:314–318.

148. Omeir, R.L. et al. (1995) Arachidonoyl ethanolamide-[1,2-^{14}C] as a substrate for anandamide amidase. *Life Sci.* 56:1999–2005.

149. Wilson, S.J., Lovenberg, T.W., and Barbier, A.J. (2003) A high-throughput-compatible assay for determining the activity of fatty acid amide hydrolase. *Anal. Biochem.* 318:270–275.

150. Boldrup, L. et al. (2004) A simple stopped assay for fatty acid amide hydrolase avoiding the use of a chloroform extraction phase. *J. Biochem. Biophys. Methods* 60:171–177.

151. Patterson, J.E. et al. (1996) Inhibition of oleamide hydrolase catalyzed hydrolysis of the endogenous sleep-inducing lipid cis-9-octadecenamide. *J. Am. Chem. Soc.* 118:5938–5945.

152. Thumser, A.E., Voysey, J., and Wilton, D.C. (1997) A fluorescence displacement assay for the measurement of arachidonoyl ethanolamide (anandamide) and oleoyl amide (octadecenoamide) hydrolysis. *Biochem. Pharmacol.* 53:433–435.

153. Patricelli, M.P. and Cravatt, B.F. (2001) Characterization and manipulation of the acyl chain selectivity of fatty acid amide hydrolase. *Biochemistry* 40:6107–6115.

154. DeBank, P.A., Kendall, D.A., and Alexander, S.P. (2005) A spectrophotometric assay for fatty acid amide hydrolase suitable for high-throughput screening. *Biochem. Pharmacol.* 69:1187–1193.

155. Ramarao, M.K. et al. (2005) A fluorescence-based assay for fatty acid amide hydrolase compatible with high-throughput screening. *Anal. Biochem.* 343:143–151.

156. Kage, K.L. (2007) A high throughput fluorescent assay for measuring the activity of fatty acid amide hydrolase. *J. Neurosci. Methods* 161:47–54.

157. Hasegawa-Sasaki, H. (1985) Early changes in inositol lipids and their metabolites induced by platelet-derived growth factor in quiescent Swiss mouse 3T3 cells. *Biochem. J.* 232:99–109.

158. Gammon, C.M. et al. (1989) Diacylglycerol modulates action potential frequency in GH3 pituitary cells: Correlative biochemical and electrophysiological studies. *Brain Res.* 479:217–224.

159. Cox, J.W. and Horrocks, L.A. (1981) Preparation of thioester substrates and development of continuous spectrophotometric assays for phospholipase A, and monoacylglycerol lipase. *J. Lipid Res.* 22:496–505.

160. Strosznajder, J., Singh, H., and Horrocks, L.A. (1984) Monoacylglycerol lipase: Regulation and increased activity during hypoxia and ischemia. *Neurochem. Pathol.* 2:139–147.

161. Farooqui, A.A. et al. (1984) Spectrophotometric determination of lipases, lysophospholipases, and phospholipases. *J. Lipid Res.* 25:1555–1562.

162. Bisogno, T. et al. (1997) Biosynthesis, release and degradation of the novel endogenous cannabimimetic metabolite 2-arachidonoylglycerol in mouse neuroblastoma cells. *Biochem. J.* 322:671–677.

163. Bisogno, T. et al. (2006) Development of the first potent and specific inhibitors of endocannabinoid biosynthesis. *Biochim. Biophys. Acta* 1761:205–212.

164. Rouzer, C.A., Ghebreselasie, K., and Marnett, L.J. (2002) Chemical stability of 2-arachidonylglycerol under biological conditions. *Chem. Phys. Lipids* 119:69–82.

165. Saario, S.M. et al. (2004) Monoglyceride lipase-like enzymatic activity is responsible for hydrolysis of 2-arachidonoylglycerol in rat cerebellar membranes. *Biochem. Pharmacol.* 67:1381–1387.

166. Ghafouri, N. et al. (2004) Inhibition of monoacylglycerol lipase and fatty acid amide hydrolase by analogues of 2-arachidonoylglycerol. *Br. J. Pharmacol.* 143:774–784.

167. Tornqvist, P. et al. (1976) Purification and some properties of a monoacylglycerol-hydrolysing enzyme of rat adipose tissue. *J. Biol. Chem.* 251:813–819.

168. Vandevoorde, S. et al. (2005) Influence of the degree of unsaturation of the acyl side chain upon the interaction of analogues of 1-arachidonoylglycerol with monoacylglycerol lipase and fatty acid amide hydrolase. *Biochem. Biophys. Res. Commun.* 337:104–109.

169. Vandevoorde, S. and Fowler, C.J. (2005) Inhibition of fatty acid amide hydrolase and monoacylglycerol lipase by the anandamide uptake inhibitor VDM11: Evidence that VDM11 acts as an FAAH substrate. *Br. J. Pharmacol.* 7:885–893.

170. Vandevoorde, S. et al. (2007) Lack of selectivity of URB602 for 2-oleoylglycerol compared to anandamide hydrolysis *in vitro. Br. J Pharmacol.* 2:186–191.

171. Laval, G. et al. (2007) Disulfiram is an inhibitor of human purified monoacylglycerol lipase, the enzyme regulating 2-arachidonoylglycerol signaling. *Chembiochem* 11:1293–1297.

172. Savinainen, J.R. et al. (2001) Despite substantial degradation, 2-arachidonoylglycerol is a potent full efficacy agonist mediating Cβ(1) receptor-dependent G-protein activation in rat cerebellar membranes. *Br. J. Pharmacol.* 134:664–672.

173. Saario, S.M. et al. (2005) Characterization of the sulfhydryl-sensitive site in the enzyme responsible for hydrolysis of 2-arachidonoyl-glycerol in rat cerebellar membranes. *Chem. Biol.* 12:649–656.

174. Saario, S.M. et al. (2006) URB754 has no effect on the hydrolysis or signaling capacity of 2-AG in the rat brain. *Chem. Biol.* 8:811–814.

175. Ueda, N., Okamoto, Y., and Morishita, J. (2005) *N*-acylphosphatidylethanolamine-hydrolyzing phospholipase D: A novel enzyme of the β-lactamase fold family releasing anandamide and other *N*-acylethanolamines. *Life Sci.* 77:1750–1758.

176. Tsou, K. et al. (1998) Fatty acid amide hydrolase is located preferentially in large neurons in the rat central nervous system as revealed by immunohistochemistry. *Neurosci. Lett.* 254:137–140.

177. Van Sickle, M.D. et al. (2001) Cannabinoids inhibit emesis through Cβ1 receptors in the brainstem of the ferret. *Gastroenterology* 121:767–774.

178. Szabo, B. et al. (2006) Depolarization-induced retrograde synaptic inhibition in the mouse cerebellar cortex is mediated by 2-arachidonoylglycerol. *J. Physiol.* 577:263–280.

179. Starowicz, K.M. et al. (2008) Endocannabinoid dysregulation in the pancreas and adipose tissue of mice fed with a high-fat diet. *Obesity* 16:553–565.

180. Egertova, M. et al. (1998) A new perspective on cannabinoid signalling: Complementary localization of fatty acid amide hydrolase and the CB1 receptor in rat brain. *Proc. Biol. Sci.* 265:2081–2085.
181. Egertova, M., Cravatt, B.F., and Elphick, M.R. (2003) Comparative analysis of fatty acid amide hydrolase and Cβ(1) cannabinoid receptor expression in the mouse brain: Evidence of a widespread role for fatty acid amide hydrolase in regulation of endocannabinoid signaling. *Neuroscience* 119:481–496.
182. Egertova, M. et al. (2004) Fatty acid amide hydrolase in brain ventricular epithelium: Mutually exclusive patterns of expression in mouse and rat. *J. Chem. Neuroanat.* 28:171–181.
183. Starowicz, K. et al. (2006) Endovanilloid degradative enzymes: Comparative analysis of FAAH, COMT and TRPV1 expression in the mouse brain. In *16th Annual Symposium on the Cannabinoids*, Burlington, VT, International Cannabinoid Research Society, p. 114.
184. Cristino, L. et al. (2006) Immunohistochemical localization of cannabinoid type 1 and vanilloid transient receptor potential vanilloid type 1 receptors in the mouse brain. *Neuroscience* 139:1405–1415.
185. Yazulla, S. et al. (1999) Immunocytochemical localization of cannabinoid Cβ1 receptor and fatty acid amide hydrolase in rat retina. *J. Comp. Neurol.* 415:80–90.
186. Glaser, S.T. et al. (2005) Endocannabinoids in the intact retina: 3 H-anandamide uptake, fatty acid amide hydrolase immunoreactivity and hydrolysis of anandamide. *Vis. Neurosci.* 6:693–705.
187. Parker, R.M. and Barnes, N.M. (1999) mRNA: Detection by *in situ* and northern hybridization. *Methods Mol. Biol.* 106:247–283.
188. Hod, Y. (1992) A simplified ribonuclease protection assay. *Biotechniques* 13:852–854.
189. Saccomanno, C.F. et al. (1992) A faster ribonuclease protection assay. *Biotechniques* 13:846–850.
190. Weis, J.H. et al. (1992) Detection of rare mRNAs via quantitative RT-PCR. *Trends Genet.* 8:263–264.
191. Bucher, P. (1999) Regulatory elements and expression profiles. *Curr. Opin. Struct. Biol.* 9:400–407.
192. Melton, D.A. et al. (1984) Efficient *in vitro* synthesis of biologically active RNA and RNA hybridization probes from plasmids containing a bacteriophage SP6 promoter. *Nucleic Acids Res.* 12:7035–7056.
193. Morrison, T.B., Weis, J.J., and Wittwer, C.T. (1998) Quantification of low-copy transcripts by continuous SYBR Green I monitoring during amplification. *Biotechniques* 24:954–962.
194. Karge, W.H., III, Schaefer, E.J., and Ordovas, J.M. (1998) Quantification of mRNA by polymerase chain reaction (PCR) using an internal standard and a nonradioactive detection method. *Methods Mol. Biol.* 110:43–61.
195. Bustin, S.A. (2002) Quantification of mRNA using real-time reverse transcription PCR (RT-PCR): Trends and problems. *J. Mol. Endocrinol.* 29:23–39.
196. Curtiss, N.P. et al. (2005) Isolation and analysis of candidate myeloid tumor suppressor genes from a commonly deleted segment of 7q22. *Genomics* 85:600–607.
197. Spoto, B. et al. (2006) Human adipose tissue binds and metabolizes the endocannabinoids anandamide and 2-arachidonoylglycerol. *Biochimie* 88:1889–1897.
198. Guo, Y. et al. (2005) *N*-acylphosphatidylethanolamine-hydrolyzing phospholipase D is an important determinant of uterine anandamide levels during implantation. *J. Biol. Chem.* 280:23429–23432.
199. Morishita, J. et al. (2005) Regional distribution and age-dependent expression of *N*-acylphosphatidylethanolamine-hydrolyzing phospholipase D in rat brain. *J. Neurochem.* 94:753–762.

200. Hansson, A.C. et al. (2007) Genetic impairment of frontocortical endocannabinoid degradation and high alcohol preference. *Neuropsychopharmacology* 32:117–126.
201. Strausberg, R.L. et al. (2002) Generation and initial analysis of more than 15,000 full-length human and mouse cDNA sequences. *Proc. Natl. Acad. Sci. USA* 99:16899–16903.
202. Gjerstorff, M.F. et al. (2006) Identification of genes with altered expression in medullary breast cancer vs. ductal breast cancer and normal breast epithelia. *Int. J. Oncol.* 28:1327–1335.
203. Jung, K.M. et al. (2007) A key role for diacylglycerol lipase-α in metabotropic glutamate receptor-dependent endocannabinoid mobilization. *Mol. Pharmacol.* 72:612–621.
204. Giang, D.K. and Cravatt, B.F. (1997) Molecular characterization of human and mouse fatty acid amide hydrolases. *Proc. Natl. Acad. Sci. USA* 94:2238–2242.
205. Helliwell, R.J. et al. (2004) Characterization of the endocannabinoid system in early human pregnancy. *J. Clin. Endocrinol. Metab.* 89:5168–5174.
206. Engeli, S. et al. (2005) Activation of the peripheral endocannabinoid system in human obesity. *Diabetes* 54:2838–2843.
207. Bluher, M. et al. (2006) Dysregulation of the peripheral and adipose tissue endocannabinoid system in human abdominal obesity. *Diabetes* 55:3053–3060.
208. Tanaka, M. et al. (2007) The mRNA expression of fatty acid amide hydrolase in human whole blood correlates with sepsis. *J. Endotoxin Res.* 13:35–38.
209. Siegmund, S.V. et al. (2006) Fatty acid amide hydrolase determines anandamide-induced cell death in the liver. *J. Biol. Chem.* 281:10431–10438.

Part II

Mass Spectrometry Methods for Lipid Analysis

6 Mass Spectrometry Methods for the Analysis of Lipid Molecular Species: A Shotgun Lipidomics Approach

Xianlin Han, Hua Cheng, Xuntian Jiang, and Youchun Zeng

CONTENTS

6.1   INTRODUCTION TO MULTIDIMENSIONAL MASS SPECTROMETRY–BASED SHOTGUN LIPIDOMICS

Lipidomics or metabolomics of lipids, defined as the large-scale study of cellular lipids, is a rapidly expanding research field [1–3]. Although it has emerged as a distinct field only since the past few years [1–3], numerous new discoveries and advances have already been made [3–12]. Its essential roles in identifying the biochemical mechanisms of lipid metabolism, investigating the functions of an individual gene of interest, identifying novel biomarkers, and evaluating drug efficacy among others are becoming increasingly appreciated. One important task in lipidomics is the large-scale identification and quantitation of individual lipid molecular species in each cellular lipidome.

One of the major new developments in current lipidomics is the multidimensional mass spectrometry (MS)-based shotgun lipidomics [13–15]. This approach comprises multiple simple steps to maximally exploit differences in the physical and chemical properties of the classes of lipids present in biological samples to maximize the coverage and dynamic range of analyses. These steps include (1) the multiplexed extractions for sample preparation, (2) intrasource separation to resolve lipid classes based on their electrical properties, (3) the multidimensional MS identification of individual lipid molecular species, and (4) two-step quantitation processing with internal standards to extend the linear dynamic range as described previously [7]. The work flow of these steps is schematically illustrated (Scheme 6.1).

In brief, the lipids of each biological sample (commonly containing 100–2000 µg of protein content from cell, tissue, or biologic fluid) can be extracted by solvent(s) under acidic, basic, and/or neutral conditions (i.e., multiplexed extractions). The differential solubility of the different lipid classes in various solvents under varying pH conditions is exploited in this step, which is a critical step to maximally separate and enrich the lipid class(es) of interest. For example, many lipid classes (e.g., sphingosine-1-phosphate, lysophosphatidic acid (lysoPtdOH), acylcarnitine, etc.) can be efficiently extracted under acidic conditions [4]. Gangliosides and acyl-CoA are highly soluble in polar solvents and are partitioned into the aqueous phase during chloroform extraction [16–18]. Thus, these lipid classes can be reverse-extracted by

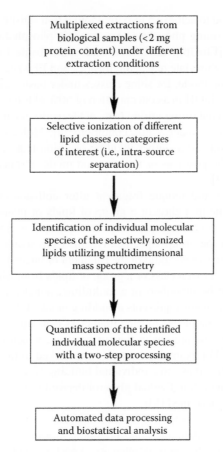

SCHEME 6.1 A work flow of multidimensional mass spectrometry–based shotgun lipidomics.

using butanol or other solvents under acidic conditions. Moreover, very hydrophobic lipid classes (e.g., cholesterol and its esters, triacylglycerol [TAG], free fatty acid [FFA], etc.) can be extracted and enriched with hexane. Fluorenylmethoxylcarbonyl (Fmoc) chloride can be added to quickly tag the amine-containing lipids and increase the sensitivity for the analysis of these lipids through neutral-loss (NL) scanning of the tagged Fmoc moiety [19].

The electrospray ion source behaves like an electrophoretic cell and can selectively separate different charged moieties under high electrical potential (typically ~4 kV) [20,21]. Since different lipid classes possess different electrical properties, largely depending on the nature of their polar head groups [1,4], the electrospray ion source can be used to resolve lipid classes in a crude lipid extract based upon the intrinsic electrical properties of each lipid class (now termed "intrasource separation of lipids") [4,13,15]. In shotgun lipidomics, the differential acidic or basic properties of lipid classes in a lipid solution prepared under a specific pH condition are exploited to selectively ionize different lipid classes in the positive- or negative-ion modes and to achieve a maximal ionization sensitivity [22]. Therefore, the lipid classes containing phosphate (e.g., anionic phospholipids, ethanolamine glycerophospholipid [EtnGpl], acyl-CoA, and sphingosine-1-phosphate), sulfate (e.g., sulfatide), and carboxylate (e.g., gangliosides and FFA) can be selectively ionized in the negative-ion mode, for some classes under basic conditions (i.e., in the presence of NH_4OH or LiOH in a concentration of 50% of total lipid concentration), whereas lipid classes containing amine (e.g., acylcarnitine) can be readily ionized in the positive-ion mode under acidic conditions [4]. Molecular species of other lipid classes can be ionized as either alkaline or anion (e.g., chloride, acetate, or formate) adducts in the positive- or negative-ion mode, respectively, as discussed previously in details [4].

Finding a sensitive and unique fragment after collision-induced dissociation (CID), which is specific to a class or a group of lipids of interest, is the third key step for successfully identifying, profiling, and quantifying individual molecular species in the class or the group. Specifically, either NL scanning or precursor-ion (PI) scanning at the mass or m/z ratio of the fragment of interest, respectively, can be performed to "isolate" a given class or a group of lipids from which individual lipid molecular species can be identified in a multidimensional array analysis fashion. Here, each of these fragments represents a building block of the class or the group of lipids, and all the building blocks of each lipid class constitute an additional dimension to the molecular ions present in the survey scan, which is referred to as the first dimension [7,13]. For example, three moieties linked to the hydroxyl groups of glycerol can be recognized as three individual building blocks, and if each building block is identified then each individual glycerol-derived lipid molecular species in a given sample can be determined [13].

Finally, quantitation by shotgun lipidomics is performed in a two-step procedure [13,23,24]. First, the abundant and nonoverlapping molecular species of a class are quantified by comparing the ion peak intensity of each individual identified molecular species with that of the preselected internal standard of the class after ^{13}C deisotoping [4,14] from a survey scan. Next, some or all of these determined molecular species of the class (plus the preselected internal standard) are used as standards to determine the content of other low-abundance or overlapping molecular species using one or multiple NL and/or PI scans, which are specific to the building blocks (e.g., head group) of the lipid class of interest [4]. Multiple standards are necessary in this second step since the fragmentation kinetics of different molecular species may be different [25,26]. It should be pointed out that such an approach by using tandem MS spectrum along with at least two internal standards for quantitation has been broadly employed in the field [26–29]. Through this second step in the quantitation process, the linear dynamic range of quantitation can be dramatically extended by eliminating background noise and by filtering the overlapping molecular species through a multidimensional mass spectrometric approach [4].

Through lipid class-selective intrasource ionization and subsequent two-dimensional MS analyses, shotgun lipidomics, at its current stage, enables us to fingerprint and quantify individual molecular species of most major and many minor lipid classes in cellular lipidomes, which collectively represent >95% of the total lipid mass (composed of as many as 2000 molecular species), directly from their $CHCl_3$

extracts. These classes of lipids include choline glycerophospholipid (ChoGpl), EtnGpl, phosphatidylinositol (PtdIns), phosphatidylglycerol (PtdGro), phosphatidyl-serine (PtdSer), phosphatidic acid (PtdOH), sphingomyelin (CerPCho), monohexosyl-ceramide (i.e., galactosylceramide and/or glucosylceramide, HexCer), sulfatide, FFA, TAG, lysoChoGpl, lysoEtnGpl, lysoPtdOH, acylcarnitine, cholesterol and cholesteryl esters, and ceramide (Cer) (including dihydroceramide). Cardiolipin (Ptd$_2$Gro) [30], sphingosine-1-phosphate/dihydrosphingosine-1-phosphate [31], sphingosine, psycho-sine, and lysoCerPCho [32] are the newest lipid classes that have been added to the list of lipids identified and quantified by shotgun lipidomics. In this chapter, the pro-tocols for identification and quantitation of some representative lipid classes of mouse brain samples are described. However, these methods are readily applicable to lipids extracted from other biological samples.

6.2 SAMPLE PREPARATION IN SHOTGUN LIPIDOMICS

6.2.1 MATERIALS

Synthetic phospholipids including 1,2-dimyristoleoyl-*sn*-glycero-3-phosphocholine (14:1–14:1 PtdCho), 1,2-dipalmitoleyl-*sn*-glycero-3-phosphoethanolamine (16:1–16:1 PtdEtn), 1,2-dipentadecanoyl-*sn*-glycero-3-phosphoglycerol (15:0–15:0 PtdGro), 1,2-dipentadecanoyl-*sn*-glycero-3-phosphoserine (15:0–15:0 PtdSer), lauroryl sphin-gomyelin (C12:0 CerPCho), and heptadecanoyl ceramide (C17:0 Cer) were purchased from Avanti Polar Lipids, Inc. (Alabaster, Alabama). Semisynthetic palmitoyl sul-fatide (C16:0 sulfatide) and pentadecanoyl GalCer (C15:0 GalCer) were obtained from Matreya, Inc. (Pleasant Gap, Pennsylvania). Deuterated cholesterol (d$_7$-cholesterol) was purchased from Cambridge Isotope Laboratories, Inc. (Cambridge, Massachusetts). All the solvents were obtained from Burdick and Jackson (Honeywell International Inc., Burdick and Jackson, Muskegon, Michigan). All other chemicals were purchased from Sigma-Aldrich (St. Louis, Missouri).

6.2.2 INTERNAL STANDARDS AND GENERAL LIPID EXTRACTION

Mice (male, C57 BL/J background, 4 months of age) were purchased from The Jackson Laboratory (Bar Harbor, Maine). All animal procedures were performed in accordance with the Guide for the Care and Use of Laboratory Animals and were approved by the Animals Studies Committee at Washington University. Mice were killed by asphyxiation with CO$_2$. Mouse brain tissues (including cortex, cerebellum, brain stem, and spinal cord or as wished) were dissected quickly. The collected brain tissues were immediately freeze-clamped at the temperature of liquid nitrogen. The tissue wafers were pulverized into a fine powder with a stainless-steel mortar and pestle cold to liquid nitrogen temperatures.

A sample (approximately 10 mg of tissue or equivalent to that containing approximately 1 mg of protein) from each dissected brain tissue was weighed and homogenized in 0.3 mL of ice-cold, diluted (10×) phosphate-buffered saline with a Potter-Elvehjem tissue grinder. Protein assays on each individual homogenate were performed using a bicinchoninic acid protein assay kit (Pierce, Rockford, Illinois) with bovine serum albumin as standard. After a certain volume of homogenate (~0.25 mL) of each sample was transferred to a disposable culture borosilicate glass tube (16 × 100 mm), 3 mL of $CHCl_3$/MeOH (1:1, by vol.) and proper volume of LiCl solution (50 mM) to make up 1.4 mL of final LiCl solution were added to each individual test tube. Internal standards (all in nmol/mg of protein) in a premixed solution for global lipid analysis including 15:0–15:0 PtdGro (3), 14:1–14:1 PtdCho (25), 16:1–16:1 PtdEtn (22.5), C12:0 CerPCho (2.5), C16:0 sulfatide (5.0), C15:0 GalCer (10), C17:0 Cer (0.8), and d_7-cholesterol (165) as well as the necessary internal standards for quantitation of individual molecular species of other lipid classes were also added to each brain tissue sample based on protein content prior to the extraction of lipids. For the myelin-dominated brain tissue samples (e.g., brain stem and spinal cord), 1.5 times of internal standards described above were added. These internal standards allowed us to normalize the final quantified lipid content to the protein content present and to eliminate the effects of potential losses from incomplete sample recovery during processing and were selected because they represent ≪0.1% of the endogenous cellular lipid content as demonstrated by electrospray ionization/ mass spectrometry (ESI/MS) lipid analysis.

The extraction mixtures were vortexed and centrifuged at 2500 rpm for 5 min. The $CHCl_3$ layer of each extract mixture was carefully removed and saved. An additional 1.5 mL of $CHCl_3$ was added into the MeOH/aqueous layer of each test tube. After centrifugation, the $CHCl_3$ layer was removed from each individual sample and combined with the first $CHCl_3$ fraction and then dried under a nitrogen stream. Each individual residue was then resuspended in 3 mL of $CHCl_3$/MeOH (1:1, by vol.), re-extracted against 1.4 mL of LiCl aqueous solution (10 mM) as previously described [4], and the extract was dried as described above. Each individual residue was resuspended in ~1 mL of chloroform and filtered with a 0.2 μm polytetrafluoroethylene (PTFE) syringe filter. Finally, each individual residue was reconstituted with a volume of 200 μL/mg of protein (the original protein content of the samples as determined from protein assays) in $CHCl_3$/MeOH (1:1, by vol.). The lipid extracts were finally flushed with nitrogen, capped with Teflon coated caps, and stored at −20°C for ESI/MS analyses (typically conducted within 1 week).

6.2.3 SAMPLE PREPARATION FOR THE ANALYSES OF COMMON BRAIN SPHINGOLIPIDS

For analysis of sphingolipids, a small portion of each individual lipid extract (20 μL, which is equivalent to the lipid extract from a tissue sample containing 0.1 mg of protein) as prepared above is transferred to a conical centrifuge glass test tube and the solvent is evaporated under the stream of nitrogen. A small volume of ice-cold LiOMe solution (1 M, 50 μL) in MeOH is added to the test tube at 0°C. The reaction mixture is vortexed for 15 s, kept in an ice bath for 1 h, and then quenched with 2 mL of 0.4% acetic acid solution. The pH of the quenched reaction solution should be adjusted to 4–5 by the addition of acetic acid, if necessary. The aqueous phase is washed with hexane (2 mL × 3 times), and the hexane solution is discarded. The lipids in the aqueous phase are extracted by the modified Bligh and Dyer method as described above. The combined $CHCl_3$ phase is dried under a stream of nitrogen. The residue is then re-extracted with a 10 mM aqueous LiCl solution. Each individual extract is reconstituted in 100 μL of $CHCl_3$/MeOH (1:1, by vol.) after filtering through a 0.2 μm polytetrafluoroethylene membrane filter, flushed with nitrogen, capped, and stored at −20°C for ESI/MS analysis of sphingolipids as described previously [32].

6.2.4   Sample Preparation for Quantitation of Cholesterol and Cholesteryl Esters

An aliquot (10 μL) of a stock solution of mixed dimethyl aminopyridine (DMAP, 1 M) and methoxyacetic acid (1 M) in CH_2Cl_2 and 20 μL of a stock solution of 1-(3-dimethyl aminopropyl)-3-ethylcarbodiimide in CH_2Cl_2 (0.142 M) are sequentially added to a brain lipid extract (10 μL of an individual lipid extract as prepared above). The mixed solution (made to a final volume of 100 μL with CH_2Cl_2) is stirred at room temperature overnight (approximately 14 h) to derivatize the cholesterol with methoxyacetic acid. To each reaction mixture, 1 mL of milli-Q purified water is added prior to extraction of derivatized cholesterol with ethyl ether (2 mL × 3 times). The combined organic extract was filtered through a 0.2 μm polytetrafluoroethylene syringe filter and evaporated under a nitrogen stream. Each individual residue was resuspended in 1 mL of $CHCl_3$/MeOH (1:1, v/v), which is further diluted 10-fold prior to MS analysis. It should be pointed out that other alternative derivative methods can also be employed for the purpose of cholesterol quantitation [33,34].

For quantitation of total esterified cholesterols, cholesteryl esters were first converted to free cholesterol in the presence of the selected internal standard (d_7-cholesterol). Specifically, a brain lipid extract (20 μL of the lipid extract as prepared above) is dissolved in tetrahydrofuran (25 μL), followed by the addition of working buffer (0.5% (w/v) taurocholate in phosphate buffer, pH 7.0, 1 mL). The mixture is sonicated for 10 min followed by the addition of cholesterol esterase stock solution (200 U/mL working buffer, 10 μL). The resulting mixture is incubated at 37°C overnight (about 16 h), cooled to room temperature, and then extracted with ethyl ether (2 mL × 3 times). The combined organic extract was evaporated under a nitrogen stream, and the residue containing cholesterol and d_7-cholesterol was directly derivatized with methoxyacetic acid and quantitated as described above.

6.2.5   Mass Spectrometry and Parameter Settings

A triple-quadrupole mass spectrometer (ThermoElectron TSQ Quantum Ultra, San Jose, California) equipped with an electrospray ion source and Xcalibur system software was utilized in the study under conditions as previously described [15]. The first and third quadrupoles serve as independent mass analyzers using a mass resolution setting of peak width 0.7 Th while the second quadrupole serves as a collision cell for tandem mass spectrometry. The diluted lipid extract is directly infused into the ESI source at a flow rate of 4 μL/min with a syringe pump. Typically, a 1 min period of signal averaging in the profile mode was employed for each MS spectrum. For tandem mass spectrometry, a collision gas (nitrogen) pressure was set at 1.0 mtorr but the collision energy varied with the classes of lipids as described previously [4,15]. Typically, a 2–5 min period of signal averaging in the profile mode is employed for each tandem MS spectrum. All the MS spectra and tandem MS spectra are automatically acquired by a customized sequence subroutine operated under Xcalibur software. Data processing of MS analyses, including ion peak selection, data transferring, peak intensity comparison, and quantitation, is conducted using self-programmed MicroSoft Excel macros [15].

6.3 ANALYSES OF INDIVIDUAL MOLECULAR SPECIES OF REPRESENTATIVE LIPID CLASSES

6.3.1 ANIONIC GLYCEROPHOSPHOLIPIDS

The lipid content of brain tissue samples varies from approximately 1200 nmol/mg of protein in cortex [35], where gray matter is predominant, to approximately 1800 nmol/mg of protein in spinal cord and PNS nerve bundles, where white matter is the major component [36]. Therefore, when the lipid extracts are reconstituted in 200 μL/mg of protein (see above for sample preparation), the lipid concentration is in the range of 5–10 nmol/μL. The highest concentration of lipids to have a minimal effect of lipid aggregation on quantitation is approximately 100 pmol/μL in 2:1 $CHCl_3$/MeOH, 50 pmol/μL in 1:1 $CHCl_3$/MeOH, and 10 pmol/μL in 1:2 $CHCl_3$/MeOH (see further discussion later). To avoid potential aggregation, brain lipid extracts are diluted 100- to 200-fold with 1:1 $CHCl_3$/MeOH for accurate analyses prior to direct infusion of a lipid solution into a mass spectrometer.

Negative-ion ESI/MS mass spectrum is acquired after direct infusion of a diluted brain lipid extract at a flow rate of 4 μL/min (Figure 6.1A). Since anionic lipids were selectively ionized in the negative-ion mode, the acquired mass spectrum displays abundant ion peaks that correspond to the molecular species of anionic lipids (e.g., PtdGro, PtdSer, PtdIns, and sulfatide) along with low abundance chloride adducts of neutral but polar lipids (e.g., ChoGpl) (Figure 6.1A). In this section, only the analyses of anionic phospholipids are described and analysis of sulfatide molecular species will be described separately.

The individual molecular species corresponding to the ion peaks displayed in Figure 6.1A are identified by two-dimensional ESI/MS analyses (Figure 6.1). The second dimension of this 2D spectrum represents the building blocks of phospholipid classes shown in Figure 6.1A. These building blocks represent either a specific component of the head group moiety of each class or the acyl chain(s) of each molecular species. Specifically, building blocks of NL87.0 (neutral loss [NL] of serine from PtdSer molecular species), PI153.1 (a glycerophosphate derived from anionic phospholipids), and PI241.2 (a phosphoinositol derivative ion arising from PtdIns molecular species) [4] identified the individual molecular species of PtdSer, anionic phospholipids, and PtdIns, respectively. In addition, the analysis of the NL50.0 building block (NL of chloromethane from molecular species containing a phosphocholine head group) shows the chloride adducts of ChoGpl or CerPCho molecular species, whose analysis will be described in Section 6.3.2. The acyl chain moieties

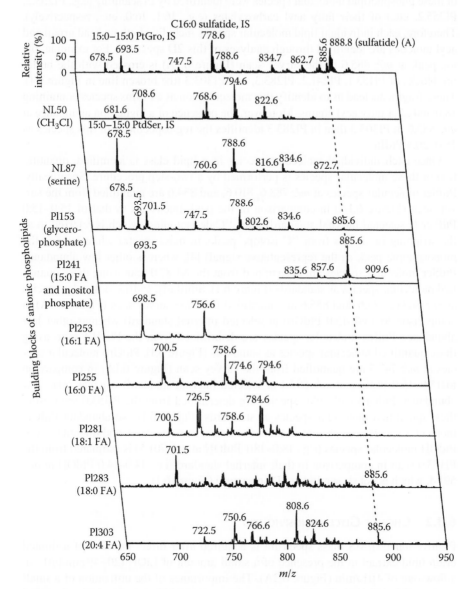

FIGURE 6.1 Representative 2D ESI/MS analyses of anionic phospholipids in a lipid extract of mouse cortex. The lipid extract is prepared by a modified Bligh and Dyer procedure and properly diluted prior to direct infusion into a mass spectrometer as described in the text. The survey scan or MS/MS scans of the 2D ESI mass spectrum are acquired in the negative-ion mode by sequentially programmed customized scans operating utilizing Xcalibur software. "PI" denotes precursor-ion scanning, "NL" stands for neutral loss scanning, and "IS" represents internal standard. All scans are displayed after normalization to the most intense peak (base peak) in each individual scan.

of these phospholipid molecular species were identified by PI scanning (e.g., PI253.2, PI255.2, etc.) of their fatty acyl carboxylates (e.g., 16:1, 16:0, etc., respectively). Therefore, each individual lipid molecular species including both its head group and acyl moieties are identified through analyses of this 2D spectrum. For example, the ion peak at m/z 885.6 in the survey scan (Figure 6.1A) is crossed with the building blocks of PI153.1, PI241.2, PI283.3, and PI303.3 (the broken line in Figure 6.1). These fragments lead us to identify this molecular ion as a PtdIns species containing 18:0 and 20:4 fatty acyl chains. Moreover, the presence of a higher ion abundance at m/z 885.6 in PI303.3 than in PI283.3 identifies the regiospecificity of this species as 18:0–20:4 PtdIns.

Once each individual molecular species in a lipid class is identified, quantitation of these molecular species is performed by a two-step procedure. Specifically, PtdSer molecular species at m/z 788.6, 810.6, and 834.6 are quantified from the survey scan (Figure 6.1A) in comparison of the peak intensities to that of 15:0–15:0 PtdSer (a selected internal standard) after ^{13}C de-isotoping (which is referred to as the stripping of signals from ^{13}C isotope peaks in mass spectra while leaving the monoisotopic peak as the representative signal) [4], whereas other low abundance PtdSer molecular species are determined from the NL87.0 scan using these quantified molecular species as standards (Figure 6.1). Similarly, PtdIns molecular species at m/z 863.6, 883.6, and 885.6 are quantified from the survey scan (Figure 6.1A) in comparison to 15:0–15:0 PtdGro (a selected internal standard) whereas other low abundance PtdIns molecular species are determined from the PI241.1 scan using these quantified molecular species as standards (Figure 6.1). PtdGro molecular species at m/z 747.5 are quantified from the survey scan (Figure 6.1A) in comparison to 15:0–15:0 PtdGro at m/z 693.4 (a selected internal standard) whereas other low abundance PtdGro molecular species are determined from the PI153.0 scan using these quantified molecular species as standards (Figure 6.1). No abundant PtdOH molecular species are present in the survey scan in some cases and quantitation of PtdOH molecular species (e.g., 18:0–18:1 PtdOH at m/z 701.5) is estimated from the PI153.0 scan in comparison to their internal standard (i.e., 14:0–14:0 PtdOH at m/z 591.4, which is not shown in the 2D spectrum) (Figure 6.1).

6.3.2 CHOLINE GLYCEROPHOSPHOLIPID

Positive-ion ESI/MS mass spectrum is acquired after direct infusion of a diluted brain lipid extract in the presence of a small amount of LiOH (20–40 pmol/μL) at a flow rate of 4 μL/min (Figure 6.2A). The importance of the utilization of a small amount of LiOH in the infusion solution has previously been extensively discussed [4]. The acquired mass spectrum displays abundant ion peaks that correspond to the lithiated ChoGpl and CerPCho molecular species in the mass region between m/z 650 and 850, whereas lithiated TAG molecular species are in the mass region between m/z 800 and 900 and lithiated HexCer molecular species are in the mass region between m/z 780 and 850, both in low abundance (Figure 6.2A). Only the analyses of ChoGpl molecular species are exemplified here, and other three lipid classes will be individually discussed below.

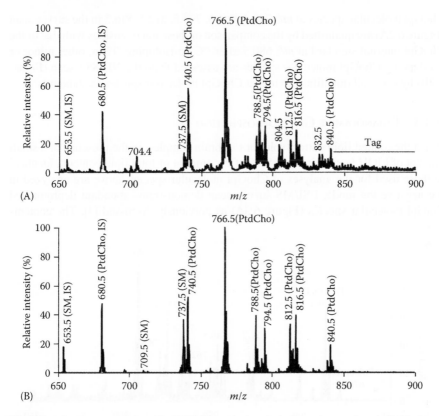

FIGURE 6.2 Representative 2D ESI/MS analysis of choline glycerophospholipids in a diluted lipid extract of mouse cortex in the positive-ion mode. A positive-ion ESI mass spectrum (Panel A) was acquired from a diluted mouse cortex lipid extract after direct infusion in the presence of LiOH. Analyses of the building blocks corresponding to ChoGpl in the second dimension (e.g., Panel B) were performed as described in the text. "IS" denotes internal standard. All mass spectral scans were displayed after normalization to the base peak in each individual scan.

Positive-ion ESI/MS survey scan after direct infusion of the diluted brain lipid extract in the presence of LiOH shows abundant ion peaks corresponding to lithiated ChoGpl molecular species (Figure 6.2A), which are identified through the characterization of the building blocks corresponding to ChoGpl head groups via NL of 183.1 u (i.e., phosphocholine) [4] (Figure 6.2B). Abundant lithiated CerPCho molecular species also appear in the NL183.1 scan, but in odd numbers of m/z according to "the nitrogen rule" [37], which are readily recognized and the effect of any CerPCho overlapping with a ChoGpl ion on the ChoGpl quantitation is minimized through ^{13}C de-isotoping. Identification of the fatty acyl chains of ChoGpl molecular species were repeatedly performed through NL of all potentially naturally occurring fatty acids (scans are not shown) in addition to the 2D MS analyses of chloride adducts of ChoGpl molecular species as shown in Figure 6.1. The identified, abundant lithiated

ChoGpl molecular species at m/z 740.5, 766.5, 794.5, 812.5, 816.5 in the survey scan (Figure 6.2A) are quantified by the comparison of these ion intensities with that of the ChoGpl internal standard at m/z 680.5 after ^{13}C de-isotoping. Then, other minor or overlapping ChoGpl molecular species are assessed from the NL183.1 scan (Figure 6.2B) by using all quantified abundant ChoGpl molecular species as standards.

6.3.3 ETHANOLAMINE GLYCEROPHOSPHOLIPID

When the diluted lipid extract solutions of brain samples, in the presence of a small amount of LiOH (approximately 50% of the amount of total lipid content), identical to those used for the analyses of ChoGpl molecular species above are analyzed in the negative-ion mode, ESI/MS survey scan demonstrates abundant deprotonated EtnGpl molecular species (Figure 6.3A) as previously discussed [4]. The contents

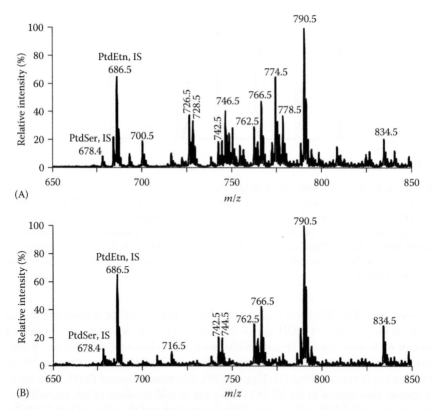

FIGURE 6.3 Representative ESI/MS analysis of ethanolamine glycerophospholipids in a diluted lipid extract of mouse cortex in the negative-ion mode. A mass spectrum (A) was acquired in the negative-ion mode from the diluted lipid extract of mouse cortex in the presence of approximately 25 pmol LiOH/μL. A mass spectrum (B) was taken in the negative-ion mode after the diluted lipid solution used in spectrum A was treated with acid vapor and a small amount of LiOH (approximately 25 pmol LiOH/μL) was added to the infused solution. "IS" denotes internal standard.

of molecular ions at m/z 700.5, 726.5, 746.5, 766.5, 774.5, 778.5, and 790.5 are determined by ratiometric comparison of these ion peak intensities with that of the selected internal standard (i.e., 16:1–16:1 PtdEtn at m/z 686.5) after ^{13}C de-isotoping. To distinguish PlsEtn molecular species from the potentially overlapped plasmenylethanolamine molecular species, a small portion (10 μL of the prepared lipid extract) of each lipid extract of a brain sample was treated with acidic vapor. Each residue was analyzed again in the negative-ion mode in the presence of LiOH (Figure 6.3B). This mass spectrum demonstrates that the major ion peaks around m/z 700.5, 726.5, 746.5, and 774.5 are all absent in Figure 6.3B, suggesting that these ions correspond to PlsEtn molecular species.

To identify individual EtnGpl molecular species in the lipid solution and quantify the low-abundance EtnGpl species, Fmoc derivatization is used [19]. After derivatization and MS analysis, NL222.1 (NL of the Fmoc moiety from Fmoc-EtnGpl molecular species) unambiguously demonstrates the presence of all EtnGpl molecular ions (Figure 6.4) with a dynamic range over four orders of magnitude [19]. The identities of the aliphatic chains of these molecular species were efficiently identified through a 2D ESI/MS analysis (Figure 6.4). The building blocks of the EtnGpl molecular ions included fatty acyl carboxylates and lysoPlsEtn derivatives (Figure 6.4). From these 2D MS analyses, many isomeric EtnGpl molecular species are readily identified. For example, the Fmoc-EtnGpl ion peak at m/z 922.6 (i.e., deprotonated EtnGpl molecular ion at m/z 700.5) comprises at least four PlsEtn molecular species (i.e., 16:0–18:1, 16:1–18:0, 18:0–16:1, and 18:1–16:0 PlsEtn) (see the left broken line in Figure 6.4). Quantitation of the low abundance ions displayed in NL222.1 of Figure 6.4 is carried out using those quantified EtnGpl molecular species (see above) as standards. The mouse brain lipidome is highly enriched with PlsEtn molecular species. Moreover, PlsEtn molecular species present in white matter predominantly contain monounsaturated fatty acids at the sn-2 position whereas PlsEtn molecular species containing polyunsaturated fatty acyl chains at the sn-2 position are predominant in the gray matter, which is similar to that found in human brain samples as previously described [38].

6.3.4 SULFATIDE

The content of sphingolipids in most biological samples is relatively lower than that of phospholipids, thus they only account for a few percent of total lipids in a lipid extract. However, the composition of sphingolipids in brain white matter samples can be as high as 30 mol% of total lipids in the lipid extracts (in which gangliosides are not included). Analyses of sphingolipids by shotgun lipidomics are much more challenging in comparison to phospholipids. However, with our new improvement made in sample preparation for sphingolipid analyses [32] as described above, the capability of sphingolipid analyses by shotgun lipidomics is comparable to the analyses of phospholipids and can penetrate into the much lower abundance molecular species of sphingolipids. Lipid samples prepared for sphingolipid analyses as described above are in the concentration of approximately 500 pmol/μL which should still be diluted to 10-fold before direct infusion for ESI/MS analyses of sphingolipids.

When this diluted lipid solution is analyzed by ESI/MS after direct infusion in the negative-ion mode, ESI mass spectrum displays very abundant ion peaks in the

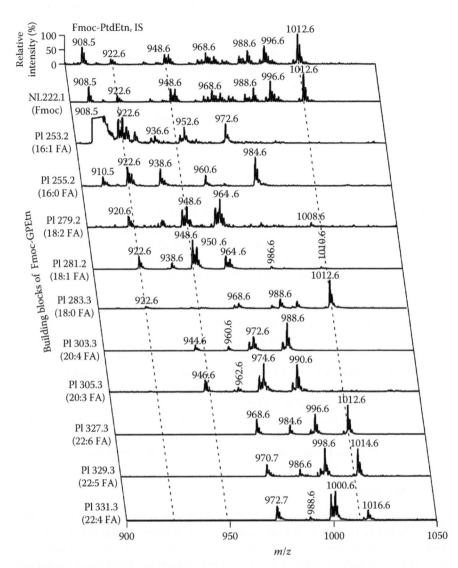

FIGURE 6.4 Representative negative-ion 2D ESI mass spectrum of a diluted chloroform extract of mouse cortex after treatment *in situ* with Fmoc chloride. An ESI mass spectrum of Fmoc-EtnGpl was acquired in the negative-ion mode directly from a diluted mouse cortex lipid extract after derivatization with Fmoc chloride [19]. Analyses of Fmoc-EtnGpl building blocks in the second dimension including the Fmoc moiety, fatty acyl carboxylates, and lysoPlsEtn ions by PI scanning and NL scanning were performed. "IS" denotes internal standard; *m:n* indicates an acyl chain containing *m* carbons and *n* double bonds; "FA" stands for fatty acyl chain. Each of the broken lines indicates the crossing peaks (fragmental ions) of a molecular ion with Fmoc-EtnGpl building blocks. All mass spectral scans were displayed after normalization to the base peak in each individual spectrum.

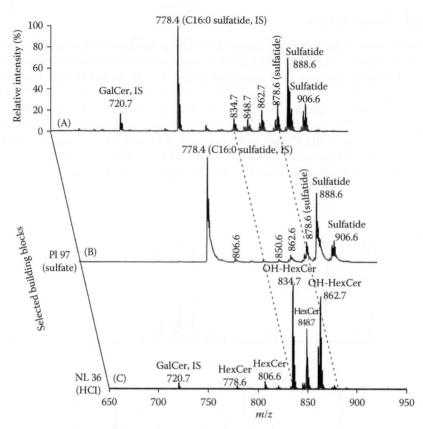

FIGURE 6.5 Shotgun lipidomics analyses of sphingolipid molecular species after treatment of a mouse cortex lipid extract with LiOMe in the negative-ion mode. Negative-ion ESI mass spectrum (A) is acquired directly from a lipid extract of mouse cortex after treatment with LiOMe. Tandem mass spectrometric analyses of the lipid extract after treatment with LiOMe are acquired by using PI scanning (B) of m/z 97 (i.e., sulfate) and NL scanning (C) of 36 u (loss of HCl from HexCer chloride adducts), respectively. "IS" denotes internal standard. Each spectrum displayed is normalized the base peak in the spectrum.

region of m/z 700–950 (Figure 6.5A). These ions correspond to deprotonated sulfatide molecular species which are identified by PI scanning of m/z 97.0 (i.e., sulfate) and HexCer chloride adducts which can be identified by NL scanning of 36 u (i.e., HCl) (Figure 6.5). The sphingoid bases of sulfatides with or without a hydroxyl group can be identified by product-ion analyses as previously described [39,40]. Thus, the acyl amide chains with or without a hydroxyl group at the α position can be derived from the identified sulfatide molecular species. After identification, sulfatide molecular species at m/z 878.6, 888.6, and 906.6 are determined from the survey scan (Figure 6.5A) whereas other low abundance or overlapped sulfatide molecular species are assessed from PI97.0 scan using these quantified molecular species as standards (Figure 6.5B). Other ion peaks present in the survey scan and those in the NL36.0 scan will be discussed in the analyses of HexCer molecular species.

6.3.5 SPHINGOMYELIN

When the diluted lipid solution (prepared for sphingolipid analysis) in the presence of a small amount of LiOH (40 pmol/μL) was analyzed by ESI/MS in the positive-ion mode, ESI mass spectrum shows abundant lithiated CerPCho and HexCer molecular species in the mass region of m/z 650–850 (Figure 6.6A). Lithiated CerPCho molecular species were identified by the NL183.1 (i.e., phosphocholine), which is specific to CerPCho molecular species in this mass region under the experimental conditions (Figure 6.6B). Comparison of this NL183.1 scan with the one shown in Figure 6.2 demonstrates that many additional CerPCho molecular species also occur in low abundance, which indicates that the new improvement in sample preparation

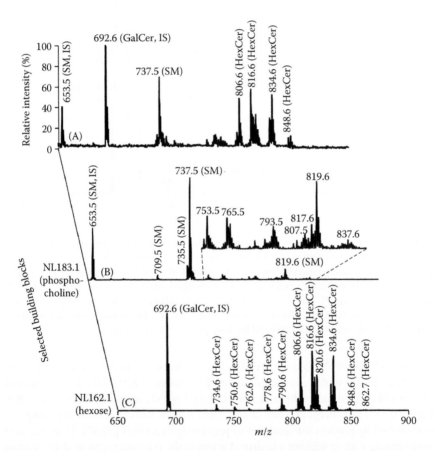

FIGURE 6.6 Shotgun lipidomics analyses of sphingolipid molecular species after treatment of a mouse cortex lipid extract with LiOMe in the positive-ion mode in the presence of a small amount of LiOH. The mass spectrum (Panel A) is acquired directly from a lipid extract of mouse cortex after treatment with LiOMe. The mass spectra (Panels B and C) are acquired from the NL scanning of 183.1 u (i.e., phosphocholine) and 162.1 u (i.e., monohexose), respectively. "IS" denotes internal standard. The ion peaks in mass spectrum B represent lithiated SM molecular species.

successfully leads to the penetrance of our shotgun lipidomics approach into the low abundance molecular species region. The sphingoid base of major CerPCho molecular species is determined as previously described [41]. The acyl amide moieties are derived based on the sphingoid base and *m/z* of each individual molecular species. The quantities of lithiated C18:1 CerPCho at *m/z* 735.5, C18:0 CerPCho at *m/z* 737.5, and C24:1 CerPCho at *m/z* 819.6 are accurately determined in comparison with that of C14:0 CerPCho at *m/z* 653.5 after ^{13}C de-isotoping (Figure 6.6A). The contents of all other low-abundance molecular species of CerPCho (see inset in Figure 6.6B) are determined through the second step of quantitation by using C14:0, C18:0, and C24:1 CerPCho as internal standards.

6.3.6 HEXOSYLCERAMIDE

Positive-ion ESI mass spectrum of the diluted lipid solution for sphingolipid analyses in the presence of LiOH (Figure 6.6A) also shows lithiated HexCer molecular species, which are identified by NL162.1 (corresponding to monohexose) (Figure 6.6C) [42]. Moreover, the molecular species of HexCer containing a hydroxyl group can be readily distinguished from nonhydroxy HexCer molecular species since the peak intensity ratios of the hydroxy to nonhydroxy HexCer molecular species in the NL36.0 scan of lipid solution in the negative-ion mode are enhanced three times in comparison to those in the survey scan [42]. Accordingly, identification of hydroxy HexCer molecular species is achieved by the identification of the enhanced ion peaks in the NL36.0 scan (Figure 6.5C) in comparison to the survey scan (Figure 6.5A) of a lipid solution (prepared for sphingolipid analyses) in the negative-ion mode. For example, the ion clusters at *m/z* 806.6, 834.7, and 862.7 are hydroxy HexCer molecular species whereas other ion clusters are nonhydroxy ones.

A large number of HexCer molecular species could be quantified in the first step of the quantitation procedure from the survey scan (e.g., ions at *m/z* 806.6, 816.6, 820.6, and 834.6) in comparison of their ion intensities to that of an internal standard (C15:0 GalCer at *m/z* 692.6) after ^{13}C de-isotoping (Figure 6.6A). The contents of other HexCer molecular species are determined from the NL162.2 scan (Figure 6.6C) by ratiometric comparison of each of their peak intensities to those of the quantified HexCer molecular species. The monohexose moiety of HexCer may be determined by profile analyses of HexCer chloride adducts as previously described [42]. Alternatively, the composition of galactosylceramide and glucosylceramide in the determined HexCer can be assessed through a conventional TLC approach [43].

6.3.7 CERAMIDE

Since the Cer content in biological samples is quite low, assessment of Cer content is heavily dependent on tandem MS spectra which can be carried out in the lipid solution with or without the treatment of LiOMe as described previously [15,44]. Specifically, we have found that to the sphingosine-based Cer molecular species, NL scan of 256.2 u is sensitive to those without a hydroxyl group at the α position, NL scan of 327.3 u is specific to those containing a hydroxyl group at the α position, and NL scan of 240.2 u is essentially equally sensitive to all Cer molecular

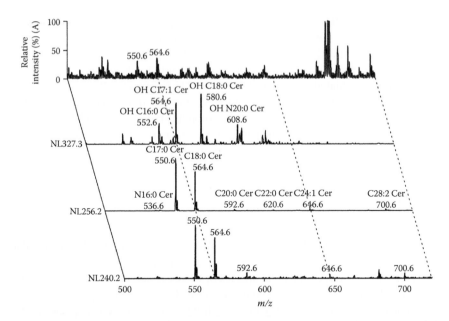

FIGURE 6.7 Representative shotgun lipidomics analysis of ceramide (Cer) molecular species in a lipid extract of mouse cortex in the presence of a small amount of LiOH in the negative-ion mode. A survey scan (A) in the negative-ion mode is acquired prior to the analysis of the building blocks of Cer molecular species by NL scanning. "IS" denotes internal standard. All mass spectral scans are displayed after normalization to the base peak in each individual spectrum.

species in both subclasses within experimental errors [44]. Therefore, NL256.2 and NL327.3 scans are used to identify sphingosine-based Cer molecular species with or without a hydroxyl group at the α position whereas NL240.2 scan was used to assess the contents of these molecular species (Figure 6.7). To the sphinganine or other sphingoid base-containing Cer molecular species, other sets of building blocks were respectively analyzed for identification and quantitation. For example, Cer molecular species containing sphinganine are identified and quantified by analyses of NL258.2, NL329.3, and NL242.2 scans whereas identification and quantification of Cer molecular species containing a sphingoid base with 20 carbons are conducted by analyses of NL284.3, NL355.3, and NL268.2 scans as previously demonstrated in human brain samples [7]. However, the contents of these sphinganine-containing and C20-sphingoid base-containing Cer molecular species in mouse brain samples are very minor.

6.3.8 Cholesterol and Cholesteryl Esters

To determine the contents of cholesterol and its esters in lipid extracts of brain tissue samples, a small portion of individual lipid extract is derivatized as described in Section 6.2. The diluted solution after derivatization (see above) in the presence of a small amount of LiOH (50 pmol/μL) was directly infused to the ESI mass

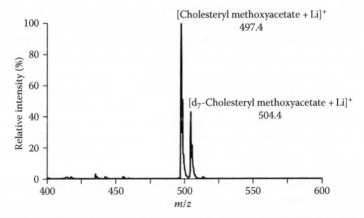

FIGURE 6.8 Quantitation of cholesterol in mouse cortex lipidomes by ESI/MS/MS analyses after derivatization. Mouse cortex lipid extracts were prepared and each of the diluted lipid extract solutions was individually modified with methoxyacetic acid. The reaction-workup solution of each lipid extract was analyzed in the positive-ion mode in the presence of LiOH. PI scanning of m/z 97.1 was acquired in the positive-ion mode and used for quantitation of cholesterol in the lipid extract.

spectrometer by a Harvard syringe pump at a flow rate of 4 μL/min. A PI scan of m/z 97.1 (corresponding to a fragment ion of [methoxyacetate + MeOH + Li]$^+$, resulting from the cholesterol derivative) was acquired in the positive-ion mode (Figure 6.8) as previously described [36]. Quantitative analysis was achieved by direct comparison of the peak intensity of cholesterol derivative to that of the internal standard derivative from the PI scan. The content of total esterified cholesterol in the lipid extracts of brain samples was obtained by subtraction of the content of free cholesterol from that of total cholesterol in the sample as previously described [36].

6.4 KEYNOTES FOR ACCURATE LIPID ANALYSES BY SHOTGUN LIPIDOMICS

6.4.1 Careful Extraction of Lipids Is Essential for Successful Analyses of Lipids by Shotgun Lipidomics

Lipid extraction is the first key step to the successful analyses of cellular lipidomes by shotgun lipidomics since the presence of a residual amount of inorganic contaminants during the extraction has substantial impacts on the quality of lipid analyses by using this approach [4]. Great efforts are taken to minimize this amount of carryover from solvents and solutions used in sample preparation. An effective way to achieve this is to back-extract the lipid extracts with aqueous solution containing very low concentration of salts (e.g., 10 mM LiCl in most of our experiments) or volatile components (e.g., acetic acid or ammonium acetate). Alternatively, significant dilution of the prepared lipid extract prior to infusion of a lipid solution will also reduce the concentration of inorganic components.

Besides the role of extraction in the recovery and cleanup of lipids, proper extraction may help to separate different categories of lipids as outlined in the above. For example, nonpolar lipids such as cholesterol, cholesteryl esters, TAG, FFA (under acidic conditions), etc., can be effectively extracted using hexane/methanol solution, thereby achieving separation of these classes of lipids from other polar lipid classes. During the analyses of brain lipids, this method is successfully used to minimize the effects of cholesterol (which is abundantly present in neuronal lipid extracts) on the analyses of other polar lipids (e.g., phospholipids). In contrast, very polar lipids such as acyl-CoA and gangliosides are largely present in aqueous phase during solvent extraction of other less polar lipids and can be recovered by using either acetonitrile [18] or butanol [16] if necessary.

It is emphasized that although acidic conditions of the aqueous phase can improve the extraction efficiency of acidic lipids such as PtdOH, acyl-CoA, acyl carnitine, etc., plasmalogen molecular species are very labile under these conditions [36,45]. Thus, if analysis of plasmalogen molecular species (which are very abundant in brain lipid extracts) is included in a study, acidic conditions during sample preparation should be avoided. Similarly, if analysis of lysolipids is the primary goal of a study, acidic conditions during sample preparation should also be avoided since degradation of plasmalogen under acidic conditions yields *sn*-1 acyl lysophospholipids in which the fatty acid esterified at the *sn*-2 position can be readily isomerized to the *sn*-1 position. Detergents complicate the analyses of lipid molecular species by ESI/MS and thus, detergents should be avoided when possible. If detergents are used, great care needs to be exercised in removing as much of the added detergents as possible.

6.4.2 ANALYSIS IN THE LOW CONCENTRATION REGION IS THE KEY FOR ACCURATE AND SIMPLE QUANTIFICATION OF INDIVIDUAL LIPID MOLECULAR SPECIES

It is still debatable whether ionization efficiencies of individual molecular species of a polar lipid class are identical or different within experimental error. In our understanding, as also demonstrated by others [46], the answer to this question is dependent on the concentration. In other words, in the low concentration range as we use, the differences of the ionization efficiencies of different molecular species in a class are minimal and is comparable with the experimental errors of biological sample preparation after ^{13}C de-isotoping which will be discussed later. However, when the lipid concentration increases, lipid interactions also increase and lipids tend to form aggregates. As well known, lipid aggregation is a process that depends on the chemical properties of each individual molecular species including the polarity of the head group, the length of aliphatic chains, and the degree of unsaturation. Therefore, when analyses of lipids in the concentration region higher than the threshold of aggregation, ionization efficiencies of different molecular species become different since it is very difficult to ionize the lipid in its aggregate state and most of the aggregated particles will end up in the waste. Thus, as the lipid concentration increases, the ionization efficiency of lipid molecular species, particularly the ones containing long and saturated fatty acyl chains, dramatically decreases as previously demonstrated [46]. In this case, a standard curve for each individual molecular species becomes necessary.

Therefore, it should be emphasized that the lipid concentration in shotgun lipidomics as well as other ESI/MS techniques must be lower than that favoring aggregate formation to avoid this complication. The upper limits of the total lipid concentrations of a biological extract is approximately 100 pmol/μL in $CHCl_3$:MeOH 2:1, 50 pmol/μL (up to 100 for multidimension MS due to intrasource separation) in $CHCl_3$:MeOH 1:1, and 10 pmol/μL in $CHCl_3$:MeOH 1:2. These concentration limits may be varied in the presence of other modifiers in the solution as well as the composition of lipid extracts. Whatever the case, analysis of lipids in the low concentration region is always favored. The quick improvements in the sensitivity of mass spectrometers greatly facilitates the expansion of lipid analyses into low abundance molecular species as previously demonstrated [30–32].

6.4.3 THE CONTRIBUTION OF ^{13}C ISOTOPOLOGUES ON THE ANALYSES OF LIPID MOLECULAR SPECIES SHOULD BE RECOGNIZED

It is well known that the relative ion peak intensities shift from monoisotopologue to other ^{13}C isotopologues as the number of carbons of an analyte increases. Therefore, the differences between the distributions of ^{13}C isotopologues in different individual molecular species of a lipid class may greatly affect the accurate quantitation of the molecular species of interest in comparison to the selected internal standard. This factor has been more and more recognized [4,14,24,47,48], but has still not fully been considered. Indeed, there are two types of effects of ^{13}C isotopologues on lipid quantitation by using shotgun lipidomics as previously described [4,14]. The first type of effect is resulting from the differences of carbon numbers between the individual lipid molecular species of interest and the selected internal standard for these species. The second type of effect is resulting from M + 2 isotopologue of a molecular species on the quantitation of another molecular species in the same class which contains one less double bond. The magnitudes of these effects are varied and the correction factors of these effects can be calculated as described previously [4,14]. It should be pointed out that the isotopologues from other atoms such as ^{18}O, ^{15}N, and ^{2}H can be ignored in most cases in comparison to the inherent experimental errors and biological variability, but can be similarly considered if one would like to include these effects as well. An extensive discussion on this topic can be found in our review article [4].

6.4.4 AT LEAST ONE INTERNAL STANDARD FOR A CLASS OF LIPIDS IS ESSENTIAL IN MASS SPECTROMETRY–BASED LIPIDOMICS

The absolute intensity determined from many detectors (e.g., a UV detector or NMR) after calibration is directly proportional to the concentration of an analyte in the linear dynamic range which is so stable and less sensitive to the environment (e.g., the presence of "insert" substances) that independent researchers can essentially repeat a measurement. In contrast, the absolute ion counts of an analyte in the MS analysis depend on many factors including the sample preparation, the ionization conditions, the tuning conditions for the analyzer, and the analyzer and the detector used in the mass spectrometer, among others. Therefore, it is very difficult to reproduce the measurement of the absolute ion counts of an analyte (e.g., an individual lipid molecular species) at a certain concentration. However, if we include an internal standard for the lipid molecular species of interest, the reproducibility of the relative ratio of the lipid molecular species to its internal standard at the low concentration region as discussed above is very high. Therefore, it is always advised to determine the relative ratios of lipid molecular species with their selected internal standard(s). Thus, unexpected changes during lipid analyses are internally controlled. Any results obtained from sample analyses without any internal control are quite doubtful. The question is when the internal standard should be included. Some investigators like to determine the absolute content of individual lipid molecular species in a lipid solution, thereby adding the standard to a lipid solution prior to the injection of the sample into an instrument [12,24]. We recommend one to add the internal standard(s) prior to lipid extraction, thereby eliminating the presence of any incomplete recovery during lipid extraction.

6.5   CONCLUSIONS

In this chapter, we presented the protocols for sample preparations and for the analyses of representative brain lipid classes by using shotgun lipidomics. Several keynotes for successful analyses of brain lipids as well as other lipid samples were also emphasized. It was demonstrated that shotgun lipidomics allows us not only to quickly analyze abundant lipid molecular species, but also to broadly assess those lipid molecular species in low abundance and those found to be minor lipid classes. We believe that the application of the shotgun lipidomics approach will take us to a new level of understanding lipid metabolism, trafficking, and homeostasis, revealing biochemical mechanisms underlying many lipid-related diseases, and identifying new lipid biomarkers.

ACKNOWLEDGMENTS

This work was supported by National Institute on Aging Grants R01 AG23168 and R01 AG31675, and the Neurosciences Education and Research Foundation. The authors are grateful to Dr. Kui Yang and Shaoping Guan for their technical help. XL serves as a consultant for LipoSpectrum LLC.

LIST OF ABBREVIATIONS

Cer	(dihydro)ceramide
CerPCho	sphingomyelin
ChoGpl	choline glycerophospholipid
CID	collision-induced dissociation
ESI	electrospray ionization

EtnGpl	ethanolamine glycerophospholipid
FFA	free fatty acid
Fmoc	fluorenylmethoxylcarbonyl
HexCer	monohexosylceramide (i.e., galactosylceramide and/or glucosylceramide)
MS	mass spectrometry
NL	neutral loss
PI	precursor ion
Ptd$_2$Gro	cardiolipin
PlsEtn	plasmenylethanolamine (i.e., ethanolamine plasmalogen)
PtdCho	phosphatidylcholine
PtdEtn	phosphatidylethanolamine
PtdGro	phosphatidylglycerol
PtdIns	phosphatidylinositol
PtdOH	phosphatidic acid
PtdSer	phosphatidylserine
TAG	triacylglycerol

REFERENCES

1. Han, X. and Gross, R.W. (2003) Global analyses of cellular lipidomes directly from crude extracts of biological samples by ESI mass spectrometry: A bridge to lipidomics. *J. Lipid Res.* 44:1071–1079.
2. Lagarde, M. et al. (2003) Lipidomics is emerging. *Biochim. Biophys. Acta* 1634:61.
3. Wenk, M.R. (2005) The emerging field of lipidomics. *Nat. Rev. Drug Discov.* 4:594–610.
4. Han, X. and Gross, R.W. (2005) Shotgun lipidomics: Electrospray ionization mass spectrometric analysis and quantitation of the cellular lipidomes directly from crude extracts of biological samples. *Mass Spectrom. Rev.* 24:367–412.
5. Walker, J.M. et al. (2005) Targeted lipidomics: Fatty acid amides and pain modulation. *Prostaglandins Other Lipid Mediat.* 77:35–45.
6. Serhan, C.N. (2005) Mediator lipidomics. *Prostaglandins Other Lipid Mediat.* 77:4–14.
7. Han, X. (2007) Neurolipidomics: Challenges and developments. *Front. Biosci.* 12:2601–2615.
8. Ivanova, P.T. et al. (2004) LIPID arrays: New tools in the understanding of membrane dynamics and lipid signaling. *Mol. Interv.* 4:86–96.
9. Welti, R. et al. (2007) Plant lipidomics: Discerning biological function by profiling plant complex lipids using mass spectrometry. *Front. Biosci.* 12:2494–2506.
10. Schiller, J. et al. (2007) MALDI-TOF MS in lipidomics. *Front. Biosci.* 12:2568–2579.
11. Albert, C.J. et al. (2007) Myocardial lipidomics. Developments in myocardial nuclear lipidomics. *Front. Biosci.* 12:2750–2760.
12. Ejsing, C.S. et al. (2006) Automated identification and quantification of glycerophospholipid molecular species by multiple precursor ion scanning. *Anal. Chem.* 78:6202–6214.
13. Han, X. and Gross, R.W. (2005) Shotgun lipidomics: Multi-dimensional mass spectrometric analysis of cellular lipidomes. *Expert Rev. Proteomics* 2:253–264.
14. Han, X. and Gross, R.W. (2001) Quantitative analysis and molecular species fingerprinting of triacylglyceride molecular species directly from lipid extracts of biological samples by electrospray ionization tandem mass spectrometry. *Anal. Biochem.* 295:88–100.

15. Han, X. et al. (2004) Towards fingerprinting cellular lipidomes directly from biological samples by two-dimensional electrospray ionization mass spectrometry. *Anal. Biochem.* 330:317–331.
16. Tsui, Z.C. et al. (2005) A method for profiling gangliosides in animal tissues using electrospray ionization-tandem mass spectrometry. *Anal. Biochem.* 341:251–258.
17. Kalderon, B. et al. (2002) Modulation by nutrients and drugs of liver acyl-CoAs analyzed by mass spectrometry. *J. Lipid Res.* 43:1125–1132.
18. Golovko, M.Y. and Murphy, E.J. (2004) An improved method for tissue long-chain acyl-CoA extraction and analysis. *J. Lipid Res.* 45:1777–1782.
19. Han, X. et al. (2005) Shotgun lipidomics of phosphoethanolamine-containing lipids in biological samples after one-step in situ derivatization. *J. Lipid Res.* 46:1548–1560.
20. Ikonomou, M.G., Blades, A.T., and Kebarle, P. (1991) Electrospray-ion spray: A comparison of mechanisms and performance. *Anal. Chem.* 63:1989–1998.
21. Gaskell, S.J. (1997) Electrospray: Principles and practice. *J. Mass Spectrom.* 32:677–688.
22. Han, X. et al. (2006) Factors influencing the electrospray intrasource separation and selective ionization of glycerophospholipids. *J. Am. Soc. Mass Spectrom.* 17:264–274.
23. Han, X. et al. (2004) Caloric restriction results in phospholipid depletion, membrane remodeling and triacylglycerol accumulation in murine myocardium. *Biochemistry* 43:15584–15594.
24. Schwudke, D. et al. (2006) Lipid profiling by multiple precursor and neutral loss scanning driven by the data-dependent acquisition. *Anal. Chem.* 78:585–595.
25. Han, X. and Gross, R.W. (1995) Structural determination of picomole amounts of phospholipids via electrospray ionization tandem mass spectrometry. *J. Am. Chem. Soc.* 6:1202–1210.
26. Brugger, B. et al. (1997) Quantitative analysis of biological membrane lipids at the low picomole level by nano-electrospray ionization tandem mass spectrometry. *Proc. Natl. Acad. Sci. USA* 94:2339–2344.
27. Lehmann, W.D. et al. (1997) Characterization and quantification of rat bile phosphatidylcholine by electrospray-tandem mass spectrometry. *Anal. Biochem.* 246:102–110.
28. Welti, R. et al. (2002) Profiling membrane lipids in plant stress responses. Role of phospholipase Dα in freezing-induced lipid changes in Arabidopsis. *J. Biol. Chem.* 277:31994–32002.
29. Welti, R. and Wang, X. (2004) Lipid species profiling: A high-throughput approach to identify lipid compositional changes and determine the function of genes involved in lipid metabolism and signaling. *Curr. Opin. Plant Biol.* 7:337–344.
30. Han, X. et al. (2006) Shotgun lipidomics of cardiolipin molecular species in lipid extracts of biological samples. *J. Lipid Res.* 47:864–879.
31. Jiang, X. and Han, X. (2006) Characterization and direct quantitation of sphingoid base-1-phosphates from lipid extracts: A shotgun lipidomics approach. *J. Lipid Res.* 47:1865–1873.
32. Jiang, X. et al. (2007) Alkaline methanolysis of lipid extracts extends shotgun lipidomics analyses to the low abundance regime of cellular sphingolipids. *Anal. Biochem.* 371:135–145.
33. Sandhoff, R. et al. (1999) Determination of cholesterol at the low picomole level by nano-electrospray ionization tandem mass spectrometry. *J. Lipid Res.* 40:126–132.
34. Jiang, X., Ory, D.S., and Han, X. (2007) Characterization of oxysterols by electrospray ionization tandem mass spectrometry after one-step derivatization with dimethylglycine. *Rapid Commun. Mass Spectrom.* 21:141–152.
35. Han, X. et al. (2003) Novel role for apolipoprotein E in the central nervous system: Modulation of sulfatide content. *J. Biol. Chem.* 278:8043–8051.

36. Cheng, H., Jiang, X., and Han, X. (2007) Alterations in lipid homeostasis of mouse dorsal root ganglia induced by apolipoprotein E deficiency: A shotgun lipidomics study. *J. Neurochem.* 101:57–76.
37. McLafferty, F.W. and Turecek, F. (1993) *Interpretation of Mass Spectra*, University Science Books, Sausalito, CA.
38. Han, X., Holtzman, D.M., and McKeel, D.W. Jr. (2001) Plasmalogen deficiency in early Alzheimer's disease subjects and in animal models: Molecular characterization using electrospray ionization mass spectrometry. *J. Neurochem.* 77:1168–1180.
39. Hsu, F.-F., Bohrer, A., and Turk, J. (1998) Electrospray ionization tandem mass spectrometric analysis of sulfatide. Determination of fragmentation patterns and characterization of molecular species expressed in brain and in pancreatic islets. *Biochim. Biophys. Acta* 1392:202–216.
40. Hsu, F.F. and Turk, J. (2005) Analysis of sulfatides, in *The Encyclopedia of Mass Spectrometry*, Caprioli, R.M. (ed.), pp. 473–492, Elsevier, New York.
41. Hsu, F.F. and Turk, J. (2005) Analysis of sphingomyelins, in *The Encyclopedia of Mass Spectrometry*, Caprioli, R.M. (ed.), pp. 430–447, Elsevier, New York.
42. Han, X. and Cheng, H. (2005) Characterization and direct quantitation of cerebroside molecular species from lipid extracts by shotgun lipidomics. *J. Lipid Res.* 46:163–175.
43. Abe, T. and Norton, W.T. (1974) The characterization of sphingolipids from neurons and astroglia of immature rat brain. *J. Neurochem.* 23:1025–1036.
44. Han, X. (2002) Characterization and direct quantitation of ceramide molecular species from lipid extracts of biological samples by electrospray ionization tandem mass spectrometry. *Anal. Biochem.* 302:199–212.
45. Ford, D.A., Rosenbloom, K.B., and Gross, R.W. (1992) The primary determinant of rabbit myocardial ethanolamine phosphotransferase substrate selectivity is the covalent nature of the *sn*-1 aliphatic group of diradyl glycerol acceptors. *J. Biol. Chem.* 267:11222–11228.
46. Koivusalo, M. et al. (2001) Quantitative determination of phospholipid compositions by ESI-MS: Effects of acyl chain length, unsaturation, and lipid concentration on instrument response. *J. Lipid Res.* 42:663–672.
47. Han, X. and Gross, R.W. (1994) Electrospray ionization mass spectroscopic analysis of human erythrocyte plasma membrane phospholipids. *Proc. Natl. Acad. Sci. USA* 91:10635–10639.
48. Liebisch, G. et al. (2004) High-throughput quantification of phosphatidylcholine and sphingomyelin by electrospray ionization tandem mass spectrometry coupled with isotope correction algorithm. *Biochim. Biophys. Acta* 1686:108–117.

7 Mass Spectrometry Methods for the Analysis of Bioactive Sphingolipids: A High-Performance Liquid Chromatography/ Tandem Mass Spectrometry Approach

Jacek Bielawski, Motohiro Tani, and Yusuf A. Hannun

CONTENTS

7.1   INTRODUCTION

7.1.1   OVERVIEW AND BACKGROUND

Sphingolipids (SPL), the general name for lipids containing a sphingoid long-chain base, are one of the most chemically and functionally diverse classes of biomolecules [1]. Sphingomyelin (SM) and glycosphingolipids (GSPL) constitute the major complex SPL in mammalian cells, both of which contain Cer as a lipid portion and phosphorylcholine or sugars, respectively, as a hydrophilic moiety. Cer, composed of a sphingoid base (SB) and an amide-linked fatty acid (FA), serves as the precursor for almost all SPL [1].

Although SPL had long been known to function as structural constituents of the plasma membranes, in the past two decades research into the SPL has progressed along two areas. First, SPL influence membrane structure, where they have been proposed to exist in clusters and form microdomains containing cholesterol at the plasma membrane, the so-called lipid rafts [2]. These lipid microdomains are thought to function as platforms required for effective signal transduction and for correct protein sorting. Second, many SPL act as both first and second messengers, as well as bioactive mediators, in a variety of signaling pathways. Thus, the SPL metabolites—Cer, Cer 1-phosphate (Cer-1P), sphingosine (Sph), and Sph 1-phosphate (Sph-1P)—have emerged as a new class of lipid biomodulators that influence cell functions through multiple signaling pathways [3–6]. The first bioactive SPL to be appreciated was Sph, and it exhibits a multitude of cellular activities through regulation of various targets such as protein kinase C [7], phospholipase D [8], and others. In yeast, SB play key roles in stress responses, probably by regulating the protein kinases, Pkh1/2 [9] and Ypk1/2 [10], while Cer induces growth arrest and apoptosis in many systems, and Sph-1P stimulates growth and suppresses apoptosis [5]. Cer mediates its effects by modulating the activity of several intracellular targets, including protein phosphatases [11], protein kinase C zeta [12], and the protease, cathepsin D [13].

Sph-1P primarily functions as an extracellular messenger that regulates cell motility and morphology through a G-protein-coupled receptor, the endothelial differentiation gene receptors S1PRs$_{1-5}$ [14]. Sph-1P may also regulate intracellular pathways through modulating Ca^{2+} release [5]. In addition, some studies suggest the regulation of cell growth and intracellular Ca^{2+} mobilization by sphingosylphosphorylcholine (a lyso-form of SM). The pathways involved in the generation of this lipid metabolite, however, remains unclear [15]. Cer 1-phosphate produced by ceramide Cer kinase was also shown to have a role in some biological functions, such as phagocytosis, degranulation, and inflammatory responses, through its direct activation of phospholipase A2 (cPLA2) [6].

Therefore, it is evident that these SPL metabolites exert distinct functions in various cellular systems. Moreover, the individual subspecies of these lipids, differing in the structure of the SB and/or the fatty acyl chain demonstrate differing function, for example, loss of function by lacking the 4–5 double bond in SB [3] and specific regulation of C18:0-Cer or C24:0 and C24:1-Cer in cell growth in human head and neck squamous cell carcinomas or breast cancer cells, respectively [16,17]. Thus, it has become critical to investigate SPL biology from the point of view of individual molecular species and not simply at the level of an entire class or subclass of molecules. Thus, the need has arisen to develop methods to more clearly identify and quantify specific molecular species.

The recent accelerated development of research in the SPL field has been driven by the development of mass spectrometry (MS) methods for lipid analysis. Specific advances include greater quantitation methods and methods that allow molecular identification and functional characterization of SPL-metabolizing enzymes. This chapter discusses specific liquid chromatography/mass spectrometry (LC/MS) methods for quantitation of bioactive SPL. A brief introduction into SPL enzymology is provided. However, there are several outstanding reviews on the topic of SPL metabolism [1,3–6,18].

7.1.1.1 Sphingolipid Metabolism: A Foundation of Structural Complexity

The biochemical pathways for the synthesis and degradation of SPL are defined, and most of the enzymes involved have been identified at the molecular level (Figure 7.1). Briefly, SB, which are initially synthesized by the condensation of L-serine and palmitoyl-CoA by serine palmitoyl-CoA transferase, are acylated to form dihydroCers (dhCer) [1]. It should be noted that the acylation of dihydroSph (dhSph) is catalyzed by individual members of the LASS family of proteins, which exhibit distinct substrate specificities for the fatty acyl-CoA that results in the synthesis of dhCers having N-linked FA with distinct chain lengths [19]. Subsequently, the *cis* 4–5 double bond is inserted to generate Cer in the ER, which in turn serves as the precursor for SM and glycosphingolipids (GlSPL) in the Golgi apparatus [1].

The breakdown of GSPL proceeds in a stepwise degradation pattern by the action of exoglycosidases that results in the formation of Cer. SM is also initially degraded through the hydrolysis of the phosphodiester bond between Cer and phosphorylcholine through the action of sphingomyelinases. Subsequently, Cer is deacylated by ceramidases to generate Sph. Sph is considered to be exclusively generated from Cer

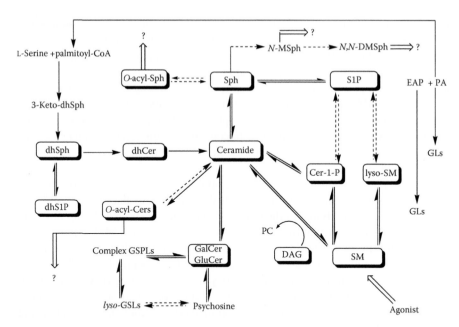

FIGURE 7.1   SPL biosynthesis and metabolism pathways; metabolomic profiling of SPL.

through the action of ceramidases, and *de novo* synthesis of Sph has been ruled out since Δ4 desaturase specifically acts on dhCer but not on dhSph [20]. In the next step, which occurs mainly in the cytoplasmic region of the cells, Sph is phosphorylated by Sph kinases into Sph-1P. In most cells, the generated Sph-1P is rapidly degraded either by a specific lyase to hexadecenal and phosphoethanolamine or by specific phosphatases to Sph (Figure 7.1).

For many of the key enzymes of SPL metabolism, such as sphingomyelinase, ceramidase, and Sph kinase, multiple forms of the enzymes have been identified that differ in their catalytic pH optimum, primary structure, and/or localization. These isoforms arise primarily as products of distinct genes, which appear to be involved in distinct pathways, regulating the generation of their respective SPL products [3–5,18]. For example, three different ceramidases have very different subcellular localization: endosome/lysosome (acid ceramidase), plasma membrane, extracellular secretion, mitochondria (neutral ceramidase), and ER/Golgi (alkaline ceramidases) [18]. This diversity of Cer-metabolizing enzymes underscores the complexity of SPL signaling, and it allows for the regulation of compartment-specific metabolism and SPL-mediated cell regulation.

In contrast to these remarkable achievements in the molecular identification of enzymes of SPL metabolism, the modes and regulation of SPL transport remain mostly to be elucidated. The hydrophobic nature of SPL prevents their spontaneous transfer through the cytosol, and therefore their transport requires facilitated mechanisms such as vesicular transport and/or the action of lipid-transfer proteins. The discovery of CERT, which transports Cer between the ER and the Golgi apparatus, provided important information concerning the intracellular transport mechanism of SPL [21]. Other evidence suggests that SPL are found in the nucleus and in mitochondria [22,23]. Unfortunately, little is known about the dynamic interplay between SPL metabolism in the ER/Golgi "classical" pathway and the other organelles not connected to the ER/Golgi complex.

7.1.1.2 LC/MS Methods for the Detection and Analysis of Bioactive Sphingolipids

Technological advances in lipid detection, analysis, and quantitation have played a key role in the development of the SPL research. Traditional lipid analytical methods, such as thin-layer chromatography (TLC), are hampered by limited sensitivity, selectivity, and resolution. Metabolic labeling using lipid precursors (such as serine or palmitic acid) have been widely used for selective labeling of certain classes of lipids, which are then typically separated using TLC and visualized by autoradiography. However, this method is affected by the incorporation of radioactive substrates, and this does not always reflect the primary lipid contents of cells. Furthermore, TLC has low resolution, sensitivity, and is affected by environmental conditions, making it difficult to identify the subspecies of individual SPL. The development of advanced mass-spectrometry-based technologies has allowed the simultaneous assessment of several SPL species as well as the measurement of individual molecular subspecies. To understand the physiological function of SPL metabolites, it is important to elucidate the metabolic changes of several SPL and their individual molecular species. Here, we review methods that allow for the simultaneous analysis of SPL and their individual molecular species using liquid chromatography/tandem mass spectrometry (LC-MS/MS).

7.1.2 LIPID DEFINITION

Historically, lipids have been defined based on their physical properties as a class of natural organic compounds that are highly soluble in organic solvents. Christie [24] proposed a more precise definition based on their molecular structure and functionality: "Lipids are fatty acids (FA) and their derivatives, and substances related biosynthetically or functionally to these compounds." Thus, FA are defined as compounds synthesized in nature via condensation of malonyl coenzyme A. Usually, mammalian FA contain even number of carbon atoms (C_{14}–C_{26}), although odd and branched chain FA are also found in many other organisms. Other structural features such as double bonds and hydroxyl groups are subsequently incorporated into the acyl chains by different enzyme systems [24].

7.1.3 GENERAL CLASSIFICATION

In addition to the metabolic classification of lipids into glycerolipids, SPL, and sterols, other structural classifications have been employed. Han and Gross [25] provide more analytically relevant classification of cellular lipids, dividing them into three large groups: nonpolar lipids, polar lipids, and metabolites, based on the relative polarity of their headgroup. Nonpolar lipids include predominantly cholesterol and triacylglycerols (TAG). Polar lipids predominantly comprise phospholipids (PL), sphingolipids (SPL), and glycolipids. Common metabolites contain unesterified FA,

FA esters, Cers (Cer), lysolipids, eicosanoids, diacylglycerols (DAG), Sph (Sph), Sph-phosphates (Sph-1P), and others.

7.1.3.1 Current Lipids' Classification

Prevalent complex SPL, phosphosphingolipids (PSPL) and GSPL, are found in all eukaryotes, some prokaryotes and viruses, mainly as components of the plasma membrane and related organelles or membranes. SPL constitute about 30% of the total lipid of plasma membranes.

7.1.3.2 Sphingolipids

General structures and nomenclature for SPL cited and described in this chapter are shown in Figures 7.1 and 7.2. SPL constitute one of the most structurally diversified classes of amphipathic lipids abundant in all living organisms. Variations in the nature of the headgroup attached to the primary hydroxyl group (carbohydrates, phosphorylcholine, phosphate, or phosphoinositol), N-acyl group, and sphingoid-base backbone result in a great number of chemically distinct SPL, where Sph, dhSph, or phytoSph (phytoSph) are the core structural moieties. Some 3-O-alkyl, 3-O-alkenyl (vinyl ether linkage), and 1- or 3-O-acyl-analogs of natural components have also been identified. Thousands of natural complex SPL have been isolated based on almost 60 distinct species of SB, although most of them are very minor components. SB, the backbone of all SPL, encompass a wide array of (2S, 3R, 4E)-2-amino-1, 3-dihydroxyalkenes (Sph), (2S, 3R)-2-amino-1,3-dihydroxy-alkanes (dhSph), and (2S,3S,4R)-2-amino-1,3,4-trihydroxyalkanes (phytoSph) with alkyl chain lengths from 14 to 22 carbon atoms and variations in the number and position of the double bonds, hydroxyl groups, and branching methyl groups. Mammalian SPL are predominantly composed of 2-amino-1,3-dihydroxyocta-decene (18Sph, abbreviated here as Sph) and 2-amino-1,3-dihydroxyoctadecane (18dhSph, abbreviated here as dhSph). Yeast and plant SB are mainly composed of 2-amino-1,3,4-trihydroxyoctadecane (18phytoSph), 18dhSph and their eicosa-homologs (20phytoSph and 20dhSph). Additionally, some SPL contain SB having a double bond in position 8 or have double bonds in positions 4 and 8 (which can be found in plant SPL).

Cers are N-acyl-derivatives of SB. Combinations of different SB with different FA (including their hydroxy-analogs) generate a huge variety of Cer, dhCer, and phytoCer. These basic SPL are modified at the 1-hydroxyl group to (1) phosphates (e.g., Sph-1P and Cer-1P), (2) phosphocholine-analogs (e.g., sphingomyelin, SM, and lysosphingomyelin, lyso-SM), and (3) glucosyl-and galactosyl-analogs (e.g., glucosylcer and galactosylceramide, known as cerebrosides, and their lyso-form, psychosine). Members of the latter group also serve as precursors to hundreds of different species of complex GSPL, with lactosylceramide (LacCer), containing only two sugar residues, being the simplest. Additionally, some Cer and GSPL can be modified on their hydroxyl groups (e.g., O-acyl-Cer), and some SB can be N-methylated.

SPL constitute the second major category of polar lipids; for example, they represent approximately 5%–10% of the total lipid mass in the mammalian brain [26–28]. Abnormal SPL metabolism could lead to their accumulation and deposition in multiple tissues, especially neural tissues, and this results in potentially severe clinical

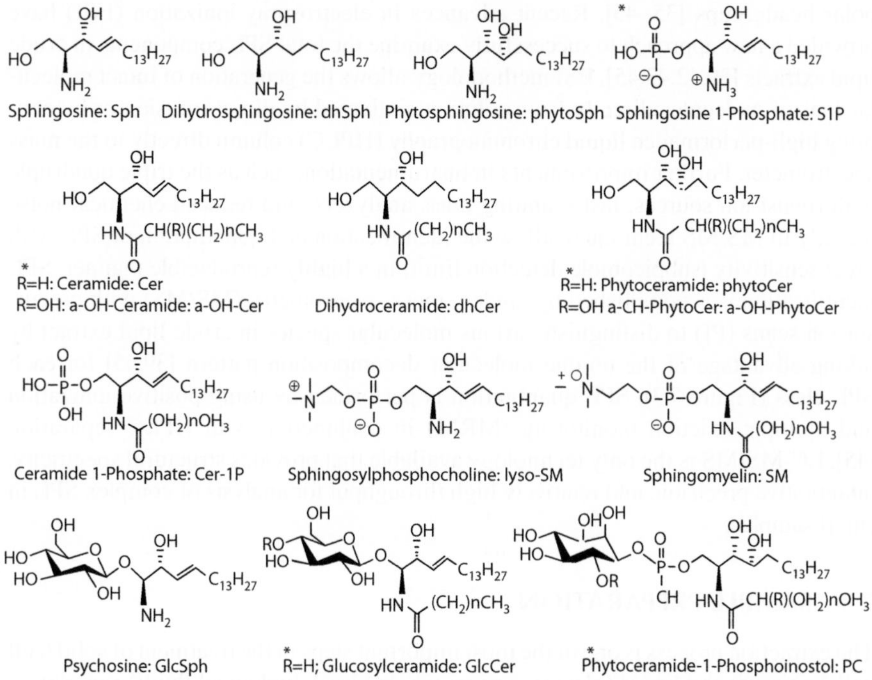

FIGURE 7.2   General structures, nomenclature and abbreviations for SPL cited and described in this presentation. Cn indicates chain length of *N*-acyl part of SPL; 18SB indicates SPL containing 2-amino-1,3-dihydroxy-octadecene-4E or 2-amino-1,3-dihydroxy-octadecane.

manifestations, known as the sphingolipidoses [26]. The structural diversity of SPL, and their implication in diseases, dictates that every step in analysis of these natural products must be carefully evaluated.

7.2   LIPIDOMICS APPROACH AND MASS SPECTROMETRY

The term "lipidomics" has recently emerged [29–33] in relation to genomics and proteomics. Thus, lipidomics can be defined as the full characterization of lipid molecules and their functions.

To understand how SPL biosynthesis and turnover regulates cell behavior under normal and abnormal conditions, how perturbations in SPL of one type may enhance or interfere with the action of another, and where and how all these SPL are made and removed, we must be able to establish the metabolomic profile of SPL. MS methodology offers an efficient tool to monitor changes in the composition of all these bioactive species under different environments.

A variety of sample preparation, ionization modes, and instrumental designs have been developed to analyze particular SPL classes by MS technology [34]. Design for this methodology has been based on the fact that different SPL subclasses dissociate into structurally distinctive patterns corresponding to their SB, *N*-acyl chains, and polar headgroups [35–45]. Recent advances in electrospray ionization (ESI) have provided a new approach to successfully examine the total SPL components in crude lipid extracts [39,42,43,45]. ESI methodology allows the generation of intact molecular ions of molecules directly from solution, delivered by direct infusion or by coupling high-performance liquid chromatography (HPLC) column directly to the mass spectrometer. Further improvements in instrumentation, such as the triple quadruple with robust ion sources, fast scanning mass analyzers, and reduced chemical noise (mainly in MS/MS technique), allow the identification and quantitation of SPL with great sensitivity (subpicomole detection limit) in a highly reproducible manner. SPL identification is accomplished by tandem mass spectrometry (MS/MS) with precursor ion scans (PI) to distinguish various molecular species in crude lipid extract by taking advantage of the unique molecular decomposition pattern [39,45] for each SPL class (Figure 7.3). SPL quantitation is performed by using positive ionization and multiple reaction monitoring (MRM) in conjunction with HPLC separation [45]. LC-MS/MS is the only technology available that provides structural specificity, quantitative precision, and relatively high throughput for analysis of complex SPL in small samples.

7.3 SAMPLE PREPARATION

The extraction process is one of the most important steps in the treatment of solid (cell pellets, tissue) and liquid (plasma, serum, whole blood, biological fluids) samples.

Chloroform:methanol 2:1 (by vol) extraction, developed in 1956 for fish tissue [46], further improved in 1959 by Bligh and Dyer [47], became a gold standard procedure, known as the "Bligh & Dyer" (B&D) method, and it is still commonly used for lipid extraction from all biological matrices. It involves a two-step extraction employing chloroform:methanol:water at well-defined ratios of 1:2:0.8 (by vol.) and 2:2:1.8 (by vol.), respectively [47]. According to the originators, the upper (methanol:water) phase contains all the "nonlipid" substances, while most lipids remain in the lower (chloroform) phase. This virtually unchanged procedure is commonly applied to most SPL sample preparation, regardless of the analytical procedure subsequently used, for example, TLC, HPLC, or MS, although the extraction efficiency, particularly for the most polar SPL components such as Sph-1P or lyso-SM is questionable.

Over time, some modifications to the B&D have been incorporated in the isolation of SPL, mostly intended to remove the bulk of the major coextracted components, especially the glycerolipids, by subjecting the initial chloroform extract to a mild alkaline hydrolysis to cleave ester-linked FA [45,48–56]. However, 1-O-acyl-Cers and related compounds [57] will also be hydrolyzed, thus artificially increasing the level of Cer. Comparison of the Cer level calculated from lipid extracts that were prepared with and without the base hydrolysis step can provide important data about the level of O-acyl Cers. Our results show a 20%–40% increase in Cer after this step. Nevertheless, this simple approach is recommended in the preparation of samples for SM analysis to reduce contamination by phosphatidyl choline (PtdCho) from the lipid extract that can interfere with SM determination. Due to close masses

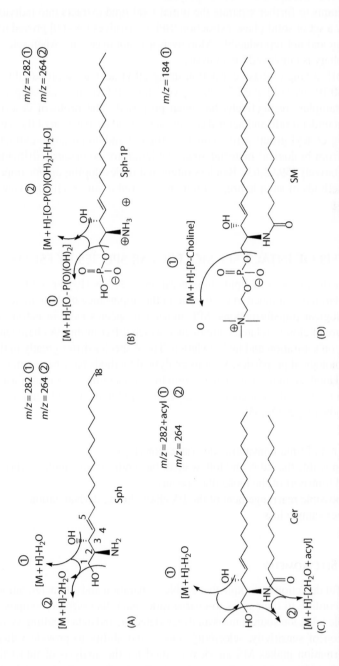

FIGURE 7.3 MS/MS fragmentation patterns of the initial positive molecular ions, generated upon ESI of Sph (A), Sph-1P (B), Cer (C), and SM (D). The specific common fragment ion of $m/z = 264$ (2, panel C) for Cers, and $m/z = 184$ (1, panel D) for SM are used in the parent ion scan experiments for determination of molecular species composition prior to quantitative analysis with the MRM experiments.

and fragmentation pattern, it is recommended to use this hydrolysis step, even if the highly specific LC-MS/MS method is employed.

Several attempts to further separate the initial total lipid extracts into individual lipid classes by a set of solid-phase extraction (SPE) cartridges [58–60] proved to be time consuming and not reproducible. Moreover, it is not necessary when selective LC-MS technology is employed for analysis.

Recently [61], a single-phase extraction using ethyl acetate:isopropanol:water system at 60:30:10 (by vol.) and 85:15:0 (by vol.) for solid (tissue and cell pellets) and aqueous samples, respectively, has been proposed. The protocol describes lipid extraction under a safe and neutral condition to avoid destruction of the parent SPL containing O-acyl group while assuring efficient and quantitative extraction of the SB-1P from biological material since the latter are notoriously difficult to recover quantitatively [59,62,63]. Readers interested in developing and/or improving existing methods of sample preparation are referred to an excellent review by McDowall [64].

7.4 ANALYSIS OF INTACT SPL MOLECULAR SPECIES BY ESI/MS

The interest in the analysis of lipids in general and SB, SB-1P, Cer, GLCer, and SM, in particular, has continued to evolve due to the importance of these molecules in various biological transformations. SPL molecular species exist in nature as a complex mixture of closely related components, which differ in the FA chain length and the degree on saturation and hydroxylation. These species differ greatly in their chemical and biological properties. Various analytical methods have been employed to separate and analyze individual species from their intact (underivatized) form, of which ESI/MS is the method of choice. Analyzing a lipid in its intact form provides the following advantages [24,65]:

- Elimination of time-consuming derivatization steps
- Making possible the study and follow up of biosynthesis, metabolism, turn-over, and transport of the molecular species
- Protect possible rearrangement of the FA chain during derivatization
- Minimizes sample loss

7.4.1 MASS SPECTROMETRY

MS is a powerful detection technique that enables separation and characterization of compounds according to their mass-to-charge ratio (m/z). Its essential components include a sample inlet, ion source, mass analyzer, detector, and data handling system. The combination of sensitivity, selectivity, speed, and ability to provide valuable structural information makes MS an ideal method for the analysis of intact lipid molecular species. The sample can be introduced either by direct infusion or through separation devices such as liquid chromatography (LC).

7.4.1.1   Mechanism of Electrospray/Mass Spectrometry

ESI/MS, invented in the 1960s, was put into practice by Fenn et al. [66,67]. It involves transformation of ions from the liquid to the gas phase. It is a method that operates at atmospheric pressure and ambient temperature. Initially, a solution containing the analytes of interest is introduced to the ESI ion source through capillary tubing. The narrow orifice at the end of the capillary and the dynamic forces facilitate the formation of sprayed small droplets in the ionization chamber. The application of electric potential (approximately 2–5 kV) causes ionization. Consequently, the droplets carry a net charge. The charged droplets are then directed into the mass analyzer by the applied electric field. The applied potential may be positive or negative depending on the physicochemical properties of the analytes. Passing through the ionization chamber, the droplets dissolve, and this effect dramatically increases the coulombic forces between the ions. Once the force exceeds the surface tension of the solvent, the droplets explode to form a fine mist of smaller droplets. This cycle is repeated until molecular ions are generated prior to their entrance into the mass analyzer [25,68].

7.4.1.2   Basics of ESI/MS for SPL Analysis

The soft ionization MS techniques such as fast atom bombardment (FAB) [69–71], field desorption (FD) [72], thermospray (TS) [73], matrix-assisted laser desorption ionization (MALDI) [74–77], and particularly ESI [38,39,45,78,79] can ionize lipid molecules without causing extensive fragmentation. Recently, ESI has become a preferred and widely used method to ionize biologically important lipid molecules. In a recent review, Han and Gross [25] assigned lipids into four classes based on their electrical propensities that are reflected in their ESI/MS analyses:

- Anionic lipids: containing net negative charges at neutral pH, such as sulfatide, phosphatidyl serine (PtdSer), phosphatidyl inositol (PtdIns), phosphatidyl glycerol
- Weak anionic lipids: carrying net negative charges at alkaline pH, such as phosphatidyl ethanolamine (PtdEtn), lyso-phosphatidyl ethanolamine (lysoPtdEtn), Cer
- Neutral polar lipids: polar but electrically neutral lipids at both neutral and alkaline pH, such as PtdCho, lysoPtdCho, SM, DAG, TAG
- Special lipids: varying in their electrical propensity, which include cholesterol, cholesteryl esters, sterols

7.4.2   MS Scan Modes

A number of mass analyzers are available, for example, quadruple, ion trap, time of flight, ion cyclotron resonance, or sector instruments, which separate charged molecules in vacuum depending on their m/z ratio. In the so-called full scan (FS) mode, a spectrum of primary, mostly molecular, ions is identified. This is the least specific mode with low sensitivity, and it is mostly used for a rough assessment of major components of biological material when no or very limited information about SPL composition exist. However, interferences from other compounds present can either suppress ionization or cause such a high chemical noise making detection of SPL virtually impossible.

Moreover, the mass analyzers can also be used for fragmentation, predominantly in the triple quadruple instrument. In this instrument, the middle (Q2) field-free quadruple either focuses and transmits all ions, or can be used as a collision cell for controlled fragmentation, called collision-induced dissociation (CID). As a result of a collision with an inert gas introduced into the cell, the internal energy of the ions increases through the conversion of kinetic energy out of breaking specific bonds, depending on the collision energy applied [74,80]. The fragment ions are then analyzed in the second mass analyzer (third quadruple Q3). The degree of fragmentation is affected by the choice of collision gas, its pressure, and particularly the applied collision energy. When a single quadruple instrument is used, partial fragmentation can be induced in the source by elevating the cone-to-skimmer potential difference. Protonated molecules desorbed from the ESI droplets are accelerated between the cone and skimmer, undergoing CID upon collision with residual carrier gas molecules such as N_2 [80].

Initial "soft" ionization of a biological extract results in numerous SPL molecular ions, either positive $(M + H)^+$ or negative $(M - H)^-$. When the precursor ion fragments, it generates secondary (daughter) distinctive pattern of ions related to the headgroup, SB and FA. This provides a wealth of structural information, enabling identification of SPL in particular biological materials. The positive ionization fragmentation can be enforced by incorporation of alkali metal ions $(M + Me)^+$, where $Me = Li^+$, Na^+, K^+, Rb^+, or Cs^+ [38,81]. In addition to the structural information, tandem MS provides a higher sensitivity, specificity, and greatly reduced chemical background due to selected mode of monitored masses.

7.4.2.1 Specific Scan Modes for MS–MS Instrumentation

7.4.2.1.1 Product Ion Scan

In product ion scan, the first mass analyzer (Q1) allows a single ion with a set m/z value to pass, and this is then further fragmented by CID in the second quadruple (Q2), the secondary (daughter) ions are then scanned over a defined mass range by the third quadruple (Q3) and passed to the detector. The relative abundance of the product ions depends on the dissociation dynamic; therefore, on changing the CID collision energy, a fragmentation pattern is observed, which is specific for each class of SPL compounds.

7.4.2.1.2 Neutral Loss

In a neutral loss (NL) scan, Q3 is offset from Q1 by fixed m/z, corresponding to specific NL, for example, 18 Da for loss of a water molecule. Both Q1 and Q3 scan over specified ranges of m/z values. In this mode, the detector records only those precursor ions that decompose, losing the specified neutral fragment. This type of MS experiments highly decreases chemical noise and is very helpful in the identification of unknown SPL.

7.4.2.1.3 Precursor Ion Scan

In a PI scan mode, the Q3 is set to pass specific m/z value, characteristic of a defined secondary ion. The Q1 scans across m/z range, recording only those primary ions which decompose to the specified product ion of interest. This highly specific scan mode eliminates or at least greatly reduces chemical noise, and it constitutes a very useful identification tool since each class of SPL yields at least one common product ion. Thus, setting Q3 to this specific daughter ion and scanning Q1 over the expected parent ion mass ranges, a spectrum of molecular species for an unknown biological sample may be identified.

7.4.2.1.4 Multiple Reaction Monitoring

In an MRM experiment, the Q1 is set to pass specific precursor ion m/z and Q3-specific daughter ion m/z only. This makes the MRM the most specific and sensitive MS–MS experiment allowing the analysis of even very minor components of a complex mixture with great precision and sensitivity. Such measurement practically eliminates chemical noise, thus making it an ideal tool for quantitative analyses, particularly if coupled with HPLC physical separation. Multiple mass transitions, specific for a particular compound, may be monitored sequentially; therefore, a large number of compounds may be analyzed together. Optimization of CID parameters for each compound of interest results in best sensitivity and specificity.

7.4.3 SPL IDENTIFICATION

Due to the complexity of SPL, which usually constitute minor components of crude lipid extracts, identification of individual molecular species is necessary before attempting any quantitative determination. This task can only be achieved with the application of MS, particularly with PI experiments. Although direct infusion FS MS has been attempted [50,56], reliable results may be obtained only for negative mode, in which a limited number of SPL, such as free FA and sulfates, are ionized. In FS positive mode, high chemical background makes any identification virtually impossible.

Qualitative analysis of SPL from crude extracts is best accomplished by analysis of their unique molecular decomposition products using a PI scan of common fragment ions, characteristic for the particular class of SPL (Figure 7.3) [39,52,53]. Sullard [38] and Shevchenko and coworkers [82] performed comprehensive studies on fragmentation patterns for mammalian and yeast SPL, respectively. Readers are directed to these excellent papers for detailed information. Briefly, for mammalian SPL, the focus of this review, m/z 264 and 266 are the common fragment ions used for identification of Sph and its saturated counterpart, dihydroSph derivatives, respectively.

Considering the complexity of SPL composition as well as the presence of many other lipid-related compounds in biological material extracts, it is advisable to confirm initial identification, derived from PI scan, in order to avoid false identification. This may be accomplished by other, more compound-specific MS experiments, such as product ion scan of the newly identified molecular ion, or in an MRM experiment, with mass transition unique for the particular molecule, for example, single or double dehydration for Cer or NL of sugar moiety for glycoCers (GlcCer).

Ionization conditions and collision energy are optimized for individual molecular species to achieve maximum sensitivity and quantitative accuracy. SPL composition has to be established for every new matrix.

Composition of Cer, Cer-1P, and GlcCer (C18-SB) is established by the PI scan, performed for the common product ion (*m/z* 264.2 and 266.1 for Sph and dhSph derivatives, respectively) at a high collision energy (35–55 eV) operating in positive ionization mode (Figure 7.3). A representative sample extract is infused directly into the ESI source, and it is then scanned for molecular ions of the potential SPL. Further confirmation of identity is achieved through MRM analysis with "soft" fragmentation (15–30 eV). Running samples through the HPLC system also confirms a reasonable retention time. Only SPL that satisfy identification criteria in both analyses should be considered truly present in the sample.

SM and dhSM composition (18C-SB): Identification of the SM and dhSM components is performed similarly, employing common product ion (*m/z* 183.9) at 40 eV collision energy (Figure 7.3).

Note: It is important to optimize the ionization conditions for each class of SPL and collision energy for each individual molecular subspecies to be applied for quantitative MRM analysis.

7.4.4 HIGH-PERFORMANCE LIQUID CHROMATOGRAPHY/MASS SPECTROMETRY METHODOLOGY

HPLC is often employed for the separation of intact lipid molecules using various detectors. SPL lack chromophores that enable direct specific spectrophotometric detection. Attempts have been made using UV [83–87] and evaporative light scattering detection (ELSD) [88–90]. Both detectors, however, lack specificity and impose additional limitations. With UV detection, it is very difficult to select a working mobile phase since underivatized SPL absorb close to the 200–210 nm range, depending upon the degree of saturation of the FA moiety, and most of the commonly used solvents strongly absorb in this region [87,91].

In the ELSD detector, the HPLC column effluent is evaporated, leaving the solute components as fine droplets, which are illuminated by laser, and the scattered light is measured. This is indiscriminatory detection since any compound that does not evaporate is detected [91].

An alternative technique that overcomes most of the above problems and provides both compound specificity and quantitative sensitivity is the use of HPLC coupled with ESI/MS. It is one of the most powerful technologies for the analysis of intact polar lipid molecules. The physical separation power of HPLC into either various lipid classes or individual molecular species within the class, together with MS highly selective detection, makes possible simultaneous determination of protonated/deprotonated molecules, providing also invaluable structural information [92]. Incorporation of analytical separation through an HPLC column provides for much better sensitivity as compared to a direct infusion by greatly reducing ion suppression, an effect caused by other components in the extract competing for MS ionization power when entering the mass spectrometer all at the same time.

Both normal and reverse phase HPLC have been employed, and it is important to select a solvent system that renders chromatographic resolution while still having compatibility with the ESI/MS to achieve maximum sensitivity [38,45,52,53,61,93–95].

Recently, the most powerful technique applied for SPL molecular species is the HPLC/MS/MS instrumentation with the MRM scanning mode where each target analyte is uniquely identified by the precursor–product ion mass transition and the specific retention time.

7.4.5 QUANTIFICATION BY HPLC-MS/MS USING MULTIPLE REACTION MONITORING

Quantitation of SPL in biological extracts has been a least developed segment of the HPLC/MS/MS analysis due to structural diversity, wide range of lipid concentrations, and a very limited supply of commercially available individual standards. Only laboratories that have access to custom-made synthetic standards [61] were able to set up reliable comprehensive quantitative protocols for various SPL classes. Recently, however, Avanti Polar Lipids Inc. (Alabaster, Alabama) and Matreya LLC (Pleasant Gap, Pennsylvania) significantly expanded their offer of synthetic standards so that the major difficulties can be partially overcome. To achieve reliable quantitation of all molecular species, calibration curves should be generated for as many representative components of SPL as possible, due to diversified MS responses, as reflected by calibration curve slopes (Figure 7.4A and B).

7.4.5.1 Selection of Internal Standards

Selection of a representative set of internal standards (IS), which serve as a reference for both identification and quantitation, is critical for the analysis of a complex mixture of SPL. IS should be as close as possible to the target analytes, providing similar MS fragmentation pattern as well as physicochemical properties reflected by similar solubility, extraction efficiency, and mobile–stationary phase relationship during the HPLC separation. The best IS is chemically identical to target analyte labeled with a stable isotope, usually ^2H or ^{13}C. However, considering the large number of SPL molecular species, such an approach is impractical; therefore, some compromise must be applied to overcome the limited number of the ^2H or ^{13}C compounds. Often one IS per SPL class is used, mostly an SPL with unnatural, usually lower number of carbon atoms in the FA moiety [38,45,51–54,94,95].

Bielawski et al. [61] selected derivatives of C17-Sph and C17-dhSph (SB with unnatural 17 carbon atom backbone) for use as IS to the specific classes of SPL with natural C18-Sph derived compounds. These IS have those physicochemical properties such as the elution order and mass fragmentation pattern that accurately reflect natural SPL, but are not present in the mammalian samples. Moreover, since these IS are introduced into the sample prior to extraction, incomplete extraction efficiencies are corrected, rendering quantitation of the target SPL more precise.

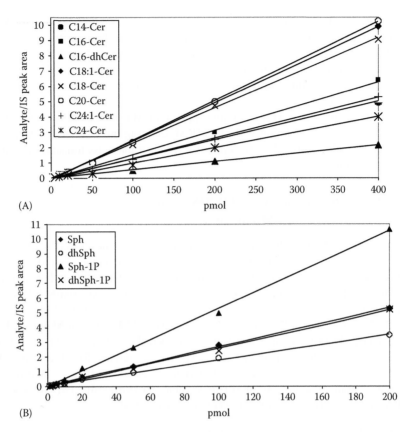

FIGURE 7.4 Calibration curves of SB and Cers. The MS response varies for molecular species even within the particular SPL class, as indicated by the calibration curve slops; therefore, individual calibrations should be generated for as many target analytes as possible. A linear instrument response (R^2 value of 0.99) is obtained for the typical calibration ranges: 1.0–200.0 pmol for SB and SB-1P (B) as well as for all Cer species (A).

7.4.5.2 Calibration Curves

Generating a calibration mechanism for each target SPL in a class greatly improves the quality of the quantitative results. However, due to the limited availability of authentic standards, calibration should be generated for as many representative standards as is practical. This way, calibration devised for the synthetic standard can be also used for few structurally closely related analytes.

Sometimes, quantitation is performed using one IS as a single point calibration [38,39,45,51–54,94,95]. This is not a very good practice since it assumes equal MS response to all molecular species in the class. Unfortunately, MS responses vary widely, depending on both structural features (number of carbon atoms, branching, unsaturation, etc.) and mobile phase composition, which changes over time, particularly when gradient elution is employed, resulting in poor quality data.

Based on the above considerations, we have adopted the following quantitative approach to SPL analysis [61]. In this approach, quantitative analyses of SPL are based on the eight-point calibration curves generated for each target analyte. The synthetic standards along with a set of IS are spiked into an artificial matrix, then subjected to the identical extraction procedures as the "biological" samples, and then analyzed by the HPLC/MS/MS system operating in positive MRM mode, employing gradient elution. Peaks for the target analytes and IS are recorded and processed using the instrument software system. Plotting the analyte/internal standard peak area ratios against analyte concentration generates the analyte-specific calibration curves. Any SPL for which no standard is available is quantitated using the calibration curve of its closest counterpart. The detailed information on HPLC and MS parameters can be found in the original publication [61].

7.4.6 DATA HANDLING/NORMALIZATION

Results from the MS analysis represent the mass level of particular SPL (in pmol) per total sample used for lipid extract preparation and quantitative analysis. In general, treatment with exogenous agents causes changes in SPL levels and compositions. For the final data presentation, MS results should be normalized to some stable parameters (which are considered not affected by that particular treatment). Total protein (mg) or PL phosphate (nmol) present in the B&D extract [47], which corresponds to the amount of the biological material used for MS analysis, can be used as the normalization parameters [49,96,97]. Normalization to the total cell number, wet or dry weight for tissue or volume for liquid samples, is also used [98]. Final results may be expressed as changes in relation to the control (% control). Each of the end users should develop his/her own data normalization strategy that will be consequently carried over the entire course of the experimental program for best data integrity.

7.5 ALTERNATIVE METHODOLOGY

A variety of different techniques (mostly radiolabeling, HPLC analysis of fluorescent analogs, and enzymatic methods) in addition to MS methodology are used for SPL analysis. The enzymatic method employing diacylglycerol kinase and [^{32}P] ATP has been the most commonly used procedure for total Cer quantitation in the range of 25 pmol to 2 nmol [97]. Cellular SB are most often analyzed by the HPLC technique developed for their fluorescent derivatives [107]. Cellular SB-1P are analyzed via their derivatization to [^{3}H] C2-Cer phosphate by an enzymatic method (employs alkaline phosphatase) followed by action of recombinant Sph kinase and [^{32}P ATP] after TLC separation of Sph-1P from the cellular Sph or by employing HPLC analysis of OPA-derived Sph-1P [99–104]. These methods have the benefit of being less expensive and do not require MS. SM may be determined by several different approaches including TLC analysis, GC analysis of silylated derivatives, and MS techniques [78,79,96,98,105,106]. Total Cer and SM can be determined following hydrolysis and analysis of the liberated and derivatized SB by means of HPLC

[102,107,108] and the liberated FA by means of GC [109] or GC/MS [110]. However, the methods are of very limiting application, providing quantitative information on total SPL class, for example, Cer, but not composition of individual molecular species within the class.

ACKNOWLEDGMENTS

This work was supported in part by NIH grants P01-CA97132 and P20 RR0176677.

LIST OF ABBREVIATIONS

B&D	Bligh & Dyer method
Cer-1P	Cer 1-phosphate
CID	collision-induced dissociation
DAG	diacylglycerols
dhCer	dihydroCer
dhSph	dihydroSph
ELSD	evaporative light scattering detection
ESI	electrospray ionization
FA	fatty acids
FAB	fast atom bombardment
FD	field desorption
FS	full scan
GC	gas chromatography
GlcCer	glycoCer
GSPL	glycosphingolipids
HPLC	high-performance liquid chromatography
IS	(selection of) internal standards
LC	liquid chromatography
LC/MS	liquid chromatography/mass spectrometry
LC-MS/MS	liquid chromatography/tandem mass spectrometry
lysoPtdCho	lysophosphatidyl choline
lysoPtdEtn	lysophosphatidyl ethanolamine
lysoSM	lysosphingomyelin
MALDI	matrix-assisted laser desorption ionization
MRM	multiple reaction monitoring
MS	mass spectrometry
m/z	mass-to-charge ratio
NL	neutral loss (scan)
PI	precursor ion scan
PL	phospholipids
PSPL	phosphosphingolipids
PtdCho	phosphatidyl choline
PtdEtn	phosphatidyl ethanolamine
PtdIns	phosphatidyl inositol
PtdSer	phosphatidyl serine

SB	sphingoid bases
SM	sphingomyelin
SPE	solid-phase extraction
Sph	sphingosine
Sph-1P	Sph 1-phosphate
SPL	sphingolipids
TAG	triacylglycerol
TLC	thin-layer chromatography
TS	thermospray

REFERENCES

1. Merrill, A.H. Jr. (2002) *J Biol Chem* **277**, 25843–25846.
2. Simons, K. and Ikonen, E. (1997) *Nature* **387**, 569–572.
3. Pettus, B.J., Chalfant, C.E., and Hannun, Y.A. (2002) *Biochem Biophys Acta* **1585**, 114–125.
4. Cuvillier, O. (2002) *Biochem Biophys Acta* **1585**, 153–162.
5. Spiegel, S. and Kolesnick, R. (2002) *Leukemia* **16**, 1596–1602.
6. Lamour, N.F. and Chalfant, C.E. (2005) *Mol Interv* **5**, 358–367.
7. Hannun, Y.A. and Bell, R.M. (1989) *Science* **235**, 670–674.
8. Lavie, Y. and Liscovitch, M. (1990) *J Biol Chem* **265**, 3868–3872.
9. Friant, S., Lombardi, R., Schmelzle, T., Hall, M.N., and Riezman, H. (2001) *EMBO J* **20**, 6783–6792.
10. Liu, K., Zhang, X., Lester, R.L., and Dickson, R.C. (2005) *J Biol Chem* **280**, 22679–22687.
11. Chalfant, C.E., Kishikawa, K., Mumby, M.C., Kamibayashi, C., Bielawska, A., and Hannun, Y.A. (1999) *J Biol Chem* **274**, 20313–20317.
12. Wang, G., Sliva, J., Krishnamurthy, K., Tran, E., Condie, B.G., and Bieberich, E. (2005) *J Biol Chem* **280**, 26415–26424.
13. Heinrich, M., Wickel, M., Schneider-Brachert, W., Sandberg, C., Gahr, J., Schwandner, R., Weber, T. et al. (1999) *EMBO J* **18**, 5252–5263.
14. Hla, T. (2003) *Pharmacol Res* **47**, 401–407.
15. Meyer zu Heringdorf, D., Himmel, H.M., and Jakobs, K.H. (2002) *Biochem Biophys Acta* **1582**, 178–189.
16. Koybasi, S., Senkal, C.E., Sundararaj, K., Spassieva, S., Bielawski, J., Osta, W., Day, T.A. et al. (2004) *J Biol Chem* **279**, 44311–44319.
17. Marchesini, N., Osta, W., Bielawski, J., Luberto, C., Obeid, L.M., and Hannun, Y.A. (2004) *J Biol Chem* **279**, 25101–25111.
18. Tani, M., Ito, M., and Igarashi, Y. (2007) *Cell Signal* **19**, 229–237.
19. Mizutani, Y., Kihara, A., and Igarahi, Y. (2005) *Biochem J* **390**, 263–271.
20. Michel, C., van Echten-Deckert, G., Rother, J., Sandhoff, K., Wang, E., and Merrill, A.H. Jr. (1997) *J Biol Chem* **272**, 22432–22437.
21. Hanada, K., Kumagai, K., Yasuda, S., Miura, Y., Kawano, M., Fukasawa, M., and Nishijima, M. (2003) *Nature* **426**, 803–809.
22. Watanabe, M., Kitano, T., Kondo, T., Yabu, T., Taguchi, Y., Tashima, M., Umehara, H. et al. (2004) *Cancer Res* **64**, 1000–1007.
23. Malisan, F. and Testi, R. (2002) *Biochim Biophys Acta* **1585**, 179–187.
24. Christie, W.W. (2003) *Lipid Analysis: Isolation, Separation, Identification and Structural Analysis of Lipids*, 3rd edn., The Oily Press, Bridgwater, U.K.
25. Han, X. and Gross, R.W. (2005) *Mass Spectrom Rev* **24**, 367–412.

26. Sastry, P.S. (1985) *Proc Lipid Res* 24, 69–176.

27. Vos, J.P., Lopez-Gordozo, M., and Gadella, B.M. (1994) *Biochim Biophys Acta* **1211**, 125–149.

28. Merrill, A.H., Schmelz, E.M., Dillehay, D.L., Spigel, S., Shayman, J.A., Schroeder, J.J., Riley, R.T. et al. (1997) *Toxicol Appl Pharmacol* **142**, 208–225.

29. Han, X. and Gross, R.W. (2003) *J Lipid Res* **44**, 1071–1079.

30. Han, X., Yang, J., Chang, H., Ye, H., and Gross, R.W. (2004) *Anal Biochem* **330**, 317–331.

31. Spener, F., Lagarde, M., and Record, M. (2003) *Eur J Lipid Sci Technol* **105**, 481–482.

32. Lagarde, M., Geloen, A., Record, M., Vance, D., and Spener, F. (2003) *Biochim Biophys Acta* **1634**, 61–69.

33. Hannun, Y.A. (2005) Functional lipidomics: Lessons and examples from the Sphingolipids in *Functional Lipidomics*, L. Feng and G. Prestwich (eds.), CRC Press, Boca Raton, FL, Chapter 7, pp. 147–158.

34. Adams, J. and Ann, Q. (1993) *Mass Spectrom Rev* **12**, 51–85.

35. Adams, J. and Ann, Q. (1992) *J Am Soc Mass Spectrom* **3**, 260–263.

36. Ann, Q. and Adams, J. (1993) *Anal Chem* **22**, 7–13.

37. Ann, Q. and Adams, J. (1993) *Biol Mass Spectrom* **22**, 285–294.

38. Sullards, M.C. (2000) *Methods Enzymol* **312**, 32–45.

39. Gu, M., Kerwin, J.L., Watts, J.D., and Aebersold, R. (1997) *Anal Biochem* **244**, 347–356.

40. Van Veldhoven, P.P., De Ceuster, P., Rozenberg, R., Mannaerts, G.P., and Hoffman, E. (1994) *FEBS Lett* **350**, 91–95.

41. Couch, L.H., Churchwell, M.I., Doerge, D.R., Tolleson, W.H., and Howard, P.C. (1997) *Rapid Commun Mass Spectrom* **11**, 504–512.

42. Mano, N., Oda, Y., Yamada, K., Asakawa, N., and Katayama, K. (1997) *Anal Biochem* **244**, 291–300.

43. Liebisch, G., Derobnik, W., Reil, M., Trumbach, B., Arnecke, R., Olgemoller, B., Roscher, A. et al. (1999) *J Lipid Res* **40**, 1539–1546.

44. Pettus, B.J., Kroesen, B., Szulc, Z.M., Bielawska, A., Bielawski, J., Hannun, Y.A., and Busman, M. (2004) *Rapid Commun Mass Spectrom* **18**, 577–583.

45. Sullard, M.C. and Merrill, A.H. (2001) *Science's stke* **67**, 1–11.

46. Folch, J., Lees, M., and Sloane, H.S. (1956) *J Biol Chem* **196**, 497–509.

47. Bligh, E.G. and Dyer, W.J. (1959) *Can J Biochem Physiol* **37**, 911–917.

48. Lin, T., Genestier, L., Pinkoski, M.J., Castro, A., Nicholas, S., Mogil, R., Paris, F. et al. (2000) *J Biol Chem* **275**, 8657–8663.

49. Bose, R., Verheij, R., Haimovitz-Friedman, A., Scotto, K., Fuks, Z., and Kolesnick, R.N. (1995) *Cells* **82**, 405–414.

50. Prasad, T., Saini, P., Guar, N.A., Vishwakarma, R.A., Khan, L.A., Haq, Q.M.R., and Prasad, R. (2005) *Antimicrob Agents Chemother* **49**, 3442–3452.

51. Monick, M.M., Mallampalli, R.K., Bradford, M., McCoy, D., Gross, T.J., Flaherty, D.M., Powers, L.S. et al. (2004) *J Immunol* **173**, 123–135.

52. Adams, J.M., Pratipanawatr, T., Berria, R., Wang, E., DeFronzo, R.A., Sullard, M.C., and Mandarino, L. (2004) *Diabetes* **53**, 25–31.

53. Zheng, W., Kollmeyer, J., Symolon, H., Momin, A., Munter, E., Wang, E., Kelly, S. et al. (2006) *Biochim Biophys Acta* **1758**(12), 1864–1884.

54. Maceyka, M., Sankala, H., Hait, N.C., LeStunff, H., Liu, H., Toman, R., Collier, C. et al. (2005) *J Biol Chem* **280**, 37118–37129.

55. VanDer Luit, A.H., Budde, M., Zerp, S., Caan, W., Klabenbeek, J.B., Verheij, M., and Van Bltterswijk, W.J. (2007) *Biochem J* **401**, 541–549.

56. Guan, X.L., He, X., Ong, W.Y., Yeo, W.K., Shui, G., and Wenk, M.R. (2006) *FASEB J* **20**, 1152–1161.

57. Abe, A. and Shayman, J.A. (1998) *J Biol Chem* **273**, 8467–8474.
58. Brodennec, J., Koul, O., Aguado, I., Brichon, G., Zwingelstein, G., and Portoukalian, J. (2000) *J Lipid Res* **41**, 1524–1531.
59. Bodennec, J., Brichon, G., Zwingelstein, G., and Portoukalian, J. (2000) *Methods Enzymol* **312**, 101–114.
60. Rizov, I. and Doulis, A. (2001) *J Chromatogr A*, **922**, 347–354.
61. Bielawski, J., Szulc, Z.M., Hannun, Y.A., and Bielawska, A. (2006) *Methods* **39**, 82–91.
62. Kralik, S.F., Du, X., Patel, C., and Walsh, J.P. (2001) *Anal Biochem* **294**, 190–193.
63. Lowry, O.H. (1951) *J Biol Chem* **193**, 265–275.
64. McDowall, R.D. (1989) *J Chromatogr* **492**, 3–58.
65. Olsson, N.U. and Salem, N. (1997) *J Chromatogr B*, **692**, 245–256.
66. Fenn, J.B., Mann, M., Meng, C.K., Wong, S.F., and Whitehouse, C.M. (1989) *Science* **246**, 64–72.
67. Fenn, J.B., Mann, M., Meng, C.K., Wong, S.F., and Whitrhouse, C.M. (1990) *Mass Spectrom Rev* **9**, 37–70.
68. Cech, N.B. and Enke, C.G. (2001) *Mass Specrom Rev* **20**, 362–387.
69. Griffiths, W.J., Brown, A., Reimendal, R., Yang, Y., Zhang, J., and Sjovall, J. (1996) *Rapid Commun Mass Spectrom* **10**, 1169–1174.
70. Tavana, A.M., Drucker, D.B., Hull, P.S., and Boote, V. (1998) *FEMS Immunol Med Mocrobiol* **21**, 57–64.
71. Korachi, M., Blinkhorn, A.S., and Drucker, D.B. (2002) *Eur J Lipid Sci Technol* **104**, 50–56.
72. Murray, H.E. and Schulten, H.R. (1981) *Chem Phys Lipids* **29**, 11–29.
73. Ma, Y.C. and Kim, H.Y. (1995) *Anal Biochem* 226, 293–301.
74. Hoffman, E.D. (1996) *J Mass Spectrom* **31**, 129–137.
75. Jackson, S.N., Wang, H.J., and Woods, A.S. (2005) *J Am Soc Mass Spectrom* **16**, 133–138.
76. Suzuki, Y., Suzuki, M., Ito, E., Goto-Inoue, N., Miseki, K., Iida, J., Yamazaki, Y. et al. (2006) *J Biochem* **139**, 771–777.
77. Ugarov, M., Egan, T., Koomen, J., Gillig, K.J, Fuhrer, K., Gonin, M., and Schultz, J.A. (2004) *Anal Chem* **76**, 2187–2195.
78. Vieu, C., Chevy, F., Rolland, C., Barbaras, R., Chap, H., Wolf, C., Perret, B. et al. (2002) *J Lipid Res* **43**, 510–522
79. Isaac, G., Bylund, D., Masson, J.E., Markides, K.E., and Bergquist, J. (2003) *J Neurosci Method* **XX**, 1–9.
80. Kuksis, A. and Muher, J.J. (1995) *J Chromatogr B* **671**, 35–70.
81. Samuelsson, K. and Samuelsson, B. (1970) *Chem Phys Lipids* **5**, 44–48.
82. Eising, C.S., Moehring, T., Bahr, U., Duchoslav, E., Karas, M., Simons, K., and Shevchenko, A. (2006) *J Mass Spectrom* **41**, 372–389.
83. Junganwala, F.B., Hayssen, V., Pasquini, J.M., and McCluer, R.H. (1979) *J Lipid Res* **20**, 579–587.
84. Smith, M. and Junganwala, F.B. (1981) *J Lipid Res* **22**, 697–704.
85. Patton, G.M., Fasulo, J.M., and Robins, S.J. (1982) *J Lipid Res* **23**, 190–196.
86. Abidi, S.L. and Mounts, T.L. (1992) *J Chromatogr A* **598**, 209–218.
87. McHowat, J., Jones, J.H., and Creer, M.H. (1996) *J Lipid Res* **37**, 2450–2460.
88. Christi, W.W. (1988) *J Chromatogr A* **454**, 273–284.
89. Brouwers, J.F., Gadella, B.M., van Golde, L.M., and Tielens, A.G. (1998) *J Lipid Res* **39**, 344–353.
90. Brouwers, J.F., Vernooij, E.A., Tielens, A.G., and van Golde, L.M. (1999) *J Lipid Res* **40**, 164–169.
91. Christie, W.W. (1992) *Advance in Lipid Methodology*, The Oily Press, Bridgwater, U.K., pp. 239–271.

92. Karlsson, A.A., Michelsen, P., and Odham, G. (1998) *J Mass Spectrom* **33**, 1192–1198.

93. Cole, R.B. (1997) *Electrospray Ionization Mass Spectrometry*, Wiley, New York.

94. Fox, T.E., Han, X., Kelly, S., Merrill, A.H., Martin, R.E., Anderson, R.E., Gardner, T.W. et al. (2006) *Diabetes* **55**(12), 3573–3580.

95. Dragusin, M., Wehner, S., Kelly, S., Wang, E., Merrill, A.H., Kalff, J.C., and van Echten-Deckert, G. (2006) *FASEB J* **20**, 1930–1939.

96. Luberto, C. and Hannun, Y.A. (1998) *J Biol Chem* **273**, 14550–14559.

97. Bielawska, A., Perry, D.K., and Hannun, Y.A. (2001) *Anal Biochem* **298**, 141–150.

98. Sullards, M.C., Wang, E., Peng, Q., and Merrill, A.H. Jr. (2003) *Cell Mol Biol* **49**, 789–797.

99. Perry, D., Bielawska, A., and Hannun, Y.A. (1999) *Methods in Enzymology*, vol. 312, Academic Press, New York, pp. 22–31.

100. Edsal, L.C. and Spiegel, S. (1999) *Anal Biochem* **272**, 80–86.

101. Caligan, T.B., Peters, K., Ou, J., Wang, E., Saba, J., and Merrill, A.H. Jr. (2000) *Anal Biochem* **281**, 36–44.

102. He, Q., Suzuki, H., Sharma, N., and Sharma, R.P. (2006) *Toxicol Sci* **94**(2), 388–397.

103. Tani, M., Kihara, A., and Igarashi, Y. (2006) *Biochem J* **394**, 237–242.

104. Carroll, J.L. Jr., McCoy, D.M., McGowan, S.E., Salome, R.G., Ryan, A.J., and Mallampalli, R.K. (2002) *Am J Physiol Lung Cell Mol Physiol* **282**(4), 735–742.

105. Tserng, K. and Griffin, R. (2003) *Anal Biochem* **323**, 84–93.

106. Byrwell, W.C. and Perry, R.H. (2006) *J Chromatogr A* **1133**, 149–171.

107. Merrill, A.H. Jr., Wang, E., Mullins, R.E., Jamison, W.C.L., Ninkar, S., and Liotta, D. (1998) *Anal Biochem* **171**, 373–381.

108. Jungalvala, B., Evans, J.E., Bremer, E., and McCluer, R.H. (1983) *J Lipid Res* **24**, 1380–1388.

109. Gaver, R.C. and Sweeley, C.C. (1965) *J Am Oil Chem Soc* **42**, 294–298.

110. Samuelsson, K. and Samuelsson, B. (1970) *Chem Phys Lipids* **5**, 44–79.

8 Mass Spectrometry Methods for the Analysis of Oxidant Stress in Biological Fluids and Tissues: Quantification of F_2-Isoprostanes and F_4-Neuroprostanes Using Mass Spectrometry

*Ginger L. Milne and Jason D. Morrow**

CONTENTS

---

* This chapter is dedicated in memory of Jason D. Morrow whose untimely passing is a great loss to us all.

8.1 INTRODUCTION

Free radicals, largely derived from molecular oxygen, are implicated in a variety of human conditions and diseases including atherosclerosis and associated risk factors, cancer, neurodegenerative diseases, and aging. Damage to tissue biomolecules, such as lipids, proteins, or DNA, by free radicals is postulated to contribute importantly to the pathophysiology of oxidative stress [1,2]. Measuring oxidative stress in humans requires accurate quantification of either free radicals or damaged biomolecules. A number of methods exist to quantify free radicals and their oxidation products, although many of these techniques suffer from a lack of sensitivity and specificity, especially when used to assess oxidant stress status *in vivo*. In a recent multi-investigator study, termed the biomarkers of oxidative stress (BOSS) study, sponsored by the National Institute of Health, it was found that the most accurate method to assess oxidant stress *in vivo* status is the quantification of plasma or urinary isoprostanes (IsoP) [3]. IsoP, a series of prostaglandin (PG)-like compounds produced by the free radical–catalyzed peroxidation of arachidonic acid independent of cyclooxygenases, were first discovered by our laboratory in 1990 [4]. The mechanism of formation of these compounds has been intensely studied and is reviewed in the literature [5,6].

F_2-IsoP (structures shown in Figure 8.1) are stable, robust molecules and are detectable in all human tissues and biological fluids analyzed, including plasma, urine, bronchoalveolar lavage fluid, cerebrospinal fluid (CSF), and bile [4]. The quantification of F_2-IsoP in urine and plasma, however, is most convenient and least invasive. Based on the available data, quantification of these compounds in either plasma or urine is representative of their endogenous production and thus gives a highly precise and accurate index of *in vivo* oxidative stress.

More recently, our laboratory has identified a series of F_2-IsoP-like molecules, termed F_4-neuroprostanes (F_4-NP, structures shown in Figure 8.2), which are formed from the oxidation of docosahexaenoic acid (DHA) *in vivo*. Unlike other tissues, the brain contains significant amounts of this polyunsaturated fatty acid. DHA is particularly enriched in neuronal membranes where it accounts for 25%–35% of the total fatty acids in the gray matter. Thus, quantification of these compounds in the brain tissue appears to be a highly sensitive marker of oxidative neuronal injury in comparison to F_2-IsoP, which is more of a global measure of oxidant stress in the central nervous system.

The methodologies used to measure F_2-IsoP and F_4-NP are summarized in Figure 8.3. To quantify these compounds, our laboratory uses gas chromatographic/ negative ion chemical ionization mass spectrometry (GC/NICI-MS) employing stable isotope dilution [7,8]. The advantages of this technique over other approaches, for example, immunoassay, include its high sensitivity and specificity, that yields quantitative results in the low picogram range. However, these methods are labor intensive and require considerable expenditure on equipment. The methodologies used to quantify F_2-IsoP and F_4-NP are detailed herein. Further, procedures are outlined for the analysis of these compounds both as unesterified fatty acids and esterified in phospholipids. Specific examples are given demonstrating the utility of these assays.

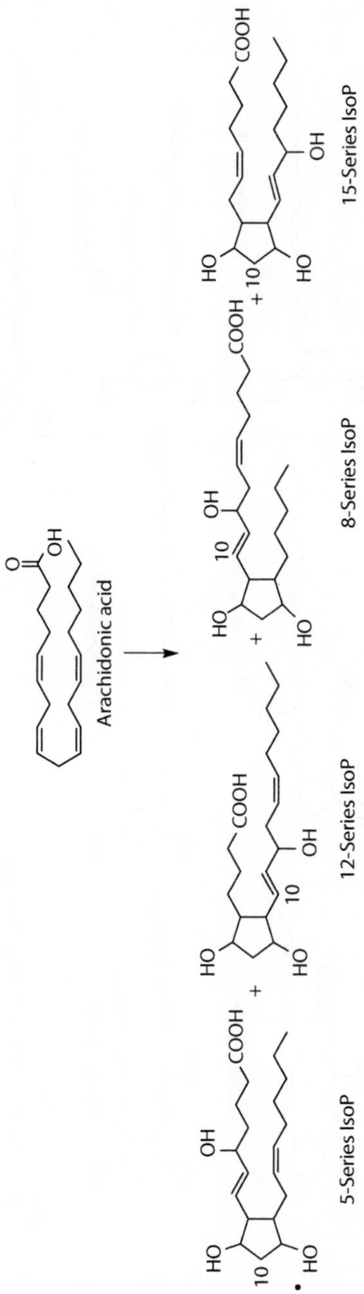

FIGURE 8.1 Structure of F$_2$-IsoP regioisomers derived from the oxidation of arachidonic acid (C20:4, *n*-6).

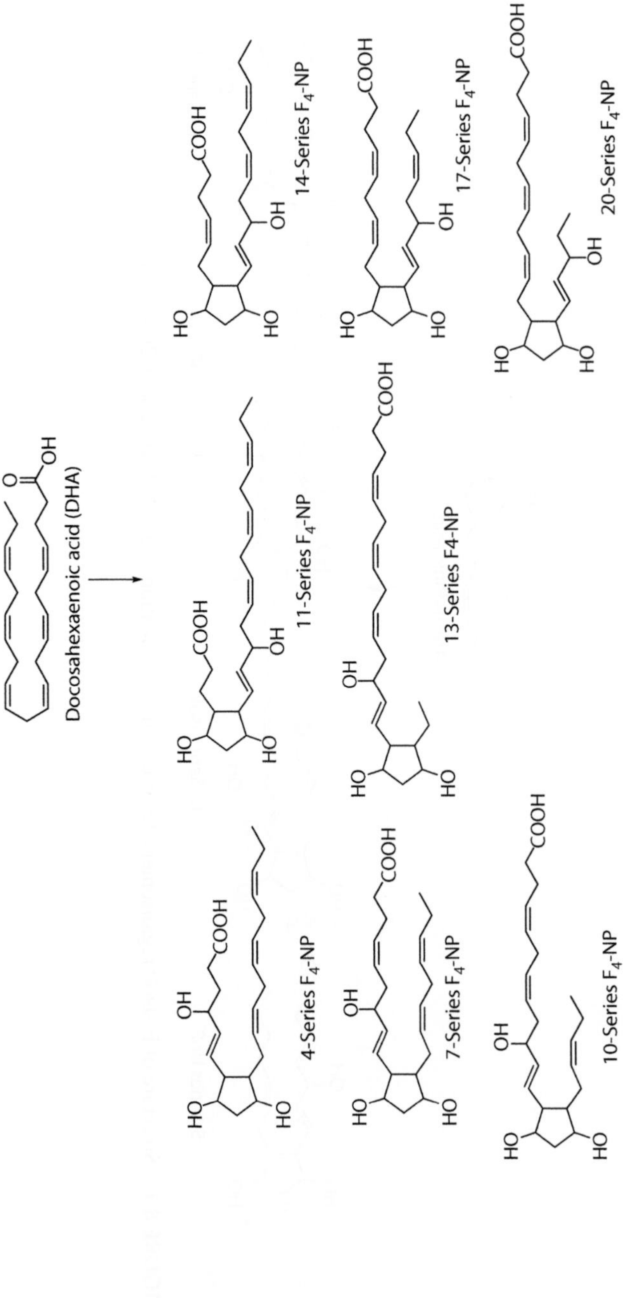

FIGURE 8.2 Structure of F_4-NP regioisomers derived from the oxidation of DHA (C22:6, *n*-3).

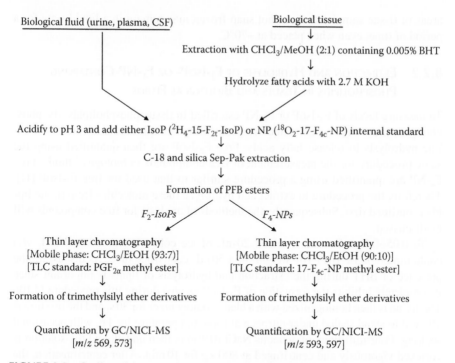

FIGURE 8.3 Outline of the procedures used for the extraction, purification, derivatization, and mass spectrometric analysis of F_2-IsoP from biological sources.

8.2 GENERAL PROCEDURES USED FOR THE MEASUREMENT OF F_2-ISOP AND F_4-NP

8.2.1 HANDLING AND STORAGE OF BIOLOGICAL FLUIDS AND TISSUES

F_2-IsoP are detected in all biological fluids and tissues examined to date. F_4-NP are primarily found in brain tissue, but their formation can also be assessed in other DHA-rich tissues, such as heart and liver, and in plasma, urine, and CSF. A potential drawback to measuring these molecules as an index of endogenous lipid peroxidation is that they can be readily generated *ex vivo* in tissues or biological fluids such as plasma in which arachidonic acid- or DHA-containing lipids are present. This occurs not only when biological fluids or tissues are kept at room temperature but also when they are stored at −20°C [9]. However, we have found that the formation of F_2-IsoP or F_4-NP does not occur if biological fluids or tissues are processed immediately after procurement and if the agents, including butylated hydroxytoluene (BHT, a free radical scavenger) and/or triphenylphosphine (a reducing agent), are added to the organic solvents during the extraction and hydrolysis of phospholipids [9,10]. Thus, samples to be analyzed for these markers should either be processed immediately or stored at −70°C. If samples are to be stored at −70°C, it is ideal for them to be rapidly frozen in liquid nitrogen prior to placement at −70°C. This is especially important for tissue samples as it is known that inner areas of tissue samples that are not snap frozen may remain in a liquid state for a period of time, even when placed at −70°C.

8.2.2 Extraction and Hydrolysis of F_2-IsoP- or F_4-NP-Containing Phospholipids in Tissues and Biological Fluids

To measure levels of F_2-IsoP or F_4-NP esterified in tissue phospholipids, the phospholipids must first be extracted from the tissue sample and subjected to alkaline hydrolysis to release fatty acids. Free F_2-IsoP are then quantified using the same procedure for the measurement of free compounds in biological fluids. Free F_4-NP are quantified using a procedure similar to that used for free F_2-IsoP [11]. Therefore, the procedure to extract and hydrolyze these molecules from tissue lipids is outlined first. Subsequently, the methods of analysis for free compounds will be discussed.

To 0.05–0.5 g of tissue is added 20 mL of ice cold $CHCl_3$/methanol (2:1, v/v) containing 0.005% BHT (by weight) in a 50 mL centrifuge tube. As discussed, the presence of BHT during the extraction and hydrolysis of lipids is important since it completely inhibits the formation of F_2-IsoP *ex vivo* during this procedure [4,10]. The tissue is then homogenized with a blade homogenizer for 30 s, and the mixture is allowed to stand sealed under nitrogen at room temperature for 1 h, with occasional shaking. Four milliliters of aqueous NaCl (0.9%) is then added, and the solution is vortexed vigorously and centrifuged at 800 × g for 10 min. After centrifugation, the upper aqueous layer is discarded and the lower organic layer is carefully separated from the intermediate semisolid proteinaceous layer.

The organic phase containing the extracted lipids is then transferred to a second 50 mL centrifuge tube and evaporated to dryness under a stream of N_2. Two to four milliliters of methanol containing BHT (0.005%) and an equal volume of aqueous KOH (2.7 M) are then added to the residue. The mixture is then vortexed and incubated at 37°C for 30 min to hydrolyze and release F_2-IsoP/F_4-NP. It is then acidified to pH 3 with 1 M HCl and diluted to a final volume of at least 40 mL with pH 3 water in preparation for extraction of free F_2-IsoP/F_4-NP (see discussion below). Dilution of the methanol in the solution with water to 5% or less is necessary to ensure a proper column extraction of F_2-IsoP in the subsequent purification procedure.

For the extraction of lipids containing esterified F_2-IsoP in biological fluids such as plasma, as opposed to tissue, a different method is used [8,10]. We have found that the addition of BHT alone to the Folch solution used to extract plasma lipids does not entirely suppress lipid peroxidation *ex vivo*. On the other hand, the addition of both triphenylphosphine and BHT to the organic solvents used to extract plasma lipids entirely prevents autoxidation. Thus, to 1 mL of a biological fluid, such as plasma, in a 40 mL centrifuge tube is added 20 mL of ice-cold $CHCl_3$/methanol (2:1, v/v) containing 0.005% BHT and triphenylphosphine (5 mg). The mixture is shaken for 2 min and then 10 mL of 0.043 wt % aqueous $MgCl_2$ is added. The mixture is shaken again for 2 min and then centrifuged at 800 × g for 10 min. After centrifugation, the upper aqueous layer is discarded, and the lower organic layer is carefully separated from the intermediate semisolid proteinaceous layer. The organic layer is then dried under nitrogen and subsequently subjected to hydrolysis using the same procedure outlined above for tissue lipids.

8.3 PURIFICATION, DERIVATIZATION, AND QUANTIFICATION OF FREE F$_2$-ISOP

Quantification of F$_2$-IsoP by GC/NICI-MS is extremely sensitive with a lower limit of detection in the range of 1–5 pg using a deuterated internal standard with a [^{2}H$_0$] blank of less than 5 ppt. Thus, it is usually not necessary to assay more than 1–3 mL of a biological fluid or a lipid extract from more than 100 mg of tissue. Further, because urinary levels of F$_2$-IsoP are high (typically greater than 1 ng/mL), we have found that 250 µL of urine is more than adequate to quantify urinary F$_2$-IsoP. The following assay procedure is the method used for analysis of free F$_2$-IsoP in plasma but is equally adaptable for quantifying F$_2$-IsoP in urine, other biological fluids such as CSF, and hydrolyzed lipid extracts of tissues.

Following acidification of 3 mL of plasma to pH 3 with 1 M HCl, 1 ng of deuterated internal standard [^{2}H$_4$]-15-F$_{2t}$-IsoP (8-iso-PGF$_{2\alpha}$) (Cayman Chemical; Ann Arbor, Michigan) is added to the sample. The mixture is then vortexed and applied to a C$_{18}$ Sep-Pak® column (Waters Associates, Milford, Massachusetts) preconditioned with 5 mL methanol and 5 mL of water (pH 3). The sample and subsequent solvents are eluted through the Sep-Pak using a 10 mL sterile plastic syringe. The column is then washed successively with 10 mL of water (pH 3) and 10 mL heptane. The F$_2$-IsoP are then eluted with 10 mL ethyl acetate/heptane (50:50, v/v).

The ethyl acetate/heptane eluate from the C$_{18}$ Sep-Pak is next dried over anhydrous Na$_2$SO$_4$ and applied to a silica Sep-Pak, prewashed with 5 mL ethyl acetate. The cartridge is then washed with 5 mL of ethyl acetate followed by elution of the F$_2$-IsoP with 5 mL of ethyl acetate/methanol (50:50 v/v). The ethyl acetate/methanol eluate is evaporated under a stream of nitrogen.

The F$_2$-IsoP are then converted to pentafluorobenzyl (PFB) esters by treatment with a mixture of 40 µL of 10% PFB bromide in acetonitrile and 20 µL of 10% N,N-diisopropylethylamine in acetonitrile at 37°C for 30 min. The reagents are then dried under nitrogen. This residue is then dissolved in 50 µL chloroform/methanol (3:2, v/v) and subjected to thin layer chromatography (TLC) (LK6D silica plates, marked lanes, glass backed; Whatman, Maidstone, United Kingdom) using the solvent chloroform/ethanol (93:7, v/v). As a TLC standard, approximately 2–5 µg of the methyl ester of PGF$_{2\alpha}$ is chromatographed on a separate plate simultaneously. (The methyl ester of PGF$_{2\alpha}$ rather than the PFB ester is used as the TLC standard because any contamination of the sample being quantified with the methyl ester of PGF$_{2\alpha}$ will not interfere with the analysis owing to the fact that the F$_2$-IsoP are analyzed as PFB esters.) After chromatography, the standard is visualized by spraying the plate with a 10% solution of phosphomolybdic acid in ethanol followed by heating. Compounds migrating in the region of the methyl ester of PGF$_{2\alpha}$ ($R_f \sim 0.15$) and the adjacent areas 1 cm above and below are scraped and extracted from the silica gel with ethyl acetate.

The ethyl acetate is dried under a stream of nitrogen and the PFB esters of the F_2-IsoP are converted to trimethylsilyl (TMS) ether derivatives by adding $20\,\mu L$ N,O-bis-(trimethylsilyl)trifluoroacetamide (BSTFA) and $10\,\mu L$ dimethylformamide and incubating at 37°C for 20 min. The reagents are dried under a stream of nitrogen and the derivatized F_2-IsoP are redissolved in $10\,\mu L$ of undecane, which has been dried over calcium hydride, for analysis by GC/MS.

For quantification of F_2-IsoP by GC/MS, we routinely use an Agilent 5973 mass spectrometer with a computer interface. The F_2-IsoP are separated on a 15 m DB1701-fused silica capillary column (i.d. $0.25\,\mu m$, film thickness $0.25\,\mu m$) because we have found that this column gives excellent separation of individual regioisomers compared to other columns. The column temperature is programmed from 190°C to 300°C at 20°C/min. Methane is used as the carrier gas for NICI. The ion source temperature is 200°C. The ion monitored for endogenous F_2-IsoP is the carboxylate anion m/z 569 (M-181, loss of $CH_2C_6F_5$). The corresponding carboxylate anion for the deuterated internal standard is m/z 573. Each day, the sensitivity of the mass spectrometer is checked by injecting a standard consisting of $PGF_{2\alpha}$ and $[^2H_4]$-15-F_{2t}-IsoP (40 pg each).

Shown in Figure 8.4 is a selected ion monitoring (SIM) chromatogram obtained from the analysis of F_2-IsoP in human plasma. The series of peaks in the upper m/z 569 ion current chromatogram represent different endogenous F_2-IsoP. This pattern of peaks is virtually identical to that obtained from all other biological fluids and tissues that we have examined to date. In the lower m/z 573 chromatogram, the single peak represents the $[^2H_4]$-15-F_{2t}-IsoP internal standard that was added to the plasma sample. For quantification purposes, the peak denoted by the star (*), which co-elutes with the 15-F_{2t}-IsoP internal standard, is routinely measured. Using the ratio of the intensity of this peak to that of the internal standard, the concentration of F_2-IsoP was calculated to be 83 pg/mL. Normal concentrations of F_2-IsoP in human plasma have been calculated to be 35 ± 6 pg/mL (mean ± 1 SD) while normal concentrations in human urine are 1.6 ± 0.6 ng/mL (mean ± 1 SD). The reader is referred to previously published data for normal levels of F_2-IsoP in other biological fluids and tissues [10].

Quantification of the F_2-IsoP based on the intensity of the starred (*) peak shown in Figure 8.4 is highly precise and accurate. The precision is ±6% and the accuracy is 96%. It should be noted herein that $PGF_{2\alpha}$, which is derived primarily from cyclooxygenase, does not co-chromatograph with the peak being quantified by this assay. Rather, it elutes with the peak noted by a plus (+). The intensity of this peak varies depending on the extent to which $PGF_{2\alpha}$ is generated enzymatically via the cyclooxygenase versus via free radical–catalyzed nonenzymatic arachidonic acid peroxidation. Therefore, we have chosen to quantify only the starred peak in our analysis and not the subsequently eluting peaks.

8.4 PURIFICATION, DERIVATIZATION, AND QUANTIFICATION OF FREE F_4-NP

In our laboratory, we primarily quantify F_4-NP in the lipid extract of tissues such as brain, heart, and liver, which contain substantial amounts of DHA. The

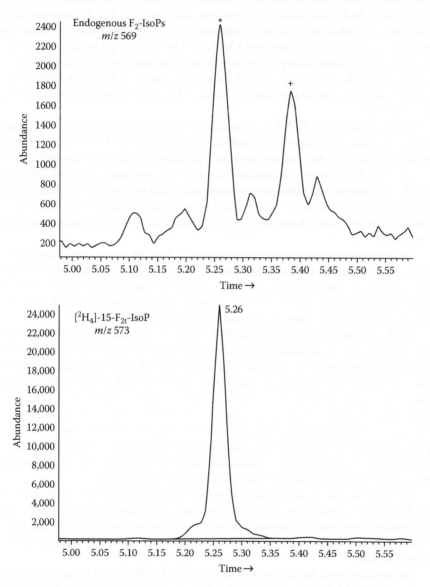

FIGURE 8.4   Analysis of F$_2$-IsoP in plasma obtained from a rat 4 h after treatment with CCl$_4$ (2 mL/kg orogastrically) to induce endogenous lipid peroxidation. The m/z 573 ion current chromatogram represents the [^{2}H$_4$]-15-F$_{2t}$-IsoP internal standard. The m/z 569 ion current chromatogram represents endogenous F$_2$-IsoP. The peak in the upper chromatogram represented by the star (*) is the one routinely used for the quantification of F$_2$-IsoP. The peak represented by the plus (+) can be comprised of both F$_2$-IsoP- and cyclooxygenase-derived PGF$_{2\alpha}$. The concentration of F$_2$-IsoP in the plasma using the starred (*) peak for quantification was calculated to be 83 pg/mL.

methodology described herein can, however, be applied to measurement of these molecules in biological fluids (CSF, plasma, and urine) [11,12]. To measure these molecules, we utilize an ^{18}O-labeled standard, $[^{18}O_2]$-17-F_{4c}-NP. 17-F_{4c}-NP was synthesized as described by Quan and Cha [13]. The ^{18}O-labeled molecule was prepared in our laboratory using the method of Murphy and Clay [14], which involves methylation of the carboxylic acid followed by alkaline hydrolysis with $Li^{18}OH$. The resulting internal standard produced a blank of less than 15 ppt when analyzed by GC/NICI-MS.

The analysis of free F_4-NP is quite similar to that described above for F_2-IsoP. After extraction and hydrolysis of the tissue sample, 1 ng of $[^{18}O_2]$-17-F_{4c}-NP internal standard is added to the resulting mixture. The sample is then vortexed and applied to a C_{18} Sep-Pak column preconditioned with 5 mL methanol and 5 mL of water (pH 3). The sample and subsequent solvents are eluted through the Sep-Pak using a 10 mL sterile plastic syringe. The column is then washed successively with 10 mL of water (pH 3) and 10 mL heptane. F_4-NP is eluted with 10 mL ethyl acetate/heptane (50:50, v/v).

The ethyl acetate/heptane eluate from the C_{18} Sep-Pak is next dried over anhydrous Na_2SO_4 and applied to a silica Sep-Pak prewashed with 5 mL ethyl acetate. The cartridge is then washed with 5 mL of ethyl acetate/heptane (75:25, v/v) followed by elution of F_2-IsoP with 5 mL of ethyl acetate/methanol (50:50, v/v). The ethyl acetate/methanol eluate is evaporated under a stream of nitrogen.

The F_4-NP are then converted to PFB esters as described above for F_2-IsoP. The reagents are dried under nitrogen. The residue is then dissolved in 50 μL chloroform/methanol (3:2, v/v) and subjected to TLC (LK6D silica plates, marked lanes, glass backed; Whatman, Maidstone, United Kingdom) using the solvent chloroform/ethanol (90:10, v/v). The TLC plate is run to 15 cm. As a TLC standard, approximately 2–5 μg of the methyl ester of 17-F_{4c}-NP is chromatographed on a separate plate simultaneously and then visualized by spraying the plate with a 10% solution of phosphomolybdic acid in ethanol followed by heating. Compounds migrating in the region from immediately below to 2 cm above the TLC standard are scraped and extracted from the silica gel with ethyl acetate.

The ethyl acetate is dried under a stream of nitrogen and the PFB esters of the F_2-NP are converted to TMS ether derivatives, as described for F_2-IsoP. The reagents are dried under a stream of nitrogen and the derivatized F_4-NP are redissolved in 10 μL of undecane, which has been dried over calcium hydride, for analysis by GC/MS.

The ion monitored for the analysis of F_4-NP by GC/NICI-MS is the carboxylate anion m/z 593 (M-181, loss of $CH_2C_6F_5$). The corresponding carboxylate anion for the ^{18}O-labeled internal standard is m/z 597. Shown in Figure 8.5 is a SIM chromatogram obtained from the analysis of F_4-NP in the brain tissue from a normal mouse. The series of peaks in the upper m/z 593 chromatogram represents different endogenous F_4-NP while the single peak in the lower m/z 597 chromatogram represents the $[^{18}O_2]$-17-F_{4c}-NP internal standard. For quantification purposes, the area of the peaks denoted by the star (*) is compared to the area of the internal standard peak. Amounts of F_4-NP in this sample were calculated to be 93.3 ng/g wet brain tissue. Levels of F_4-NP in normal human brain tissue average 4.9 ± 0.6 ng/g.

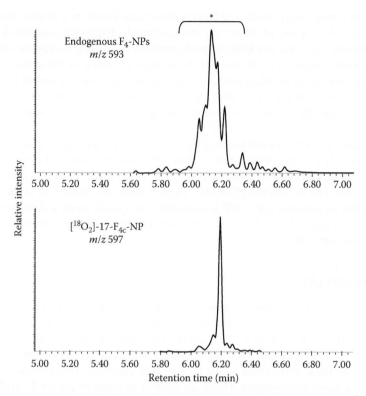

FIGURE 8.5   Analysis of F$_4$-NP in basal ganglia brain tissue from a normal mouse. The m/z 597 ion current chromatogram represents the [^{18}O$_2$]-17-F$_{4c}$-NP internal standard. The m/z 593 ion current chromatogram represents endogenous F$_4$-NP. The area of the peaks indicated in the upper chromatogram by the star (*) is measured for quantification of the F$_4$-NP. The concentration of F$_4$-IsoP in this tissue sample was calculated to be 93.3 ng/g wet brain tissue.

8.5   F$_2$-ISOP AND F$_4$-NP AS INDEXES OF OXIDANT STRESS *IN VIVO*

The true utility of the F$_2$-IsoP and F$_4$-NP is in the quantification of lipid peroxidation and thus oxidant stress status *in vivo*. As discussed, these molecules are stable and robust. In addition, F$_2$-IsoP have been detected in all human tissues and biological fluids that have been analyzed, including plasma, urine, bronchoalveolar lavage fluid, CSF, and bile [7,10]. Based on available data, quantification of these compounds in either plasma or urine is representative of their endogenous production, and thus gives a highly precise and accurate index of *in vivo* oxidant stress. While measurement of F$_2$-IsoP in plasma is indicative of their endogenous formation at a specific point in time, analysis of these compounds in urine is an index of systemic or "whole body" oxidant stress integrated over time.

Importantly, normal levels of F$_2$-IsoP in healthy humans have been defined [7,10,15]. These normal determinations thus allow for the assessment of the effects of diseases on endogenous oxidant tone and for the determination of the extent to which various therapeutic interventions affect levels of oxidant stress. Elevations of IsoP in human body fluids and tissues have been found in a diverse array of human disorders, some of which are atherosclerosis, hypercholesterolemia, diabetes, obesity, cigarette smoking, neurodegenerative diseases, rheumatoid arthritis, and many others [16–28]. Further, treatments for some of these conditions, including antioxidant supplementation, antidiabetic treatments, cessation of smoking, and even weight loss, have been shown to decrease the production of F_2-IsoP [17,21,29]. Thus, the clinical utility of measuring the molecules is great and continues to grow.

As discussed above, normal levels of F_4-NP have been defined in brain tissue from humans [12,14]. Elevated levels of these molecules have been detected in hippocampal and temporal lobe tissues from Alzheimer's disease patients as well as in the CSF from patients with the disease [27,30,31]. These findings are significant as they suggest that the measurement of F_4-NP is a valuable and specific analytical tool that can be used to explore the role of oxidant stress and possible therapeutic interventions in neurodegenerative diseases.

8.6 SUMMARY

This chapter outlines methods to assess lipid peroxidation associated with oxidant injury *in vivo* by quantifying concentrations of either free F_2-IsoP in biological fluids or levels of F_2-IsoP or F_4-NP esterified in tissue lipids. Our GC/MS methodologies described herein are highly precise and accurate. A potential shortcoming with these approaches is that they require expensive instrumentation, for example, a mass spectrometer. However, several immunoassays for F_2-IsoP and 15-F_{2t}-IsoP have become available from commercial sources. At this time, the accuracy and reliability of these assays for quantifying F_2-IsoP in biological fluids have not been fully validated by mass spectrometry. If these immunoassays prove to be a reliable measure of F_2-IsoP, however, this should greatly expand the use of F_2-IsoP to assess oxidant stress. In conclusion, studies carried out over the past several years have shown that measurement of either F_2-IsoP or F_4-NP has overcome many of the limitations associated with the other methods to assess oxidant status, especially when applied to the measurement of oxidant stress *in vivo* in humans.

ACKNOWLEDGMENTS

Supported by NIH grants DK48831, ES000267, ES13125, GM42056 and GM15431.

LIST OF ABBREVIATIONS

BHT	butylated hydroxytoluene
CSF	cerebrospinal fluid
DHA	docosahexaenoic acid
GC	gas chromatography

IsoP   isoprostanes
NICI-MS   negative ion chemical ionization mass spectrometry
NP   neuroprostanes
PFB   pentafluorobenzyl

REFERENCES

1. Chisholm, G.M. and Steinberg, D. (2000) The oxidative modification hypothesis of atherogenesis: An overview. *Free Radic. Biol. Med.* 18:1815–1826.
2. Halliwell, B. and Gutteridge, J.M. (1990) Role of free radicals and catalytic metal ions in human disease: An overview. *Methods Enzymol.* 186:1–85.
3. Kadiiska, M.B. et al. (2005) Biomarkers of oxidative stress study II: Are oxidation products of lipids, proteins, and DNA markers of CCl4 poisoning? *Free Radic. Biol. Med.* 38:698–710.
4. Morrow, J.D. et al. (1990) A series of prostaglandin F2-like compounds are produced *in vivo* in humans by a non-cyclooxygenase, free radical-catalyzed mechanism. *Proc. Natl. Acad. Sci. USA* 87:9383–9387.
5. Morrow, J.D. et al. (1999) The isoprostanes: Unique prostaglandin like products of free radical-catalyzed lipid peroxidation. *Drug Metab. Rev.* 31:117–139.
6. Famm, S.S. and Morrow, J.D. (2003) The isoprostanes: Unique products of arachidonic acid oxidation—A review. *Curr. Med. Chem.* 10:1723–1740.
7. Morrow, J.D. et al. (1999) Quantification of the major urinary metabolite of 15-F2t-isoprostane (8-iso-PGF2alpha) by a stable isotope dilution mass spectrometric assay. *Anal. Biochem.* 269:326–331.
8. Milne, G.L. et al. (2007) Quantification of F$_2$-isoprostanes as a biomarker of oxidative stress. *Nat. Protoc.* 2:221–226.
9. Morrow, J. D., Harris, T.M., and Roberts II, L.J. (1990) Noncyclooxygenase oxidative formation of a series of novel prostaglandins: Analytical ramifications for measurement of eicosanoids. *Anal. Biochem.* 184:1–10.
10. Morrow, J.D. and Roberts, L.J. (1999) Mass spectrometric quantification of F$_2$-isoprostanes in biological fluids and tissues as a measure of oxidant stress. *Methods Enzymol.* 300:3–12.
11. Musiek, E.S. et al. (2004) Quantification of F-ring isoprostane-like compounds (F4-neuroprostanes) derived from docosahexaenoic acid *in vivo* in humans by a stable isotope dilution mass spectrometric assay. *J. Chromatogr. B Analyt. Technol. Biomed. Life Sci.* 799:95–102.
12. Roberts, L.J. II et al. (1998) Formation of isoprostane-like compounds (neuroprostanes) *in vivo* from docosahexaenoic acid. *J. Biol. Chem.* 273:13605–13612.
13. Quan, L.G. and Cha, J.K. (2002) Preparation of isoprostanes and neuroprostanes. *J. Am. Chem. Soc.* 124:12424–12425.
14. Murphy, R.C. and Clay, K.L. (1982) Preparation of 18O derivatives of eicosanoids for GC-MS quantitative analysis. *Methods Enzymol.* 86:547–551.
15. Liang, Y. et al. (2003) Quantification of 8-iso-prostaglandin-F(2alpha) and 2,3-dinor-8-iso-prostaglandin-F(2alpha) in human urine using liquid chromatography-tandem mass spectrometry. *Free Radic. Biol. Med.* 34:409–418.
16. Milne, G.L., Musiek, E.S., and Morrow, J.D. (2005) F2-isoprostanes as markers of oxidative stress *in vivo*: An overview. *Biomarkers* 10(Suppl 1):S10–23.
17. Morrow, J.D. (2005) Quantification of isoprostanes as indices of oxidant stress and the risk of atherosclerosis in humans. *Arterioscler. Thromb. Vasc. Biol.* 25:279–286.

18. Gniwotta, C. et al. (1997) Prostaglandin F2-like compounds, F2-isoprostanes, are present in increased amounts in human atherosclerotic lesions. *Arterioscler. Thromb. Vasc. Biol.* 17:3236–3241.
19. Gross, M. et al. (2005) Plasma F2-isoprostanes and coronary artery calcification: The CARDIA study. *Clin. Chem.* 51:125–131.
20. Davi, G. et al. (1997) *In vivo* formation of 8-Epi-prostaglandin F2 alpha is increased in hypercholesterolemia. *Arterioscler. Thromb. Vasc. Biol.* 17:3230–3235.
21. Davi, G. et al. (2002) Platelet activation in obese women: Role of inflammation and oxidant stress. *JAMA* 288:2008–2014.
22. Gopaul, N.K. et al. (1995) Plasma 8-epi-PGF2 alpha levels are elevated in individuals with non-insulin dependent diabetes mellitus. *FEBS Lett.* 368:225–229.
23. Keaney, J.F. Jr. et al. (2003) Obesity and systemic oxidative stress: Clinical correlates of oxidative stress in the Framingham study. *Arterioscler. Thromb. Vasc. Biol.* 23:434–439.
24. Morrow, J.D. et al. (1995) Increase in circulating products of lipid peroxidation (F2-isoprostanes) in smokers. Smoking as a cause of oxidative damage. *N. Engl. J. Med.* 332:1198–1203.
25. Montine, K.S. et al. (2004) Isoprostanes and related products of lipid peroxidation in neurodegenerative diseases. *Chem. Phys. Lipids* 128:117–124.
26. Montine, T.J. et al. (1999) Cerebrospinal fluid F2-isoprostanes are elevated in Huntington's disease. *Neurology* 52:1104–1105.
27. Montine, T.J. et al. (1998) Cerebrospinal fluid F2-isoprostane levels are increased in Alzheimer's disease. *Ann. Neurol.* 44:410–413.
28. Basu, S. et al. (2001) Raised levels of F(2)-isoprostanes and prostaglandin F(2alpha) in different rheumatic diseases. *Ann. Rheum. Dis.* 60:627–631.
29. Dietrich, M. et al. (2002) Antioxidant supplementation decreases lipid peroxidation biomarker F(2)-isoprostanes in plasma of smokers. *Cancer Epidemiol. Biomarkers Prev.* 11:7–13.
30. Montine, T.J. et al. (2002) Lipid peroxidation in aging brain and Alzheimer's disease. *Free Radic. Biol. Med.* 33:620–626.
31. Reich, E.E. et al. (2001) Brain regional quantification of F-ring and D-/E-ring isoprostanes and neuroprostanes in Alzheimer's disease. *Am. J. Pathol.* 158:293–297.

Part III

Lipid-Mediated Regulation
of Gene Expression

9 Methods for Assessing the Role of Lipid-Activated Nuclear Receptors in the Regulation of Gene Transcription

Jenny Kaeding, Jocelyn Trottier, and Olivier Barbier

CONTENTS

9.1 INTRODUCTION

9.1.1 CLASSIFICATION OF NUCLEAR RECEPTORS

Nuclear receptors (NRs) function as ligand-activated transcription factors that regulate the expression of target genes to affect processes such as reproduction, development, and general metabolism. Forty-eight members of the NR family have so far been identified in the human genome.

Based on their ligand affinity, NRs can be divided into three categories: the endocrine receptors, the adopted orphan receptors, and the orphan receptors (Figure 9.1) [1]. The lipid sensors, peroxisome proliferator-activated receptors (PPARs), liver X receptors (LXRs), farnesoid X receptor (FXR), pregnane X receptor (PXR) (also named steroid and xenobiotic receptor in mice; SXR), constitutive androstane receptor (CAR) and retinoid X receptor (RXR), belong all to the second class. The endogenous ligands identified so far consist of low-affinity (compared to those of classical

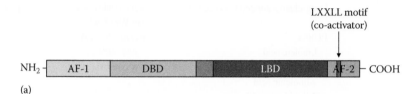

(a)

	Ligands	NR	
Endocrine receptors	Hormonal lipids	ER α,β	RAR α,β
		PR	TR α,β
	(high-affinity)	AR	VDR
		GR	EcR
		MR	
Adopted orphan receptors	Dietary lipids	RXR α,β	FXR
		PPAR α,β	PXR/SXR
	(low-affinity)	LXR a,b	CAR
Orphan receptors	Unknown	SF-1	ROR α,β
		LRH-1	ERR α,β
		DAX-1	RVR α,β
		SHP	GCNF
		TLX	TR 2,4
		PNR	HNF-4
		NGFI-B α,β	COUP-TF α,β

(b)

FIGURE 9.1 (a) NR structure. (b) NR classification. (a) NR contain a ligand-independent activation function domain (AF-1), a DNA-binding domain (DBD), a ligand-binding domain (LBD), and a ligand-dependent activation function domain (AF-2). Many co-activators interact with the LXXLL motif localized in the AF-2, when the ligand is bound to the NR to induce a conformation change. (b) Human NR are divided into three groups: (1) Endocrine receptors activated by hormonal receptor; (2) adopted orphan receptors activated by dietary lipids; (3) orphan receptors that are activated by unknown endogenous ligands.

steroid hormone receptors) dietary lipids and their metabolites (Table 9.1). In addition, a multitude of synthetic high-affinity agonists (Table 9.2) and antagonists (Table 9.3) specific for each lipid sensor have been created. Lipid sensors bind to DNA as heterodimers with RXR as an obligate (permissive) partner to regulate transcription of genes implicated in various metabolic pathways in different tissues. In particular, lipid sensors regulate a metabolic gene network controlling lipid synthesis, metabolism, storage, transport, and elimination [1,2].

TABLE 9.1
Endogenous Ligands of Lipid Sensors

Nuclear Receptor	Ligand	EC50 (µM)	Ref.
PPARα (NR1C1)			
	8-HEPE		
	8(S)-HETE	0.2	[218]
	Long-chain saturated FAs (≥C12)	Very weakly: ~30% of Wy14643	[218]
	PUFAs	Relatively weak: 30%–40% of	[218]
	Linoleic acid	Wy14643	
	α-linolenic acid		
	γ-linolenic acid		
	Arachidonic acid		
	Docosahexanoic acid		
	Eicosapentaenoic acid		
PPARγ (NR1C3)			
	13-HETE		
	9-HODE		
	13-HODE		
	15d-PGJ$_2$ (15-deoxy prostaglandin J$_2$)		[219]
	Triterpenoids		[1]
PPARδ (NR1C2)			
	Carboprostacyclin		[220]
	PUFAs		[218]
	Linoleic acid		
	Arachidonic acid		
	Docosahexanoic acid		
	Eicosapentaenoic acid		
LXRβ (NR1H2)			
	24(S),25-epoxycholesterol	3.0	[221]
	D-glucose	3141	[222]
	D-glucose-6-phosphate		[222]
	24(S)-hydroxycholesterol	3.0	[221]
	22(R)-hydroxycholesterol	3.0	[221]
	27-hydroxycholesterol	0.071	[2]

TABLE 9.1 (continued)
Endogenous Ligands of Lipid Sensors

Nuclear Receptor	Ligand	EC50 (µM)	Ref.
LXRα (NR1H3)			
	24(S),25-epoxycholesterol	4.0	[2]
	D-glucose	308	[222]
	D-glucose-6-phosphate		[222]
	24(S)-hydroxycholesterol	4.0	[2]
	22(R)-hydroxycholesterol	5.0	[2]
	27-hydroxycholesterol	0.085	[2]
FXR (NR1H4)			
	CA	>10	[2]
		586	[223]
	CDCA	7.0	[224]
		5.0	[2]
		17	[223]
	DCA	100	[2]
		131	[223]
	6-ECDCA	0.099	[10]
	6α-ethyl-chenodeoxycholic acid		
	22(R)-hydroxycholesterol	>3µM	[2]
	LCA	5.0	[2]
		3.0	
	UDCA	185	[223]
PXR (NR1I2)			
	5β-cholestane-3α,7α,12α-triol	3–5	[2]
	LCA	9–15	[2]
	Pregnenolone-16α-carbonitrile	0.3	[2]
	Vitamin K		[2]
CAR (NR1I3)			
	Pregnanedione	0.67	[2]
RXRα (NR2B1)			
	Methoprenic acid	2.0	[8]
	9-*cis* retinoic acid	0.0067–0.073	[8]
RXRβ (NR2B2)			
	9-*cis* retinoic acid	0.0061–0.117	[8]
RXRβ (NR2B3)			
	9-*cis* retinoic acid	0.0097–0.085	[8]

TABLE 9.2

Synthetic Ligands of Lipid Sensors

Nuclear Receptor	Ligand	EC50 (µM)	Ref.
PPARα (NR1C1)			
	Bezafibrate	50	[225]
	Ciprofibrate	20	[226]
	Clofibrate	55	[225]
	Fenofibrate	30	[225]
	GI 262570	0.45	[225]
	GW2433	0.17	[225]
	GW7647	0.006	[3]
	GW7845	3.5	[225]
	GW9578	0.05	[225]
	JTT-501	1.9	[225]
	KRP-297	0.85	[225]
	L-796449	0.0041	[225]
	L-165041	10	[225]
	SB 213068	0.74	[225]
	TAK-559	0.067	[3]
	Wy14643	5.0	[225]
PPARγ (NR1C3)			
	Bezafibrate	60	[225]
	Clofibrate	~500	[225]
	Fenofibrate	300	[225]
	GI 262570	0.00034	[225]
	GW0207	0.044	[225]
	GW2433	2.5	[225]
	GW1929	0.0062	[225]
	GW7845	0.00071	[225]
	GW9578	1.0	[225]
	JTT-501	0.083	[225]
	KRP-297	0.083	[225]
	L-165041	5.5	[225]
	L-796449	0.0052	[225]
	Pioglitazone	0.58	[225]
	Rosiglitazone	0.043	[225]
	SB 213068	0.066	[225]
	TAK-559	0.031	[3]
	Troglitazone	0.55	[225]
	Wy14643	60	[225]
PPARδ (NR1C2)			
	Bezafibrate	20	[225]
	GW0742	0.001	[3]

TABLE 9.2 (continued)
Synthetic Ligands of Lipid Sensors

Nuclear Receptor	Ligand	EC50 (μM)	Ref.
	GW2433	0.19	[225]
	GW9578	1.4	[225]
	GW501516	0.0012	[70]
	KRP-297	9.1	[225]
	L-165041	0.53	[225]
	L-796449	0.0079	[225]
	Wy14643	35	[225]
LXRβ (NR1H2)			
	Acetyl-podocarpic dimmer	0.001	[2]
	F3methylAA	Kd: 0.007	[2]
	GW3965	0.03	[227]
	Paxillone	4.0	[2]
	T091317	0.02	[228]
		0.05	[2]
LXRα (NR1H3)			
	GW3965	0.19	[227]
	F3methylAA	Kd: 0.013	[2]
	Paxilline	4.0	[2]
	T091317	0.02	[228]
		0.05	[2]
FXR (NR1H4)			
	AGN29	2.0	[229]
	AGN31	2.9	[229]
	Fexaramine	0.25	[2]
	GW4064	0.08	[10]
		0.015	[2]
PXR (NR1I2)			
	CITCO	~3	[18]
	Clotrimazole	0.8–1.6	[18]
	Dexamethasone	0.8	[2]
	Hyperforin	0.027	[2]
	Lovastatin	1–5	[18]
	Nifedipine	4.3	[18]
	Paclitaxel	5.0	[18]
	5β-Pregnane-3,20-dione	3.1	[230]
	Rifampicin	0.8–8	[2,18]
	RU486	5.5	[230]
	(+)-S20	0.4	
	Schisandris A and B	1.25–2	[2]

(continued)

TABLE 9.2 (continued)
Synthetic Ligands of Lipid Sensors

Nuclear Receptor	Ligand	EC50 (µM)	Ref.
	SR12813	0.12	[230]
		0.2	[2]
	Taxol	5.0	[2]
	TCPOBOP 1,4-bis[2-(3,5-dichlorpyridyloxyl)]-benzene	3.9	[230]
	Troglitazone	~3	[18]
CAR (NR1I3)			
	CITCO	0.049	[2]
		0.025	[18]
	Meclizine (also antagonist)	0.025	[2]
	TCPOBOP 1,4-bis[2-(3,5-dichlorpyridyloxyl)]-benzene	0.02	[2]
RXRα (NR2B1)			
	CD3254	0.003	[8]
	LGD1069	0.036	[8]
	LG100268	0.0032	[8]
RXRβ (NR2B2)			
	LGD1069	0.021	[8]
	LG100268	0.003–0.0068	[8]
RXRγ (NR2B3)			
	LGD1069	0.029	[8]
	LG100268	0.003–0.0097	[8]

9.1.2 ROLE OF LIPID SENSORS

Each isoform of the PPAR NR family has a specific, nonredundant role. PPARα functions as a global regulator of fatty acid metabolism. Its target genes coordinate the metabolic changes necessary to conserve energy during fasting and feeding such as fatty acid β-oxidation [3]. PPARγ is an important regulator of adipogenesis. It promotes adipocyte differentiation and has been shown to be essential for the development of adipose tissue *in vivo* [3,4]. PPARδ was the last PPAR isoform characterized and has been shown to play a role in epidermal maturation. In addition, emerging data suggest that PPARδ may also affect lipid metabolism and energy utilization in peripheral tissues [4,5].

Both LXRα and LXRβ act as cholesterol sensors and regulate fatty acid metabolism. In response to elevated sterol concentrations, LXR transactivate genes that control transport, catabolism, and elimination of cholesterol [2,6]. FXRα is the only isoform of this NR expressed in humans. FXR is a bile acid sensor responsible for repressing bile acid synthesis and regulating enterohepatic recirculation [2]. In this context, the NR

TABLE 9.3

Antagonists of Lipid Sensors

Nuclear Receptor	Ligand	IC$_{50}$ [μM]	Ref.
PPARα (NR1C1)			
	MK886	4.6	[3]
PPARγ (NR1C3)			
	BADGE	5.0	[3,231]
	Diclofenac		
	GW9662	8.48	[3]
	LG-100641	6.36	[3,232]
	PD068235	6.1	[3]
	SR-202	3.85	[3]
PPARδ (NR1C2)			
	—		
LXRβ (NR1H2)			
	—		
LXRα (NR1H3)			
	—		
FXR (NR1H4)			
	(Guggulsterone) [not specific]	1–5	[233]
		10	[2]
PXR (NR1I2)			
	Ecteinascidin 743	0.003	[2]
CAR (NR1I3)			
	Androstanol	0.4	[2]
	Androstenol	0.4	[2]
	Clotrimazole	0.69	[2]
	Meclizine	0.069	[2]
RXRα (NR2B1)			
	LG100754	0.0034	[8]
RXRβ (NR2B2)			
	LG100754	0.01	[8]
RXRγ (NR2B3)			
	LG100754	0.0122	[8]

is acting as an opponent to LXR [7]. PXR and CAR function as regulators of detoxification and elimination of xenobiotics and toxic endogenous lipids by controlling the expression of phase I and II enzymes as well as drug transporters [2,7].

RXRs are predominantly known as obligate partners of the other lipid sensors. However, the ability of RXR to form homodimers and to bind DR-1 elements *in vitro* raised the question of the existence of an independent RXR signaling pathway, which needs further investigation [8].

The main expression site of all lipid sensors is the liver; however, in concert with their specific regulatory role, their expression has also been identified in many peripheral tissues (Table 9.4). A considerable number of endogenous ligands have been identified, which consist mainly of dietary lipids and their metabolites (Table 9.1). However, additional, not-yet-identified endogenous ligands may exist. Furthermore, the physiologic importance of a subgroup of ligands remains to be proven because their receptor affinity is considerably lower than their circulating concentrations [9].

9.1.3 TRANSCRIPTIONAL MECHANISMS OF LIPID SENSORS

NRs mainly share the same structural organization (Figure 9.1): they possess a ligand-independent N-terminal transcriptional activation function domain (AF-1), a core DNA-binding domain (DBD) consisting of two highly conserved zinc-finger motifs, a hinge region that allows flexibility of the NR to dimerize and bind to DNA, and a large C-terminal region containing the ligand-binding domain (LBD), a dimerization interface and a ligand-dependent activation function domain (AF-2). Upon ligand binding, lipid sensors undergo allosteric changes within the LBD that coordinately dissociates co-repressors (Table 9.5) and facilitates co-activator recruitment (Table 9.5) to enable transcriptional activation [1].

TABLE 9.4

Tissue Distribution of Lipid Sensors

Nuclear Receptor	Tissue	Ref.
PPARα	Liver, kidney, heart, brown fat, skeletal muscle, large intestine, macrophages, SMCs, endothelial cells, T lymphocytes	[3,5,234]
PPARγ	Liver, adipose tissue, skeletal muscle, colon, macrophages, SMCs, endothelial cells, lymphoid tissues	[3,5,234]
PPARδ	Wide range of tissues, high levels in brain, adipose tissue, skin	[3,5,234]
LXRβ	Ubiquitous	
LXRα	Liver, adipose tissue, kidney, intestine, lung, adrenals, macrophages, spleen	[2]
FXR	Liver, kidney small intestine, colon, adrenals, SMC	
PXR	Liver, intestine, kidney, colon, lung, lymphocytes	[2]
CAR	Liver, intestine, kidney, stomach	[2]
RXRα	Liver, lung, muscle, kidney, epidermis, intestine	[8]
RXRβ	Ubiquitous	[8]
RXRγ	Brain, muscle	[8]

TABLE 9.5

Co-Repressors and Co-Activators of Lipid Sensors

Nuclear Receptor	Co-Activators	Co-Repressors
PPARα (NR1C1)	PPARBP	NRIP1
	NCOA6	NCOR1
	BFE	
	CREBBP	
	CITED2	
	NCOA1	
	NCOA3	
	SWI2/SNF2	
	PGC-1α	
	PPARGC1B	
PPARγ (NR1C3)	PGC-2	NRIP1
	ARA-70	SAF-B
	PGC-1α	TAZ
	PPARGC1B	NCOR1
	CREBBP	NCOR2
	P300	
	CITED2	
	ERAP140	
	PPARBP	
	PRMT-2	
	PIMT	
	NCOA1–3,6	
	SWI/SNF	
	PDIP	
PPARδ (NR1C2)	NCOA1	NCOR1
	NCOA3	NCOR2
	NCOA6	
	PGC-1α	
LXRβ (NR1H2)	NCOA1	NCOR1
	P300	NCOR2
LXRα (NR1H3)	NCOA1	NCOR1
	P300	NCOR2
	TRRAP	
	GRIP1/TIF2	
	PGC1α	
	PGC1β	
FXR (NR1H4)	NCOA1	—
	TRRAP	
	DRIP205	
	PGC1α	

(continued)

TABLE 9.5 (continued)
Co-Repressors and Co-Activators of Lipid
Sensors

Nuclear Receptor	Co-Activators	Co-Repressors
PXR (NR1I2)	NCOA1	SHP
	NRIP1	NCOR2
	PGC-1	
	FOXO1	
	GRIP1	
CAR (NR1I3)	NCOA1	—
	PPARBP	
	PGC-1	
RXRα (NR2B1)	NCOA1	—
	NCOA2	
	NCOA3	
	PGC-1α	
	TBP	
	TAFII110	
	TAFII28	
	CREBBP	
	P300	
RXRβ (NR2B2)	NCOA1	—
	NCOA2	
	NCOA3	
RXRγ (NR2B3)	NCOA1	—
	NCOA2	
	NCOA3	

Sources: Moore, D.D. et al., *Pharmacol. Rev.*, 58, 742, 2006;
Michalik, L. et al., *Pharmacol. Rev.*, 58, 726, 2006;
Germain, P. et al., *Pharmacol. Rev.*, 58, 760, 2006.

Members of the NR family regulate transcription by several mechanisms. They can activate or repress target genes by binding directly to response elements (REs) within the promoter/enhancer regions of target genes. Those REs are composed of two hexanucleotides with the sequence AGGTCA separated by a varying number of spacer nucleotides, which are characteristic for each lipid sensor (Table 9.6). Some selected "classical" target genes are listed in Table 9.7. NRs can also act by binding to other classes of DNA-bound transcription factors such as NF-κB or AP-1 [5]. In addition, negative regulation also includes more complex mechanisms like displacement or inhibition of positive regulators [10–12].

The ability of NRs to initiate or suppress the transcription is mediated by numerous accessory proteins, classified as co-activators and co-repressors. These cofactors constitute a large and heterogenous family which has, in common, to interact with NRs in a ligand-dependent manner. Without ligand binding, heterodimerized NR

TABLE 9.6

Consensus Sequences of NR RE

Nuclear Receptor	Response Element	Ref.
PPARα	DR-1, DR-2	[3]
PPARγ	DR-1	[3]
PPARδ	DR-1	[3]
LXRβ	DR-1, DR-4	[2]
LXRα	DR-4	[2]
FXR	IR-1	[2]
PXR	DR-3, ER-6, DR-4, ER8, IR-0, PBRE	[2]
CAR	DR-4, DR-5	[2]
RXRα	DR-1, -2, -3, -4, -5	[8]
RXRβ	DR-1, -2, -3, -4, -5	[8]
RXRγ	DR-1, -2, -3, -4, -5	[8]

Note: DR, direct repeat; ER, everted repeat; IR, inverted repeat; PBRE, pheno-barbital response element.

TABLE 9.7

Examples of Well-Established "Classical" Target Genes of Lipid Sensors

Nuclear Receptor	Target Gene	Ref.
PPARα	ApoAI, LPL, LXRα	[234]
PPARγ	LPL, FATP, ApoA2	[3]
PPARδ	Adipophilin-related protein (ADRP)	
LXRβ	ABCA1, ABCG1, ApoE, GLUT4	[2]
LXRα	ABCA1, ABCG1, ApoE, LPL, GLUT4	[2]
FXR	SHP, BSEP, IBABP, MDR3, MRP-2	[2]
PXR	CYP3A4, CYP2C9, SULT2A1, UGT1A1	[2]
CAR	CYP3A4, CYP2B6, UGT1A1, SULT1A1	[2,18]
RXRα	—	
RXRβ	—	
RXRγ	—	

associate with co-repressors (Table 9.5). An important characteristic of co-repressors is their histone deacetylase activity, which leads to transcriptional inhibition [13]. Contrariwise, co-activators can only bind to NR after ligand-induced changes in the NRs conformation (Table 9.5). Many of them interact with the repositioned AF-2 helix through so-called LXXLL motifs, known as NR boxes. A common characteristic is their histone acetyltransferase (HAT) activity, which facilitates chromatin remodeling and subsequent access of the transcriptional machinery to promoters. Another feature of co-activators is to modify and recruit basal factors and RNA polymerase II [13].

The diversity of cofactors suggests that transcriptional activation occurs through recruitment of multiple factors acting sequentially or in combination.

9.2 POSSIBLE EXPERIMENTAL APPROACHES

To investigate the role of lipid sensors in the regulation of gene transcription, two different general strategies are possible: the candidate gene approach or a genome-spanning strategy. In the past, most studies about lipid sensors were based on the candidate gene approach, i.e., starting with the hypothesis that a specific gene may be regulated by the lipid sensor of interest. However, this approach has its limitations because it is laborious, and unexpected roles of lipid sensors are difficult to discover [14].

With the rising use of high-throughput methods in molecular analysis, it has become more and more popular to systematically screen lipid sensor target genes or their cofactors on a genome-wide basis. Likewise, experimental approaches have been developed to more systematically screen endogenous metabolites for their receptor-binding capacity. These methods are generally more expensive because they usually rely on automated systems and/or chip technologies. Furthermore, adequate algorithms have to be defined for the analysis of the large number of data generated by these techniques.

At present, both approaches are used in parallel to elucidate the role of lipid sensors in gene regulation, and important techniques will be presented below. They are summarized in Tables 9.8 through 9.13. Most of the techniques described below

TABLE 9.8

Overview of Technical Approaches Used for RNA Analysis

Technique	Advantages	Limitations	Exemplary References
ISH	Intratissular and cellular localization possible No RNA extraction required	Use of radioactive isotopes	[125]
Northern blot	Simple Low cost Good specificity Analysis of multiple genes at the same time	Use of radioactive isotopes Low sensibility	[128]
RT-PCR	Simple Fast Low cost Good specificity	Low sensitivity Only semiquantitative Interference with DNA contamination	
Real-time PCR	Simple Fast High sensitivity Quantitative	Cost Need for individual optimization	[108,134]

TABLE 9.9

Overview of Technical Approaches Used for Protein Analysis

Technique	Advantages	Limitations	Exemplary References
IHC	Intratissular and cellular localization possible No extraction of proteins required	High background activity possible Highly specific antibodies required	[137]
Western blot (WB)	Simple Low cost	Highly specific antibodies required Only semiquantitative	[140]

TABLE 9.10

Overview of Technical Approaches Used for Lipid Sensor–DNA Binding Analysis

Technique	Advantages	Limitations	Exemplary References
Reporter gene assay	Fast Flexible High sensitivity	Optimization for different cell lines required Cloning to generate plasmid constructs may be time consuming	[35,39,144,147–150]
EMSA	Simple Fast Low cost	Use of radioactive isotopes Now high-throughput possible	[11,111]
Footprinting	Sensitive Localization of REs possible	Use of radioactive isotopes Time consuming Now high-throughput possible Laborious analysis of results Requires laborious optimization DNA conformation may be affected by chemical treatment	[119,144,146,157,159]
ChIP	Analysis in living cells	Laborious optimization Laborious protocol Requires highly specific antibodies Requires hypothesis about localization of potential binding site	[11,135,164]
FRET	No use of radioactive isotopes High sensitivity Real-time monitoring in living cells possible	High cost Labeling of proteins required Possible interference of label with protein structure or binding interactions High proximity of interacting molecules required Suboptimal ratios between binding partners may lead to false-negative results	

TABLE 9.11

Overview of Technical Approaches Used for High-Throughput Methods to Identify Lipid Sensor Target Genes

Technique	Advantages	Limitations	Exemplary References
Microarray	High-throughput analysis possible High flexibility due to choice of chips	High cost Low reproducibility High noise Statistical approach and filtering largely influence results Analysis limited to genes represented by the chip	[176]
ChIP-on-chip	High-throughput analysis possible No hypothesis for binding site required (see ChIP)	High cost Low reproducibility High noise Statistical approach and filtering largely influence results Analysis limited to genes represented by the chip	[180]
SAGE	Analyses the complete transcriptome	High cost Large quantities of RNA required Demanding technique Problems with tag identification	[182]
Gene trapping	Monitoring of gene expression in living cells Generation of transgenic (TG) animals based on trapped ES cells No extraction of RNA necessary	High cost Demanding technique	

TABLE 9.12

Overview of Techniques Used for Lipid Sensor–Protein Binding Analysis

Technique	Advantages	Limitations	Exemplary References
Reporter gene assays	Fast Flexible High sensitivity	Optimization for different cell lines required Cloning to generate plasmid constructs may be time consuming	[184]
ChIP	Analysis in living cells	Laborious optimization Laborious protocol Requires highly specific antibodies Requires hypothesis about localization of potential binding site	[185]

TABLE 9.12 (continued)
Overview of Techniques Used for Lipid Sensor–Protein Binding Analysis

Technique	Advantages	Limitations	Exemplary References
GAL4 two-hybrid system	Easy Analysis in living cells Sensitive High-throughput scale possible	Use of chimeric proteins may cause disturbances in protein conformation Large number of false-positive results	[39,189]
GST pull-down assay	Easy Sensitive No problems associated with cell culture	High tendency of artifactual binding Isolated proteins required	[39,191]
Co-IP	Relatively easy	Weak interactions are difficult to detect Highly specific (monoclonal) antibodies against both proteins required Possible problems with high nonspecific background	[39,194]
FRET	No use of radioactive isotopes High sensitivity Real-time monitoring in living cells possible	High cost Labeling of proteins required Possible interference of label with protein structure or binding interactions High proximity of interacting molecules required Suboptimal ratios between binding partners may lead to false-negative results	[196]
BRET	No use of radioactive isotopes High sensitivity Less expensive than FRET Lower background than FRET Real-time monitoring in living cells possible	Labeling of proteins required Possible interference of label with protein structure or binding interactions High proximity of interacting molecules required Suboptimal ratios between binding partners may lead to false-negative results	[197]
FCS	Analysis in living cells Distinction between different cellular compartments possible Quantification possible	High cost Demanding technique	[198]

TABLE 9.13
Overview of Technical Approaches Used for Lipid Sensor–Ligand Interactions

Technique	Advantages	Limitations	Exemplary References
Radioligand-binding assay	Easy to use High sensitivity High specificity Analysis in intact cells possible High-throughput applications possible	Use of radioactive isotopes Laborious and slow Requires isolated proteins Separation step of bound and unbound ligand required	[201]
SPA	High reliability when automated No separation of bound and unbound ligand required	Use of radioactive isotopes Potential problems with receptor immobilization	[202,203]
FRET	No use of radioactive isotopes No separation of bound and unbound ligand required High sensitivity Real-time monitoring in living cells possible High-throughput applications possible	High cost Labeling of donor and acceptor molecules required Possible interference of label with protein structure or binding interactions High proximity of interacting molecules required Suboptimal ratios between binding partners may lead to false-negative results	[205]
FP	No use of radioactive isotopes No separation of bound and unbound ligand required High sensitivity Relatively low cost Relatively simple to perform Only one labeling step required High-throughput applications possible	Low sensitivity when ligand affinity and/or receptor concentration are low	[206]
AlphaScreen	Detects also ligand–receptor interactions with elevated distances (≤ 200 nM) Very high sensitivity High-throughput applications possible		[209]
GAL4 one-hybrid system	Easy Analysis in living cells Sensitive High-throughput applications possible	Use of chimeric proteins may cause disturbances in protein conformation Large number of false-positive results	[211,212]

are not specifically associated with lipid metabolism and are also used in studies about genes related to other metabolic pathways. However, particular challenges and problems that may occur in the context of studying lipid sensors have been emphasized.

9.3 CELLULAR MODELS AND EXPERIMENTAL STRATEGIES

9.3.1 CELLULAR MODELS

In vivo experiments in cell culture or animals are usually aimed at analyzing the function of the lipid sensor of interest. To compare gene expression between an inactive and an active state of the lipid sensor, the receptor can be activated by a specific ligand (Tables 9.1 and 9.2), and changes in gene expression levels are subsequently determined on the RNA and/or protein level by methods outlined below. In fundamental research, cell culture is the only way to study NR signaling in humans in an *in vivo* context, and a large number of cellular models have been developed.

Treatment of established continuous cell lines derived from the tissue of interest (e.g., HepG-2 from the liver or Caco2 from the colon) with specific lipid sensor agonists is the most convenient experimental approach. Cancer or immortalized cell lines are readily available from culture collections such as the American Type Culture Collection (ATCC) or the European Collection of Cell Cultures (ECACC), as well as from commercial suppliers, and can also be obtained through other investigators [15]. They are relatively easy to maintain, requiring uncomplicated culture and media conditions. Usually, they can be used within a wide range of passages and can proliferate rapidly so that they can be amplified to required cell numbers for experiments [16]. However, they have limitations. Cancer cell lines may display changes in their karyotype or even less evident DNA mutations, leading to changes in the phenotype compared to the nonmalignant cells from the same tissue. HepG-2 cells, for example, only express a subpattern of proteins found in normal hepatocytes; e.g., they do not express the NR, PXR [17]. Furthermore, those differences have often not been characterized in detail and may reduce the value/significance of results obtained with them. Another aspect is that a certain interlaboratory variability in the phenotype of cells continuously cultured for a longer period of time has been observed. This may result in variation of biological responses [18].

Another option is to use cells in primary culture which have been isolated from the respective normal or cancerous tissues [19,20]. Their characteristics resemble the original tissue more, which makes results more reliable. However, their availability is relatively limited and costs from commercial sources are more elevated compared to cell lines. They usually grow only for a limited number of passages, e.g., human umbilical vein epithelial cells (HUVEC), or do not proliferate at all in cell culture, e.g., human hepatocytes, so that the number of experiments is relatively limited. In addition, they display a larger variability, especially interbatch variability, in their biological response, making it more difficult to obtain significant results and to compare data from different donors. It is also to be considered that those cells— especially human hepatocytes—often respond to (imperfect) culture conditions with changes in protein expression and deterioration of cell-specific functions [21].

These limitations can partially be improved by modifying culture conditions with extracellular matrix components such as collagen gel sandwich culture systems. This technique facilitates the preservation of certain tissue characteristics even after longer *in vitro* culture [22]. Precision-cut tissue slices represent another important cellular model. They are frequently used in toxicological, pharmacologic, and metabolic studies, because this *in vitro* system is highly reminiscent of the *in vivo* situation, mainly due to the maintenance of complex cell-to-cell interactions [23]. Furthermore, a large number of different components such as enzyme systems, cofactors, and transporters are present in a physiological context [24]. The preparation of precision-cut slices overcomes initial problems with nutritional and oxygen distribution as observed in the culture of larger tissue pieces [17]. Tissue slice preparation and maintenance of liver, kidney, lung, and heart is well established and protocols are summarized in [23]. More recently, the technique has further been expanded to tissues which are more difficult to cultivate such as heart [25], small intestine, and colon [26]. However, the loss of specific biotransformation functions and tissular differentiation features still have to be considered in those alternative models [27], and care has to be taken when data obtained with cell culture experiments should be extrapolated to an *in vivo* situation.

To conserve cells for longer periods, they are frequently cryopreserved. Techniques are considered to be optimized to minimize sudden intracellular formation of ice crystals that could result in ultrastructural damage. Generally, cell viability, attachment, and metabolic activity on thawing are well maintained [19]. Zhu et al. [28] reported that cryopreserved, transiently transfected HepG-2 cells had a comparable performance in a transactivation assay compared to freshly transfected cells. In contrast, another study found out that the metabolic capability of hepatocytes after cryopreservation, in particular, with respect to phase II metabolism was diminished [29].

In summary, cell lines—if carefully chosen for a specific study—may be used for initial screening experiments. If an appropriate cellular model is available, the results should be validated in a more physiologic context, i.e., cells in primary culture. When the study is aimed at investigating an intracellular signal transduction pathway, these models will usually be sufficient. For more complex studies, which require, e.g., the inclusion of cellular import or export systems or intercellular signaling, more sophisticated cellular models such as sandwich culture or tissue slices may be necessary.

9.3.2 TRANSFECTION OF CULTURED CELLS

Several experimental approaches such as heterologous cell lines, small interfering RNA (siRNA), or reporter gene assays (see Section 9.6.1.1) require transient or stable transfection of mammalian cells, i.e., introduction of exogenous DNA into cells to study recombinant gene expression. Several techniques have been established to transport DNA across the cell membrane and can generally be divided into chemical, physical, and viral methods [30].

Chemical methods, including calcium phosphate, cationic lipids, and cationic polymers, are most commonly used for transient transfections [30]. The ability of these transfection reagents to introduce DNA molecules into cells relies on the fusion of membrane systems. The negative charge of the plasmids to be transfected is masked by the transfection reagent. When added to the cell culture medium, they associate spontaneously with and cross the negatively charged, hydrophobic cell membrane and are subsequently internalized by endocytosis [31]. DNA molecules are then transported by a largely unknown mechanism from the cytosol into the nucleus, where gene expression is achieved [30]. The cells are incubated together with the DNA–transfection reagent complexes for at least 5h to up to overnight. Afterward, the medium is changed and cells are incubated an additional day or two to allow expression of the transgene.

In general, all different types of cells, adherent cell lines, primary cultures as well as suspension cultures (here, the cells are added to the wells after the preparation of the DNA–transfectant complexes) can be transfected by chemical methods. The transfection efficiency of both cationic lipids and polymers is comparably high but varies significantly with the type of cells used, and the transfection conditions always have to be optimized depending on the cells used as expression system: the number of cells, the transfectant (commercially available from different companies), the ratio of transfectant versus DNA, and the transfection duration are parameters to be considered. Transfection efficiency can be monitored by using green fluorescent protein (GFP, see Section 9.6.3.1) as reporter gene, which allows optical examination and estimation of transfection efficiency in the living cells under the microscope [32].

The most critical parameter for successful transfection is cell health. Transfection is most efficient when the cells are maintained in mid-log growth [33]. Therefore, cells which do not divide in cell culture are generally not transfectable by chemical methods. In general, cationic polymers like polyamines are more toxic compared to cationic lipids [30]. It has also to be considered that complexes between the cationic liposomes and the DNA must not contain serum, because it contains sulfated proteoglycans and other proteins, which compete with the DNA for binding to the cationic lipids. For most transfection reagents, the medium should also not contain antibiotics, to avoid additional cell toxicity [34].

In most cases, cells will be transfected in a satisfying manner using chemical-based methods. Particular lipid-based transfection reagents have even been developed for cells which are considered to be difficult to transfect, suspension cells, and cells in primary culture. A large variety of transfection protocols using lipid sensors can be found in the literature. As examples, the transfection of hepatoma HepG2 cells with expression plasmids for LXR and RXR as well as luciferase reporter plasmid [35]; difficult-to-transfect MCF-7 breast cancer cells with an inducible PPARα expression vector [36]; or primary rat hepatocytes with a reporter gene [37] have been described.

However, if sufficient transfection efficiency cannot be achieved using chemical methods, cell toxicity is too high, or highly efficient stable transfection is required, alternative approaches can be tried. In contrast to chemical-based transfection methods which rely on endocytosis, viral vectors use natural cell receptors to enter the cells, which significantly reduces cell toxicity and increases transfection (or better, transduction) efficiency. This makes them particularly suitable for stable gene transfer into difficult-to-transfect cells such as cells in primary culture. Different recombinant viral vectors have already successfully been established. Retroviral vectors are most commonly used. They possess the natural ability to stably integrate into eukaryotic genomes. Lentiviral vectors, which are another class of retroviruses, have the additional characteristic to be even able to transduce nondividing cells. By combining the right viral vector, envelope, and promoter, efficient transduction using retroviruses is possible for virtually any cell type. For example, Rizzo et al. [38] described the successful infection of 3T3-L1 preadipocytes with an FXR expression plasmid and an FXR responsive reporter plasmid by the aid of retrovirus, and Brendel et al. [39] infected mouse hepatic BNL-CL.2 cells with an LRH-1 expression vector. The insert length is restricted to approximately 10 kb, but this is sufficient for cloning cDNA of any lipid sensor (of approximately 1.5 kb) or relatively large parts of gene promoters [40]. Adenovirus-based vectors are an alternative to lentiviruses, because they also allow transfer of genetic information into nondividing cells. Furthermore, adenoviral vectors have an enlarged capacity of approximately 35 kb. However, only transient transfections are possible, because adenovirus do not integrate into the host genome [41]. For example, the technique has been applied to deliver siRNA into 3T3-L1 preadipocytes to silence PPARγ in theses cells [42]. Herpes simplex viruses (HSV) have a large transgene capacity and can deliver plasmids of up to 150 kb into the host cells. Similar to retroviruses, HSV can also be targeted to specific cells by modifying envelope proteins [43]. Koldamova et al. [44] described the use of HSV-mediated gene delivery for primary mouse neurons which were infected with an expression plasmid for the amyloid precursor protein to study the effects of LXR agonism on ABCA1 expression and amyloid β secretion. The major concern about the use of viral methods is safety, as some vectors are of pathogen origin and may be reactivated; their potential for tumorigenicity and the possibility of allergic reactions to proteins still remain within these vectors [45]. Due to the high safety requirements, laboratory costs are relatively high. In addition, the production of viral vectors is generally laborious and time consuming [46].

Physical methods include conventional needle and microinjection, particle bombardment, electroporation, hydrodynamics, encapsulated microspheres, and ultrasound. Generally, some type of physical force is used to overcome the barrier of a cell membrane [45]. Their major disadvantage is the low transfection efficiency of primary cells. Particularly in the case of microinjection, the number of transfected cells is very low. Particle bombardment and electroporation are also associated with high cell mortality [46]. Furthermore, physical methods require additional expenditures with regard to instrumentation. These methods are rarely used to transfect cells in NR research. However, microinjection is an important technique in the generation of TG animals (see Section 9.4.2). With regard to the additional financial, instrumental, and experimental effort required for viral and physical DNA delivery methods, they can only be recommended when chemical-based approaches are not sufficiently working.

9.3.3 HETEROLOGOUS CELL LINES

Another interesting tool is the establishment of heterologous cell lines overexpressing the lipid sensor of interest. By expressing a regulator of gene expression at high and commonly supraphysiological levels, the biological response is expected to be amplified. If cell lines with low endogenous expression levels are used, interference with competing or redundant pathways can be limited. To create heterologous cell lines, an appropriate plasmid expression vector has to be chosen. It is important that the plasmid has a strong promoter for transcription (such as the major immediate early promoter of the human cytomegalovirus; CMV) and directs expression of a selectable marker such as the neomycin phosphotransferase gene, which confers resistance to the antibiotic, G418 [47]. The cells are transfected following the guidelines outlined in Section 9.3.2. After a sufficient incubation time which should allow the plasmid to stably insert into the genome (usually 2–3 days), the appropriate antibiotic is added to the culture medium. Successfully transfected cells which express the gene of interest and the antibiotic-resistance gene (encoded by the transfected plasmid) will selectively survive in an environment containing the respective antibiotic. Several weeks have to be allowed for the successful selection of stable expressing clones. For example, Hosono et al. [42] described the construction of stable mCAR expressing NIH3T3 cells and Rizzo et al. [38] described the creation of stable FXR expressing 3T3-L1 adipocytes. Unfortunately, only one in 10^4 transfected cells will stably integrate DNA. Furthermore, stably transfected cells integrate only a few DNA copies, and consequently, the expression level of the transfected gene is generally lower than in transiently transfected cells.

Another variant is inducible heterologous cell lines. Inducible systems allow flexible temporospatial control of lipid sensor expression in a host system. This might be of advantage because the host cell response to the presence of varying quantities (i.e., induced) or absence (i.e., noninduced) of the lipid sensor can be studied in detail. Furthermore, constitutive overexpression may also lead to general disturbances in cellular metabolism, while a limited induction for defined periods of time allows the cells to recover after induction. Several systems are commercially available: the ecdysone-, the estrogen-, the progesterone-, the chemical inducers of dimerization (CID)-, and tetracycline-based systems [48].

Using the ecdysone-inducible mammalian expression system, cells are sequentially transfected with the effector plasmid, pIND, which contains the cDNA of the lipid sensor of interest or any other gene to be inducibly expressed under the control of tandem repeats of the ecdysone RE and the transactivator plasmid, pVgRXR, which encodes for the ecdysone and RXR. The selection of stable transfectants is achieved by neomycin and zeocin resistance genes in the respective vectors. NR expression in stably expressing cells is induced by treatment with the ecdysone analogue, pronasterone A [49]. The tetracycline inducible system exists in two variants, the tet-off and tet-on systems. The tet-off transactivator plasmid is composed of the DBD of the tetR gene and the transactivation domain of the VP-16 herpes simplex virus under the control of a tissue-specific promoter. Activation of this promoter allows the expression of the tTA fusion protein in a tissue-specific manner. The tef-off effector plasmid consists of the gene of interest driven by the minimal human CMV promoter under the control of tetO. tTA protein can only bind to tetO and activate target gene expression in the absence of doxycycline, which antagonizes tTA binding to tetO. In the tet-on system, the tetR DBD is mutated, producing a protein that only binds tetO and activates target gene expression in the presence of doxycycline [50]. An interesting example for the use of the inducible pIND system has been reported by Pai et al. [51], who explored the individual roles of different sterol regulatory element-binding proteins (SREBP) by preparing lines of CHO cells that expressed graded amounts of each SREBP. Faddy et al. [36] generated a tetracycline-dependent (tet-off) conditionally expressing PPARα MCF-7 cell line to study its role in the cellular response to sodium butyrate.

The establishment of cell lines stably expressing both, the inducible lipid sensor expression vector and the regulatory plasmid, is more time consuming than constitutively overexpressing cell lines. Furthermore, the concentrations of the inducing agent have to be optimized for each inducible heterologous cell line. Therefore, inducible cell lines should only be considered if the experimental approaches require conditional lipid sensor expression or if additional important information about the lipid sensor under investigation is thought to be obtained with this system.

9.3.4 CELLULAR GENE KNOCKDOWN BY siRNA

The opposite approach to gene overexpression comprises gene knockdown. Specific gene silencing in cell culture can be achieved by the use of siRNA. siRNA are 19–25 bp double-stranded (ds) RNA molecules that lead to sequence-specific inhibition of target mRNA in mammalian cells in culture without causing any immune response [52]. The technique is based on the highly conserved eukaryotic pathway, RNA interference, which represents a cellular line of defense directed against invading viral genomes or serves as a method to clear a cell of aberrant transcription products [53]. Two different experimental approaches are possible: Cells can either be directly transfected with siRNA that are synthesized *in vitro* or be transfected with expression vectors to express short hairpin RNA (shRNA) that can be converted into siRNA *in vivo*. The latter approach can also be used to stably express siRNA in cell lines [54] and even be extended to an inducible system by the use of tetracycline operator/repressor interaction [55] to benefit from the particular advantages of the respective models (see Section 9.3.3).

siRNA delivery into mammalian cells can be achieved by different transfection methods as described above (see Section 9.3.2). Chemical-based siRNA transfection reagents are most widely used (Table 9.8). An important issue for successful gene knockdown is the posttransfection incubation time. First, the size of the mRNA pool and the rate of transcription of the target gene are correlated with the velocity of gene knockdown. Second, the half-life of the target protein has to be considered and third, differences in activity levels of RNAi pathways in different cell lines may also contribute to variations [56]. Another key step is represented by the assessment of the level of target gene knockdown. On the mRNA level, RT-PCR, Northern blotting, RNase protection, and branched DNA assays can be used. The cellular level of the target protein can also be analyzed in total cell lysates of the transfected population using techniques such as Western blotting. To ensure good experimental quality and to allow troubleshooting if no sufficient gene knockdown is observed, several controls should be included in the experimental design [56]:

1. Positive control: A siRNA targeting a well-characterized, constitutively expressed housekeeping gene such as lamin A/C or cyclophilin B should be transfected in parallel experiments to monitor the efficiency of siRNA transfection into cells.
2. Negative control: A nonspecific siRNA that does not target any endogenous gene in the cell should be used as a negative control. This helps to distinguish between sequence-specific silencing and nonspecific effects.
3. Transfection control: A fluorescently labeled siRNA can be transfected to determine optimal transfection conditions. The fluorescent signal in the nucleus can be directly observed in the intact cell by fluorescent microscopy and gives evidence if transfection efficiency is too low (approximately less than 20%).
4. Untreated control: Cells without any siRNA treatment should be cultured in parallel to determine baseline cell viability, phenotype, and target gene level.

siRNA is increasingly used as an analytic tool in functional genomics. For example, the technique has been applied to silence the aryl hydrocarbon receptor (AhR) by the aid of an expression vector to study the role of polycyclic aromatic hydrocarbons and AhR signaling on LXRα-regulated genes and atherosclerosis in HepG2 cells [57]. Direct transfection of siRNA has been used to silence PPARγ and LXRα in THP-1 macrophages to demonstrate their involvement in the modification of cholesterol efflux by angiotensin type 1 receptor blockers [58]. The same approach was used to silence the Niemann Pick C (NPC) proteins, NPC1 and NPC2, in the context of a study about the influence of LXR activation on intracellular cholesterol trafficking in primary human macrophages [59] and to confirm the negative regulation of hepatic lipase by FXR [60].

By silencing the lipid sensor of interest, its specific involvement in the regulation of a particular gene can be convincingly demonstrated. The technique also has its limitations: The first problem is the design of effective siRNA sequences. Even though a multitude of algorithms have been developed by commercial and noncommercial suppliers to consider important sequence requirements [61], only approximately 25% of selected target siRNA are functional [62]. Therefore, several siRNA have to be tested in time-consuming screening experiments to select one with sufficient silencing efficiency. Another limiting factor is transfection efficiency: the delivery conditions have to be optimized for each target gene and each cell line to be used. Finally, off-target effects also have to be considered. This phenomenon describes the nonspecific silencing of not functionally related genes which have poor sequence identity with the siRNA and are probably caused by several mechanisms [63]. Off-target effects are difficult to determine, because of their non–sequence-specific occurrence, but may significantly bias the results of siRNA experiments when randomly affected genes interfere with the signaling pathway under investigation [64]. Recent results suggest that a 6 bp core sequence within the siRNA, called seed region, may be responsible for off-target effects, and special modifications lead to significant reductions in off-target effects [65,66]. A possible solution might be to order predesigned siRNA because commercial suppliers try to incorporate the new knowledge into their design algorithms and testing strategies. To circumvent the problem of off-target effects on the level of experimental design, results should be validated by using several efficient siRNA targeted against different regions of the mRNA of interest and/or by performing cell treatments with a specific NR antagonist.

In comparison with other techniques, transiently transfected siRNA is the technique of choice when specific lipid sensor agonists (e.g., RXR) or antagonists (e.g., PPARδ, LXRs, and FXR) are not available or results with antagonist treatment should be confirmed by another experimental approach. Stable transfection of shRNA might be an interesting approach when a more complex investigation of biological responses to the absence of the lipid sensor of interest is planned (such as effects on cell growth or viability), which potentially interferes with a simple antagonist treatment.

9.3.5 CELL TREATMENT STRATEGIES

In order to characterize the role of the lipid sensor of interest in cellular models, several treatment strategies are possible: treatment with specific endogenous or exogenous agonists allow specific activation of the lipid sensor of interest and therefore analysis of the biologic response in a given cellular model. This approach is very flexible, because depending on the analysis strategy (see Section 9.2) it allows the investigation of either specific genes or a systematic analysis of a genome-wide response (high-throughput approaches like DNA-microarray; see Section 9.6.2.1). Treatment with a single dose of a given agonist (Tables 9.1 and 9.2) is often not sufficient and does not allow comparison of ligands [18]. The potency, selectivity, and pharmacokinetic properties of a ligand used for the study will largely influence the results. Therefore, it is advisable to sequentially use multiple ligands for one lipid sensor or to perform dose–response curves to increase the predictive power of an experiment [14]. Dose–response curves should be based on previously determined IC_{50}-values, because they allow demonstrating possible saturating doses and the specificity of the effect (increasing agonist doses should lead to an increasing effect). This can be further extended by conducting co-treatment experiments with specific antagonists of the lipid sensor under investigation (Table 9.3) or by the redundant approach of treating receptor knockout cells by using siRNA (see Section 9.3.4). Selected examples for each lipid sensor are summarized in Table 9.14.

The duration of the experiment is also an important parameter: the appropriate time to yield a determinable effect on target gene expression has to be elaborated [14]. Time course experiments not only reveal the velocity of the biological response after lipid sensor activation but also give the first idea if the observed effect is more likely of direct (more rapid response) or indirect nature (delayed response). This question can further be analyzed by pre- or co-treatment experiments using actinomycin D (an RNA synthesis inhibitor; approx. 1 μg/mL) and cycloheximide (a protein synthesis inhibitor; approx. 10 μg/mL), respectively [67]. If the effect is still observed in the presence of both inhibitors, the gene of interest is most likely directly regulated by the NR. An inhibition of the observed effect in the presence of cycloheximide indicates that the signal transduction cascade involved requires *de*

TABLE 9.14

Selected Examples for Cell Treatment with NR Agonists

NR	Treatment	Objective of the Study	Ref.
PPARα/PPARγ	Fenofibric acid, 100–500 µM, 24 h and time course Rosiglitazone (BRL 49653), 0.1–10 µM, 24 h	Evaluation in AML-12 mouse hepatocytes and 3T3-L1 preadipocytes if PPARα and PPARγ (see below) regulate LPL	[159]
PPARα	Fenofibric acid, 100 µM, 24 h Wy 14643, 10–500 µM, 24 h	Analysis of the regulation of Rev-erbα by fibrates in human hepatocytes in primary culture and human hepatoma HepG2 cells	[235]
PPARγ	Rosiglitazone, 1.0 nM–10 µM, 2–48 h Pioglitazone, 1 µM, 24 h	Study about the regulatory effects of PPAR on hormone-sensitive lipase (HSL) gene expression in human SMMC-7721 hepatoma cells and rat preadipocytes	[236]
PPARδ PPARδ/RXR	GW501516, 1 µM, 24 h LG101305, 0.1 µM, 24 h	Investigation of gene regulation involved in skeletal muscle lipid and carbohydrate metabolism by PPARδ in C2C12 skeletal muscle cells	[237]
LXR	T0901317, 1 µM, 4 days 22-R-HC, 10 µM, 4 days 22-S-HC 10 µM, 4 days	Evaluation of the role of natural versus synthetic LXR ligands in lipid metabolism in human skeletal muscle cells in primary culture	[238]
LXR	T091317, 1 µM, 72 h	Exploration of the effects of LXR activation on insulin secretion and insulin content in the insulinoma cell line INS-1E	[239]
LXR	T091317, 1–2 µM, 12–48 h	Test of the effect of LXR ligands on APOA5 expression in human hepatoma HepG2 and Huh-7 cells	[240]
LXR/RXR	9-cis-retinoic acid, 10 µM (cotreatment with T091317)		
LXR/PPARγ	GW3965 GW7845 3–9 days	Evaluation of the function of LXR in preadipocyte cell lines 3T3-L1 and 3T3-F442A	[241]
FXR	CDCA, 250µM, 24 h	Verification in Caco2 cells that FXR negatively regulates LXR via induction of SHP	[39]
FXR	CDCA, 25–75 µM, 6–48 h GW4064, 1–5 µM, 6–48 h	Investigation in HepG2 cells if FXR regulates hepatic lipase (HP) expression	[60]

(*continued*)

TABLE 9.14 (continued)
Selected Examples for Cell Treatment with NR Agonists

NR	Treatment	Objective of the Study	Ref.
PXR	Rifampicin, 50 µM, 16 h	Evaluation of the antifibrinogenic capacity of PXR in human hepatic stellate cells	[242]
PXR	Rifampicin, 5 µM, 12–48 h clotrimazole, 5 µM, 48 h Dexamethasone, 10 nM– 1 µM, 12–48 h CHX, 25 µM, 2 h (pretreatment)	Evaluation of PXR and GR in the regulation of CYP3A4 in human hepatocytes in primary culture	[243]
RXR/RAR	9-*cis* retinoic acid, LG1069, all trans retinoic acid, all at 1 nM–10 µM, 24 h	Elucidation of signal transduction pathways involved in control of apo C-III expression by retinoids in human hepatoma HepG2 cells	[244]

novo protein synthesis of a mediator protein while an inhibition in the presence of actinomycin D indicates that *de novo* RNA synthesis is necessary. These approaches are very frequently used; however, one has to keep in mind that these substances are nonspecific synthesis blockers that may largely disturb the intracellular homeostasis. High toxicity is frequently observed, and results have to be validated with additional experiments, e.g., direct effects by DNA-NR-binding studies using electrophoretic mobility shift assay (EMSA) (see Section 9.6.1.2) or indirect effect by identifying a mediator protein using co-immunoprecipitation (Co-IP; see Section 9.6.3.5). Actinomycin D treatment only leads to conclusive results if the original effect consists of a positive regulation, that is, an increase in target gene expression. Because all lipid sensors form permissive heterodimers with RXR, it might also be of interest to study the cellular response by treating or co-treating them with specific RXR agonists (Tables 9.1 and 9.2).

Selected examples for cell treatment in lipid sensor research are summarized in Table 9.6. In general, it should be taken into account when preparing stock solutions that the concentration of solvent should never exceed 0.1% (v/v) of the culture medium in which the cells are treated. Furthermore, many biologic responses may be disturbed by factors present in serum. It might therefore be advisable to use reduced serum concentrations (e.g., 0.2% to maintain vital functions) or charcoal- or resin-stripped serum during cell treatments.

9.4 ANIMAL MODELS AND EXPERIMENTAL STRATEGIES

9.4.1 POSSIBLE ANIMAL MODELS

In addition to cells, animals can also provide useful *in vivo* information about lipid-activated NR signaling. Other than cells which are usually derived from only one cell type, the complex interactions between different cell types, tissues, and systems remain intact in animals. Therefore, animal studies can provide information about treatment effects on the entire organism. Animals used in lipid sensor research include nonhuman primates, dogs, rabbits, and rodents. Numerous experimental animal models have been established to study particular human diseases or pathological states. Nonhuman primates are genetically and physiologically more similar to humans in comparison with other animals. They are frequently used as experimental models for diseases including infectious, neurological, and heart diseases. Because of their phylogenetical closeness to humans and their similar metabolism, nonhuman primates are also used to analyze effects of NR activators [68]. For example, they often serve as animal models to collect preclinical data about newly developed synthetic lipid sensor agonists including pharmacokinetics and dynamics as well as drug-associated toxicity, as reported in [69,70]. However, they live very long, thus requiring lengthy experiments for longitudinal studies, are expensive to maintain, and may also be carriers of viral, potentially dangerous zoonoses. In addition, the use of primates is also limited by ethical considerations [71]. Therefore, their use is relatively restricted. Dogs are often used in investigations about heart diseases, because their cardiovascular and respiratory systems bear strong analogy to those of humans. In addition, they are frequently used for pharmacologic studies, for example, the examination of the pharmacologic interaction between the lipid lowering drugs, simvastatin and gemfibrozil [72]. While they are relatively easy to maintain, there are problems associated with the status of dogs in the Western culture and with anthropomorphic attitudes toward them [73]. The rabbit has the advantage of being small and relatively inexpensive, but large enough to permit physiological experiments. It has been widely used in studies of atherosclerosis. Mild cholesterol supplementation leads to hypercholesterolemia and associated vascular damage in these animals. For example, the role of C-reactive protein in atherosclerotic lesions has been studied in a rabbit model [74]. However, the highly abnormal diet for an herbivore and the major differences in lipid metabolism between rabbits and humans raise the question of the validity of the rabbit model [71]. The Watanabe rabbit is a heritable hyperlipidemic rabbit strain, which has extreme hypercholesterolemia and associated advanced atherosclerosis. It is based on a mutation of the LDL receptor, thus providing an excellent animal model of the genetic defect underlying familial hypercholesterolemia in humans [71]. An example of the use is the examination of taurine as an antiatherosclerotic agent in the Watanabe rabbit [75].

Rodents such as rats and mice are the most frequently (about 90%) used animal models due to strong resemblances in metabolism and physiology as well as small place requirements, short generation time, and low cost of housing. Moreover, like the human genome, the mouse and the rat genome are fully characterized [76,77]. This facilitates interspecies comparisons and molecular biology experimental work. Several important experimental rodent models have been developed which can be used for lipid sensor signaling research. For example, a mouse model of cholestasis is operative bile duct ligation [78]. Feeding rodents with a "westernized" diet represents a model for obesity and metabolic syndrome [79]. However, rats and mice are relatively resistant to atherosclerosis and even high dietary cholesterol contents are not sufficient to create significant hypercholesterolemia. Therefore, the use of thiopropyluracil to create hypothyroidism is often required [71]. In addition, several spontaneously genetically modified rodent models have been developed in recent years. The fatty Zucker rat is a classical rat model and carries a spontaneous mutation in the leptin receptor gene. Homozygous animals develop a variant of metabolic syndrome, becoming obese, moderately insulin-resistant, and hypertriglyceridemic [80], making them a widely used model for studies in obesity. The Zucker diabetic fatty rat is a variant developed from the original fatty Zucker rat which shows many of the complications of the hyperglycemic diabetic state [81]. In addition, several rat models exist for atherosclerosis research [71]. Like fatty Zucker rats, db/db mice carry a spontaneous mutation in the leptin receptor gene, leading to obesity, insulin resistance, hyperinsulinemia, and hypertriglyceridemia [82]. In addition, those mice show impaired endothelial and vascular function, abnormal cardiac metabolism and function, retinal damage, and glomerular sclerosis [71]. Similarly, ob/ob mice produce a structurally defective leptin that makes them obese, insulin-resistant, hypertriglyceridemic, but normoglycemic. They also show vascular dysfunction, but no atherosclerosis [83]. A new mouse model of obesity-induced diabetes has recently been generated by combining quantitative trait loci from obese and nonobese nondiabetic mice [84].

Despite the existence of a multitude of rodent animal models, consensus about which one is most effective for a particular disease is missing. When a particular model has to be chosen, care should be taken that the respective model is well characterized and that the associated phenotype resembles strongly the phenotype of the human disease under investigation. This might be difficult, i.e., for complex multifactorial and polygenetic diseases such as cardiovascular disease.

9.4.2 TRANSGENIC ANIMALS

Due to the short generation time and low cost of housing of mice, genetic modification is usually done in them. Three different approaches are very frequently used: gene knockout, gene knockin (KI), and TG animals [84]. Methods to produce TG mice include

1. *DNA microinjection.* This method consists of the injection of the gene construct into the pronucleus of a fertilized ovum. It is then transferred into the oviduct of a pseudopregnant female mouse. The introduction of DNA into the genome with this method is random. Therefore, it can even inactivate another gene, if the insert is placed within the coding region [85]. It is also possible that even if the gene is inserted, it will not be expressed.
2. *Embryonic-stem-cell-mediated gene transfer.* The DNA sequence is first introduced into embryonic stem cells (ES) in *in vitro* culture by homologous recombination. Afterward, the ES are incorporated into an embryo at the blastocyst stage of development to result in a chimeric animal [85].
3. *Retrovirus-mediated gene transfer.* The sequence of interest is cloned into the virus genome. Retroviruses have the ability to infect host cells by transferring their genetic material, which is inserted into the host genome.

However, because every cell will not be infected by viruses, the resulting animals are chimeric [85]. This technique requires several generations before homozygous animals are obtained.

TG mice are genetically modified by overexpressing a gene from another species, such as humans. This is of advantage, because native animal proteins, including lipid sensors, are not 100% homologous to human proteins and can therefore show differences in ligand-binding affinities or gene regulation compared to their human orthologues. For example, using wild-type PXR$^{-/-}$ and TG hPXR mice, it was demonstrated that rifaximin is a specific activator of the human orthologue. Furthermore, murine *mrp2*, a well-known PXR target gene in humans, was not induced in hPXR mice after rifaximin treatment, suggesting differences between mouse and human *mrp2*-controlling elements [86]. Because of the ethical limitation of *in vivo* studies in humans, TG mouse models seem to be an appropriate alternative to study human gene regulation *in vivo*. The TG approach is straightforward and relatively inexpensive. In addition, TG mice often demonstrate an obvious phenotype which permits a profound understanding of the function of the transgene [87]. Lipid receptor TG mice have been generated for PPARα [88], PXR [89], and CAR and give new insights in their metabolic regulatory role [89,90].

As previously outlined for cell culture methods (see above), TG expression can also be controlled by a chemical inducer, thus limiting the expression of the transgene. In mice, the same systems are available as in mammalian cell culture (see Section 9.3.3) [91]. However, the tetracycline system is the most frequently used one in lipid sensor research. For example, using the tet-off system, a mouse conditionally expressing PPARα was generated to examine the species differences in response to peroxisome proliferator-induced hepatocarcinogenesis [92].

A possible limitation of TG mouse models is the so-called "position effect": the random integration of the transgene into the mouse genome can seriously affect the levels of transgene expression. Furthermore, it may also lead to an unintentional inactivation of a native mouse gene, thus creating a phenotype which does not solely reflect the presence of the transgene. When using inducible TG models, it also has to be considered that inducer treatment may have significant effects on the phenotype. Despite those limitations, TG mouse models can provide important insights into normal human gene function *in vivo*.

9.4.3   Gene Knockout in Animals

A more recent technique is gene knockout, i.e., silencing of the expression of an endogenous gene in mice by genetic modification. KO mice are used to analyze the physiological function of a gene, the effect of a drug in the absence of a gene, or to simulate a human disease which is caused by a lack of a specific protein. Homologous recombination is used to replace the gene with a noncoding or inactive gene DNA sequence, resulting in the elimination of the targeted gene.

Gene knockout has extensively been used to characterize the function of lipid sensors. For example, the generation of PXR$^{-/-}$ mice resulted in the discovery of its function in biosynthesis, transport, and metabolism of bile acids and its protective role against severe liver damage induced by lithocholic acid (LCA) [93]. Similarly, by comparing FXR$^{-/-}$ mice to their wild-type littermates, the critical role of FXR for bile acid and lipid homeostasis was established [94]. In a more specific model, the PPARγ$_2$ was knocked out in genetically obese mice to investigate the role of this PPARγ variant on adipose tissue metabolism under conditions of excess nutrients [95]. More detailed data about available knockout models and their phenotype characterization can be found online in the NIH Deltagen and Lexicon Knockout Mice and Phenotypic Data Resource.

Even though the generation of knockout mice is becoming more and more routine, this approach is still more time consuming and expensive compared to the TG approach. Furthermore, global knockout of a gene essential for embryonic development will result in embryonic or neonatal lethality, thus prohibiting the investigation of its function in postnatal life [87]. For example, RXRα$^{-/-}$ embryos die *in utero* [96]. Alternative approaches are the use of heterozygous animals, the application of siRNA or the generation of conditional KO mice as described below [87]. Another possible problem may be the absence of a detectable phenotype or different phenotype characteristics between the knockout model and the human disease under investigation [97]. Finally, it might be difficult to distinguish between direct effects of the gene knockout and secondary consequences potentially arising through indirect effects of the absence of the targeted gene [87].

In KI models, the endogenous gene is replaced by a mutant variant of the same gene or another coding gene, such as a reporter gene or a human gene using homologous recombination. In this way, subtle changes can be introduced into lipid sensors, their cofactors, or target proteins. Other than in classic TG mice, this approach allows tight control of the inserted gene and position effects are avoided [98]. For example, a KI mice bearing a naturally occurring dominant negative mutation in PPARγ was generated to study new therapeutic strategies and to establish correlation between the mutation and diseases like metabolic syndrome [99].

Another modification of the TG approach is to block the expression of the targeted lipid sensor (or other protein) by using RNA interference. Similar to approaches used in cell culture (see Section 9.3.4), siRNA can be directly delivered by viral or other techniques to silence gene expression in particular organs. Alternatively, long-term expression can be achieved by generating TG animals expressing siRNA [87]. For example, the technique has been used to transiently silence the transporter, ABCA1, in the liver of mice [100]. Like in cells (see Section 9.3.4), the siRNA approach bears some limitations, such as a limited effect of sequences, unsatisfying degree of knockdown, varying levels of knockdown depending on the tissue and off-target effects [87]. However, they may be a more rapid and straightforward alternative to conditional knockout systems (see below) if global knockout is not working.

The *Cre/lox* system controls the deletion of the targeted gene in a tissue- and/or time-specific manner. The *Cre/lox* systems use two mice lines: The first one carries the targeted gene for deletion flanked by two *lox P* sequences, the so-called "floxed gene." The second line carries the transgene for the *Cre* recombinase under the control of an inducible or tissue-specific promoter. By breeding the two mouse lines, *Cre* recombinase expression leads to recognition of the floxed targeted gene and its deletion. *Cre/lox* is a highly flexible system. The floxed target mouse line can be breeded with different *Cre* mouse lines carrying either an inducible or different tissue-specific promoters. This allows the characterization of the targeted gene in numerous different contexts [50]. The system is particularly useful for the knockout of genes that are essential for embryonic development or that lead to infertility. For example, *Cre/lox* has been used to generate PPARγ knockout mice to determine its role in cholesterol homeostasis. Because general disruption of this lipid sensor leads to embryonic lethality, a mouse line was generated bearing the floxed PPARγ and *cre* under the control of an α/β interferon-inducible promoter which was subsequently activated by dIdC treatment [101].

Phenotype characterization has to be carried out for both KO and TG animals, because transgenesis has only an extremely low success rate. The first generation (F1) of animals can be chimeric depending on the technique used. It is generally anticipated that at least five generations are needed until stable homozygous TG animals are obtained [87]. Successful gene delivery/gene KO has to be confirmed by genomic DNA analysis using techniques such as PCR or Southern blot (see Section 9.5.3.3). The expression/absence of expression is usually analyzed by techniques such as RT-PCR, Northern blot, RNase protection, branched DNA assays, or WB (see Sections 9.5.3 and 9.5.4). Furthermore, phenotype characterization includes parameters such as weight, viability, fertility, developmental characteristics, or histological analysis. For example, the characterization of FXR$^{-/-}$ mice has been described in detail [94].

9.4.4 ANIMAL TREATMENT STRATEGIES

Independent of the particular animal model used, animals can be treated by specific lipid sensor activators (Tables 9.1 and 9.2) to analyze their role in an *in vivo* context. There are multiple ways for animal treatment to determine the role of lipid-activated NR, depending on the characteristics of the molecules used and the purpose of the research (Table 9.15).

1. *Oral administration:* Substances are added to food or water. For example, CDCA and WY-14643 are generally administrated in this way, mixed with corn-oil [102,103]. Oral administration is an easy way to treat animals. However, the pharmacologically relevant doses may vary between animals because of variations in food consumption and intestinal absorption rate.
2. *Gavage:* The activator is directly injected into the stomach. This technique allows giving an exact quantity of substance to each animal via the digestive system. However, elevated danger exists of hurting the animal during the administration. Many protocols using lipid receptor activators use gavage, for example, rifampicin and T0901317 [104,105].
3. *Microinjection:* The substance is administrated parenterally which can occur by multiple ways, such as intravenous, intraperitoneal, intramuscular,

TABLE 9.15

Selected Examples for Animal Treatment with NR Agonists

Receptor	Treatment	Administration	Ref.
PPARα	Clofibrate	i.p.	[245]
	Wy-14643	Gavage	[246]
PPARγ	Ciglitazone	i.p.	[247]
	Rosiglitazone	p.o.	[248]
	Rosiglitazone,	i.p.	[249]
	Pioglitazone	i.p.	
PPARδ	L-165041	i.p.	[250]
	GW501516	p.o.	[251]
LXRβ	GW3965	Gavage	[252]
	T0901317	Gavage	[253]
	T0901317	i.p.	[254]
LXRα	T0901317	Gavage	[253]
	T0901317	i.p.	[254]
FXR	GW4064	i.p.	[255]
	CA	p.o.	[256]
PXR	Paclitaxel	i.v.	[257]
	PCN	i.p.	[258]
	PCN	i.p.	[259]
	RIF	p.o.	
CAR	TCPOBOP	i.p.	[260]
	TCPOBOP	i.p.	
RXR	Bexarotene	p.o.	[261]
	LG100268	p.o.	[262]

Note: i.p., intraperitoneal; p.o., oral.

intradermal, or subcutaneous administration. This technique can be used to avoid early degradation of the substance by digestion and/or faster absorption in target tissues. For example, intraperitoneal injection is the most frequent way for the administration of PCN and GW4064 [101,106,107].

4. *Implant:* A subcutaneous implant, generally placed at the dorsal neck, allows a slow and constant release of the activator into an animal. Also, the release can be extended to several days, even weeks, without the need to administer further doses. Treatment via implant is not commonly used for lipid-activated lipid sensors.

Concentrations used depend on the molecule and the purpose of the research; for example, an analysis of toxic effects requires more elevated doses than a gene activation–effect analysis. However, activators are often given at supraphysiological concentrations to amplify the biologic response. Time-course and dose–response experiments can be performed to determine optimal treatment conditions. Selected examples are summarized in Table 9.7.

9.5 ANALYSIS OF CELLULAR AND ANIMAL SAMPLES

9.5.1 SAMPLE PREPARATION

DNA, mRNA, and proteins can be used to characterize cells or TG animals and to analyze the effects of lipid sensor activation. Trizol, a monophasic solution of phenol and guanidine isothiocyanate, is widely used because it allows to concomitantly extract DNA, RNA, and total protein from the same biological samples. The protocol is based on cell/tissue lysis by the reagent, followed by liquid–liquid extraction, precipitation, and washing steps [108]. Other commercially available RNA extraction methods use silica-based microspin or vacuum columns and allow rapid RNA extraction even from small quantities of cell or tissue. In addition, special kits have been developed to purify RNA from difficult-to-extract sources such as fiber-rich or lipid tissues.

RNA extraction using trizol yields total RNA, while mRNA only represents approximately 1% of the extract. Column-based methods are often designed to more specifically enrich mRNA during the purification. However, if the gene of interest is only expressed at very low levels, mRNA can also specifically be isolated by using commercially available column-based kits. They are based on affinity purification of poly A^+ mRNA. The disadvantage of these methods is that commercially available kits are more expensive. Also, due to the considerable loss of RNA during the purification process, more starting material is needed to obtain sufficient extraction yield. In addition, mRNA that does not contain a poly A^+ signal will be lost by these methods [109].

When handling RNA, high precaution needs to be taken to avoid RNA degradation by RNase. In addition, good quality of the RNA samples should be ensured, because low-quality RNA may strongly compromise the results of downstream applications [110]. Using a UV/VIS spectrophotometer, the RNA quantity can be determined by the aid of the optical density (OD) 260, while the ratio of the OD 260/280 informs about the quality and OD 260/240 or OD 260/320 about the purity and the extraction performance. An OD 260/280 ratio greater than 1.8 is usually considered an acceptable indicator of good RNA quality. Furthermore, the ratio between 28S (5 kb) and 18S (2 kb) rRNA can be used to determine mRNA integrity, because these two species constitute more than 80% of total RNA. The sample is separated by agarose gel electrophoresis and visualized with ethidium bromide (EtBr). A ratio of 2:1 (28s:18s) is considered to reflect intact mRNA. More recently developed commercial spectrometric methods allow evaluating both RNA concentration and integrity by using very small sample volumes. In addition, these technologies are less laborious while being more sensitive and highly specific for RNA [110]. Analysis of the efficiency of mRNA extraction (poly A^+ RNA) is performed by denaturing agarose gel electrophoresis, wherein the absence of 18S and 28S rRNA on indicates a successful purification.

9.5.2 Labeling of Probes and Protein Preparation

Several techniques such as ISH, Northern blot, EMSA, or footprinting require the use of labeled probes. Mainly short ds DNA sequences serve as probes. The most common label is radioactive ^{32}P, which can be prepared in different ways: One possibility is the use of polynucleotide kinase (PNK). This enzyme adds a phosphate from $[\gamma\text{-}^{32}P]ATP$ to the 5′ end of the probe. This technique is often used for hybridization probes. Only the 5′ nucleotide contains a ^{32}P. The other possibility is Klenow fragment (KF). KF of DNA polymerase I has a DNA polymerase and 3′–5′ exonuclease activities. A ds DNA is synthesized from a single-stranded (ss) template. Using radioactive dCTP, a radiolabeled probe is produced, where every cytosine contains a radioactive phosphate [111].

Radioactive labels have considerable disadvantages including disposal of radioactive waste, limited half-life, costs, health hazards, and requirement for special licenses [112]. To counteract those problems, several nonradioactive labels have been developed, such as biotin- and fluorescent labels. In biotinylated probes, the 5′ end CTP is replaced by biotin-14-CTP. They are detected by streptavidin covalently linked to reporter molecules, such as fluorophores or enzyme reactants (e.g., peroxidase, alkaline phosphatase). A possible limitation is that the relatively big biotin label may block the binding between the probe and the sample. In addition, the sample must be transferred onto a membrane after migration and blocked with a blocking reagent [113]. In fluorescent probes, the 5′ end CTP is replaced by dCTP-texas red or another fluorescent dye. When using fluorescent probes, proteins are generally tagged with GFP [114]. Because they can be stored for long periods of time without decreases in signal intensity, large amounts of probe can be prepared and stored. This, in turn increases the reproducibility between experiments performed with the same batch of labeled probe.

Some techniques also require the use of isolated proteins that can be obtained from multiple sources:

Cellular extracts: Because lipid sensors are predominantly found in the cellular nucleus, nuclear extracts (NE) are prepared to increase the concentration. NE may be prepared by different methods described in [115–117]. The advantage of this preparation is that the extract also contains possible lipid sensor cofactors and thus better reflects the *in vivo* situation. However, the large variety of proteins present in the extract may increase nonspecific binding.

Expression systems: Bacteria or mammalian cells can be used to produce lipid sensor proteins. Bacteria or mammalian cells are transformed or transfected, respectively, with an appropriate expression vector coding for a lipid sensor or another protein of interest. The protein is overexpressed by the cellular machinery and can be then purified. In bacteria, overexpression often causes problems, because the solubility of large quantities of protein is restricted, leading to protein aggregation and loss of activity [118]. Another disadvantage is that the overexpressed protein needs to be purified by sometimes laborious methods [119,120].

Purified proteins: Proteins such as lipid sensors can be partially or entirely purified from NE or cellular lysate by using affinity resin or chromatography [121].

Cell-free systems: Proteins can also be produced by cell-free *in vitro* translation systems. Eukaryotic systems are prepared from crude extracts, such as rabbit reticulocyte lysate or wheat germ extract, which contain the required components for translation of RNA templates. They can also be combined with a prokaryotic phage RNA polymerase to utilize DNA (such as a gene cloned into a plasmid vector or a PCR product) as a template for protein synthesis. Both kinds of assay kits are commercially available from different companies. The advantage of cell-free systems is that a defined template can be used. In addition, they are less time consuming and show less variability than *in vivo* systems.

9.5.3   RNA ANALYSIS

One approach to study lipid sensor expression or lipid sensor–induced changes in target gene expression is the analysis of mRNA. Different frequently used techniques have been established, which are summarized in Table 9.8.

9.5.3.1   *In Situ* Hybridization

In situ hybridization (ISH) is used to directly analyze RNA in tissues or cells, i.e., unlike other techniques described below, RNA extraction is not required for ISH. The location and expression of the gene of interest can be tracked [122,123]. Furthermore, ISH allows identification and quantification of mRNA sequences. Frozen or fresh tissue is embedded in paraffin and cut to form light sections. These sections are fixed on microscopic slides by agents like paraformaldehyde. Then, labeled nucleic acid probes are used to hybridize with complementary mRNA sequences in the fixed tissues [124]. Finally, detection is carried out, depending on the type of label, using either autoradiographic films (radioactive probes) or microscopy (FISH).

Four different types of probes can be used: ds DNA, ss DNA, RNA (riboprobe), or oligonucleotides. While the use of ds DNA is easy and allows the production of long sequences the choice of ss DNA, riboprobes, or oligonucleotides avoids self-annealing. In addition, riboprobes and oligonucleotides have considerably higher hybridization efficiency and riboprobes provide a high specific activity, yielding a better signal-to-noise ratio [122]. In lipid receptor analysis, riboprobes are most frequently used. As an example, Nakamura et al. [125] studied the regulation of the murine cholesterol transporter ABCG1 by LXR and demonstrated its expression in different tissues by ISH in mouse embryos.

Possible probe labels include radioactive isotopes, biotin, digoxigenin (DIG) or fluorochromes [123,126,127]. Radiolabeled probes offer a high sensitivity, which allows even detection of lowly abundant mRNA. However, in addition to the general disadvantages of radioactivity (see Section 9.5.2), the limitations of ^{32}P labeled probes are the decays of the signal over time, low resolution, and often long exposure times. These problems can be overcome by using ^{35}S radiolabel. This isotope has a better resolution and a longer half-life. Besides, several types of nonradioactive probes can be used. FISH is safer and faster than its radioactive counterpart. The fluorochrome can be easily detected by fluorescent microscopy [127], which even allows simultaneous detection of multiple fluorochromes [126]. However, signal enumeration is more difficult with fluorescent than radioactive labels. A possible disadvantage of biotin labels is its possible interference with endogenous biotin, which can lead to false positive results. The inclusion of negative controls such as RNase-treated sections is of major importance. Detection of DIG labeled probes can be performed by IgG conjugated to a reporter molecule such as alkaline phosphatase, FITC, or Texas red. This method offers increased sensitivity because multiple labels can be incorporated into the probe and detection with IgG allows antibody bridging, which results in considerable signal amplification [123,126,127].

9.5.3.2   Northern Blot

This technique is mainly used to detect mRNA and to estimate mRNA size [108]. In addition, cells of animals or treatment effects on gene expression can be characterized. Basically, extracted RNA samples are separated by agarose gel electrophoresis under denaturing conditions. Migration is followed by the capillarity-driven transfer of RNA onto a nylon or nitrocellulose membrane. The membrane is then incubated with a labeled probe. This can be cloned cDNA of the gene of interest, oligonucleotides, or antisense RNA, usually labeled with ^{32}P or biotin [108,122].

Visualization of radiolabeled blots may be done by autoradiographic films or, better, by phosphoimaging. The latter method is 10–100 times more sensitive than the films. Therefore, it is faster and can even detect relatively weak signals. In addition, the signal range for phosphoimaging is larger (100,000:1), making this technique more appropriate for quantification. For example, Northern blot has been used to analyze the effect of LXR and different PPAR isoforms on the expression of the sulfotransferase, SULT2B1b, in human keratinocytes [128].

Northern blot is relatively simple and is associated with low cost, unless phosphoimaging is required for detection. The technique also has a good specificity, and it is possible to detect two or more genes in one hybridization step. However, high abundance of mRNA is required because of its low sensibility [108]. In addition, the use of radioactivity may be considered a disadvantage (see above). This can be avoided by using nonradioactive probes.

9.5.3.3   PCR-Based RNA Analysis

PCR-based techniques require reverse transcription of extracted RNA to form ds complementary DNA (cDNA) before sample analysis. For this purpose, samples are incubated with the reverse transcriptase enzyme and RNase inhibitor at 42°C to allow elongation of cDNA [129].

Different types of primers can be used: Specific primers can be used to generate cDNA of restricted RNA. The use of specific primers generally decreases background signal in downstream analysis. Another option is oligo(dT) primers which only allow reverse transcription of mRNA. However, most reverse transcriptase enzymes only efficiently synthesize cDNA of 1–2 kb of length. Therefore, very long mRNA may be difficult to be determined by this method. Finally, random hexamers, which represent all possible hexameric nucleic acid sequences, can be used as primers. Using this strategy, virtually all RNA present in a sample is reverse transcribed into cDNA. By using random hexamers or oligo(dT) primers, quantitatively more RNA is reverse transcribed than by methods using specific primers. Therefore, several genes can be analyzed with the same sample [129]. For the experimental planning, it also has to be considered that RT-PCR gene expression measurements are only comparable when the same priming strategy and reaction conditions are used in all experiments and the samples contain the same total amount of RNA [130].

The polymerase chain reaction (PCR) is a basic molecular biology method that allows amplification of DNA. RT-PCR is a PCR performed on cDNA. For example, this technique is frequently used to qualitatively and semiquantitatively analyze the expression of lipid receptors and/or their (potential) target genes. PCR on genomic DNA is also suitable to initially screen TG animals for the presence of a transgene. Furthermore, this technique is required for the detection of lipid sensor protein or lipid sensor DNA interactions in the classical ChIP approach (see Section 9.6.1.4).

Primers are designed to allow specific amplification of the gene of interest. It is recommended to design primers that anneal in two different exons to ensure selective cDNA amplification and avoid interference with possible genomic DNA contamination. If it is not possible to design exon-spanning primers, non–reverse-transcribed controls should be included and/or RNA samples be digested with DNase prior to reverse transcription.

RT-PCR is a fast and simple method that allows semiquantitative determination of changes in gene expression between treated and untreated cell or animal samples. To further precise the analysis, agarose gel electrophoresis can also be followed by Southern blot [122]. However, RT-PCR also has its limitations: The technique is only semiquantitative, thus results cannot be numerically expressed and its sensitivity is relatively low. Due to their similar chemical properties, DNA contamination can easily occur during RNA extraction. This may bias the results.

9.5.3.4 Real-Time RT-PCR

Like RT-PCR, real-time PCR is based on the analysis of cDNA. However, this technique is more sensitive, so only low mRNA abundance is required for the analysis. Different from RT-PCR, which only allows detection at the end of thermal cycling, real-time PCR determines sample amplification at the end of each cycle owing to the use of fluorescence. Four main types of detection have been developed [129]:

1. *Hydrolysis probes:* They are based on the 5′-nuclease activity of the DNA polymerase. Two specific primers are used to start the amplification and define the endpoints of amplification. In addition, a third oligonucleotide, the probe, binds to the amplicon between both primers. The probe is linked to a reporter dye and a quencher at the 5′ and 3′ ends respectively. When the probe is intact and unbound, emission by the reporter dye is absorbed by the quencher due to the close proximity of both. However, when the probe binds the template, it is cleaved by the enzymatic activity during amplification and the quencher becomes too distant from the reporter dye, so the fluorescence is emitted and can be quantified [129,131].
2. *DNA-binding dyes:* Unbound dye only emits a little fluorescence, while molecules bound by ds DNA emit high fluorescence. The emitted light is proportional to the quantity of DNA in a sample. The specificity of the signal is only determined by the primers.

3. *Molecular beacons:* These probes have a stem-and-loop structure that contains specific sequences for the target gene (loop) flanked by self-complementary stem sequences, which are covalently linked to a fluorochrome and a quencher at the 5′ and 3′ ends, respectively [132]. In the absence of a complementary sequence, the probe keeps its stem-loop structure and, therefore, the fluorochrome and the quencher are in a proximity that allows the absorption of the fluorescence emitted from the fluorochrome by the quencher. However, in the presence of the complementary sequence, hybridization of the probe with the sequence leads to the separation of the fluorochrome and the quencher, and fluorescence can be detected [129,132].

4. *Hybridization probes:* A technique which is based on the fluorescence resonance energy transfer (FRET; see Section 9.6.1.5). Two hybridization probes are used in addition to 3′-end-specific primers. One probe is linked to a green fluorescent dye at its 3′ end. The second probe carries an acceptor fluorochrome at its 5′ end. While they are in solution, no fluorescence is emitted by the second probe. During annealing step, both probes hybridize the template and the 3′ end of the first probe becomes close to the 5′ end of the second one. The emitted energy by the first probe excites the second dye and a red fluorescence can be detected.

Primers used with DNA-binding dye technique have to be designed to produce an amplicon of 75–250 bp in length. This ensures amplification efficiency and reproducibility. Care has to be taken to design specific primers that allow specific amplification of only the gene of interest. Probe-based real-time techniques have a higher specificity compared to DNA-binding dyes. However, required primers and probes have to be purchased commercially, making these techniques more expensive in regular use. Also, the sequences of primers and probes are usually not disclosed, which prohibits specificity controls such as sequencing of PCR products. It is also possible to use self-designed hydrolysis probes but they are accompanied with elevated cost compared to predesigned commercial probes. Most studies about the role of lipid receptors are using hydrolysis probes or DNA-binding dyes for real-time PCR analysis as for example in [133].

Results of real-time PCR are expressed as threshold cycle (C_T), which indicates the fractional cycle number at which the amount of amplified target reaches a fixed threshold. The more template present in the sample, the fewer cycles needed to reach the C_T. Results can be quantified by two principal methods:

1. *Standard curve method:* A nucleic acid with known concentration and length is first used to construct a standard curve. For relative quantification, the C_T of the gene of interest is directly compared to the curve. This method can also be used for absolute quantification, i.e., to precisely determine the mRNA copy number per cell for the gene of interest. This requires an absolute standard curve for each individual amplicon [129].

2. *Comparative C_T method:* This method is used for relative quantification. All values are normalized to an endogenous reference, i.e., an internal control

gene which is constitutively expressed, such as GAPDH, β-actin, or rRNA to correct results for differing amounts of input RNA (ΔC_T sample or ΔC_T reference). Finally, ΔC_T values of a calibrator such as untreated control cells or normal tissue are subtracted from values of any sample ΔC_T ($\Delta\Delta C_T = \Delta C_T$ sample – ΔC_T calibrator). The amount of target is then given by $2^{-\Delta\Delta C_T}$ [134]. Values represent the changes in expression of the gene of interest normalized to an endogenous reference and relative to a calibrator. For the comparative C_T method, it has to be ensured that the amplification efficiency for the gene of interest and the housekeeping gene are approximately equal to obtain valid results [134].

The standard curve method serves to quantify and compare the expression of one gene, e.g., a lipid sensor, in different tissues or cell lines. However, the comparative C_T method is more frequently used, because it allows comparing levels of gene expression in samples from the same source in an activated versus an inactivated state. Therefore, this method is particularly interesting to use in lipid sensor signaling by analyzing cell or animal samples with and without lipid sensor activator treatment for changes in target gene expression. An example of the use of the comparative C_T method is a study in which the relative effect of rifampicin on CYP27A1 mRNA expression in hepatic, colon, and kidney cells was compared [135]. Also, real-time PCR is frequently used in primary validation of microarray data [108].

In addition to its above-mentioned high sensitivity, the advantages of real-time PCR are its safety and simplicity for use. Furthermore, this technique allows the quantification of the observed effects. However, lightcycler and fluorescent probes are relatively expensive to purchase. In addition, only RNA of excellent quality can be used for real-time PCR analysis to ensure efficient amplification. When DNA-binding dyes are used, the method for the quantification of a specific gene needs to be optimized for each tissue or cell line before using it for analysis to ensure efficient amplification. This includes parameters such as primer concentration, template dilution, and amplification temperature and may be laborious [108,129].

Over the last years, Northern blot has been more and more replaced by real-time PCR due to its increased sensitivity, better quantification, and rapidity of analysis. RT-PCR is the method of choice for fast and simple qualitative analysis requirements. Because of its different technical approach, ISH is not redundant with the other RNA analyzing techniques and will be compared in the context of immunohistochemistry (IHC; see Section 9.5.4.2).

9.5.4 PROTEIN ANALYSIS

Exclusive RNA analysis is not sufficient to allow final conclusions about any effects of lipid sensor activation on gene expression because RNA is not necessarily translated into protein. Therefore, changes in mRNA levels do not always result in similar changes of protein levels. This implies that observed treatment effects on mRNA expression have to be confirmed by protein analysis. Important techniques in lipid sensor research are summarized in Table 9.9.

9.5.4.1 Antibodies

Proteins from treated cells or animals can be analyzed by two ways: WB and IHC. The common point between these two techniques is the use of specific antibodies which target the protein of interest. Two types of antibodies can be distinguished:

Polyclonal antibodies (PCA) can bind to multiple epitopes of an antigen. Because of this, they are more tolerant to small changes in antigen expression. However, some PCA may show more unspecific binding. PCA are often the preferred choice to detect denatured proteins, so they are more likely used in WB.

Monoclonal antibodies (MCA) only bind to a single epitope. MCA produce generally less background than PCA. In constant experimental conditions, reproducibility between experiments will be high when MCA are used. They are excellent to detect antigens in tissues or during affinity purification. So, they are more likely used in Co-IP and IHC.

9.5.4.2 Immunohistochemistry

IHC combines anatomical, immunological, and biochemical techniques for the identification of tissular proteins by means of a specific antigen/antibody reaction tagged with a visible label. Using IHC, it is possible to analyze the localization of proteins such as lipid sensors or their target genes in different types of cells or tissue sections and even their subcellular distribution. IHC can also be used to semiquantitatively analyze changes in expression levels of target proteins thought to be regulated by a lipid sensor.

Samples are fixed on a microscopic slide mainly by formaldehyde. The protein of interest is then detected using specific antibodies coupled to a fluorescent dye, a reporter enzyme such as peroxidase, or a radioactive isotope. Like ISH, visualization is carried out by microscopy [136]. An illustrating example of the use of IHC is the study of Watanabe et al. [137], who determined the exact localization of LXR protein expression in components of atherosclerotic lesions.

IHC allows analysis and identification of proteins directly in the tissue of interest, which distinguish this technique from WB, where proteins have to be extracted and are analyzed in a denatured state (see Section 9.5.4.3). However, IHC may show high background activity due to disturbances such as endogenous peroxidase activity, nonspecific antigen diffusion, or pigmentation of tissues [136]. In order to control these problems, several controls should be included in the experiment: (1) use of preimmune serum from the antibody-producing animal, (2) use of the secondary antibody alone, (3) no use of antibody, use of an antibody directed against a gene that is known not to be expressed in the cells/tissues to be studied, (4) use of a known positive antibody, such as anti-actin, and (5) use of a limited dilution series for the primary antibody [138].

9.5.4.3 Western Blot

WB or immunoblot is a very frequently used technique to characterize the expression of proteins in cell or animal samples. In addition, techniques such as Co-IP (see Section 9.6.3.5) require final detection by WB. Proteins to be analyzed may be extracted from cell or tissue samples using trizol (see Section 9.5.1); however, depending on the protein to be analyzed, simple cell or tissue homogenate as well as purified subcellular content such as NE or microsomes are also frequently used [139–141].

Proteins are separated by their molecular weight in denaturing conditions using polyacrylamide gel electrophoresis. Afterward, they are transferred to a membrane, usually nitrocellulose, which is incubated with a specific antibody for the protein of interest [142]. A second antibody, linked to a detectable label, such as biotin or a reporter enzyme like horseradish peroxidase or alkaline phosphatase, binds specifically to the first antibody and can be detected using autoradiographic films. The use of a second antibody increases the signal strength and its specificity. Changes in protein expression can be detected by comparing the band intensities from treated versus untreated samples. WB is a widely used technique which is relatively easy to perform and does not require expensive equipment. It is frequently applied to identify lipid sensors as well as to analyze their target gene expression in cells and tissues. Numerous examples can be found, which exceed the capacities of this chapter; one example might be the confirmation of the functionality of siRNA directed against the constitutive active/androstane receptor retention protein (CCRP), which lead to inhibition of its protein levels, while PXR (which was supposed to be associated with CCRP) was not affected by these siRNA [140].

Depending on the analysis, several parameters have to be adjusted: The percentage of acrylamide used for protein separation depends on the size of the protein of interest. For example, the molecular weights of lipid sensors are between 40 and 60 kDa; thus optimal separation in this weight range is achieved by using 10% polyacrylamide gels. The type of membrane, the blocking agent to reduce nonspecific binding of the antibody, as well as the supplier and the dilution of the antibody can also be modified. For example, even a difference between lots can change the specificity of antibodies. Bands appearing on the film can also be quantified to more precisely compare protein expression levels between control and treated samples by phosphoimaging. It has to be considered that specificity of the antibodies is highly important to allow a valid quantification. A poor quality antibody will show nonspecific bands that render detection of the protein of interest difficult.

WB is an irreplaceable technique to analyze levels of protein expression. In this context, it is an advisable approach to confirm results obtained by real-time PCR using WB. IHC is a more qualitative approach, more redundant to ISH than to WB. In NR research, IHC is much more frequently used than its RNA counterpart, due to increasing availability of specific antibodies, better stability of proteins compared to RNA, and easier experimental design.

9.6 PURE *IN VITRO* APPROACHES

While cell and animal models (see Sections 9.3.1 and 9.4.1) can reveal the specific involvement of a given lipid sensor in the regulation of a gene or a pathway under investigation, those approaches are not sufficient to exactly localize the putative lipid sensor RE on the gene promoter or to give information about interactions between lipid sensors and other proteins such as coregulators. Different *in vitro* techniques have been developed, which allow to investigate direct interactions of lipid sensor with DNA as well as with other proteins and to characterize more precisely lipid sensor activation by their ligands. Most of them work in a completely cell-free environment. Even though several techniques such as reporter gene assays or ChIP also include cell treatments, they are considered as *in vitro* approaches, because the biological samples undergo strong *in vitro* modification during analysis.

9.6.1 ANALYSIS OF LIPID SENSOR–DNA BINDING

Several techniques exist for the analysis of lipid sensor–DNA binding which are partly redundant and partly complementary. An overview is given in Table 9.10.

9.6.1.1 Reporter Gene Assays

Signal transduction can routinely be studied by coupling transcriptional elements such as RE to the expression of a reporter gene. The latter encodes proteins that can easily be measured and quantified by associated reactions producing color, light, or fluorescence upon modulation of their expression. This makes the assay suitable for high-throughput screenings (HTS), for example, for the examination of multiple compounds for their ability to transactivate an already identified RE. The effectiveness and sensitivity of a reporter gene depends upon a number of factors including the level of expression in a cell, stability of the expressed protein in the intercellular or intracellular milieu, background due to endogenous protein, and sensitivity of detection of the expressed protein [143]. Commonly used reporters are listed in Table 9.16. The most convenient reporter gene is firefly luciferase, because of its high sensitivity and easy determination.

The DNA sequence under investigation (RE) is cloned into a plasmid, which contains the coding region of a reporter gene such as firefly luciferase and which will express the reporter gene product under the transcriptional direction of the RE. When reporter plasmids are transfected into mammalian cells (see Section 9.3.2), luciferase expression is regulated through the transcriptional activation of upstream

---

TABLE 9.16
Examples of Commonly Used Reporter Genes

Reporter Gene	Detection Method
Alkaline phosphatase:	Colorimetry: Nitro blue tetrazolium
Chloramphenicol acetyltransferase (CAT)	Autoradiography: [^{14}C]chloramphenicol
	Scintillation: [^{3}H]acetyl-CoA
	Immunoassay
Firefly/renilla luciferase [217]	Bioluminescence
β-Galactosidase (lac Z)	Colorimetry: X-GAL
	Fluorescence: Fluorescein digalactoside,
	4-methylumbelliferyl β-D-galactopyranoside

Source: Bronstein, I. et al., *Anal. Biochem.*, 219, 169, 1994.

---

promoter elements. Expression levels of the bioluminescent enzyme luciferase are monitored by introducing its substrate, beetle luciferin, according to the reaction shown in Figure 9.2. Frequently, another plasmid encoding a second reporter gene under constitutive noninducible control is co-transfected, to serve as an internal reference for transfection efficiency [143].

Depending on the results already obtained by other experimental approaches and on the specific interest, different strategies may be used in transient transfections.

1. *"Full"-length promoter sequence.* There is no consensus in the literature about which length comprises a complete promoter. It is important to incorporate sufficient flanking sequence to ensure that regulatory mechanisms function as similar as possible to the *in vivo* situation [18]. Generally, at least 2–3 kb of the promoter sequence upstream of the transcription initiation site of the gene under investigation is used. However, this might not always be sufficient. For example, the CYP3A4 promoter contains a distal PXR RE at approximately −7.7 kb from the transcriptional start site [144]. Even if cloning of long promoter fragments easily cause difficulties, it might be preferable to increase the sequence length. In addition, more and more evidence is arising that introns also contain transcriptional regulatory elements such as NREs [145,146]. This might also be important to consider.

An increase in reporter gene activation will give evidence that the gene is regulated by the lipid sensor tested in the assay. Different transfection conditions with and without expression plasmids that encode the lipid sensor of interest and RXR (to ensure a high level constitutive expression of encoded

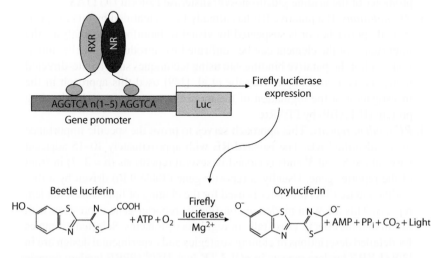

FIGURE 9.2 The luciferase reaction. Firefly luciferase expression is induced or repressed through the binding of ligand-activated NR to RE within the promoter region cloned in front of its cDNA. Firefly luciferase catalyses the conversion of beetle luciferin into oxyluciferin. This reaction also produces light (560 nm) that can be read. Produced light correlated with the firefly luciferase expression level (NR: nuclear receptor).

cDNA) are usually used. After an appropriate transfection incubation (see Section 9.3.2), the transfected cells are treated with a specific lipid sensor activator (Tables 9.1 and 9.2) for 24–48 h before reporter gene expression is analyzed. A lipid sensor positively regulated gene will respond to activator treatment with an induction in reporter gene expression. Because of the strong overexpression of the reporter gene, endogenous lipid sensor expression may be a limiting factor [18]. In this case, co-transfection of lipid sensor expression plasmids may additionally increase the response. For example, Chen et al. [147] used CYP7A1 promoter reporter genes in their investigation of the regulation of this gene by FXR. Goodwin et al. [144] described a strategy of cloning a 13 kb fragment of the CYP3A4 promoter.

2. *Serial promoter deletions.* To more specifically localize the RE, serial deletion constructs can be generated. Unless suitable restriction sites are directly available within the promoter sequence, this is easily achieved by PCR amplification of the desired regions. Primers have to be engineered containing restriction in their 5′ ends to allow their cloning into a reporter gene plasmid using the same restriction sites. Transient transfections of promoter deletions will give information about which sequence is minimally necessary to achieve transcriptional reporter gene activation. By generating, in the beginning, a few constructs with low resolution and then, depending on the results obtained with those constructs, more constructs with higher resolution, putative RE can be localized relatively precisely. For example, deletion constructs have been used to localize an LXRE on the SREBP-1c gene promoter [148]. Another example is the study of Verreault et al. [35], where the deletion constructs yielded the identification of an LXRE on the promoter of the uridine glucuronosyltransferase isoform UGT1A3.

3. *RE mutations.* If a putative RE has already been identified by other experimental approaches or is suspected by visual or bioinformatic analysis, the importance of the element can be confirmed by introducing specific mutations within the putative binding site using techniques such as site-directed mutagenesis. For example, Kawabe et al. [149] used this approach in the investigation of the regulation of the mouse heart-type fatty acid-binding protein (H-FABP) by PPARα.

4. *RE tandem repeats.* This approach serves to prove the specific importance of the identified RE. The isolated RE with approximately 10–15 adjacent bases at the 5′ and 3′ ends is cloned in several repetitions ($n = 3–7$) in front of the reporter gene. Usually, a reporter gene (Table 9.16) driven by a thymidine kinase (TK) promoter is used for the cloning of lipid sensor tandem repeats. This allows a moderate constitutive reporter gene expression with minimal trans effects, cross-talk, or regulatory problems. Selected examples for detailed descriptions of cloning strategies and experimental design are in [39] (LXRE tandem repeats in pGL3-TK-luc), [150] (PPRE tandem repeats in pGL3-TK-luc), and [149] (PPRE tandem repeats in pGL3-TK-luc). The introduction of mutations in the RE tandem repeats further confirms the specific binding of the lipid sensor to the identified RE.

Reporter gene assays are widely used to analyze lipid sensor-mediated regulation of gene expression. However, valid conclusions are only possible when results obtained by reporter gene assays are confirmed by other methods: Full-length promoter transfections precede or confirm *in vivo* cell culture or animal experiments using specific lipid sensor activation (see Sections 9.3.5 and 9.4.4) and can be replaced or confirmed by ChIP (see Section 9.6.1.4). Serial deletion construct transfections provide similar information to footprinting techniques (see Section 9.6.1.3), and RE multicopy reporter assays are redundant with EMSA (see Section 9.6.1.2).

9.6.1.2   Electrophoretic Mobility Shift Assay

EMSA is a very commonly used technique to analyze DNA–protein interactions. In the context of this chapter, it is used to demonstrate the direct binding of the lipid sensor of interest (protein) to its potential DNA-RE in the promoter of a target gene.

Basically, the lipid sensor protein under investigation is incubated with labeled oligonucleotides representing potential RE. Usually, NE or *in vitro* translated proteins (see Section 9.5.2) are used as source for the lipid sensor protein. The mixture is then separated in a nondenaturing acrylamide gel. The proteins linked to DNA have a higher weight and therefore migrate more slowly in the polyacrylamide gel than the free probe. Nondenaturing conditions are important in EMSA to conserve the three-dimensional structures of the protein that determine its capacity to bind to DNA [151].

The results should be validated using some of the possible assay variations:

1. Use of a mutated RE as probe to demonstrate that the specific (nonmutated) sequence is needed for lipid sensor binding.
2. Competition with unlabeled "cold" probe: If the unlabeled probe competes with the labeled probe, no shift will appear [152]. Very frequently, competition experiments are carried out using the RE under investigation (homologous competition). However, to validate the specificity of the binding, another previously described RE should be used as competitor (heterologous competition). This allows the comparison of the binding strength of the lipid sensor under investigation to different REs and may be an indicator for the degree of their transcriptional activation capacity. Competition results can be further strengthened by using mutated competitor sequences.
3. Supershift: The incubation of RE and protein is carried out in the presence of a specific antibody directed against the lipid sensor of interest. When the antibody specifically binds the protein, the complex becomes larger than the protein–DNA complex alone and will migrate more slowly. Therefore, this shift will appear higher on the gel/film than the original complex [152]. The supershift assay confirms the identity of the protein that binds the RE.

Because lipid sensors, especially *in vitro* translated proteins, already show a strong ligand-independent binding *in vitro*, it may be difficult to use the addition of a specific ligand as another assay condition. Several parameters can be adapted to increase the strength and specificity of binding. The concentration of the acrylamide gel may be modified to increase the separation between specific and nonspecific bands. To decrease nonspecific binding, competitors such as salmon sperm DNA and/or poly(dIdC) are almost always added to the incubation mix. Incubation temperatures, pH, and salt concentration of the incubation buffer also largely influence binding specificity. In addition, incubation on ice is suitable when NE is used to avoid premature protein degradation [153]. Migration time and voltage also influence the results.

For example, a negative PXR RE on the CYP7A1 promoter was identified using EMSA [111]. In their investigation about the negative regulation of apo AI by LXR, Huuskonen et al. [11] confirmed by EMSA that the NR HNF4 was displaced from its binding site on the apo AI promoter in the presence of activated LXR.

EMSA is an easy and relatively affordable technique that does not require expensive equipment. However, the assay is not suitable for high-throughput analysis. The use of radioactive compounds may also be considered as a negative aspect (see Section 9.5.2) [112]. To circumvent this problem, nonradioactive probes, such as fluorescent labels, can be used [154].

9.6.1.3 Footprinting Methods

Footprinting methods work by using an enzymatic or chemical reagent that will cut the DNA backbone. The conditions of treatment are adjusted such that each DNA molecule receives statistically one lesion, thereby producing a statistical mixture of singly-modified DNA molecules each including a cleavage site at a random position along the DNA chain. A protein such as lipid sensors specifically bound to the DNA will inhibit the cutting reaction at sites on the DNA to which it is bound. The degradation products are separated by denaturing polyacrylamide electrophoresis and visualized by autoradiography. The presence of a specific binding by a nuclear protein will leave a blank in the cutting pattern, which is aptly dubbed the "footprint" of the protein. By comparison of the bands representing bound DNA with bands representing unbound DNA, a lack of bands in the protein-bound fraction and presence of bands in the protein-unbound fraction is indicative of protein binding at that region (Figure 9.3) [155].

> *DNase I footprinting:* This technique was originally developed by Galas and Schmitz [156] and is the most often used footprinting technique in research about lipid-activated lipid sensors. A 5′-end-labeled, ds DNA segment is incubated in the presence and absence of NE to allow potential binding of nuclear proteins. Subsequently, the nonprotected parts of the DNA segment are fragmented by DNase I digestion. The enzyme DNase I nonspecifically hydrolyzes DNA sequences in the presence of divalent cations such as Mg^{2+} or Ca^{2+} [155]. For example, the technique has extensively been used to investigate the regulation of the drug metabolizing enzyme, CYP3A4, by PXR [117,144] or to study the basis of the regulation of LXR expression [157]. When applying the technique, it should be considered that GC-rich regions are poor DNase I substrates. Thus, footprinting may

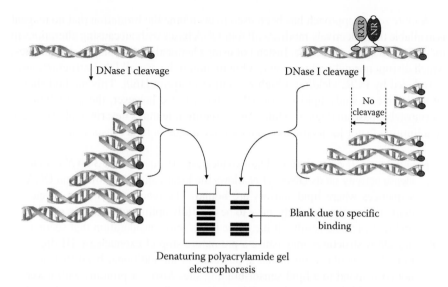

FIGURE 9.3   The principle of footprint. DNase I cleave DNA molecule randomly. NR specifically bound to DNA inhibits cleavage by DNase I at the binding site. Degradation products are separated by denaturing polyacrylamide gel electrophoresis. Blank in the cutting pattern indicates a specific binding (NR: nuclear receptor).

be less successful if the lipid sensor under investigation binds to one of these poorly cut regions. Because of sterical requirements of the enzyme, the accuracy for estimating the preferred binding sites is limited and the ligand site size is overestimated by about three base pairs. Furthermore, the footprint is staggered in the 3′-direction by about 2–3 bases which requires that the protection patterns on each strand of the duplex be compared in separate experiments [155].

Methylation interference: The methylation interference assay involves a chemical modification of guanines and adenines with dimethylsulfate (DMS) to produce N7-methylguanine or N3-methyladenine residues. The probe is treated to generate about 1%–2% modified purines. If the base which was modified on a given DNA molecule was critical for binding, that molecule will be left unbound after the addition of lipid sensors. Bound and unbound populations of DNA are separated on mobility shift assay gels. The two populations are then treated by piperidine to cleave the DNA at the modified bases and run on a denaturing PAGE gel. This method is very efficient to precisely localize guanine residues that are directly in contact with the NR; however, it only works poorly for adenine residues [158]. While being precise, it is only applicable when the lipid-sensor-binding sequence is already localized on the DNA, thus it requires preliminary analysis by other techniques. For example, Schoonjans et al. [159] combined methylation interference with EMSA (see Section 9.6.1.2) to identify a PPRE on the LPL promoter.

A more recent approach has been used to overcome the limitation that no reagent is available that selectively modifies all four DNA bases without causing alterations in the DNA secondary structure. Instead of using chemical treatments, base analogues, which disrupt normal protein contacts but maintain the overall DNA structure, were introduced by PCR, yielding a high-resolution footprint [160]. This method should be applicable to study lipid sensor–DNA interactions; however, the availability of thermostable base analogues exhibiting the required features described above might be a limiting factor for its use as a standard technique in lipid sensor research.

> *Exonuclease III footprinting:* The exonuclease III possesses a $3' \rightarrow 5'$ exonuclease activity on ds DNA. This allows the localization of $3'$ limits on DNA sequences where lipid sensors are bound. The reaction conditions, especially the enzyme quantity, have to be carefully optimized to obtain reliable and interpretable results. It has to be taken into consideration that secondary DNA structures may cause a premature stop of exonuclease III digestion. This would result in the appearance of an additional band that does not correspond to a lipid-sensor-binding site. Also, for proteins with lower binding affinity, exonuclease III sometimes nibbles into a protein-bound DNA segment or even displaces the protein from the DNA [161]. This technique appears to be rarely used in research about lipid sensors.
>
> *Hydroxy radical formation:* This method was originally developed by Tullius and Dombroski [162]. Hydroxyl radicals, generated by iron(II)-promoted reduction of dioxygen or hydrogen peroxide, attack the deoxyribose sugars arrayed along the DNA surface with virtually no dependence on the base sequence. Hydroxyl radical footprinting provides higher resolution footprints than DNase I footprinting, from which it is much easier to identify the exact ligand-binding sites. In addition, hydroxyl radicals often detect weaker, secondary binding sites that are not evident in DNase I footprints [155]. However, it is less sensible: it requires an occupation rate of at least 75%, while DNase I footprinting only requires 50%. Like with DNase I, the hydroxyl radical footprint is staggered in the $3'$-direction by about 2–3 bases. In order to precisely localize the RE, the protection patterns on each strand of the duplex have to be compared in separate experiments. Another possible disadvantage of the method is that glycerol concentrations beyond 0.5%—as commonly used for NE—inhibit the reaction [155].

Because all footprinting techniques involve the attachment of a chemical moiety to DNA, their general limitation is that larger-scale phenomena such as DNA conformation might be affected, leading to false-negative or false-positive results [155]. In addition, footprinting is generally laborious, time consuming, and requires the utilization of radioactive isotopes (see Section 9.5.2), which hinders this technique from more extensive applications and HTS of various potential target genes at the same time. In terms of those limitations, other techniques providing similar results, such as EMSA or reporter gene assays using serial promoter deletions (see Section 9.6.1.1), are generally preferred for the investigation of lipid sensor–DNA interactions.

9.6.1.4 Chromatin Immunoprecipitation

Chromatin immunoprecipitation (ChIP) is another technique that can be performed to analyze interactions between lipid sensors and DNA. More precisely, ChIP analyses of a DNA sequence bound by transcription factors or chromatin-associated proteins in living cells.

Receptor–DNA complexes formed in living cells by receptor agonist treatment are stabilized by covalent crosslinking [163]. An antibody targeting the lipid receptor of interest is used to precipitate the protein with the DNA bound to it. The coprecipitated DNA is then isolated and analyzed by techniques like semiquantitative PCR, real-time PCR, or Southern blot (Figure 9.4).

For example, Li et al. [135] demonstrated by ChIP that rifampicin-activated PXR induced the recruitment of steroid receptor co-activator-1 (SRC-1) to the CYP27A1 promoter. In a toxicological study about the molecular effects of phthalates on adipogenesis, Huuskonen et al. [11] demonstrated by ChIP that treatment with a phthalate derivative results in the recruitment of a specific subset of PPARγ coregulators leading to selective activation of several PPARγ target genes. Furthermore, Claudel et al. [164] demonstrated by ChIP that activated FXR/RXR heterodimers displace the NR, HNF4α, from the apo AIII promoter to explain the negative regulation of this apolipoprotein by FXR.

ChIP is a complementary technique to EMSA to analyze binding of lipid receptor to DNA. However, unlike EMSA, ChIP allows the analysis of DNA–protein interaction in living cells or tissues making this experimental approach more physiologic. Conversely, ChIP often requires long and laborious optimization of conditions such as crosslinking, immunoprecipitation, and DNA fragmentation. ChIP is also based on large cell numbers: for each treatment condition and antibody to be used, approximately one million of cells have to be considered. Furthermore, highly specific antibodies have to be used to avoid false-positive results. The final analysis by PCR or real-time PCR requires the use of predefined amplification strategies, but this approach does not allow the detection of unexpected REs. However, this disadvantage may be circumvented by using ChIP on Chip (see Section 9.6.2.2).

Compared to ChIP, EMSA is less complex, less laborious and easier to perform. In the case of an analysis of a highly conserved RE, EMSA might be sufficient to provide a final proof to demonstrate direct lipid sensor–DNA binding. However, if the RE to be characterized is composed of a poorly conserved sequence, it might be necessary to additionally use ChIP to confirm positive results obtained with EMSA.

9.6.1.5 Fluorescence Resonance Energy Transfer

FRET has been developed as a powerful technique for the detection of various biological interactions. This technique uses the transfer of excitation energy from a donor fluorophore to an acceptor chromophore when the two dye molecules are in close proximity to each other (<100 Å) and an overlap between the donor emission and acceptor absorption spectra occurs (Figure 9.5). A major advantage of FRET is that fluorescence is used instead of radioactivity for the detection. Thus, disadvantages associated with the use of radioactivity are avoided (see Section 9.5.2) while high assay sensitivity is conserved. Additionally, fluorescence provides an immediate and

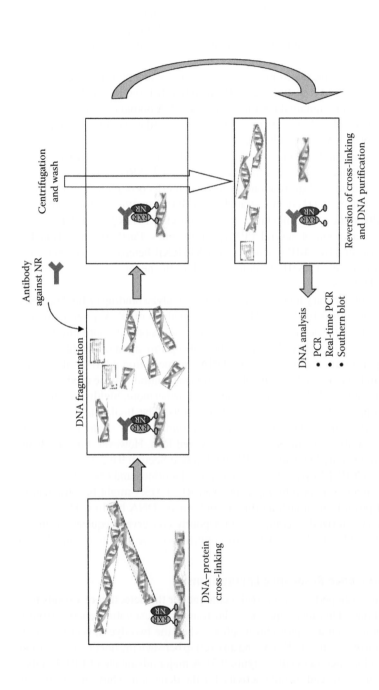

FIGURE 9.4 (See color insert following page 108.) The principle of ChIP. After cell treatment, cross-linking of protein–DNA is first done with formaldehyde. Cells are lysed and DNA is fragmented by restriction enzyme cut or sonication in 200–1000 bp fragment length. DNA-bound NR is then targeted by a specific antibody. The complex is precipitated and free fraction is washed. Cross-linking is then reversed and DNA is purified. Purified DNA can be analyzed by PCR, real-time PCR, or Southern blot (NR, nuclear receptor; PCR, polymerase chain reaction).

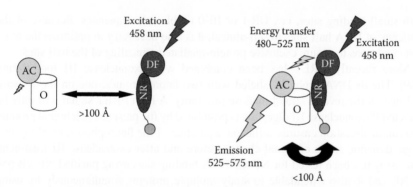

FIGURE 9.5 (See color insert following page 108.) The principle of FRET. The NR is linked with a donor fluorophore (DF) and the other (O) molecule (DNA sequence or protein) is linked with an acceptor chromophore (AC). When both molecules are far from each other, excitation of DF does not lead to an emission of AC. However, when the two dye molecules are close (less than 100 Å), excitation of DF leads to an energy transfer toward AC. AC excitation leads to an emission that can be read.

continuous signal, allowing real-time monitoring and imaging by microscopy [165]. To reduce interferences from background fluorescence, donor molecules with a long excited-state lifetime (time-resolved FRET, TR-FRET) can be used [112].

For the analysis of DNA–protein interactions, two different FRET strategies have been successfully applied. In the first one, both the DNA sequence of interest and the protein are labeled with a fluorochrome and FRET can be observed due to the formation of a DNA–protein-binding complex [166]. The technique has originally been used to study DNA conformations bound to the KF of *Escherichia coli* DNA polymerase I [166], but apparently no applications using lipid sensors have yet been published. A major disadvantage of this method is the requirement to label the protein with a fluorochrome that may affect the native features of the protein (e.g., interfere with ligand-induced conformational changes to induce DNA-binding affinity). In addition, large proteins may increase the distance between the donor and the acceptor in the DNA–protein complex and, therefore, the FRET signal may not be measurable and may give false-negative results. Furthermore, the optimum formation of DNA–protein complex often requires an excess of one of the components. This may cause difficulties in fluorescence detection, because the required 1:1 stoichiometry of donor and acceptor groups is not respected.

In the other strategy, two DNA fragments which correspond to two binding half sites of the protein are each labeled with a fluorochrome. FRET is then produced due to the protein-driven annealing of the two half sites. This technique is referred to as molecular beacons and displays a simple, rapid, homogenous, and high-throughput applicable method to study DNA–protein interactions [167]. Even though only a very limited number of proteins were tested in the original assay and no published tests with lipid sensors are available, it seems plausible that it will be applicable for lipid-activated NRs, because the labeling conditions were optimized for proteins that bind as a dimer to two distinct DNA half-sites [168] as it is also the case for lipid sensors. However, it might become difficult to design molecular beacons for lipid sensors with small binding sites, i.e., DR-1 or IR-0 (Table 9.3) sequences. Because of the tendency of DNA half sites to self-anneal, it is also necessary to optimize the reaction conditions to ensure exclusive protein-mediated annealing of the half sites.

More recently, FRET has been combined with exonuclease III footprinting [169]. The ds DNA probe is labeled with two fluorochromes, one on the forward and one on the reverse strand in close proximity. A high FRET signal will only be detected if exonuclease III digestion is prohibited by the presence of a bound protein. Enzymatic digestion ensures sufficient separation of the fluorophores, resulting in a large, dynamic range of signal change before and after exonuclease III treatment. The assay has been tested for NF-κB DNA-binding sites using purified NF-κB p50 or NE and is also extendable to study multiple proteins simultaneously by using multiple probes labeled with different FRET pairs [169]. The application of the exonuclease III/FRET assay to study lipid sensor–DNA interactions seems possible; however, the exact conditions for using lipid sensor proteins and their respective binding sites remain to be established.

9.6.2 HIGH-THROUGHPUT METHODS TO IDENTIFY LIPID SENSOR TARGET GENES

9.6.2.1 Microarray

The most frequently used high-throughput method to analyze the pattern of gene expression is the microarray. In the context of lipid sensor signaling analysis, microarrays can be used to determine the gene expression pattern after lipid sensor activation in cells or animals [170]. Based on chip technology, microarrays contain oligonucleotides or cDNA with a complementary sequence to a gene, promoter, or other genomic sequences, which are arranged in spots. Each of those spots represents one transcript.

Purified RNA from control and treated samples is reverse transcribed and labeled using different fluorochromes for control and treated samples. The labeled cDNA is then hybridized with the microarray chip. Afterward, the array is scanned and analyzed bioinformatically (Figure 9.6) [171]. Control and treated samples are initially analyzed separately by quantifying the emitted fluorescence of each spot, which is proportional to the amount of cDNA bound to it. After the individual analysis, both expression patterns are superposed, and by calculating the ratio between the two different fluorescent colorations of each spot, it is possible to quantify the relative change in gene expression between control and treated samples.

Microarrays use either probes generated from amplified cDNA [170,171] or ~25 mer oligonucleotides complementary to a selected gene or EST sequence [171]. Unlike cDNA assays where control and treated samples are analyzed on the same chip, oligonucleotide microarrays can only be used for a single sample [172]. Depending on the purpose of the analysis, probes can be selected from existing databases [173–175] or randomly chosen from any library [171]. When microarrays are used as a screening tool, probes are usually chosen randomly [170]. A particularly interesting tool for the analysis of the role of lipid sensor is the metabolic array. This oligonucleotide chip is composed of probes representing genes involved in synthesis, transport, and metabolism of lipids. As an example, Liu et al. [176] use microarray analysis

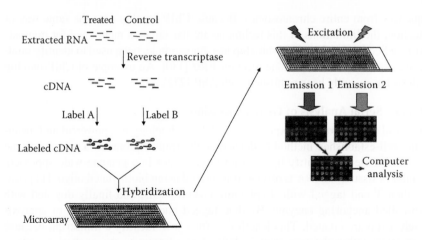

FIGURE 9.6 (See color insert following page 108.) The principle of microarray. cDNA microarray scheme. Total RNA from control and treated cells or animal tissues is first reverse transcribed and then fluorescently labeled with a different dye. Cye3 or Cye5 is usually used. Both samples are mixed and hybridized to the microarray under stringent conditions. The chip is afterward excited by laser. Emissions are scanned and computer analyzed. Images obtained are merged and ratio between control coloration and treated sample coloration is used to indicate the change in expression level. (Adapted from Duggan, D.J. et al., *Nat. Genet.* 21, 10, 1999.)

to identify transcripts in the human kidney cell line, HK-2, which could be used as surrogate markers for on-target activity of PPARα agonists.

Microarrays allow the analysis of a large quantity of genes in response to lipid sensor activation. However, it is a very expensive method due to the apparatus required to print cDNA on the matrix, the chip readers and, if required, for commercially prefabricated chips. Data analysis and filtering can dramatically change the results [177] and it is very difficult to distinguish real from false positive results because of the "noise" associated with stochastic variations in gene expression [178,179]. In addition, microarrays show a strikingly low reproducibility, and results are highly cell-model and context specific [177]. Most importantly, the number and identity of results always depend on those present on the chip [178].

In summary, the microarray, if affordable, is a very interesting technique in the field for large-scale studies of gene regulation. Because of its limitations, it is necessary to confirm the results by other analytical methods such as real-time PCR to generate an extensively validated data set.

9.6.2.2 ChIP-on-Chip

Instead of using PCR or Southern blot, protein-bound DNA which was purified by ChIP (see Section 9.6.1.4) can also be analyzed by microarray; a technique called ChIP-on-chip. In combination with microarrays, ChIP allows identification of target genes of transcription factors such as lipid sensors on a chromosome- or genome-wide scale. Four different types of arrays are used for ChIP-on-chip analysis: proximal promoter (~1 kb) arrays, large promoter (several kb) tiling arrays, arrays composed of amplified CpG islands, and the genome tiling array containing nonrepetitive sequences from entire chromosomes. Because ChIP-on-chip uses the same way of detection, the limitations of this technique are the same as mentioned for microarrays (see Section 9.6.2.1). ChIP-on-chip has, for example, been used to identify binding sites of the nuclear hormone receptor ERα [180]. An example of ChIP-on-chip analysis has recently been published with AhR [272].

9.6.2.3 Serial Analysis of Gene Expression

The serial analysis of gene expression (SAGE) technology is a powerful and meanwhile well-established method to determine the transcriptome of a cell or a tissue and can be used to identify lipid sensor target genes in a genome-wide approach. Cellular or tissular mRNA is reverse-transcribed using biotinylated oligo(dT) primers, then 3′-end tagged with streptavidin-coated beads, and finally digested with a so-called anchoring enzyme. By that, tagged nucleotide sequences of approximately 10 bp are created. This length is sufficient to identify a transcript, because its position within the transcript is defined as adjacent to the anchoring restriction enzyme that is nearest to the polyadenylation signal (because of the use of oligo(dT) primers). The tagged sequences are subcloned into an *E. coli* expression vector to form a SAGE library, which has then to be sequenced systematically to calculate the frequency of the individual tags. The tag frequency is believed to be proportional to its mRNA expression level [14]. A comparison of datasets between unactivated and activated lipid sensor in a biological model allows identifying putative target genes of a lipid sensor (Figure 9.7). For example, SAGE has been used

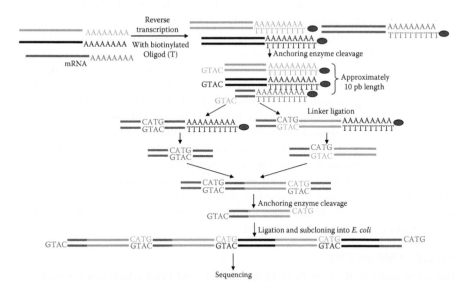

FIGURE 9.7 (See color insert following page 108.) The principle of SAGE. mRNA is reverse-transcribed using biotinylated oligo(dT) primers. A so-called anchoring enzyme cleaves cDNA approximately 10 bp before the 3′ end. A linker is added to cleaved cDNA. After removing the poly(A) sequence, cDNA are linked together. A second cleavage by the anchoring enzyme is done to allow the ligation of multiple cDNA fragments. Ligations are cloned into *E. coli* bacteria and can be sequenced.

to quantitatively describe the genomic expression profile of mice retroperitoneal adipose tissue [181] and further to analyze the transcriptomic effects of castration and androgen treatment. A large number of new transcripts regulated by androgens were thereby identified [182]. A study analyzing lipid-sensor-regulated gene expression by SAGE seems not yet published, but it would perform in a similar way to the example described above.

The major advantage of SAGE over cDNA microarrays is that it even allows identification of gene products that have not been described before. While the microarray technology is limited to genes represented on the chip used for the analysis (see Section 9.6.2.1), SAGE allows the analysis of the entire cellular transcriptome. However, due to licensing and sequencing cost, SAGE is very expensive. Furthermore, a large amount of RNA (requiring approximately 100–120 mg tissue) is needed to produce a good SAGE library. Even though a sequence length of 10 bp is theoretically sufficient to identify the transcript, a considerable number of tags often cannot be assigned to a gene or multiple tags produce a match with the same gene [14].

In conclusion, SAGE can be considered as a redundant technique to microarrays. Its advantage over the array approach is that it covers the complete genome instead of selected genes. Both techniques have similar limitations concerning their large dependence on the experimental conditions and statistical analysis, what often limits the overlaps in gene expression profiles in different studies. Because of this, both SAGE and microarrays, require additional verification by standard molecular biology approaches such as real-time PCR. However, a disadvantage of SAGE is that it is technically considerably more demanding than microarrays, making the technique probably less attractive for occasional use.

9.6.2.4   Gene Trapping

Gene trapping is a cell-based approach which allows the *in vivo* monitoring of the expression of the trapped gene through reporter monitoring. The technique is based on random insertion of a promoterless reporter gene into the host cell genome. When the reporter gene has been integrated into a target gene of a lipid sensor, a change in reporter expression is observed after lipid sensor activation. The so-identified cells are cloned to homogeneity and the identity of the corresponding targeted gene is determined by sequencing across the fusion boundary.

The production of a trapping library includes transfection of target cells with the gene trap vector, selection of positive cells by marker expression, and their isolation and further characterization. The experimental details are beyond the purpose of the chapter and are outlined elsewhere [14]. Usually a retroviral vector is used for gene delivery in the targeted cells. Two major strategies are existing to design the gene trap vector: the promoter trap (promoterless reporter gene and a gene serving as selection marker under the control of its own promoter) or the polyadenylation trap (selection marker with a constitutive promoter). The most common reporter system used for gene trapping to monitor the potential lipid sensor target gene expression is GFP. GFP is a small 27 kDa protein from the jellyfish, *Aequorea victoria*. It forms a fluorophore within itself when expressed in a cell [47] and therefore allows its direct determination in living cells without requiring substrates or other additional factors. Hence, selection of positive cells via fluorescence-activated cell sorting (FACS) is also possible using GFP. So far, mutated murine ES cell lines have been generated for the lipid sensors, PPARδ, RXRα, and RXRβ [183]. More detailed information can be found on the Web site of the International Trap Consortium [183].

Because living cells can be monitored, gene expression modulation can easily be studied over a period of time, to identify direct (immediate response) and indirect (delayed response) effects. For some cases, ectopic expression systems may also be used. A particularly interesting approach is the use of ES cells for gene trapping. These cells can be differentiated into various lineages and the role of lipid sensor target genes can be analyzed in each of them. Furthermore, trapped ES cells can be used to directly generate KO mice (see above), which allows further characterization of the identified target gene *in vivo*. Because gene expression is directly determined *in vivo*, the results are not affected by errors occurring during RNA analysis, such as quality of the extracted RNA, the cellular RNA level, kinetics of mRNA induction, mRNA stability, or cell cycle status [14].

9.6.3 ANALYSIS OF LIPID SENSOR–PROTEIN INTERACTIONS

The analysis of lipid sensor–protein interactions can be used to elucidate the role of lipid sensor cofactors. In addition, interaction with other signaling pathways can be determined. An overview of possible techniques for the investigation of lipid sensor–protein interactions is given in Table 9.12.

9.6.3.1 Reporter Gene Assays

Reporter gene assays (see Section 9.6.1.1) can also be used to investigate interactions between a lipid sensor and other proteins. Expression plasmids for the factor of interest are cotransfected with a promoter or an isolated RE in tandem repeats (see Section 9.6.1.1), which has already been proven to be regulated by the lipid sensor of interest. Changes in reporter gene activation with or without the coexpression of the protein under investigation indicate an interaction. For example, this approach has been used to demonstrate an interaction between PPARγ and tumor necrosis factor (TNFα) [184].

9.6.3.2 Chromatin Immunoprecipitation

ChIP is also widely used to study lipid sensor–cofactor interactions by using specific antibodies against the cofactors of interest. Unlike reporter gene assays that can determine any protein–protein interaction influencing transcriptional activation, ChIP is limited to lipid sensor–protein interactions, which concomitantly bind to DNA. For example, Jia et al. [185] used ChIP to study the role of the co-activator PBP for PPARα in mouse liver.

9.6.3.3 GAL4 Two Hybrid Systems

Similar to reporter gene assays (see Section 9.6.1.1), GAL4 two-hybrid systems can be used to detect protein–protein interactions in an intracellular setting. It is considered a powerful tool for this means and probably over half of the protein interaction discoveries reported in the literature originate from yeast two-hybrid experimentation [186]. GAL4 is a yeast DNA-binding transcription factor required for the activation of the GAL genes in response to galactose [187]. The native GAL4 transcription factor contains an N-terminal DBD and a C-terminal activation domain (AD). When GAL4 binds to its cognate binding site, the AD is brought close to the promoter, allowing the AD to interact with the transcription machinery and resulting in activation of transcription [187].

In the original yeast two-hybrid system, two chimeras are created: one that contains the lipid sensor fused to the GAL4 DBD and another containing a possible cofactor protein fused to the GAL4-AD. LacZ is used as a reporter gene [186]. This system allows screening of lipid sensor cofactors by preparing a recombinant DNA library in which genes for many different proteins are fused to the AD [188]. For example, the GAL4 two-hybrid system has been used with GAL4-PPARγ as bait to screen a mouse liver cDNA library and has lead to the identification of SRC-1 as a PPAR transcriptional co-activator [189].

The GAL4 two-hybrid system has been further developed for use in mammalian cells by replacing the GAL4-AD by the non-yeast herpes simplex virus, VP16, or *E. coli* B42 AD. The fusion proteins are co-transfected into mammalian cells using a GAL4-responsive luciferase reporter. The mammalian expression systems provide the advantage that protein interactions can be studied in an environment that is more similar to that *in vivo* compared to expression in yeast. Brendel et al. [39], for example, described the use of a mammalian GAL4 two-hybrid system to study interactions between LXRα (LXRα-LBD-GAL4 fusion protein) with the NR, SHP (SHP-VP16 fusion protein).

The GAL4 assay tests protein interactions in living cells without the need for isolated protein (only the gene) and is easy to perform. A single reporter construct can be used to detect responses to a range of receptor–ligand interactions. It can even be scaled up and automated for HTS of a large number of interacting proteins. Because of the signal amplification afforded by the use of enzymes as reporters, the system is also very sensitive and can detect weak protein–protein interactions. However, GAL4 is an artificial system in which the proteins under study are chimeric. This leads possibly to a loss of subtle conformational changes and interactions of regions other than the LBD [18]. Furthermore, the GAL4 system creates many false-positive results. Therefore, like in the example mentioned above [39], results always have to be verified by other techniques such as Co-IP or GST pull-down assay [186].

9.6.3.4   GST Pull-Down Assay

Another technique to determine interactions between lipid sensors and potential cofactors is GST pull-down assay. This cell-free assay uses proteins fused to glutathione-*S*-transferase (GST fusion proteins) to affinity-purify other proteins [190]. The GST fusion protein is first purified on glutathione-agarose beads. The fusion proteins bound to the beads are then used as a "bait" to test for binding to another protein which may be either a specific test protein (see Section 9.5.2; e.g., a purified protein or a protein that has been labeled with ^{35}S-methionine by *in vitro* translation) or a complex mixture of proteins such as crude NE. The beads and associated protein complexes are removed from the incubation mixture, washed, eluted in loading buffer, and analyzed.

Two methods are commonly used to detect binding of the test protein: immunoblotting using a protein-specific antibody (see Section 9.5.4.1) or measuring bound radioactivity, if a ^{35}S-methionine-labeled protein has been used. Radioactive detection is preferable, because it confers greater detection sensitivity, and it is possible to quantify the protein by phosphoimaging or scanning densitometry. Using immunoblotting, the relatively large amount of GST fusion protein bait may exhibit nonspecific cross-reactivity with the antibodies employed, making detection difficult if the GST-bait and test proteins are similar in size.

Because of the relatively high tendency of artifactual binding, a negative control for nonspecific retention should be included. This can be either a parallel assay using glutathione-agarose beads with bound GST alone, or competition with a large excess of cold target protein or the use of a variant of the bait protein which is known not to bind the test protein *in vivo*. The technique has successfully been applied to determine co-activator interactions of DRIP205 with FXR [191]. GST pulldown was also used to verify if LXRs interacted with the atypical NR small heterodimer partner (SHP). Using the same technique by generating lipid sensor-GST deletion mutant fusion proteins, the authors were also able to determine the interaction domain between the two receptors [39].

The GST pulldown technique can complement other methods for assessing protein–protein interactions such as Co-IP or GAL4-based systems. The assay is simple to perform; relatively powerful and sensitive by being capable of detecting interactions with K_d (dissociation constant) in the micromolar range. Different from the GAL4 system, the GST pull-down assay is a pure *in vitro* approach. This avoids problems associated with transformation/transfection of yeast or cells, respectively. However, it is prone to artifacts due to mass-action effects, and some care should be used in interpreting positive results. It may also happen that the bait protein is sterically inhibited because of its fusion to GST. Because a protein–protein interaction determined *in vitro* is not necessarily physiologically important *in vivo*, the assay always needs *in vivo* confirmation by other experimental approaches such as Co-IP or ChiP. Another difference to GAL4 systems is the essential requirement of proteins (because of its lack of *in vivo* reference). They can be obtained from multiple sources (see Section 9.5.2) but might be more difficult to produce, handle, and store than DNA.

9.6.3.5 Co-Immunoprecipitation

Co-IP is another technique frequently used for protein–protein interaction analysis. However, interactions between a receptor and its ligand, an enzyme and its substrate, as well as a subunit and a protein complex can also be determined by Co-IP [192]. In the case of lipid sensors, Co-IP is used to investigate interactions of receptors with cofactors (Table 9.5) and binding partners such as RXR.

Co-IP uses a specific antibody targeting the lipid sensor of interest. The antibody is incubated with the protein mixture to immunoprecipitate the targeted lipid sensor and also to co-precipitate any bound interacting proteins in the sample. Potentially formed antibody–antigen complexes are absorbed from the solution through the addition of an immobilized antibody-binding protein such as protein A- or protein G-sepharose beads [192]. After washing, the complexes are eluted. Proteins captured by immunoprecipitation are then analyzed by WB using an antibody against the suspected protein co-immunoprecipitated with the lipid sensor.

Co-IP requires highly specific antibodies for the proteins of interest without cross-reactivity with other proteins to ensure a specific binding. To ensure that the co-precipitated protein is precipitated by the antibody itself and not by a contaminating antibody in the preparation, the use of MCA (see Section 9.5.4.1) is recommended. Alternatively, PCA can be preadsorbed against extracts lacking the lipid sensor protein or prepurified with authentic antigen. It also has to be established that the antibody does not itself recognize the coprecipitated protein by using independently derived antibodies or protein mixtures lacking the lipid sensor protein as a negative control [193]. When selecting the antibodies, it also has to be considered that the antibody used for immunoprecipitation of the lipid sensor has not been raised in the same animal species than the antibody used for detection of the co-immunoprecipitated protein. This would lead to a too strong background signal during WB.

In most cases, Co-IP is performed with crude cell lysates as a protein source. This has the advantage that the experiment already has a built-in specificity control (because all proteins in the lysate potentially compete for the binding). Furthermore, the interacting proteins are present in the same relative concentrations as in the cell and in their natural state including posttranslational modifications. This avoids artificial effects in the experiment. However, to determine if the interaction between the two proteins is of direct or indirect nature, Co-IP should be repeated with the purified proteins for which the interaction has previously been observed [193]. The choice of immobilized antibody-binding protein depends upon the species that the antibody was raised in. Detailed information about the preferences is given by companies retailing Co-IP-related products.

Problems may arise if the lipid sensor or its interaction partner and the antibody heavy and light chains have similar relative molecular weights, because under reducing conditions of WB they will comigrate, resulting in an interference with the detection of the interacting protein [193]. Furthermore, Co-IP is less suitable for weak protein interactions because their binding might be not stable enough during the washing step, leading to false-negative results.

For example, Brendel et al. [39] used Co-IP to analyze the association between LXRα and SHP. Co-IP can also be performed using a chimeric protein tagged with Myc or FLAG. This strategy might be interesting, if no antibody against the studied protein is available. Ichida et al. [194] used heterologously expressed Myc-tagged PPARγ co-activator (PGC)-1α and FLAG-tagged estrogen-related receptor (ERR)-α to study the recruitment of ERRα to PGC-1α for transcriptional activation. However, it has to be kept in mind that the protein tag can inhibit the receptor–protein interaction, leading to false-negative results.

In summary, Co-IP is a suitable method to study interactions of nuclear sensors with other proteins. Once the lipid sensor of interest has been precipitated, WB can be performed with a multitude of different antibodies to study different protein interactions. However, because of its susceptibility for false-positive results, Co-IP experiments should always be confirmed by other approaches. For example using Co-IP, heat shock protein 90 (hsp90) and the hepatitis B virus X-associated protein 2 (XAP2) were identified to be associated with PPARα. A following reporter gene assay confirmed the role of XAP2 as a co-repressor of PPARα transactivation activity [195]. In a very elegant approach, it was demonstrated by Co-IP that cytosolic PXR was associated with CCRP. This association was confirmed *in vivo* in mouse liver slides by using FRET (see Section 9.6.3.6). Contrariwise, Co-IP is also frequently used to confirm results obtained by GAL4 two-hybrid screenings, as for example in [194].

9.6.3.6 Fluorescence Energy Transfer and Bioluminescence Resonance Energy Transfer

The strengths of FRET (see Section 9.6.1.5) make this technique also useful to study lipid sensor–protein interactions. An example of the application is a study about the ability of ligand-activated RXRβ to recruit co-activator [196].

Bioluminescence resonance energy transfer (BRET) is a luminescent variant of FRET, which can also be used to investigate protein–protein interactions. Energy emission of the donor is achieved by an enzymatic substrate oxidation. Thus, in contrast to FRET no external light source is needed, making the instrumentation simpler and cheaper. The absence of excitation light also results in a considerable reduction of background signal, increasing the sensitivity of BRET compared to FRET [112]. The assay has been tested to assess protein–protein interactions for the estrogen receptor and the co-activator SRC-1 in living HEK293 cells [197]. BRET has been considered an excellent method for measuring lipid sensor interactions with other proteins, but at present, no publications are available where the technique has been used for lipid-activated NR.

9.6.3.7 Fluorescence Correlation Spectroscopy

Fluorescence correlation spectroscopy (FCS) is a technology that can determine protein–protein associations by measuring fluorescence intensity fluctuations as a result of differences in diffusion rate of individual dye-labeled proteins free in solution or bound to a lipid sensor [112]. The complex physical considerations behind this technique are described in [198]. The specific challenge of FCS is the small optical observation volume, which allows probing within any location of a cell, making FCS very attractive for *in vivo* applications. Initially, the technique had been used to quantitatively study homodimer formation in cells. More recently, theory and experimental design were expended to quantify homodimer and heterodimer formation between RAR and RXR in CV-1 cells [198]. Another example of the use of FCS in combination with FRET is the determination of PPAR-RXR binding properties and PPAR nuclear mobility characteristics in living COS-7 cells [199].

9.6.4 ANALYSIS OF LIPID SENSOR–LIGAND INTERACTIONS

Another important aspect in the investigation of lipid sensor-mediated gene regulation is the identification of endogenous ligands. In addition to the molecules that are already known (Table 9.1), it seems probable that more endogenous activators exist [9]. Furthermore, lipid sensors are important targets in drug discovery because of their involvement in disease development, e.g., atherosclerosis. Several methods have been established to analyze lipid sensor–ligand interactions. They are based on either the determination of a functional response (e.g., target gene transactivation capacity) or the *in vitro* determination of ligand–receptor interactions [112]. A summary of the experimental approaches is given in Table 9.13.

9.6.4.1 Radioligand-Binding Assay

The radioligand-binding assay is a widely used method to identify lipid sensor ligands. Heterogeneously expressed protein of the receptor of interest is incubated with a radiolabeled potential ligand for an appropriate time. Afterward, the free fraction is separated from the lipid sensor-bound radioligand by centrifugation, dialysis, or filtration and the radioactivity associated with the lipid sensor is determined (Figure 9.8) [112]. If binding is observed, three further types of experiments can be carried out [200]:

Saturation: Saturation binding experiments measure specific binding at equilibrium at various concentrations (often 6–12) of the radioligand to estimate the K_d of the lipid sensor for the ligand.

Kinetic: Time-course experiments are carried out with a constant amount of lipid sensor and radiolabeled ligand. First ligand binding is initiated and then blocked, and specific binding is measured over time (typically 10–20 measurements) to determine how rapidly the ligand dissociates from the lipid sensor.

Inhibition: In this assay, increasing concentrations of unlabeled ligand are included in the incubation mixture (competition). This serves to validate the assay by using well-defined specific ligands of the lipid sensor (Tables 9.1 and 9.2) as competitors. Furthermore, different molecules can be screened if they compete with the known ligand and finally the affinity constant (K_i)

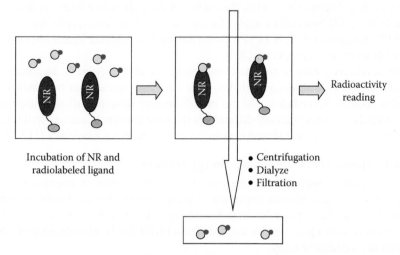

FIGURE 9.8 The principle of radioligand binding assay. NR are incubated with radiolabeled potential ligand. Free fraction is separated from NR-bound radioligand by centrifugation, dialysis, or filtration. Radioactivity is then read.

of the new ligand for the lipid sensor can be estimated. For example, Berger et al. [201] described a radioligand-binding assay for the identification of novel, non-thiazolidinedione agonists for PPARγ and PPARδ.

The assay can also be carried out in intact cells and thus, in the native environment of the lipid sensor. Receptor binding can then be studied under the same conditions as the biological responses of activated lipid sensor-signaling. This system also allows monitoring changes in lipid sensor expression following cell treatment, such as agonist-induced lipid sensor downregulation [200].

A major advantage of radioligand-binding assays is their sensitivity, specificity, and ease of use. Therefore, the technique is also applicable for high-throughput-screenings, e.g., of pharmaceutic synthetic compounds. The assay requires only labeling the ligand, which often does not reduce the affinity toward the receptor. Finally, many high-affinity receptor ligands are commercially available, allowing to easily validate an assay [112]. An important disadvantage of these assays is the need to separate free from bound ligand, which make these assays labor-intensive and relatively slow. Moreover, these assays require that the dissociation of the ligand proceeds much slower than the time to perform the separation step [112].

9.6.4.2 Scintillation Proximity Assay

Scintillation proximity assay (SPA) is another radioactive assay for the determination of lipid sensor–ligand interactions. It uses an immobilized lipid sensor on small fluomicrospheres (beads) and ligands labeled with a radioactive isotope, which may be ^{3}H or preferably ^{125}I because of its higher specific activity. The beads, on which the lipid sensor is immobilized, contain a scintillant, which emits light as a result of energy transfer from the radioactive decay of the ligand to the bead. The emission of light only proceeds if the labeled ligand and the lipid sensor are in close proximity (approximately 10 μm). Otherwise, the energy of the radioactive ligand is absorbed by the buffer [112]. Selected examples for the use of SPA are an assay for the screening of PPAR ligands [202], or an SPA-based approach to compare ligand interactions with PXR from different species [203].

SPA is relatively easy to automate, enhancing the assay reliability. Its advantage over radioligand-binding assays is that no labor-intensive separation of free and bound ligand is needed. Nevertheless, the use of radioactivity remains a disadvantage (see Section 9.5.2). Another potential difficulty is the need to immobilize the lipid sensor on a solid surface, where it should remain stable and maintain its affinity [112,204].

9.6.4.3 Fluorescence Resonance Energy Transfer

FRET not only is used for the analysis of lipid sensor–protein interactions (see Section 9.6.3.6) but is also an important nonradioactive technique for the determination of ligand–lipid sensor binding. For example, Chin et al. [205] combined the GAL4 system with a β-lactamase reporter to use FRET for the identification of LXR agonists in a cell-based assay.

Similar to SPA, the assay does not require a separation step of bound and unbound ligand. However, its convenience is limited by the requirement of labeling both the donor and acceptor molecules [112].

9.6.4.4   Fluorescence Polarization

FP is based on fluorescence anisotropy and measures the change in rotational speed of a ligand during its excited lifetime upon binding to its receptor [112]. For example, the technology has been applied to a screening assay for dual PPARα/γ agonists [206].

The advantage over FRET is that only the ligand needs to be labeled, saving one labeling step. In addition, the technology of FP is relatively simple and does not require expensive instrumentation. Conversely, the sensitivity of FP decreases considerably when the ligand affinity (≥5 nM) and/or the lipid sensor concentration (≥1 pmol/mg protein) are low [207].

9.6.4.5   AlphaScreen™

Amplified luminescence proximity homogenous assay, AlphaScreen™, is a bead-based assay which uses the principle of luminescent oxygen channeling. A donor (the lipid sensor) and an acceptor (ligand) pair of reagent-coated microbeads are brought into proximity by biomolecular interaction of binding partners immobilized to these beads. Singlet oxygen ($^1O_2$) is produced on donor beads by laser excitation and photosensitizer conversation, and a chemiluminescent reaction on the acceptor beads can only occur, when the bead pair is in sufficient proximity to allow energy transfer [208]. The determination is carried out indirectly via the generation of fluorophore excitation (by chemiluminescence). Compared to FRET, the assay displays an improved sensitivity due to low background and signal amplification [209]. Glickman et al. [209] screened FXR agonists and antagonists to directly compare the performance of the AlphaScreen method with TR-FRET. They found out that AlphaScreen was indeed superior in terms of sensitivity, dynamic range, plate reading time, and material consumption [209]. These characteristics make the assay particularly useful for high-throughput applications.

9.6.4.6   GAL4 One-Hybrid System

The GAL4 system is a cell-based reporter gene assay, which can be used to study lipid sensor–ligand interactions in addition to lipid sensor–protein interactions (see Section 9.6.3.3). In contrast to other approaches, the GAL4 one-hybrid system is based on the transactivation capacity of a potential ligand. For the investigation of ligand–lipid sensor interactions, only one-hybrid protein is created, containing the LBD of the lipid sensor of interest fused to GAL4-DBD. This chimera is then coexpressed with GAL-responsive reporter plasmid. This system allows testing molecules for their potential to activate the lipid sensor of interest in transient transfections but minimizes background caused by endogenous lipid sensor expression [210]. The application of the GAL4 one-hybrid system, including its analysis, is explained in detail in [211], where sage and rosmary extracts were tested for their potential to act as PPARγ ligands. Another example of the use of the GAL4 one-hybrid system is the test of perfluorinated fatty acid analogues and naturally occurring fatty acids to activate different lipid sensors of human, mouse, or rat origin. The application of the GAL4 assay in this context allowed a better cross-species understanding of lipid sensor activation by the respective molecules while limiting the variability usually associated with complex signaling systems [212].

In a comparative study, primary-hit identification in HTS by TR-FRET and AlphaScreen was tested using FXR co-activator recruitment. Interestingly, the overlap of identified active compounds (for the same biochemical target) was relatively low, and even the overlap of the most active compounds was generally less than 50% [213]. Even though HTS is probably not the most widely used application in fundamental lipid sensor research, it should be considered for down-scaled screenings that different assay technologies can lead to significantly different outcomes. Complementary assay formats may be needed for improvement and validation.

Furthermore, it has to be considered that the ligand-binding affinity as determined by assays, such as radioligand-binding assay or SPA, does not always correlate with the transactivation capacity of a lipid sensor ligand as determined by GAL4 assays [213]. When different PPARα ligands were tested, co-activator recruitment (determined by a quantitative FRET assay) correlated better with the transactivation capacity (determined by a GAL4-PPARα responsive reporter gene assay) when compared to *in vitro* ligand binding (determined by SPA). The most intriguing examples are gemfibrozil and fenofibrate, which did not show any measurable *in vitro* receptor binding but which have numerous PPARα-associated effects *in vivo* [214]. This underlines the necessity to apply different redundant and nonredundant experimental approaches to fully characterize a novel lipid sensor ligand. For example, Pellicciari et al. [215] combined a cell-free FRET co-activator assay and a luciferase reporter gene assay to characterize their FXR-modulating function and Ondeyka et al. [216] screened a natural product library of microbial extracts for LXR agonists using SPA and a GAL4-based transactivation assay.

9.7 *IN SILICO* APPROACHES

9.7.1 LIPID SENSOR DATABASES

Due to the considerable research dedicated worldwide, data about lipid sensor signaling is increasing rapidly. Different research groups and consortiums try to organize the available information to help the research community to more easily find relevant data for their own projects. Some selected Web resources are summarized in Table 9.17.

NuReBase is a freely accessible bioinformatic database on all classes of NR. It provides protein and DNA sequence data, alignments, and phylogenetic trees. Furthermore, data on alternative transcripts as well as expression data for human and mouse NR have been added. Its particular strength is that the database is automatically updated daily, stored in NUREBASE_DAILY. NucleaRDB is another database that provides primary and secondary data on NR, such as sequence information and chromosomal location, structural information, mutation data, and information about binding partners and ligands. The Nuclear Receptor Signaling Atlas (NURSA) is a trans NIH consortium designed to provide new findings, access to new reagents, large validated data sets, and a library of annotated prior publications about of the structure, function, and role in disease of NR.

There are also less NR-specific, but often useful online resources existing. The NCBI-related GenBank, together with the European EMBL Nucleotide Sequence

TABLE 9.17
Useful Web Resources in NR Research

Name	Web Address
NuReBase	http://www.ens-lyon.fr/LBMC/laudet/nurebase/nurebase.html
NucleaRDB	http://www.receptors.org/NR/
Nuclear Receptor Signaling Atlas	http://www.nursa.org
Biological software	http://bioweb.pasteur.fr/intro-uk.html
Biobase	http://www.gene-regulation.com/index.html
The Eukaryotic Promoter Database	http://www.epd.isb-sib.ch/
Ensembl	http://www.ensembl.org
Kyoto Encyclopedia of Genes and Genomes	http://www.genome.ad.jp/kegg/
GenBank	http://www.ncbi.nlm.nih.gov/Genbank/
EMBL	http://www.ebi.ac.uk/embl/
DDBJ	http://www.ddbj.nig.ac.jp/

Database and the DNA Database of Japan (DDBJ) are DNA databanks that contain all publicly available DNA sequences, including preliminary genome fragments, EST, and mutations. The most efficient way to access them for NR-related information is through BLAST. BIOBASE is a commercial content provider of biological databases, knowledge tools, and software for the life science industry. Their first database was TRANSFAC, a transcription factor database, which contains data on transcription factors including NR, and their experimentally proven binding sites and regulated genes. Other interesting tools include TRANSPATH, a signal transduction pathway database, and ExPlain, a tool for expression data analysis. Biological software for different NR analysis such as DNA, RNA, or protein sequences, structures, and phylogeny or sequence alignments and comparisons can also be found online. The Eukaryotic Promoter Database (EPD) is a nonredundant collection of eukaryotic POL II promoters, for which the transcription start site has been determined experimentally, including description of the initiation site mapping data, cross-references to other databases, and bibliographic references. Ensembl is a joint project to develop a software system which produces and maintains automatic annotation on selected eukaryotic genomes. Kyoto Encyclopedia of Genes and Genomes (KEGG) is a database of biological systems, consisting of genetic building blocks of genes and proteins (KEGG GENES), chemical building blocks of both endogenous and exogenous substances (KEGG LIGAND), molecular wiring diagrams of interaction and reaction networks (KEGG PATHWAY), and hierarchies and relationships of various biological objects (KEGG BRITE).

9.7.2 FUNCTIONAL BIOINFORMATICS

Because of the relative high degree of sequence conservation, possible NR RE can also be screened by computer using DNA promoter sequences. Some examples

TABLE 9.18

Freely Accessible Software for the Identification of Possible XREs

Name	Internet Address	Ref.
ConSite	http:/www.phylofoot.org/consite	[263]
jPREdictor	http://bibiserv.techfak.uni-bielefeld.de/jpredictor	[264]
MAPPER	http://bio.chip.org/mapper	[265]
Match	http://compel.bionet.nsc.ru/Match/Match.html	[266]
Mat-Inspector	http://www.genomatix.de/online_help/help_matinspector/matinspector_help.html	[267]
NUBIScan	http://www.nubiscan.unibas.ch/	[268]
PMSearch	http://www.nicemice.cn/bioinfo/PMS	[269]
PROMO	http://www.lsi.upc.es/~alggen	[270]
TESS	http://www.cbil.upenn.edu/tess/	
TFBind	http://www.hgc.ims.u-tokyo.ac. jp/service/tooldoc/TFBIND	[271]

of freely accessible software that can assist in identifying RE are listed in Table 9.18.

The computer-based analysis allows rapid analysis of large sequences; however, the current methods frequently show a large number of false-positive as well as false-negative results. Because most algorithms are based on weight matrices from data bases such as JASPAR, TRANSFAC, TFD, or IMD, which are biased toward easily recognized strong binding sites, computer prediction is especially inaccurate for relatively weak binding sites of which the sequences differ more from the consensus site [217]. Therefore, predicted RE always have to be verified in experimental approaches, such as EMSA or transient transfection.

9.8   CONCLUSION

The major roles that lipid sensors play in controlling lipids and carbohydrate homeostasis have generated an intense research effort aimed at investigating their potential as pharmaceutic targets in the treatment of a variety of diseases, such as dyslipidemias, diabetes, and cholestasis. These researches have identified novel synthetic high-affinity agonists/antagonists and have identified a subset of target genes, being regulated in a tissue-specific manner, for each receptor. Most of these findings play important roles in current and future studies on these transcription factors, since they constitute a series of tools for investigators. Furthermore, these researches have also contributed to the discovery or rediscovery of novel/forgotten technical approaches, allowing the deciphering of NR' biological actions. These techniques are now affordable and routinely used in many laboratories. Each technique having its own advantages and limitations, one should keep in mind that results obtained through one technique generally require to be confirmed using other approaches.

LIST OF ABBREVIATIONS

ABC	ATP binding cassette (transporter)
AD	activation domain
AF	activation function
AhR	aryl hydrocarbon receptor
AlphaScreen	amplified luminescence proximity homogenous assay
AP-1	activator protein
Apo	apolipoprotein
AR	androgen receptor
ARA	AR-associated protein
ATCC	American Type Culture Collection
BADGE	bisphenol diglycidyl ether
BFE	enoyl-CoA hydratase/3-hydroxyacyl-CoA dehydrogenase
BRET	bioluminescence resonance energy transfer
BSEP	bile salt export pump
CA	cholic acid
CAR	constitutive androstane receptor
CAT	chloramphenicol acetyltransferase
CCRP	constitutive active/androstane receptor retention protein
CDCA	chenodeoxycholic acid
cDNA	complementary DNA
ChIP	chromatin immunoprecipitation
CID	chemical inducers of dimerization
CITCO	6-(4-chlorophenyl)imidazo[2,1-b][1,3]thiazole-5-carbaldehyde O-(3,4-dichlorobenzyl)oxime
CITED	CBP/p300 interacting transactivator with ED-rich tail
CMV	cytomegalovirus
Co-IP	co-immunoprecipitation
CREBBP	CREB-binding protein
C_T	threshold cycle
CYP	cytochrome P450
DBD	DNA-binding domain
DCA	deoxycholic acid
DDBJ	database and the DNA Database of Japan
DEAE	diethylaminoethyl
DIG	digoxigenin
DMS	dimethylsulfate
DR	direct repeat
ds	double strand
ECACC	European collection of cell cultures
6-ECDCA	6a-Ethyl-chenodeoxycholic acid
ECL	enhanced chemiluminescence
EMSA	electrophoretic mobility shift assay
EPD	Eukaryotic Promoter Database
ER	everted repeat

ERR	estrogen-related receptor
ES	embryonic stem (cells)
EtBr	ethidium bromide
FA	fatty acid
FACS	fluorescence-activated cell sorting
FATP	fatty acid transport protein
FBS	fetal bovine serum
FCS	fluorescence correlation spectroscopy
FISH	fluorescence *in situ* hybridization
FITC	fluorescein isothiocyanate
FoxO	forkhead box, class O
FP	fluorescence polarization
FRET	fluorescence resonance energy transfer
FXR	farnesoid X receptor
GFP	green fluorescent protein
GRIP	glucocorticoid receptor-interacting protein
GST	glutathione *S*-transferase
HAT	histone acetyltransferase
HEPE	hydroxyeicosapentaenoic acid
HETE	hydroxyeicosatetraenoic acid
HODE	hydroxyoctadecadienoic acid
HTS	high-throughput screening
HSV	herpes simplex virus
HUVEC	human umbilical vein endothelial cells
IBABP	ileal bile acid-binding protein
IC	inhibitory constant
IGTC	International Trap Consortium
IHC	immunohistochemistry
IR	inverted repeat
ISH	*in situ* hybridization
kb	kilo base
K_d	dissociation constant
KEGG	Kyoto Encyclopedia of Genes and Genomes
KF	Klenow fragment
K_i	affinity constant
KO	knockout
LBD	ligand-binding domain
LCA	lithocholic acid
LDL	low-density lipoprotein
LPL	lipoprotein lipase
LPS	lipopolysaccharide
LXR	liver X receptor
MCA	monoclonal antibody
MDR	multidrug resistance
MRP	multidrug resistance protein
NCO	nuclear receptor co-activator

NCOR	nuclear receptor co-repressor
NE	nuclear extract
NF-κ	nuclear factor κ
NPC	Niemann Pick C
NR	nuclear receptor
NRIP	nuclear receptor interacting protein
OD	optical density
PAGE	polyacrylamide gel electrophoresis
PBRE	phenobarbital response element
PCA	polyclonal antibody
PCN	pregnenolone 16α-carbonitrile
PCR	polymerase chain reaction
PDIP	PPAR-DBD-interacting protein
PGC	PPARγ co-activator
PGJ$_2$	15-deoxy prostaglandin J$_2$
PIMT	PRIP-interacting protein with methyltransferase domain
PNK	polynucleotide kinase
PPAR	peroxisome proliferator-activated receptor
PPARBP	PPAR binding protein
PPARGC	PPARγ co-activator
PRIP	PPAR-interacting protein
PRMT	protein argentine methyltransferase
PUFA	polyunsaturated fatty acid
PXR	pregnane X receptor
RE	response element
RT	reverse transcription
RXR	retinoid X receptor
SAF	serum amyloid A activating factor-1
SAGE	serial analysis of gene expression
SDS	sodium dodecyl sulfate
SERBP	sterol regulatory element-binding proteins
SG	SYBR Green
SHP	small heterodimer partner
shRNA	short hairpin RNA
siRNA	small interfering RNA
SMC	smooth muscle cells
SPA	scintillation proximity assay
ss	single strand
SULT	sulfotransferase
SXR	steroid and xenobiotic receptor
TCPOBOP	1,4-bis[2-(3,5-dichlorpyridyloxyl)]-benzene
Tet	tetracycline
TG	transgenic
TIF	transcription intermediary factor
TK	thymidine kinase
TR-FRET	time-resolved FRET

TRRAP   transformation/transcription domain-associated protein
UDCA   ursodeoxycholic acid
UGT   UDP-glucuronosyltransferase
XAP   hepatitis B virus X-associated protein

REFERENCES

1. Chawla, A. et al. (2001) Nuclear receptors and lipid physiology: Opening the x-files. *Science* 294:1866–1870.
2. Moore, D.D. et al. (2006) International Union of Pharmacology. LXII. The NR1H and NR1I receptors: Constitutive androstane receptor, pregnene X receptor, farnesoid X receptor α, farnesoid X receptor β, liver X receptor α, liver X receptor β, and vitamin D receptor. *Pharmacol. Rev.* 58:742–759.
3. Michalik, L. et al. (2006) International Union of Pharmacology. LXI. Peroxisome proliferator-activated receptors. *Pharmacol. Rev.* 58:726–741.
4. Brown, J.D. and Plutzky, J. (2007) Peroxisome proliferator-activated receptors as transcriptional nodal points and therapeutic targets. *Circulation* 115:518–533.
5. Li, A.C. and Palinski, W. (2006) Peroxisome proliferator-activated receptors: How their effects on macrophages can lead to the development of a new drug therapy against atherosclerosis. *Annu. Rev. Pharmcol. Toxicol.* 46:1–39.
6. Lu, T.T., Repa, J.J., and Mangelsdorf, D.J. (2001) Orphan nuclear receptors as eLiXiRs and FiXeRs of sterol metabolism. *J. Biol. Chem.* 276:37735–37738.
7. Trottier, J. et al. (2006) Coordinate regulation of hepatic bile acid oxidation and conjugation by nuclear receptors. *Mol. Pharm.* 3:212–222.
8. Germain, P. et al. (2006) International Union of Pharmacology. LXIII. Retinoid X receptors. *Pharmacol. Rev.* 58:760–772.
9. Mata de Urquiza, A. and Perlmann, T. (2003) *In vivo* and *in vitro* reporter systems for studying nuclear receptor and ligand activities. *Methods Enzymol.* 364:463–475.
10. Goodwin, B. et al. (2000) A regulatory cascade of the nuclear receptors FXR, SHP-1, and LRH-1 represses bile acid biosynthesis. *Mol. Cell* 6:517–526.
11. Huuskonen, J. et al. (2006) Liver X receptor inhibits the synthesis and secretion of apolipoprotein A1 by human liver-derived cells. *Biochemistry* 45:15068–15074.
12. Weinhofer, I. et al. (2005) Liver X receptor α interferes with SREBP1c-mediated Abcd2 expression: Novel cross-talk in gene regulation. *J. Biol. Chem.* 280:41243–41251.
13. Glass, C.K. and Rosenfeld, M.G. (2000) The coregulator exchange in transcriptional functions of nuclear receptors. *Genes Dev.* 14:121–141.
14. Koutnikova, H. et al. (2003) Serial analysis of gene expression and gene trapping to identify nuclear receptor target genes. *Methods Enzymol.* 364:299–322.
15. Hay, R.J. (1998) Cell line availability: Where to get the cell lines you need. *Methods Cell Biol.* 57:31–47.
16. Honegger, P. (1999) Overview of cell and tissue culture techniques. In: *Current Protocols in Pharmacology*, T. Kenakin (ed.), John Wiley & Sons, New York.
17. Jung, D. et al. (2007) Analysis of bile acid-induced regulation of FXR target genes in human liver slices. *Liver Int.* 27:137–144.
18. Stanley, L.A. et al. (2006) PXR and CAR: Nuclear receptors which play a pivotal role in drug disposition and chemical toxicity. *Drug Metab. Rev.* 38:515–597.
19. Moyer, M.P. (1995) Tumor cell culture. *Methods Enzymol.* 254:153–165.
20. Helmrich, A. and Barnes, D. (1998) Animal cell culture equipment and techniques. *Methods Cell Biol.* 57:3–17.
21. LeCluyse, E.L. (2001) Human hepatocyte culture systems for the *in vitro* evaluation of cytochrome P450 expression and regulation. *Eur. J. Pharm. Sci.* 13:343–368.

22. Kono, Y. et al. (1995) Establishment of a human hepatocyte line derived from primary culture in a collagen gel sandwich culture system. *Exp. Cell Res.* 221:478–485.
23. Parrish, A.R., Gandolfi, A.J., and Brendel, K. (1995) Precision-cut tissue slices: Applications in pharmacology and toxicology. *Life Sci.* 57:1887–1901.
24. van de Kerkhof, E.G. et al. (2005) Characterization of rat small intestinal and colon precision-cut slices as an *in vitro* system for drug metabolism and induction studies. *Drug Metab. Dispos.* 33:1613–1620.
25. Stuckmann, I., Evans, S., and Lassar, A.B. (2003) Erythropoietin and retinoic acid, secreted from the epicardium, are required for cardiac myocyte proliferation. *Dev. Biol.* 255:334–349.
26. de Kanter, R. et al. (2005) A new technique for preparing precision-cut slices from small intestine and colon for drug biotransformation studies. *J. Pharmacol. Toxicol. Methods* 51:65–72.
27. Olsavsky, K.M. et al. (2007) Gene expression profiling and differentiation assessment in primary human hepatocyte cultures, established hepatoma cell lines, and human liver tissues. *Toxicol. Appl. Pharmacol.* 222:42–56.
28. Zhu, Z. et al. (2007) Use of cryopreserved transiently transfected cells in high-throughput pregnane X receptor transactivation assay. *J. Biomol. Screen.* 12:248–254.
29. Mitry, R.R., Hughes, R.D., and Dhawan, A. (2002) Progress in human hepatocytes: Isolation, culture and cryopreservation. *Semin. Cell Dev. Biol.* 13:463–467.
30. Liu, Y., Borchert, G.L., and Phang, J.M. (2004) PEA3, an Ets transcription factor, mediates the induction of cyclooxygenase-2 by nitric oxide in colorectal cancer cells. *J. Biol. Chem.* 279:18694–18700.
31. Behr, J.P. et al. (1989) Efficient gene transfer into mammalian primary endocrine cells with lipopolyamine-coated DNA. *Proc. Natl. Acad. Sci. USA* 86:6982–6986.
32. Wasungu, L. and Hoekstra, D. (2006) Cationic lipids, lipoplexes and intracellular delivery of genes. *J. Control. Release* 116:255–264.
33. Mortimer, I. et al. (1999) Cationic lipid-mediated transfection of cells in culture requires mitotic activity. *Nature* 6:403–411.
34. Azzam, T. and Domb, A.J. (2004) Current developments in gene transfection agents. *Curr. Drug Deliv.* 1:165–193.
35. Verreault, M. et al. (2006) The liver X-receptor alpha controls hepatic expression of the human bile acid-glucuronidating UGT1A3 enzyme in human cells and transgenic mice. *Hepatology* 44:368–378.
36. Faddy, H.M. et al. (2006) Peroxisome proliferator-activated receptor α expression is regulated by estrogen receptor α and modulates the response of MCF-7 cells to sodium butyrate. *Int. J. Biochem. Cell Biol.* 38:255–266.
37. Jacoby, D., Zilz, N., and Towle, H. (1989) Sequences within the 5′-flanking region of the S14 gene confer responsiveness to glucose in primary hepatocytes. *J. Biol. Chem.* 264:17623–17626.
38. Rizzo, G. et al. (2006) The farnesoid X Receptor promotes adipocyte differentiation and regulates adipose cell function *in vivo*. *Mol. Pharmacol.* 70:1164–1173.
39. Brendel, C. et al. (2002) The small heterodimer partner interacts with the liver X receptor α and represses its transcriptional activity. *Mol. Endocrinol.* 16:2065–2076.
40. Blesch, A. (2004) Lentiviral and MLV based retroviral vectors for *ex vivo* and *in vivo* gene transfer. *Methods* 33:164–172.
41. Ghosh, S.S., Gopinath, P., and Ramesh, A. (2006) Adenoviral vectors: A promising tool for gene therapy. *Appl. Biochem. Biotechnol.* 133:9–29.
42. Hosono, T. et al. (2005) RNA interference of PPARγ using fiber-modified adenovirus vector efficiently suppresses preadipocyte-to-adipocyte differentiation in 3T3-L1 cells. *Gene* 348:157–165.
43. Grandi, P. et al. (2004) Targeting HSV amplicon vectors. *Methods* 33:179–186.

44. Koldamova, R.P. et al. (2005) The liver X receptor ligand T0901317 decreases amyloid β production *in vitro* and in a mouse model of Alzheimer's disease. *J. Biol. Chem.* 280:4079–4088.

45. Chou, T.W., Biswas, S., and Lu, S. (2004) Gene delivery using physical methods. An overview. In: *Gene Delivery to Mammalian Cells, Volume 1: Nonviral Gene Transfer Techniques*, W.C. Heiser (ed.), Humana Press, Totowa, NJ.

46. Gresch, O. et al. (2004) New non-viral method for gene transfer into primary cells. *Methods* 33:151–163.

47. Gopalakrishnan, M. and Molinari, J. (1998) Expression of cloned receptors in mammalian cell lines. In: *Current Protocols in Pharmacology*, John Wiley & Sons, New York.

48. Ryding, A., Sharp, M., and Mullins, J. (2001) Conditional transgenic technologies. *J. Endocrinol.* 171:1–14.

49. No, D., Yao, T.-P., and Evans, R.M. (1996) Ecdysone-inducible gene expression in mammalian cells and transgenic mice. *Proc. Natl. Acad. Sci. USA* 93:3346–3351.

50. Maddison, K. and Clarke, A.R. (2005) New approaches for modelling cancer mechanisms in the mouse. *J. Pathol.* 205:181–193.

51. Pai, J.-T. et al. (1998) Differential stimulation of cholesterol and unsaturated fatty acid biosynthesis in cells expressing individual nuclear sterol regulatory element-binding proteins. *J. Biol. Chem.* 273:26138–26148.

52. Elbashir, S.M. et al. (2001) Duplexes of 21-nucleotide RNAs mediate RNA interference in cultured mammalian cells. *Nature* 411:494–498.

53. Sijen, T. et al. (2001) On the role of RNA amplification in dsRNA-triggered gene silencing. *Cell* 107:465–476.

54. Tuschl, T. (2002) Expanding small RNA interference. *Nat. Biotechnol.* 20:446–448.

55. van de Wetering, M. et al. (2003) Specific inhibition of gene expression using a stably integrated, inducible small-interfering-RNA vector. *EMBO Rep.* 4:609–615.

56. Brazas, R.M. and Hagstrom, J.E. (2005) Delivery of small interfering RNA to mammalian cells in culture by using cationic lipid[+45 degree rule]polymer-based transfection reagents. In: *Methods in Enzymology: RNA Interference*, J.J. Rossi and D.R. Engelke (eds.), Academic Press, San Diego, CA, pp. 112–124.

57. Iwano, S. et al. (2005) A possible mechanism for atherosclerosis induced by polycyclic aromatic hydrocarbons. *Biochem. Biophys. Res. Commun.* 335:220–226.

58. Nakaya, K. et al. (2007) Telmisartan enhances cholesterol efflux from THP-1 macrophages by activating PPARγ. *J. Atheroscler. Thromb.* 14:133–141.

59. Rigamonti, E. et al. (2005) Liver X receptor activation controls intracellular cholesterol trafficking and esterification in human macrophages. *Circ. Res.* 97:682–689.

60. Sirvent, A. et al. (2004) Farnesoid X receptor represses hepatic lipase gene expression. *J. Lipid Res.* 45:2110–2115.

61. Reynolds, A. et al. (2004) Rational siRNA design for RNA interference. *Nat. Biotechnol.* 22:326–330.

62. Singer, O., Yanai, A., and Verma, I.M. (2004) Silence of the genes. *Proc. Natl. Acad. Sci. USA* 101:5313–5314.

63. Semizarov, D. et al. (2003) Specificity of short interfering RNA determined through gene expression signatures. *Proc. Natl. Acad. Sci. USA* 100:6347–6352.

64. Scacheri, P.C. et al. (2004) Short interfering RNAs can induce unexpected and divergent changes in the levels of untargeted proteins in mammalian cells. *Proc. Natl. Acad. Sci. USA* 101:1892–1897.

65. Schwarz, D.S. et al. (2006) Designing siRNA that distinguish between genes that differ by a single nucleotide. *PLoS Genet.* 2:e140.

66. Jackson, A.L. et al. (2006) Widespread siRNA "off-target" transcript silencing mediated by seed region sequence complementarity. *RNA* 12:1179–1187.

67. Suzuki, N. et al. (1992) Effect of dexamethasone on nucleolar casein kinase II activity and phosphorylation of nucleolin in lymphosarcoma P1798 cells. *J. Steroid. Biochem. Mol. Biol.* 42:305–312.
68. Shayu, D. et al. (2005) Effects of ICI 182780 on estrogen receptor expression, fluid absorption and sperm motility in the epididymis of the bonnet monkey. *Reprod. Biol. Endocrinol.* 3:10.
69. Sierra, M.L. et al. (2007) Substituted 2-[(4-aminomethyl)phenoxy]-2-methylpropionic acid PPARα agonists. 1. Discovery of a novel series of potent HDLc raising agents. *J. Med. Chem.* 50:685–695.
70. Oliver, W.R. et al. (2001) A selective peroxisome proliferator-activated receptor delta agonist promotes reverse cholesterol transport. *Proc. Natl. Acad. Sci. USA* 98:5306–5311.
71. Russell, J.C. and Proctor, S.D. (2006) Small animal models of cardiovascular disease: Tools for the study of the roles of metabolic syndrome, dyslipidemia, and atherosclerosis. *Cardiovasc. Pathol.* 15:318–330.
72. Prueksaritanont, T. et al. (2005) Interconversion pharmacokinetics of simvastatin and its hydroxy acid in dogs: Effects of gemfibrozil. *Pharm. Res.* 22:1101–1109.
73. Labinskyy, V. et al. (2007) Chronic activation of peroxisome proliferator-activated receptor-alpha with fenofibrate prevents alterations in cardiac metabolic phenotype without changing the onset of decompensation in pacing-induced heart failure. *J. Pharmacol. Exp. Ther.* 321:165–171.
74. Sun, H. et al. (2005) C-reactive protein in atherosclerotic lesions: Its origin and pathophysiological significance. *Am. J. Pathol.* 167:1139–1148.
75. Murakami, S. et al. (2002) Taurine suppresses development of atherosclerosis in Watanabe heritable hyperlipidemic (WHHL) rabbits. *Atherosclerosis* 163:79–87.
76. Botcherby, M. (2002) Just click on the mouse! *Brief Funct. Genomic. Proteomic.* 1:226–229.
77. Consortium, R.G.S.P. (2004) Genome sequence of the Brown Norway rat yields insights into mammalian evolution. *Nature* 428:493–521.
78. Weiler-Normann, C., Herkel, J., and Lohse, A.W. (2007) Mouse models of liver fibrosis. *Z. Gastroenterol.* 45:43–50.
79. Demigne, C. et al. (2006) Mice chronically fed a westernized experimental diet as a model of obesity, metabolic syndrome and osteoporosis. *Eur. J. Nutr.* 45:298–306.
80. Amy, R.M. et al. (1988) Atherogenesis in two strains of obese rats. The fatty Zucker and LA/N-corpulent. *Atherosclerosis* 69:199–209.
81. Gealekman, O. et al. (2004) Endothelial dysfunction as a modifier of angiogenic response in Zucker diabetic fat rat: Amelioration with Ebselen. *Kidney Int.* 66:2337–2347.
82. Lee, G.-H. et al. (1996) Abnormal splicing of the leptin receptor in diabetic mice. *Nature* 379:632–635.
83. Naveilhan, P. et al. (2002) Attenuation of hypercholesterolemia and hyperglycemia in ob/ob mice by NPY Y2 receptor ablation. *Peptides* 23:1087–1091.
84. Cho, Y.R. et al. (2007) Hyperglycemia, maturity-onset obesity, and insulin resistance in NONcNZO10/LtJ males, a new mouse model of type 2 diabetes. *Am. J. Physiol. Endocrinol. Metab.* 293:E327–E336.
85. Chan, A.W. (1999) Transgenic animals: Current and alternative strategies. *Cloning* 1:25–46.
86. Ma, X. et al. (2007) Rifaximin is a gut-specific human pregnane X receptor activator. *J. Pharmacol. Exp. Ther.* 322:391–398.
87. Davey, R.A. and MacLean, H.E. (2006) Current and future approaches using genetically modified mice in endocrine research. *Am. J. Physiol. Endocrinol. Metab.* 291:E429–E438.

88. Cheung, C. et al. (2004) Diminished hepatocellular proliferation in mice humanized for the nuclear receptor peroxisome proliferator-activated receptor α. *Cancer Res.* 64:3849–3854.

89. Gong, H. et al. (2005) Animal models of xenobiotic receptors in drug metabolism and diseases. *Methods Enzymol.* 400:598–618.

90. Gonzalez, F.J. and Yu, A.M. (2006) Cytochrome P450 and xenobiotic receptor humanized mice. *Annu. Rev. Pharmacol. Toxicol.* 46:41–64.

91. Mills, A.A. (2001) Changing colors in mice: An inducible system that delivers. *Genes Dev.* 15:1461–1467.

92. Morimura, K. et al. (2006) Differential susceptibility of mice humanized for peroxisome proliferator-activated receptor alpha to Wy-14,643-induced liver tumorigenesis. *Carcinogenesis* 27:1074–1080.

93. Staudinger, J.L. et al. (2001) The nuclear receptor PXR is a lithocholic acid sensor that protects against liver toxicity. *Proc. Natl. Acad. Sci. USA* 98:3369–3374.

94. Sinal, C.J. et al. (2000) Targeted disruption of the nuclear receptor FXR/BAR impairs bile acid and lipid homeostasis. *Cell* 102:731–744.

95. Medina-Gomez, G. et al. (2007) PPAR gamma 2 prevents lipotoxicity by controlling adipose tissue expandability and peripheral lipid metabolism. *PLoS Genet.* 27:e64.

96. Kastner, P. et al. (1994) Genetic analysis of RXRα developmental function: Convergence of RXR and RAR signaling pathways in heart and eye morphogenesis. *Cell* 78:987–1003.

97. Artandi, S.E. et al. (2000) Telomere dysfunction promotes non-reciprocal translocations and epithelial cancers in mice. *Nature* 406:641–645.

98. Cohen-Tannoudji, M. and Babinet, C. (1998) Beyond 'knock-out' mice: New perspectives for the programmed modification of the mammalian genome. *Mol. Hum. Reprod.* 4:929–938.

99. Freedman, B.D. et al. (2005) A dominant negative peroxisome proliferator-activated receptor-gamma knock-in mouse exhibits features of the metabolic syndrome. *J. Biol. Chem.* 280:17118–17125.

100. Ragozin, S. et al. (2005) Knockdown of hepatic ABCA1 by RNA interference decreases plasma HDL cholesterol levels and influences postprandial lipemia in mice. *Arterioscler. Thromb. Vasc. Biol.* 25:1433–1438.

101. Akiyama, T.E. et al. (2002) Conditional disruption of the peroxisome proliferator-activated receptor gamma gene in mice results in lowered expression of ABCA1, ABCG1, and apoE in macrophages and reduced cholesterol efflux. *Mol. Cell Biol.* 22:2607–2619.

102. Houten, S.M. et al. (2007) *In vivo* imaging of farnesoid X receptor activity reveals the ileum as the primary bile acid signaling tissue. *Mol. Endocrinol.* 21:1312–1323.

103. Stienstra, R. et al. (2007) Peroxisome proliferator-activated receptor alpha protects against obesity-induced hepatic inflammation. *Endocrinology* 148:2753–2763.

104. Zhai, Y. et al. (2007) Activation of pregnane X receptor disrupts glucocorticoid and mineralocorticoid homeostasis. *Mol. Endocrinol.* 21:138–147.

105. Dai, X. et al. (2007) Effect of T0901317 on hepatic proinflammatory gene expression in ApoE-/- mice fed a high-fat/high-cholesterol diet. *Inflammation* 30:105–117.

106. Kodama, S. et al. (2007) Human nuclear pregnane X receptor cross-talk with CREB to repress cAMP activation of the glucose-6-phosphatase gene. *Biochem. J.* 407:373–381.

107. Anakk, S. et al. (2007) Gender dictates the nuclear receptor-mediated regulation of CYP3A44. *Drug Metab. Dispos.* 35:36–42.

108. Dvorak, Z., Pascussi, J.M., and Modriansky, M. (2003) Approaches to messenger RNA detection—Comparison of methods. *Biomed. Pap. Med. Fac. Univ. Palacky Olomouc Czech Repub.* 147:131–135.

109. Jacobsen, N. et al. (2004) Direct isolation of poly(A) + RNA from 4 M guanidine thiocyanate-lysed cell extracts using locked nucleic acid-oligo(T) capture. *Nucleic Acids Res.* 32:e64.

110. Fleige, S. and Pfaffl, M.W. (2006) RNA integrity and the effect on the real-time qRT-PCR performance. *Mol. Aspects Med.* 27:126–139.

111. Li, T. and Chiang, J.Y. (2005) Mechanism of rifampicin and pregnane X receptor inhibition of human cholesterol 7 alpha-hydroxylase gene transcription. *Am. J. Physiol. Gastrointest. Liver Physiol.* 288:G74–84.

112. de Jong, L.A.A. et al. (2005) Receptor–ligand binding assays: Technologies and applications. *J. Chromatogr. B* 829:1–25.

113. Ludwig, L.B., Hughes, B.J., and Schwartz, S.A. (1995) Biotinylated probes in the electrophoretic mobility shift assay to examine specific dsDNA, ssDNA or RNA–protein interactions. *Nucleic Acids Res.* 23:3792–3793.

114. Forwood, J.K. and Jans, D.A. (2006) Quantitative analysis of DNA–protein interactions using double-labeled native gel electrophoresis and fluorescence-based imaging. *Electrophoresis* 27:3166–3170.

115. Dignam, J.D., Lebovitz, R.M., and Roeder, R.G. (1983) Accurate transcription initiation by RNA polymerase II in a soluble extract from isolated mammalian nuclei. *Nucleic Acids Res.* 11:1475–1489.

116. Choi, Y.H. et al. (2003) Apoptosis and modulation of cell cycle control by synthetic derivatives of ursodeoxycholic acid and chenodeoxycholic acid in human prostate cancer cells. *Cancer Lett.* 199:157–167.

117. Bombail, V. et al. (2004) Role of Sp1, C/EBP alpha, HNF3, and PXR in the basal- and xenobiotic-mediated regulation of the CYP3A4 gene. *Drug Metab. Dispos.* 32:525–535.

118. Marston, F.A. and Hartley, D.L. (1990) Solubilization of protein aggregates. *Methods Enzymol.* 182:264–276.

119. Lechtken, A. et al. (2006) Overexpression, refolding, and purification of polyhistidine-tagged human retinoic acid related orphan receptor RORα4. *Protein Expr. Purif.* 49:114–120.

120. Peng, X. et al. (2007) Overexpression of ER and VDR is not sufficient to make ER-negative MDA-MB231 breast cancer cells responsive to 1α-hydroxyvitamin D5. *Carcinogenesis* 28:1000–1007.

121. Kim, J., Min, G., and Kemper, B. (2001) Chromatin assembly enhances binding to the CYP2B1 phenobarbital-responsive unit (PBRU) of nuclear factor-1, which binds simultaneously with constitutive androstane receptor (CAR)/retinoid X receptor (RXR) and enhances CAR/RXR-mediated activation of the PBRU. *J. Biol. Chem.* 276:7559–7567.

122. Raval, P. (1994) Qualitative and quantitative determination of mRNA. *J. Pharmacol. Toxicol. Methods* 32:125–127.

123. Warford, A. and Lauder, I. (1991) *In situ* hybridisation in perspective. *J. Clin. Pathol.* 44:177–181.

124. Collett, G.P. et al. (2000) Peroxisome proliferator-activated receptor alpha is an androgen-responsive gene in human prostate and is highly expressed in prostatic adenocarcinoma. *Clin. Cancer Res.* 6:3241–3248.

125. Nakamura, K. et al. (2004) Expression and regulation of multiple murine ATP-binding cassette transporter G1 mRNAs/isoforms that stimulate cellular cholesterol efflux to high density lipoprotein. *J. Biol. Chem.* 279:45980–45989.

126. Levsky, J.M. and Singer, R.H. (2003) Fluorescence *in situ* hybridization: Past, present and future. *J. Cell Sci.* 116:2833–2838.

127. Plummer, T.B. et al. (1998) *In situ* hybridization detection of low copy nucleic acid sequences using catalyzed reporter deposition and its usefulness in clinical human papillomavirus typing. *Diagn. Mol. Pathol.* 7:76–84.

128. Jiang, Y.J. et al. (2005) LXR and PPAR activators stimulate cholesterol sulfotransferase type 2 isoform 1b in human keratinocytes. *J. Lipid Res.* 46:2657–2666.

129. Bustin, S.A. (2000) Absolute quantification of mRNA using real-time reverse transcription polymerase chain reaction assays. *J. Mol. Endocrinol.* 25:169–193.

130. Stahlberg, A. et al. (2004) Properties of the reverse transcription reaction in mRNA quantification. *Clin. Chem.* 50:509–515.

131. Heid, C.A. et al. (1996) Real time quantitative PCR. *Genome Res.* 6:986–994.

132. Tyagi, S. and Kramer, F.R. (1996) Molecular beacons: Probes that fluoresce upon hybridization. *Nat. Biotechnol.* 14:303–308.

133. Bookout, A.L. and Mangelsdorf, D.J. (2003) Quantitative real-time PCR protocol for analysis of nuclear receptor signaling pathways. *Nucl. Recept. Signal.* 1:e012.

134. Livak, K.J. and Schmittgen, T.D. (2001) Analysis of relative gene expression data using real-time quantitative PCR and the 2(-$\Delta \Delta$ C(T)) method. *Methods* 25:402–408.

135. Li, T., Chen, W., and Chiang, J.Y. (2007) PXR induces CYP27A1 and regulates cholesterol metabolism in the intestine. *J. Lipid. Res.* 48:373–384.

136. Ramos-Vara, J.A. (2005) Technical aspects of immunohistochemistry. *Vet. Pathol.* 42:405–426.

137. Watanabe, Y. et al. (2005) Expression of the LXRalpha protein in human atherosclerotic lesions. *Arterioscler. Thromb. Vasc. Biol.* 25:622–627.

138. Watkins, S. (2000) Immunohistochemistry. In: *Current Protocols in Molecular Biology*, John Wiley & Sons, New York.

139. Senekeo-Effenberger, K. et al. (2007) Expression of the human UGT1 locus in transgenic mice by 4-chloro-6-(2,3-xylidino)-2-pyrimidinylthioacetic acid (WY-14643) and implications on drug metabolism through peroxisome proliferator-activated receptor alpha activation. *Drug Metab. Dispos.* 35:419–427.

140. Squires, E.J., Sueyoshi, T., and Negishi, M. (2004) Cytoplasmic localization of pregnane X receptor and ligand-dependent nuclear translocation in mouse liver. *J. Biol. Chem.* 279:49307–49314.

141. Talukdar, S. and Hillgartner, F.B. (2006) The mechanism mediating the activation of acetyl-coenzyme A carboxylase-alpha gene transcription by the liver X receptor agonist T0901317. *J. Lipid. Res.* 47:2451–2461.

142. Towbin, H., Staehelin, T., and Gordon, J. (1979) Electrophoretic transfer of proteins from polyacrylamide gels to nitrocellulose sheets: Procedure and some applications. *Proc. Natl. Acad. Sci. USA* 76:4350–4354.

143. Bronstein, I. et al. (1994) Chemiluminescent and bioluminescent reporter gene assays. *Anal. Biochem.* 219:169–181.

144. Goodwin, B., Hodgson, E., and Liddle, C. (1999) The orphan human pregnane X receptor mediates the transcriptional activation of CYP3A4 by rifampicin through a distal enhancer module. *Mol. Pharmacol.* 56:1329–1339.

145. Hubbert, M.L. et al. (2007) Regulation of hepatic insig-2 by the farnesoid X receptor. *Mol. Endocrinol.* 21:1359–1369.

146. Hu, T. et al. (2006) Farnesoid X receptor agonist reduces serum asymmetric dimethylarginine levels through hepatic dimethylarginine dimethylaminohydrolase-1 gene regulation. *J. Biol. Chem.* 281:39831–39838.

147. Chen, W. et al. (2001) Nuclear receptor-mediated repression of human cholesterol 7α-hydroxylase gene transcription by bile acids. *J. Lipid Res.* 42:1402–1412.

148. Yoshikawa, T. et al. (2001) Identification of liver X receptor-retinoid X receptor as an activator of the sterol regulatory element-binding protein 1c gene promoter. *Mol. Cell. Biol.* 21:2991–3000.

149. Kawabe, K. et al. (2005) Peroxisome proliferator-activated receptor α and its response element are required but not sufficient for transcriptional activation of the mouse heart-type fatty acid binding protein gene. *Int. J. Biochem. Cell Biol.* 37:1534–1546.

150. Kliewer, S.A. et al. (1992) Convergence of 9-cis retinoic acid and peroxisome prolif-erator signalling pathways through heterodimer formation of their receptors. *Nature* 358:771–774.
151. Revzin, A. (1989) Gel electrophoresis assays for DNA–protein interactions. *Bio-techniques* 7:346–355.
152. Barbier, O. et al. (2003) Peroxisome proliferator-activated receptor alpha induces hepatic expression of the human bile acid glucuronidating UDP-glucuronosyltransferase 2B4 enzyme. *J. Biol. Chem.* 278:32852–32860.
153. Xu, G. et al. (2003) FXR-mediated down-regulation of CYP7A1 dominates LXRalpha in long-term cholesterol-fed NZW rabbits. *J. Lipid Res.* 44:1956–1962.
154. Chen, W. et al. (2007) Nuclear receptors RXRα:RARα are repressors for human MRP3 expression. *Am. J. Physiol. Gastrointest. Liver Physiol.* 292:G1221–1227.
155. Hampshire, A.J. et al. (2007) Footprinting: A method for determining the sequence selectivity, affinity and kinetics of DNA-binding ligands. *Methods* 42:128–140.
156. Galas, D.J. and Schmitz, A. (1978) DNAse footprinting: A simple method for the detec-tion of protein–DNA binding specificity. *Nucleic Acids Res.* 5:3157–3170.
157. Steffensen, K.R. et al. (2003) Glucocorticoid response and promoter occupancy of the mouse LXRα gene. *Biochem. Biophys. Res. Commun.* 312:716–724.
158. Loizos, N. (2004) *Current Protocols in Protein Science*, John Wiley & Sons, New York.
159. Schoonjans, K. et al. (1996) PPARα and PPARγ activators direct a distinct tissue-specific transcriptional response via a PPRE in the lipoprotein lipase gene. *EMBO J.* 15:5336–5348.
160. Storek, M.J., Ernst, A., and Verdine, G.L. (2002) High-resolution footprinting of sequence-specific protein–DNA contacts. *Nat. Biotechnol.* 20:183–186.
161. Metzger, W. and Heumann, H. (2001) Footprinting with exonuclease III. *Methods Mol. Biol.* 148:39–47.
162. Tullius, T.D. and Dombroski, B.A. (1986) Hydroxyl radical "footprinting": High-resolution information about DNA–protein contacts and application to lambda repressor and Cro protein. *Proc. Natl. Acad. Sci. USA* 83:5469–5473.
163. Ma, H. et al. (2003) Study of nuclear receptor-induced transcription complex assembly and histone modification by chromatin immunoprecipitation assays. *Methods Enzymol.* 364:284–296.
164. Claudel, T. et al. (2003) Farnesoid X receptor agonists suppress hepatic apolipoprotein CIII expression. *Gastroenterology* 125:544–555.
165. Cardullo, R.A. (2007) Theoretical principles and practical considerations for fluorescence resonance energy transfer microscopy; Methods in cell biology. In: *Digital Microscopy*, 3rd edn., D.E.A. Wolf and G. Sulder (eds.), Academic Press, San Diego, CA, pp. 479–494.
166. Furey, W.S. et al. (1998) Use of fluorescence resonance energy transfer to investigate the conformation of DNA substrates bound to the Klenow fragment. *Biochemistry* 37:2979–2990.
167. Heyduk, T. and Heyduk, E. (2002) Molecular beacons for detecting DNA binding pro-teins. *Nat. Biotechnol.* 20:171–176.
168. Jantz, D. and Berg, J.M. (2002) Binding assays get into the groove. *Nat. Biotechnol.* 20:126–127.
169. Wang, J. et al. (2005) Exonuclease III protection assay with FRET probe for detecting DNA–binding proteins. *Nucleic Acids Res.* 33:e23.
170. Deyholos, M.K. and Galbraith, D.W. (2001) High-density microarrays for gene expres-sion analysis. *Cytometry* 43:229–238.
171. Duggan, D.J. et al. (1999) Expression profiling using cDNA microarrays. *Nat. Genet.* 21:10–14.
172. Do, J.H. and Choi, D.K. (2006) Normalization of microarray data: Single-labeled and dual-labeled arrays. *Mol. Cells* 22:254–261.

173. Benson, D.A. et al. (1997) GenBank. *Nucleic Acids Res.* 25:1–6.

174. Boguski, M.S., Lowe, T.M., and Tolstoshev, C.M. (1993) dbEST—Database for "expressed sequence tags". *Nat. Genet.* 4:332–333.

175. Schuler, G.D. et al. (1996) A gene map of the human genome. *Science* 274:540–546.

176. Liu, P.C. et al. (2003) Induction of endogenous genes by peroxisome proliferator activated receptor alpha ligands in a human kidney cell line and *in vivo. J. Steroid Biochem. Mol. Biol.* 85:71–79.

177. Shi, L. et al. (2005) Cross-platform comparability of microarray technology: Intra-platform consistency and appropriate data analysis procedures are essential. *BMC Bioinformatics* 6(Suppl 2):S12.

178. Tavera-Mendoza, L.E., Mader, S., and White, J.H. (2006) Genome-wide approaches for identification of nuclear receptor target genes. *Nucl. Recept. Signal.* 4:e018.

179. Pawitan, Y. et al. (2005) False discovery rate, sensitivity and sample size for microarray studies. *Bioinformatics* 21:3017–3024.

180. Laganière, J. et al. (2005) From the cover: Location analysis of estrogen receptor alpha target promoters reveals that FOXA1 defines a domain of the estrogen response. *Proc. Natl. Acad. Sci. USA* 102:11651–11656.

181. Bolduc, C. et al. (2004) Adipose tissue transcriptome by serial analysis of gene expression. *Obesity Res.* 12:750–757.

182. Bolduc, C. et al. (2004) Effects of dihydrotestosterone on adipose tissue measured by serial analysis of gene expression. *J. Mol. Endocrinol.* 33:429–444.

183. IGTC. (2007) International Trap Consortium, San Francisco, CA (http://www.genetrap.org/).

184. Kudo, M. et al. (2004) Transcription suppression of peroxisome proliferator-activated receptor γ2 gene expression by tumor necrosis factor α via an inhibition of CCAAT/enhancer-binding protein Δ during the early stage of adipocyte differentiation. *Endocrinology* 145:4948–4956.

185. Jia, Y. et al. (2004) Transcription coactivator PBP, the peroxisome proliferator-activated receptor (PPAR)-binding protein, is required for PPARα-regulated gene expression in liver. *J. Biol. Chem.* 279:24427–24434.

186. Hollingsworth, R. and White, J.H. (2004) Target discovery using the yeast two-hybrid system. *Drug Discov. Today: TARGETS* 3:97–103.

187. Johnston, M. (1987) A model fungal gene regulatory mechanism: The GAL genes of *Saccharomyces cerevisiae. Microbiol. Rev.* 51:458–476.

188. Fields, S. and Song, O.-K. (1989) A novel genetic system to detect protein–protein interactions. *Nature* 340:245–246.

189. Zhu, Y. et al. (1996) Cloning and identification of mouse steroid receptor coactivator-1 (mSRC-1), as a coactivator of peroxisome proliferator-activated receptor gamma. *Gene Expr.* 6:185–195.

190. Kaelin, W.G. et al. (1991) Identification of cellular proteins that can interact specifically with the T/E1A-binding region of the retinoblastoma gene product. *Cell* 64:521–523.

191. Pineda Torra, I., Freedman, L.P., and Garabedian, M.J. (2004) Identification of DRIP205 as a coactivator for the farnesoid X receptor. *J. Biol. Chem.* 279:36184–36191.

192. Qoronfleh, M.W. et al. (2003) Use of immunomatrix methods to improve protein–protein interaction detection. *J. Biomed. Biotechnol.* 2003:291–298.

193. Phizicky, E. and Fields, S. (1995) Protein–protein interactions: Methods for detection and analysis. *Microbiol. Rev.* 59:94–123.

194. Ichida, M., Nemoto, S., and Finkel, T. (2002) Identification of a specific molecular repressor of the peroxisome proliferator-activated receptor gamma coactivator-1 alpha (PGC-alpha). *J. Biol. Chem.* 277:50991–50995.

195. Sumanasekera, W.K. et al. (2003) Evidence that peroxisome proliferator-activated receptor alpha is complexed with the 90-kDa heat shock protein and the hepatitis virus B X-associated protein 2. *J. Biol. Chem.* 278:4467–4473.

196. Stafslien, D.K. et al. (2007) Analysis of ligand-dependent recruitment of coactivator peptides to RXRβ in a time-resolved fluorescence resonance energy transfer assay. *Mol. Cell. Endocrinol.* 264:82–89.

197. Koterba, K.L. and Rowan, B.G. (2006) Measuring ligand-dependent and ligand-independent interactions between nuclear receptors and associated proteins using bioluminescence resonance energy transfer (BRET). *Nucl. Recept. Signal.* 4:e021.

198. Chen, Y., Wei, L.-N., and Muller, J.D. (2005) Unraveling protein–protein interactions in living cells with fluorescence fluctuation brightness analysis. *Biophys. J.* 88:4366–4377.

199. Feige, J.N. et al. (2005) Fluorescence imaging reveals the nuclear behavior of peroxisome proliferator-activated receptor/retinoid X receptor heterodimers in the absence and presence of ligand. *J. Biol. Chem.* 280:17880–17890.

200. Bylund, D.B. and Toews, M.L. (1993) Radioligand binding methods: Practical guide and tips. *Am. J. Physiol. Lung Cell Mol. Physiol.* 265:L421–L429.

201. Berger, J. et al. (1999) Novel peroxisome proliferator-activated receptor (PPAR) gamma and PPARdelta ligands produce distinct biological effects. *J. Biol. Chem.* 274:6718–6725.

202. Sun, S. et al. (2005) Assay development and data analysis of receptor–ligand binding based on scintillation proximity assay. *Meta. Eng.—Evol. Eng.* 7:38–44.

203. Jones, S.A. et al. (2000) The pregnane X receptor: A promiscuous xenobiotic receptor that has diverged during evolution. *Mol. Endocrinol.* 14:27–39.

204. Cook, N. et al. (2002) Scintillation proximity assay (SPA) technology to study biomolecular interactions. In: *Current Protocols in Protein Science*, John Wiley & Sons: New York.

205. Chin, J. et al. (2003) Miniaturization of cell-based beta-lactamase-dependent FRET assays to ultra-high throughput formats to identify agonists of human liver X receptors. *Assay Drug Dev. Technol.* 1:777–787.

206. Seethala, R. et al. (2007) A rapid, homogeneous, fluorescence polarization binding assay for peroxisome proliferator-activated receptors alpha and gamma using a fluorescein-tagged dual PPARα/γ activator. *Anal. Biochem.* 363:263–274.

207. Gagne, A., Banks, P., and Hurt, S.D. (2002) Use of fluorescence polarization detection for the measurement of fluopeptidetm binding to G protein-coupled receptors. *J. Recept. Signal Transduct. Res.* 22:333–343.

208. Ullman, E. et al. (1994) Luminescent oxygen channeling immunoassay: Measurement of particle binding kinetics by chemiluminescence. *Proc. Natl. Acad. Sci. USA* 91:5426–5430.

209. Glickman, J.F. et al. (2002) A comparison of ALPHAScreen, TR-FRET, and TRF as assay methods for FXR nuclear receptors. *J. Biomol. Screen* 7:3–10.

210. Kliewer, S.A. et al. (1997) Fatty acids and eicosanoids regulate gene expression through direct interactions with peroxisome proliferator-activated receptors alpha and gamma. *Proc. Natl. Acad. Sci. USA* 94:4318–4323.

211. Rau, O. et al. (2006) Carnosic acid and carnosol, phenolic diterpene compounds of the labiate herbs rosemary and sage, are activators of the human peroxisome proliferator-activated receptor gamma. *Planta Med.* 72:881–887.

212. Vanden Heuvel, J.P. et al. (2006) Differential activation of nuclear receptors by perfluorinated fatty acid analogs and natural fatty acids: A comparison of human, mouse, and rat peroxisome proliferator-activated receptor-α, -β, and -γ, liver X receptor-β, and retinoid X receptor-α. *Toxicol. Sci.* 92:476–489.

213. Zhu, Z. et al. (2004) Correlation of high-throughput pregnane X receptor (PXR) trans-activation and binding assays. *J. Biomol. Screen.* 9:533–540.

214. Mukherjee, R. et al. (2002) Ligand and coactivator recruitment preferences of peroxisome proliferator activated receptor α. *J. Steroid Biochem. Mol. Biol.* 81:217–225.

215. Pellicciari, R. et al. (2006) Back door modulation of the farnesoid X receptor: Design, synthesis, and biological evaluation of a series of side chain modified chenodeoxycholic acid derivatives. *J. Med. Chem.* 49:4208–4215.

216. Ondeyka, J.G. et al. (2005) Steroidal and triterpenoidal fungal metabolites as ligands of liver X receptors. *J. Antibiot. (Tokyo)* 58:559–565.

217. Locker, J. et al. (2002) Definition and prediction of the full range of transcription factor binding sites—The hepatocyte nuclear factor 1 dimeric site. *Nucleic Acids Res.* 30:3809–3817.

218. Forman, B.M., Chen, J., and Evans, R.M. (1997) Hypolipidemic drugs, polyunsaturated fatty acids, and eicosanoids are ligands for peroxisome proliferator-activated receptors α and Δ. *Proc. Natl. Acad. Sci. USA* 94:4312–4317.

219. Forman, B.M. et al. (1995) 15-Deoxy-Δ12,14-prostaglandin J2 is a ligand for the adipocyte determination factor PPARγ. *Cell* 83:803–812.

220. Lim, H. et al. (1999) Cyclo-oxygenase-2-derived prostacyclin mediates embryo implantation in the mouse via PPARΔ. *Genes Dev.* 13:1561–1574.

221. Janowski, B.A. et al. (1999) Structural requirements of ligands for the oxysterol liver X receptors LXRα and LXRβ. *Proc. Natl. Acad. Sci. USA* 96:266–271.

222. Mitro, N. et al. (2007) The nuclear receptor LXR is a glucose sensor. *Nature* 445:219–223.

223. Lew, J.-L. et al. (2004) The farnesoid X receptor controls gene expression in a ligand- and promoter-selective fashion. *J. Biol. Chem.* 279:8856–8861.

224. Makishima, M. et al. (2002) Vitamin D receptor as an intestinal bile acid sensor. *Science* 296:1313–1316.

225. Willson, T.M. et al. (2000) The PPARs: From orphan receptors to drug discovery. *J. Med. Chem.* 43:527–550.

226. Guerre-Millo, M. et al. (2000) Peroxisome proliferator-activated receptor alpha activators improve insulin sensitivity and reduce adiposity. *J. Biol. Chem.* 275:16638–16642.

227. Collins, J.L. et al. (2002) Identification of a nonsteroidal liver X receptor agonist through parallel array synthesis of tertiary amines. *J. Med. Chem.* 45:1963–1966.

228. Schultz, J.R. et al. (2000) Role of LXRs in control of lipogenesis. *Genes Dev.* 14:2831–2838.

229. Dussault, I. et al. (2003) Identification of gene-selective modulators of the bile acid receptor FXR. *J. Biol. Chem.* 278:7027–7033.

230. Moore, L.B. et al. (2000) Orphan nuclear receptors constitutive androstane receptor and pregnane X receptor share xenobiotic and steroid ligands. *J. Biol. Chem.* 275:15122–15127.

231. Wright, H.M. et al. (2000) A synthetic antagonist for the peroxisome proliferator-activated receptor gamma inhibits adipocyte differentiation. *J. Biol. Chem.* 275:1873–1877.

232. Mukherjee, R. et al. (2000) A selective peroxisome proliferator-activated receptor-γ (PPARγ) modulator blocks adipocyte differentiation but stimulates glucose uptake in 3T3-L1 adipocytes. *Mol. Endocrinol.* 14:1425–1433.

233. Wu, J. et al. (2002) The hypolipidemic natural product guggulsterone acts as an antagonist of the bile acid receptor. *Mol. Endocrinol.* 16:1590–1597.

234. Kota, B.P., Huang, T.H.-W., and Roufogalis, B.D. (2005) An overview on biological mechanisms of PPARs. *Pharm. Res.* 51:85–94.

235. Gervois, P. et al. (1999) Fibrates increase human REV-ERBα expression in liver via a novel peroxisome proliferator-activated receptor response element. *Mol. Endocrinol.* 13:400–409.

236. Deng, T. et al. (2006) Peroxisome proliferator-activated receptor-γ transcriptionally up-regulates hormone-sensitive lipase via the involvement of specificity protein-1. *Endocrinology* 147:875–884.
237. Dressel, U. et al. (2003) The peroxisome proliferator-activated receptor β/Δ agonist, GW501516, regulates the expression of genes involved in lipid catabolism and energy uncoupling in skeletal muscle cells. *Mol. Endocrinol.* 17:2477–2493.
238. Tranheim Kase, E. et al. (2006) 22-Hydroxycholesterols regulate lipid metabolism differently than T0901317 in human myotubes. *Biochim. Biophys. Acta—Mol. Cell Biol. Lipids* 1761:1515–1522.
239. Zitzer, H. et al. (2006) Sterol regulatory element-binding protein 1 mediates liver X receptor-β-induced increases in insulin secretion and insulin messenger ribonucleic acid levels. *Endocrinology* 147:3898–3905.
240. Jakel, H. et al. (2004) The liver X receptor ligand T0901317 down-regulates APOA5 gene expression through activation of SREBP-1c. *J. Biol. Chem.* 279:45462–45469.
241. Hummasti, S. et al. (2004) Liver X receptors are regulators of adipocyte gene expression but not differentiation: Identification of apoD as a direct target. *J. Lipid Res.* 45:616–625.
242. Haughton, E.L. et al. (2006) Pregnane X receptor activators inhibit human hepatic stellate cell transdifferentiation *in vitro*. *Gastroenterology* 131:194–209.
243. Pascussi, J.-M. et al. (2000) Dexamethasone induces pregnane X receptor and retinoid X receptor-alpha expression in human hepatocytes: Synergistic increase of CYP3A4 induction by pregnane X receptor activators. *Mol. Pharmacol.* 58:361–372.
244. Vu-Dac, N. et al. (1998) Retinoids increase human apo C-III expression at the transcriptional level via the retinoid X receptor. Contribution to the hypertriglyceridemic action of retinoids. *J. Clin. Invest.* 102:625–632.
245. Moffit, J.S. et al. (2007) Differential gene expression in mouse liver associated with the hepatoprotective effect of clofibrate. *Toxicol. Appl. Pharmacol.* 222:169–179.
246. Baranowski, M. et al. (2007) PPARα agonist induces the accumulation of ceramide in the heart of rats fed high-fat diet. *J. Physiol. Pharmacol.* 58:57–72.
247. Xin, B. et al. (2007) Inhibitory effect of meloxicam, a selective cyclooxygenase-2 inhibitor, and ciglitazone, a peroxisome proliferator-activated receptor gamma ligand, on the growth of human ovarian cancers. *Cancer* 110:791–800.
248. Schaiff, W.T. et al. (2007) Ligand-activated peroxisome proliferator activated receptor gamma alters placental morphology and placental fatty acid uptake in mice. *Endocrinology* 148:3625–3634.
249. Tureyen, K. et al. (2007) Peroxisome proliferator-activated receptor-gamma agonists induce neuroprotection following transient focal ischemia in normotensive, normoglycemic as well as hypertensive and type-2 diabetic rodents. *J. Neurochem.* 101:41–56.
250. Choi, K.C. et al. (2007) Effect of PPAR-delta agonist on the expression of visfatin, adiponectin, and resistin in rat adipose tissue and 3T3-L1 adipocytes. *Biochem. Biophys. Res. Commun.* 357:62–67.
251. Nagasawa, T. et al. (2006) Effects of bezafibrate, PPAR pan-agonist, and GW501516, PPARdelta agonist, on development of steatohepatitis in mice fed a methionine- and choline-deficient diet. *Eur. J. Pharmacol.* 536:182–191.
252. Bradley, M.N. et al. (2007) Ligand activation of LXRbeta reverses atherosclerosis and cellular cholesterol overload in mice lacking LXRalpha and apoE. *J. Clin. Invest.* 117:2337–2346.
253. Riddell, D.R. et al. (2007) The LXR agonist T0901317 selectively lowers hippocampal Abeta42 and improves memory in the Tg2576 mouse model of Alzheimer's disease. *Mol. Cell Neurosci.* 34:621–628.
254. Uppal, H. et al. (2007) Activation of LXRs prevents bile acid toxicity and cholestasis in female mice. *Hepatology* 45:422–432.

255. Cariou, B. et al. (2006) The farnesoid X receptor modulates adiposity and peripheral insulin sensitivity in mice. *J. Biol. Chem.* 281:11039–11049.

256. Qin, P. et al. (2005) Bile acid signaling through FXR induces intracellular adhesion molecule-1 expression in mouse liver and human hepatocytes. *Am. J. Physiol. Gastrointest. Liver Physiol.* 289:G267–273.

257. Wang, H. et al. (2007) Activated pregnenolone X-receptor is a target for ketoconazole and its analogs. *Clin. Cancer Res.* 13:2488–2495.

258. Teng, S. and M. Piquette-Miller (2007) Hepatoprotective role of PXR activation and MRP3 in cholic acid-induced cholestasis. *Br. J. Pharmacol.* 151:367–376.

259. Ma, X. et al. (2007) The PREgnane X receptor gene-humanized mouse: A model for investigating drug–drug interactions mediated by cytochromes P450 3A. *Drug Metab. Dispos.* 35:194–200.

260. Baskin-Bey, E.S. et al. (2006) Constitutive androstane receptor (CAR) ligand, TCPOBOP, attenuates Fas-induced murine liver injury by altering Bcl-2 proteins. *Hepatology* 44:252–262.

261. Lalloyer, F. et al. (2006) The RXR agonist bexarotene improves cholesterol homeostasis and inhibits atherosclerosis progression in a mouse model of mixed dyslipidemia. *Arterioscler. Thromb. Vasc. Biol.* 26:2731–2737.

262. Yamauchi, T. et al. (2001) Inhibition of RXR and PPARgamma ameliorates diet-induced obesity and type 2 diabetes. *J. Clin. Invest.* 108:1001–1013.

263. Sandelin, A., Wasserman, W.W., and Lenhard, B. (2004) ConSite: Web-based prediction of regulatory elements using cross-species comparison. *Nucleic Acids Res.* 32:W249-W252.

264. Fiedler, T. and Rehmsmeier, M. (2006) jPREdictor: A versatile tool for the prediction of cis-regulatory elements. *Nucleic Acids Res.* 34:W546–W550.

265. Marinescu, V.D., Kohane, I.S., and Riva, A. (2005) The MAPPER database: A multi-genome catalog of putative transcription factor binding sites. *Nucleic Acids Res.* 33:D91–D97.

266. Kel, A.E. et al. (2003) MATCHTM: A tool for searching transcription factor binding sites in DNA sequences. *Nucleic Acids Res.* 31:3576–3579.

267. Cartharius, K. et al. (2005) MatInspector and beyond: Promoter analysis based on transcription factor binding sites. *Bioinformatics* 21:2933–2942.

268. Podvinec, M. et al. (2002) NUBIScan, an in silico approach for prediction of nuclear receptor response elements. *Mol. Endocrinol.* 16:1269–1279.

269. Su, G., Mao, B., and Wang, J. (2006) A web server for transcription factor binding site prediction. *Bioinformation* 1:156–157.

270. Farré, D. et al. (2003) Identification of patterns in biological sequences at the ALGGEN server: PROMO and MALGEN. *Nucleic Acids Res.* 31:3651–3653.

271. Tsunoda, T. and Takagi, T. (1999) Estimating transcription factor bindability on DNA. *Bioinformatics* 15:622–630.

272. Cui, Y.J. et al. (2009) Ontogenic expression of hepatic Ahr mRNA is associated with histone H_3K_4 di-methylation during mouse liver development. *Toxicol. Lett.* 189:184–190.

10 Fluorescence Methods to Assess the Impact of Lipid Binding Proteins on Ligand-Mediated Activation of Gene Expression

Heather A. Hostetler, Avery L. McIntosh,
Anca D. Petrescu, Huan Huang,
Barbara P. Atshaves, Eric J. Murphy,
Ann B. Kier, and Friedhelm Schroeder

CONTENTS

10.1 INTRODUCTION

Although animal studies over the past few decades have demonstrated that dietary
lipid nutrients are responsible for the regulation of downstream gene expression,
detailed mechanistic explanations for these observations have only begun to be real-
ized in the last decade. A fundamental problem in understanding the potential role
of lipids as natural ligands that activate transcriptional activity of nuclear receptors
has been the lack of knowledge regarding the presence and distribution of unesterified long-chain fatty acids (LCFA) and nonhydrolyzed long-chain fatty acyl-CoA
(LCFA-CoA) in the nuclei of living cells. Previously, very little was known regarding
the potential factors regulating the distribution and targeting of these lipids to the
nucleus or their effect on nuclear regulation. Recent work has indicated that cytoplasmic lipid binding proteins function in this role through high affinity binding of

lipidic ligands and direct interaction with nuclear receptors. Furthermore, several nuclear receptors have now been shown to exhibit high affinity (i.e., nanomolar K_d) for LCFA and LCFA-CoA, consistent with their physiological significance as endogenous ligands of these nuclear receptors. The current review is not meant to be all inclusive, but will instead focus on the use of fluorescent techniques to elucidate the role of cytoplasmic lipid binding proteins in the uptake and transport of lipidic ligands to the nucleus and their subsequent interaction with nuclear receptors.

10.1.1 CURRENT SCHEMATIC AND HYPOTHESIS

Since nutrient lipids such as fatty acids and their intracellular activated forms (i.e., fatty acyl-CoA) are poorly soluble in an aqueous environment and high concentrations result in cellular toxicity, cytoplasmic lipid binding proteins (liver fatty acid binding protein, L-FABP; acyl-CoA binding protein, ACBP; sterol carrier protein-2, SCP-2) function to increase their solubility and facilitate movement through the cytoplasm [1,2]. The control of gene expression by fatty acids and/or fatty acyl-CoA is currently thought to be mediated by nuclear transcription factors including the peroxisome proliferator-activated receptors (PPAR), hepatocyte nuclear factor-4α (HNF-4α), and others (thyroid hormone receptor, TR; glucocorticoid receptor, GR; retinoid X receptor, RXR; retinoic acid receptor, RAR; liver X receptor, LXR; sterol regulatory element binding protein, SREBP) (reviewed in [3,4]). Consequently, transport of these lipids from the cytoplasm to the nucleus would seem to be required for interaction with nuclear transcription factors.

Recent fluorescence-based assays have led to the hypothesis that transfer and channeling of LCFA and LCFA-CoA to nuclear receptors in liver nuclei is mediated by cytoplasmic lipid binding proteins (Figure 10.1). In the case of PPARα, unsaturated LCFA, branched-chain fatty acids, or fatty acyl-CoA are transported through the cytoplasm and into the nucleus by L-FABP [5,6]. Once in the nucleus, L-FABP and PPARα interact, and ligands bound to L-FABP are transferred to PPARα, which in turn heterodimerizes with RXRα to bind to DNA to affect transcription. The interaction of PPARα with activating ligands results in the displacement of corepressors and increased association with coactivators (Figure 10.1A). For HNF-4α, ACBP transports LCFA-CoA into the nucleus, HNF-4α and ACBP interact [7], and the ligand is passed to HNF-4α (Figure 10.1B). HNF-4α forms homodimers prior to DNA binding. Saturated LCFA-CoA result in enhanced HNF-4α binding to DNA, release of corepressors, and increased association with coactivators [8], while polyunsaturated LCFA-CoA result in the opposite effect, with repression of HNF-4α-mediated downstream gene expression [9].

10.1.2 RECONCILIATION OF *IN VIVO* AND *IN VITRO* FINDINGS ADDRESSING THE HYPOTHESIS

While results *in vivo* suggest that LCFA or their metabolites play a pivotal role in energy and glucose homeostasis, early studies *in vitro* using radioligand competition binding assays and subcellular fractionation techniques only poorly support the hypothesis that LCFA and/or LCFA-CoA regulate nuclear receptors.

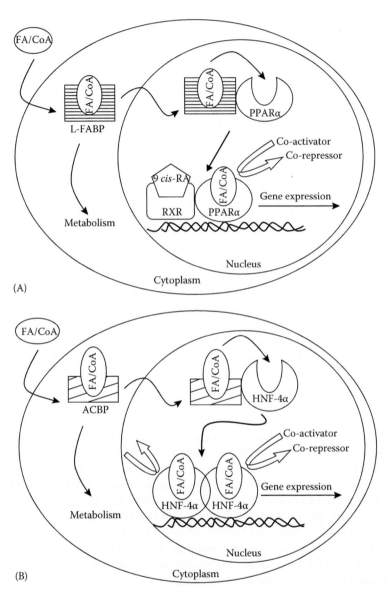

FIGURE 10.1 Schematic of the proposed pathway for selective cooperation between lipid binding proteins and nuclear receptors in LCFA and LCFA-CoA signaling to the nucleus. (A) LCFA or FA-CoA are shuttled to the nucleus for interaction with the PPARα by the L-FABP. Upon ligand binding, PPARα heterodimerizes with the retinoic X receptor-α (RXRα), which in turns binds its ligand, 9 cis-retinoic acid (9 cis-RA). (B) FA-CoA are shuttled to the nucleus for interaction with the HNF-4α by the ACBP.

First, radioligand competition binding studies show that nuclear receptors have only poor affinities for LCFA, suggesting that either LCFA are not physiologically significant endogenous ligands or LCFA metabolite(s) represent the endogenous ligand. For example, radioligand competition binding studies suggested that PPARα bound unsaturated LCFA with micromolar K_d and saturated LCFA even less well [10–13]. These radioligand-based affinities for LCFA are several orders of magnitude weaker than what PPARα exhibits for synthetic xenobiotics. Part of the difficulty in resolving this issue is the inherent complications of radioligand competition binding assays (i.e., competition of binding protein with Lipidex, charcoal, etc. for ligand) to resolve bound from free LCFA or LCFA-CoA—procedures that often underestimate LCFA and LCFA-CoA binding affinities by 2–3 orders of magnitude relative to other methods such as fluorescence and microcalorimetry [14–16]. For instance, while radioligand binding studies indicate that cytoplasmic lipid binding proteins bind LCFA and/or LCFA-CoA with moderate affinities, i.e., submicromolar to micromolar K_d [16–19], subsequent studies with fluorescence and other binding assays not requiring separation of bound from free ligand yield nanomolar K_d for L-FABP [14,16,18,20] and ACBP [21,22]. Similarly, recent fluorescence binding assays show that PPARα has high affinity (nanomolar K_d) for unsaturated LCFA, saturated LCFA-CoA, and unsaturated LCFA-CoA—all in the same range as fibrate peroxisome proliferator agents [23–26]. Thus, while early radioligand competition binding assays suggested that LCFA and LCFA-CoA are unlikely to be significant physiological ligands for PPARα, subsequent studies with more sensitive fluorescence and microcalorimetry assays not requiring separation of bound from free ligand showed that nuclear receptors bound LCFA and/or LCFA-CoA with high affinities (i.e., low nanomolar K_d), well within the range of nuclear concentrations of these ligands as indicated below.

Second, although studies in animals and cultured cells show that both saturated and unsaturated LCFA activate PPARα, the radioligand binding assays show that PPARα has only weak or no measurable affinity for saturated LCFA or their CoA thioesters (reviewed in [23,25]). Recent fluorescence binding assays have reconciled this phenomenon. Since PPARα exhibits high affinity (low nanomolar K_d) for unsaturated LCFA as well as their CoA derivatives, either or both of these ligands could contribute to PPARα activation [23,25]. In contrast, since PPARα exhibits high affinity (low nanomolar K_d) for saturated LCFA-CoA, but not for saturated LCFA, the saturated LCFA-CoA formed by intracellular fatty acyl-CoA synthases represent the high affinity endogenous ligands accounting for PPARα transactivation and activation by exogenous saturated LCFA [23,25].

Third, cellular concentrations of LCFA and LCFA-CoA are high, near 20 μM [10], suggesting that if nuclear levels are similar, then ligand-dependent nuclear receptors with nanomolar K_d for these ligands would be saturated. However, it has been postulated that because of their high affinities for LCFA and LCFA-CoA, the cytoplasmic lipid binding proteins lower the unbound/free cellular level of these lipids by 3 orders of magnitude to near 7–50 nM [15,27,28]. Indeed, nuclear LCFA and LCFA-CoA concentrations are much lower than those in the whole cell and well within the range of affinities of nuclear receptors such as PPARα (reviewed in [5,21,29,30]).

Fourth, determination of LCFA and LCFA-CoA levels in purified nuclei isolated by biochemical fractionation overestimates these levels due to (1) redistribution of

LCFA/LCFA-CoA from other organelles, (2) activation of acyl-CoA synthetases, (3) hydrolysis of LCFA-CoA to LCFA, (4) lipolytic release of LCFA and resynthesis to LCFA-CoA, or (5) cross-contamination with other membranes during the subcellular fractionation procedure [15]. Since LCFA and LCFA-CoA readily partition into membranes [31–35], analysis of purified whole nuclei does not discriminate between lipids in the nuclear envelope membranes versus the nucleoplasm. Further, such assays measure the total lipid level and do not distinguish free lipids from bound lipids. In order to overcome the limitations of biochemical fractionation techniques and to begin to resolve the differences between *in vivo* and *in vitro* data, several fluorescence-based techniques were developed.

Many of the fluorescence-based experiments investigating the impact of binding proteins on acyl-CoA activation of gene expression have utilized either naturally occurring or synthetic fluorescent lipid analogues. Such experiments have examined (1) LCFA/LCFA-CoA uptake into cells, (2) LCFA/LCFA-CoA binding to cytoplasmic binding proteins, (3) LCFA/LCFA-CoA transport into the nucleus, (4) LCFA/LCFA-CoA interaction with nuclear receptors, and (5) nuclear protein–protein interactions between cytoplasmic binding proteins and nuclear receptors. The structure of some of the more commonly used fluorescent fatty acid analogues and the structure of their corresponding natural lipids are presented in Figure 10.2. The remainder of this review focuses on the fluorescence methodology and representative findings addressing these issues.

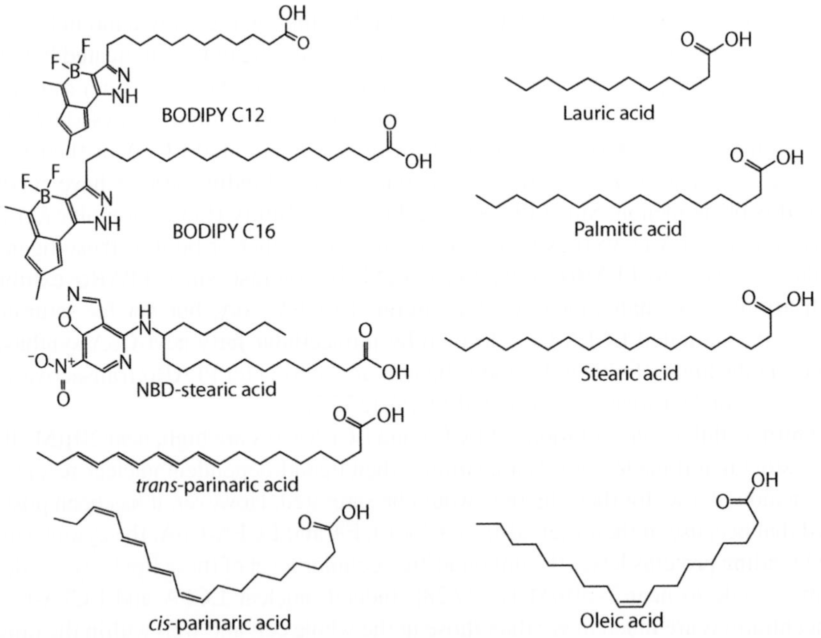

FIGURE 10.2 The chemical structure of several of the more commonly used fluorescent fatty acid analogues and the corresponding nonfluorescent, endogenous fatty acids. All structures were drawn in ChemDraw Ultra 6.0 (Cambridgesoft.com) and energy minimized.

10.2   UPTAKE OF LCFA INTO THE CELL

10.2.1   INTRODUCTION

A fundamental problem in understanding the potential role of lipids as natural ligands that activate transcriptional activity of nuclear receptors is our lack of knowledge regarding the uptake and distribution of LCFA and LCFA-CoA independent of concomitant metabolism of these ligands. Since LCFA are highly potent detergents, once taken up they are rapidly converted to LCFA-CoA, and these activated forms are even more rapidly metabolized (esterification, oxidation) within cells. Therefore, it is necessary to utilize poorly or nonmetabolizable LCFA and LCFA-CoA analogues to differentiate uptake/distribution in the absence of metabolism. Furthermore, techniques not requiring disruption of cellular equilibriums are important to accurately quantitate ligand uptake and transport in real time within living cells. Fluorescence-based assays utilizing live cells and not requiring disruption of cellular membranes have been used to examine LCFA uptake as well as LCFA and LCFA-CoA distribution in both cell suspensions and single cells. An extension of these applications has been useful in determining the extent of metabolizability of several fluorescent lipids.

10.2.2   CELL SUSPENSION METHODS—FLUORESCENCE SPECTROSCOPY

Live cell suspensions and fluorescent lipids have been utilized to provide a more accurate measurement of lipid uptake at a whole-cell level using cuvette-based assays. While flow cytometry is an alternate method for providing information on fluorescent LCFA uptake by multiple cells, it examines single cells sequentially. Since flow cytometry is better for end-point analysis, rather than kinetic analysis, of fluorescent LCFA uptake, this subject will not be covered further in this review. The cuvette-based assays are useful in resolving whether the expression of fatty acid binding proteins alters the initial rate or the total amount of LCFA uptake in a large population of cells. Furthermore, inclusion of biochemical analysis is useful for determining the extent that the fluorescent lipid is metabolized under the conditions used to study lipid uptake. In general, cell lines with no native FABP expression are transfected with one of the cytoplasmic binding proteins, and stable clones expressing that FABP are compared to mock-transfected or control cells. A suspension of these cells is incubated with a fluorescent lipid (e.g., a naturally occurring fluorescent LCFA such as *cis*-parinaric acid or a synthetic fluorescent LCFA such as NBD-stearic acid or BODIPY-palmitic acid), and fluorescence intensity is measured over a given time frame in a photon-counting fluorometer with constant stirring [36–38]. One of the unique properties of *cis*-parinaric acid is that its fluorescence is negligible in aqueous buffer but increases upon associating with a hydrophobic environment, such as the membranes or binding proteins within cells [39]. Since *cis*-parinaric acid only fluoresces upon entering the cell, this assay permits sensitive, continuous measurement of fatty acid uptake without separation of free fatty acid in solution from that taken up by the cells. Thus, an increase in fluorescence represents increased *cis*-parinaric acid uptake, and maximal fluorescence intensity indicates the extent of LCFA uptake. Following spectroscopic measurements and extensive washing to remove extracellular fluorescent lipid, cellular lipids are extracted and fluorescent fatty acid or fatty acyl-CoA metabolism is determined by the fluorescence detection of lipid fractions following lipid separation by TLC and HPLC [29,30,40].

Such assays demonstrate that *cis*-parinaric acid uptake is rapid and maximal by 1 min postincubation. Both the initial and total uptake rates of *cis*-parinaric acid are significantly increased by 50% in L-FABP-expressing cells as compared to control or I-FABP-expressing cells [37,38]. Moreover, esterification of *cis*-parinaric acid was determined to be less than 3% after 30 min of incubation [37,38]. This suggests that *cis*-parinaric acid is a poorly metabolized fatty acid and, as such, an ideal fluorescent lipid for use in uptake studies. Similar experiments have been used to examine the stability of other fluorescent fatty acid or fatty acyl-CoA analogues. With the exception of the short-chain NBD fatty acid analogue, most of the fluorescent lipids are fairly nonmetabolizable in the examined time frame (Table 10.1).

10.2.3 SINGLE CELL METHODS—FLUORESCENCE MICROSCOPY

Although the fluorescent LCFA uptake experiments in cell suspensions (cuvette-based or flow cytometry–based assays) showed that L-FABP expression enhances fluorescent fatty acid uptake, such assays do not readily distinguish between fatty acid binding to the cell surface membrane from the internalization of LCFA. To better distinguish actual lipid internalization and distribution, single-cell microscopic methods have been developed. These techniques require cultured cells to be seeded onto chambered cover glass slides at a density low enough to ensure a single monolayer. Again, cells lacking lipid binding proteins are compared with cells which express lipid binding proteins. Such experiments have used both cultured fibroblast cells overexpressing lipid binding proteins as well as cultured primary hepatocytes from mice lacking one or more of the cytoplasmic lipid binding proteins.

TABLE 10.1
Percent of Intact Fluorescent Lipid after Incubation under the Conditions Utilized to Determine Cellular Uptake

Fluorescent Lipid	Incubation Time (min)	% Nonmetabolized	Ref.
BODIPY C-5	30	100	[29]
BODIY C-12	5	91.7 ± 0.6	[29]
	30	88.9 ± 1.7	[29]
BODIPY C-16	5	98.1 ± 1.1	[29]
	30	96.9 ± 1.6	[29]
BODIPY C16-S-S-CoA	30	100	[30]
Cis-parinaric acid	30	97 ± 1	[40]
NBD-C6	30	0	[29]
NBD-C18	5	100	[29]
	30	100	[29]

10.2.3.1   Intracellular Uptake

In order to examine individual cells and visualize intracellular fluorescent lipid uptake in real time, laser scanning confocal microscopy (LSCM) and multiphoton laser scanning microscopy (MPLSM) have been employed. LSCM is generally used for imaging fluorescent molecules excited in the visible wavelength region (e.g., NBD- or BODIPY-labeled lipids), while MPLSM is used for fluorescent molecules requiring UV excitation (e.g., *cis*-parinaric acid). Cells are incubated with a nonsaturating level of fluorescent lipid, and digital images are acquired every 1–10s over a given time frame using a laser scanning imaging system to obtain fluorescence signals [29,30,41,42]. A representative LSCM or MPLSM section through the medial part of the cell (i.e., including the nucleus) is then analyzed as follows: The total area of the cell section showing increased fluorescence over time is measured and used to gauge the rate of uptake of fluorescent lipid into the cell. The maximum uptake capacity is determined as maximal fluorescence intensity (F_{max}) and is obtained by fitting the saturation curves to a rectangular hyperbola with the equation:

$$I = \frac{F_{max}t}{B+t} \tag{10.1}$$

where
  I is the fluorescence intensity
  B is a constant
  t is the time [29,30,41,42]

The initial uptake rates are determined by fitting the linear regions of the uptake curves to the linear equation:

$$y = mx + b \tag{10.2}$$

where m is the slope that corresponds to the initial rate [29,30,41,42]. Alternatively, for a short period of time, the uptake rate can be determined using the ratio F_{max}/b from the above equation.

Overall, expression of cytoplasmic binding proteins increases the initial rate and maximal uptake of fluorescent fatty acids. I-FABP expression in undifferentiated ES cells increased the initial rate and maximal uptake of NBD-stearic acid 1.7- and 1.6-fold, respectively, although there was no effect of I-FABP expression in differentiated ES cells [42]. L-FABP expression in fibroblasts increased BODIPY-C-16 uptake 1.7-fold, suggesting that fatty acid enhancement was independent of metabolism [42]. Similarly, L-FABP expression in fibroblasts enhances branched-chain fatty acid uptake [42], while L-FABP gene ablation in mouse hepatocytes decreases branched-chain fatty acid uptake [43].

10.2.3.2   Intracellular Transport/Diffusion

Fluorescence recovery after photobleaching (FRAP) is a useful technique for examining intracellular movement of fluorescent lipids. Cells are loaded with a nonsaturating amount of fluorescent lipid (e.g., NBD-stearate) for a time sufficiently long enough to establish an equilibrium distribution and washed to remove unincorporated fluorescent lipid, a medial section is taken through the cell, the desired spot within the cell (e.g., cytoplasm, nucleoplasm, etc.) is selected for photobleaching, and the recovery of fluorescence in the photobleached spot is measured to determine the intracellular diffusional mobility of the fluorescent lipid [41,43]. For example, to determine cytoplasmic diffusional mobility, a small area between the cell membrane and the nucleus is photobleached with a short laser blast for a time sufficient enough to reduce the lipid fluorescence intensity by 70%–90%. Digital fluorescence imaging is used to monitor fluorescence recovery of the fluorescent lipid at the center of the photobleached area. The intracellular mobility of the fluorescent lipid is calculated from the photobleach recovery curves as a function of time [41,43].

Although only a limited number of studies have utilized FRAP to examine the effect of cytoplasmic binding proteins in lipid transport or diffusion throughout the cell, expression of cytoplasmic binding proteins seems to increase the mobility of fluorescent fatty acids. I-FABP expression in undifferentiated ES cells increases the effective intracellular diffusion constant of NBD-stearate 1.8-fold, indicating an increase in fatty acid movement throughout the cell [41]. Similarly, L-FABP expression enhances the effective intracellular diffusion of NBD-stearate in cultured L-cell fibroblasts [15] and hepatocytes [43–46]. In contrast, the absence of L-FABP expression in L-FABP gene ablated hepatocytes reduced the cytoplasmic diffusional component of NBD-stearate, while having no effect on the membrane component [43].

10.2.4 PROS AND CONS

Such assays offer the benefits of utilizing live cells and not requiring cellular disruptions to examine both overall and individual cellular lipid uptake. Furthermore, the esterification of fluorescent lipids can be examined using standard techniques for lipid separation (HPLC, TLC) and observing each fraction for fluorescence. The single cell assays resolve binding to the plasma membrane from alterations due to intracellular fatty acid binding protein expression, intracellular membrane binding sites, intracellular diffusion, or metabolism (each of which can influence cellular fatty acid uptake). Thus, microscopic methods allow individual cells to be visualized in real time with sensitive and reliable results.

Unfortunately, the number of commercially available, nonmetabolizable, fluorescent fatty acids is limited, and the number of commercially available, nonmetabolizable, fluorescent fatty acyl-CoA is even fewer. However, recent work has resulted in several fluorescent long-chain $n - 3$ and $n - 6$ fatty acids [47], which have been used to study cellular uptake [48]. These fluorescent fatty acids are similar to the parinaric acids, in that they are fluorescent due to the presence of a conjugated tetraene group, rather than the addition of a fluorescent moiety, thus they closely resemble endogenous polyunsaturated fatty acids (Figure 10.3). Although many of the fluorescent fatty acid analogues are poorly metabolized, especially within the short time frame (minutes) required for microscopic imaging, variation in metabolism occurs and each probe must be examined to determine its level of metabolizability. Moreover, there is always the possibility that synthetic lipids will not behave the same as endogenous

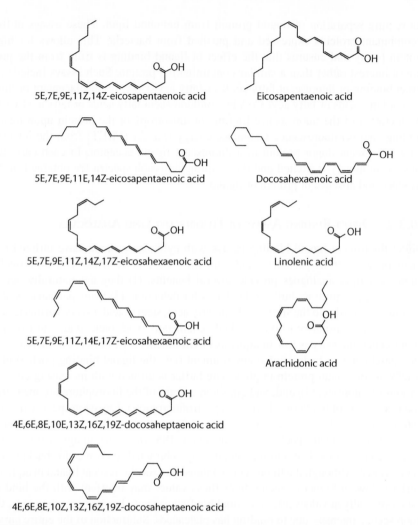

5E,7E,9E,11Z,14Z-eicosapentaenoic acid

Eicosapentaenoic acid

5E,7E,9E,11E,14Z-eicosapentaenoic acid

Docosahexaenoic acid

5E,7E,9E,11Z,14Z,17Z-eicosahexaenoic acid

Linolenic acid

5E,7E,9E,11Z,14E,17Z-eicosahexaenoic acid

Arachidonic acid

4E,6E,8E,10E,13Z,16Z,19Z-docosaheptaenoic acid

4E,6E,8E,10Z,13Z,16Z,19Z-docosaheptaenoic acid

FIGURE 10.3 The chemical structure of several newly derived fluorescent fatty acid derivatives and the chemical structures of the corresponding endogenous fatty acids. All structures were drawn in ChemDraw Ultra 6.0 (Cambridgesoft.com) and energy minimized.

ligands, thus, whenever possible, it is recommended to verify that each synthetic ligand is bound by the FABP or nuclear receptor, taken up by the cell, and distributed similarly within the cell as endogenous ligands.

10.3 LCFA AND LCFA-CoA INTERACTION WITH CYTOPLASMIC BINDING PROTEINS

10.3.1 INTRODUCTION

In studying the binding affinity of cytosolic lipid binding proteins for fatty acids and fatty acyl-CoA, several fluorescence techniques have been employed that do not require separation of bound protein from unbound lipid. These assays utilize recombinant proteins expressed and purified from bacteria. This allows for high protein purity and ensures that the effect of ligand binding is truly from the protein of interest rather than a similar contaminating protein. Such assays include (1) direct binding of fluorescent fatty acids or fatty acyl-CoA, (2) displacement of fluorescent fatty acids or fatty acyl-CoA by nonfluorescent fatty acids or fatty acyl-CoA, (3) alteration of the fluorescence lifetime or anisotropy of the protein upon ligand binding, and (4) fluorescence resonance energy transfer (FRET) from the intrinsically fluorescent donor protein to a fluorescent ligand acceptor. In each case, the fluorescent moiety is excited and the resulting emission spectra are obtained in the presence and absence of protein or ligand.

10.3.2 DIRECT BINDING ASSAYS OF FLUORESCENT LIPID ANALOGUES

Direct fluorescent ligand binding assays with cytosolic proteins have utilized *cis*-parinaric acid, *trans*-parinaric acid, and their CoA derivatives. These particular fluorescent lipid analogues provide several benefits: (1) they are naturally occurring, (2) they require no bulky side chains for detection, (3) *trans*-parinaric acid is structurally similar to the straight-chain fatty acid, stearic acid, (4) *cis*-parinaric acid has structural similarity to the "kinked" chain fatty acid, oleic acid, and (5) their fluorescence intensity is low in aqueous environments (i.e., buffer) but increases as the ligand enters a hydrophobic environment (i.e., the ligand binding pocket) [49]. Briefly, recombinant protein in phosphate buffer is titrated with increasing concentrations of fluorescent ligand, and emission spectra of the fluorophore are measured upon excitation of the fluorophore in a spectrofluorometer. Emission spectra are also obtained for the fluorophore in the absence of protein at each examined concentration. Using a software package such as Vinci (ISS, Inc., Champaign, Illinois), the spectra can be smoothed and mathematically subtracted to correct for background fluorescence. Although the fluorescence intensity of the fluorescent ligand in aqueous buffer is low, at higher concentrations these values may interfere with the binding data, especially at values after maximal binding capacity has been reached (e.g., the fluorescence intensity due to binding has plateaued). Subtraction of the entire curve, rather than the single point of maximal fluorescence intensity, is preferred since the fluorescence maxima of a fluorophore may shift upon ligand binding (reviewed in [14,49,50]). Maximal fluorescence intensities are recorded after background correction for solvent effects, protein fluorescence, and fluorescent intensities of the fluorophore in an aqueous environment. These values are used to determine the ligand binding affinity (K_d) and the number of binding sites (n) by transforming the data to a linear least-square plot of $1/(1 - \Delta F/\Delta F_{max})$ versus $C_L/(\Delta F/\Delta F_{max})$, where ΔF is the change in fluorescence intensity upon the addition of ligand, ΔF_{max} is the maximal change in fluorescence intensity, C_L is the concentration of ligand, the ordinate intercept is equal to $(nE_0)/K_d$, where E_0 is the protein concentration, and the slope is equal to $1/K_d$ [39,51,52]. It is important that data be collected from many titration points to ensure that values are recorded for concentrations before and after the linear phase of binding. For best results, a plot of the ligand concentration by the maximal fluorescence intensity should provide a hyperbolic shaped curve with several points after the plateau phase, ensuring maximal fluorescence intensity (and maximal binding) has been obtained.

For example, the maximal fluorescence emission of *cis*-parinaric acid occurs around 416 nm upon excitation at 320 nm, although a slight shift in the emission maximum may occur upon ligand binding. If the emission spectra of a protein (100 nM) titrated with 0, 2, 5, 10, 20, 50, 100, and 500 nM of *cis*-parinaric acid was recorded and corrected for the fluorescence emission from 0, 2, 5, 10, 20, 50, 100, and 500 nM of *cis*-parinaric acid in buffer, these spectra could be used to determine the binding affinity of the protein for *cis*-parinaric acid. For this example, the maximal fluorescence intensity (at 416 nm) of the protein titrated with 0, 2, 5, 10, 20, 50, 100, and 500 nM of *cis*-parinaric acid is 0, 500, 800, 900, 950, 980, and 1000 arbitrary fluorescence units (a.u.), respectively (Figure 10.4A). Since the maximal fluorescence intensity is approximately equal to 1000 a.u., this value is used for our calculations (Table 10.2). By plotting a graph of $1/(1 - \Delta F/\Delta F_{max})$ on the y-axis and $C_L/(\Delta F/\Delta F_{max})$ on the x-axis, we get a straight line where $y = 1.0121x - 1.5124$ (Figure 10.4B). From this equation, the slope of the line is 1.0121, so the K_d is equal to 1/1.0121, or approximately 1 nM, showing very strong ligand affinity.

These experiments indicate that I-FABP [39], L-FABP [14], and SCP-2 [51] bind *cis*-parinaric acid with similar high affinities; however, L-FABP has two binding sites (similar affinities) and has a much stronger affinity for *trans*-parinaric acid than either I-FABP or SCP-2 (Table 10.3). Furthermore, L-FABP has a fourfold higher affinity for *trans*-parinaric acid than for *cis*-parinaric acid, while I-FABP and SCP-2 have a threefold higher affinity for *cis*-parinaric acid than *trans*-parinaric acid (Table 10.3). This suggests that L-FABP has a higher affinity for naturally occurring straight-chain fatty acids, while I-FABP and SCP-2 prefer kinked-chain fatty acids.

SCP-2 displays stronger affinity for both acyl-CoA derivatives than the fatty acids themselves, whereas L-FABP affinity varies with ligand (Table 10.3). However,

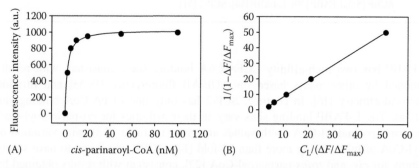

FIGURE 10.4   Protein binding to *cis*-parinaric acid. (A) A simulated plot of the maximal fluorescence intensity (at 416 nm) upon excitation of *cis*-parinaric acid at 320 nm of protein titrated with 0, 2, 5, 10, 20, 50, 100, and 500 nM of *cis*-parinaric acid. (B) A plot of the data from (A) transformed to a linear least-square plot of $1/(1 - \Delta F/\Delta F_{max})$ versus $C_L/(\Delta F/\Delta F_{max})$, where ΔF is the change in fluorescence intensity upon the addition of ligand, ΔF_{max} is the maximal change in fluorescence intensity, C_L is the concentration of ligand, the ordinate intercept is equal to $(nE_0)/K_d$ where E_0 is the protein concentration, and the slope is equal to $1/K_d$.

TABLE 10.2

Application of the Ligand Binding Equations to an Example of Direct Binding of a Fluorescent Fatty Acid

C_L (nM)	FI	ΔF	$\Delta F/\Delta F_{max}$	$1 - \Delta F/\Delta F_{max}$	$C_L/(\Delta F/\Delta F_{max})$	$1/(1 - \Delta F/\Delta F_{max})$
0	0	0	0	1	0	1
2	500	500	0.5	0.5	4	2
5	800	800	0.8	0.2	6.25	5
10	900	900	0.9	0.1	11.111111	10
20	950	950	0.95	0.05	21.052632	20
50	980	980	0.98	0.02	51.020408	50
100	1000	1000	1	0	100	Infinity

TABLE 10.3

Ligand Binding Affinity (K_d) of ACBP, I-FABP, L-FABP, and SCP-2 for Naturally Occurring Fluorescent Lipids

	ACBP K_d (nM)	I-FABP K_d (µM)	L-FABP K_d (nM)	SCP-2 K_d (nM)
Cis-parinaric acid		0.2 ± 0.1	171 ± 46	180 ± 10
Cis-parinaroyl-CoA	7.0 ± 0.9		8 ± 3	4.6 ± 0.8
			97 ± 42	
Trans-parinaric acid		0.7 ± 0.2	44 ± 8	560 ± 40
Trans-parinaroyl-CoA	4.4 ± 0.4		10 ± 1	2.8 ± 0.1
			180 ± 23	

Note: Since L-FABP has two acyl-CoA binding sites, the affinity of each site is provided in the table. ACBP [58], I-FABP [39], L-FABP [14], SCP-2 [51].

L-FABP has two high-affinity LCFA-CoA binding sites, consistent with results obtained by other assays based on ADIFAB fluorescence [18,53] and titration microcalorimetry [16]. In contrast, SCP-2 has only one LCFA-CoA binding site [51,54]. The L-FABP binding sites vary in their affinities for *cis*-parinaroyl-CoA and *trans*-parinaroyl-CoA (both values are given in Table 10.3), with variation of acyl-CoA affinities being more than 12-fold [14]. ACBP was found to have strong affinity for *cis*- and *trans*-parinaroyl-CoA [22], consistent with results obtained by titration microcalorimetry [21,55], validating the use of the fluorescence approach. Thus, ACBP's affinity for LCFA-CoA is similar to that of SCP-2 and L-FABP for the higher affinity ligand binding site. Although the ability of I-FABP to bind acyl-CoA was unable to be determined by this technique due to the unique properties of the LCFA-CoA binding site in this protein [52], this issue was subsequently resolved by an alternative approach [22,49]. These data demonstrate that most of the cytosolic binding proteins are capable of binding to LCFA-CoA with strong affinity, and in most cases, that affinity was higher than that for LCFA. Moreover, these data demonstrate the utility of direct binding assays using fluorescent fatty acids or their respective acyl-CoA derivatives.

10.3.3 DISPLACEMENT ASSAYS

In order to determine whether this high affinity binding to fluorescent ligands reflected that of typical dietary lipids, the ability of endogenous nonfluorescent lipids to displace the fluorescent lipid was examined. These assays allow for the examination of fatty acid chain length and saturation status on ligand affinity. Protein is mixed with the fluorescent ligand at the minimal concentration required to produce the maximal fluorescence intensity upon binding, as determined by the direct binding assay. Maximal fluorescence intensities are obtained from cytosolic protein bound fluorescent fatty acid or fluorescent fatty acyl-CoA, and the resulting complex is titrated with increasing concentrations of nonfluorescent lipid [14,51,56]. Fluorescence measurements are obtained as described for direct binding of fluorescent lipids. Since the fluorescence intensity of the fluorophore is environmentally dependent, the ability of the nonfluorescent lipid to serve as a competitor is measured by its ability to displace the fluorescent ligand from the binding pocket, thereby reducing the fluorescence intensity. The change in fluorescence intensity upon addition of competitor is divided by the maximal fluorescence in the absence of competitor and multiplied by 100 to obtain relative percent displacement [14,51,56]. By comparing the percent displacement of a variety of fatty acids or fatty acyl-CoA for a given concentration range, the relative affinities of these lipids can be distinguished. For example, a fatty acyl-CoA with a percent displacement value of 50% is bound with higher affinity than a fatty acyl-CoA with a percent displacement value of 10%. This allows for a relative determination of fatty acid chain length or saturation effects.

Each of the cytosolic binding proteins (L-FABP, SCP-2, ACBP) has similar strong affinity for acyl-CoA, although variation in percent displacement is noted (Table 10.4). For each of the examined cytosolic binding proteins, short-chain saturated acyl-CoA (C2:0–C8:0) were unable to displace the fluorescent lipid [14,51,56]. For each of these cytosolic proteins, increasing the chain length increased ligand affinity of saturated acyl-CoA. This trend was consistent through all of the examined chain lengths (up to C20:0, arachidoyl-CoA) for ACBP and L-FABP, while SCP-2 displayed stronger affinity for myristoyl-CoA (C14:0) and palmitoyl-CoA (C16:0) than steroyl-CoA (C18:0) or arachidoyl-CoA (C20:0). In general, unsaturation resulted in a slight decrease in ligand affinity. The major exception was SCP-2 for 18 and 20 carbon acyl-CoA, with both monounsaturation and polyunsaturation increasing binding affinity [51], suggesting that SCP-2 preferentially binds unsaturated LCFA-CoA.

10.3.4 FLUORESCENCE LIFETIME AND ANISOTROPY ASSAYS

Fluorescence lifetime and anisotropy measurements of unbound and bound protein have been used to examine the ligand binding pocket of cytosolic binding proteins as well as to confirm ligand binding. These methods utilize the intrinsically

TABLE 10.4

Percent Displacement of *cis*-Parinaroyl-CoA from ACBP or L-FABP and Percent Displacement of *cis*-Parinaric Acid from SCP-2 by Endogenous LCFA-CoA

| | % Displacement of *cis*-Parinaroyl-CoA or *cis*-Parinaric Acid* | | | |
Competitor Ligand	ACBP	L-FABP Site 1	L-FABP Site 2	SCP-2*
CoA	0	0	0	0
C2:0–C8:0-CoA	0	0	0	0
C10:0-CoA	4.9 ± 2.3	0	0	18.3 ± 3.8
C12:0-CoA	20.8 ± 3.5	43 ± 1	44 ± 1	54.1 ± 2.7
C14:0-CoA	46.6 ± 0.8	64 ± 2	60 ± 5	64.9 ± 0.8
C16:0-CoA	72.1 ± 2.9	80 ± 4	76 ± 1	61.0 ± 1.1
C18:0-CoA	79.6 ± 0.8	87 ± 3	78 ± 1	11.3 ± 2.0
C20:0-CoA	82.0 ± 1.5	91 ± 3	83 ± 3	0
C14:1-CoA	28.7 ± 3.5	39 ± 5	43 ± 5	64.3 ± 1.9
C18:2-CoA	55.9 ± 2.0	85 ± 1	80 ± 2	57.6 ± 5.3
C18:3-CoA	45.7 ± 1.0	63 ± 2	59 ± 5	62.4 ± 1.2
C20:3-CoA	71.2 ± 2.5	83 ± 1	69 ± 3	60.5 ± 0.6
C20:4-CoA	53.4 ± 2.0	80 ± 1	74 ± 7	65.0 ± 0.9

Note: ACBP [56], L-FABP [14], SCP-2 [51].

*=SCP-2 values are for displacement of *cis*-parinaric acid.

fluorescent properties of the aromatic amino acid residues, tryptophan and tyrosine. If the protein of interest contains one or more of these amino acids, then the lifetime components of these molecules can be used to determine whether they are located within the hydrophobic environment of the ligand binding pocket or on the protein's surface, exposed to the aqueous environment [49]. Further, intrinsically fluorescent amino acids within the ligand binding pocket are expected to have altered lifetime measurements upon conformation changes. This is especially useful if the protein undergoes conformational changes upon ligand binding. Anisotropy decay measurements are used to determine ligand effects on protein molecule mobility, since the mobility of a free/unbound protein is expected to differ from that of the ligand-bound protein, this method can be useful in confirming ligand binding.

In order to determine lifetime components for a given protein, fluorescence decay kinetics is first measured to obtain rotation-free results. This can be done using time-domain or frequency-domain measurements. In frequency-domain measurements, the sample fluorophore must be compared to a reference lifetime fluorescent standard, whose emission characteristics overlap with the sample of interest. The time-dependent intensity for a fluorophore with a single decay time would be related by

$$I(t) = I_0 \exp\left(-\frac{t}{\tau}\right) \tag{10.3}$$

where

I_0 is the initial intensity immediately after excitation before decay

τ is the lifetime of the fluorescence decay [57]

However, many fluorophores exhibit multiexponential decays either due to multiple spectrally overlapping fluorophores (e.g., tryptophan and tyrosine in a protein) or a complex decay of a single fluorophore. This concept can be extended so that the intensity is described by

$$I(t) = \sum_{i=1}^{n} \alpha_i \exp\left(-\frac{t}{\tau_i}\right) \qquad (10.4)$$

where α_i would be the contribution to the intensity at $t=0$ of the ith fluorescence component corresponding to the multiple discrete τ_i lifetime [22,51,58]. The decay components are related to the steady-state intensity by

$$f_i = \frac{\alpha_i \tau_i}{\sum_{j=1}^{n} \alpha_j \tau_j} \qquad (10.5)$$

where $\sum_{j=1}^{n} \alpha_j \tau_j$ is the integrated intensity over time. Practically, in a solution with a significant number of the sample fluorophores, $\sum_{j=1}^{n} \alpha_j \tau_j$ would be the integrated intensity per excitation event. The fractional contribution, f_i, can be used to determine the mean lifetime [22,51,58]:

$$\overline{\tau} = \sum_{j=1}^{n} f_j \tau_j \qquad (10.6)$$

As an example, a time-dependent decay curve with a double exponential decay simulated using experimental data involving the L-FABP in buffer without ligand was plotted (dotted points—Figure 10.5A). The data points would be acquired using a fast time-domain acquisition spectrofluorometer. In this example, the intensity was normalized so that the sum of the α values is 1. This curve was fit (solid line—Figure 10.5A), using a nonlinear fitting routine in Sigma Plot 8.0 (SPSS, Inc., Chicago, Illinois), to the multiexponential decay equation, which was simplified in the case of a two discrete lifetime component decay system [57]:

$$I(t) = \sum_{i=1}^{2} \alpha_i \exp\left(-\frac{t}{\tau_i}\right) = \alpha_1 \exp\left(-\frac{t}{\tau_1}\right) + \alpha_2 \exp\left(-\frac{t}{\tau_2}\right) \qquad (10.7)$$

The lifetime values of $\tau_1 = 3.3 \pm 0.5\,\text{ns}$ and $\tau_2 = 1.2 \pm 0.2\,\text{ns}$ with corresponding contributions of $\alpha_1 = 0.35 \pm 0.11$ and $\alpha_2 = 0.66 \pm 0.11$ were obtained from the fit. For such

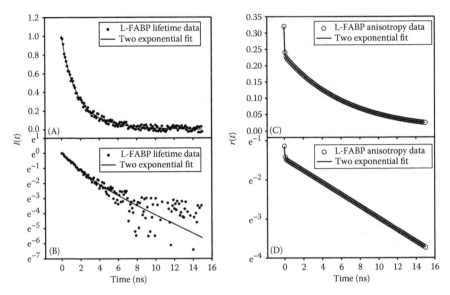

FIGURE 10.5 Lifetime and rotational correlation time measurements. (A) An example plot of time-dependent intensity decay simulated using parameters obtained from measurements of unbound L-FABP (solid circles) and the resulting fit to the two exponential equation for two component lifetimes (solid line): $I(t) = \sum_{i=1}^{2} \alpha_i \exp(-t/\tau_i) = \alpha_1 \exp(-t/\tau_1) + \alpha_2 \exp(-t/\tau_2)$.

Random noise was added to the simulated data to show its effect. (B) A plot of (A) with the intensity scaled as a function of e^x, which shows effectively the deviation from a single lifetime. (C) An example plot of time-dependent rotational anisotropy decay simulated using parameters obtained from measurements of unbound L-FABP (open circles) and the resulting fit to the two exponential equation for two component rotational correlation times (solid line): $r(t) = \sum_{i=1}^{2} r_{0i} \exp(-t/\theta_i) = r_{01} \exp(-t/\theta_1) + r_{02} \exp(-t/\theta_2)$. (D) A plot of (C) with the intensity scaled as a function of e^x that shows effectively the deviation from a single rotational correlation time.

a two lifetime system, the fractional components to the steady state intensity can be calculated using the equations

$$f_1 = \frac{\alpha_1 \tau_1}{\sum_{j=1}^{2} \alpha_j \tau_j} = \frac{\alpha_1 \tau_1}{\alpha_1 \tau_1 + \alpha_2 \tau_2} = \frac{0.35 * 3.3}{0.35 * 3.3 + 0.66 * 1.2} = 0.59$$

corresponding to the lifetime, τ_1, and

$$f_2 = \frac{\alpha_2 \tau_2}{\sum_{j=1}^{2} \alpha_j \tau_j} = \frac{\alpha_2 \tau_2}{\alpha_1 \tau_1 + \alpha_2 \tau_2} = \frac{0.66 * 1.2}{0.35 * 3.3 + 0.66 * 1.2} = 0.41$$

corresponding to lifetime, τ_2. Problems in fitting this curve occur if any two lifetimes are very similar (e.g., $\tau_1 \sim \tau_2$) or if there is significant noise in the data. This can be visualized more effectively when examining the plot of the $\ln[I(t)]$ vs. time (Figure 10.5B).

In order to examine the rotational characteristics of a fluorophore in its environment, polarizers can be placed within the excitation and emission beam path. Often, the decay intensity is measured with the polarizers in place such that the excitation is vertically oriented and the emission polarizer is oriented at the magic angle of 54.7° to minimize the scattering contributions and recover the total intensity [59]

$$I_T = I_{pl} + 2I_{pd} \tag{10.8}$$

However, with the excitation polarizer oriented vertically, the decay intensity can be monitored with the emission polarizer oriented parallel (pl) and perpendicular (pd) with the excitation polarizer. Thus, the time-dependent anisotropy can be calculated from these measured values by the equation [59],

$$r(t) = \frac{I_{pl}(t) - I_{pd}(t)}{I_{pl}(t) + 2I_{pd}(t)} \tag{10.9}$$

A simulated example of the anisotropy resolved time-dependent intensity decay with two rotational correlation times is shown in Figure 10.5C (open circles). Subsequently, the time-dependent anisotropy can be related to the rotational correlation time θ by the equation,

$$r(t) = \sum_{i=1}^{n} r_{0i} \exp\left(-\frac{t}{\theta_i}\right) \tag{10.10}$$

where the sum of r_{0i} would be equal to the limiting anisotropy of a fluorophore (e.g., $r(t=0) = r_0$), which can be fixed to a measured value [51,58]. In the example, $n=2$, so the equation was simplified as before, such that the anisotropy can be fitted to the double exponential equation

$$r(t) = \sum_{i=1}^{2} r_{0i} \exp\left(-\frac{t}{\theta_i}\right) = r_{01} \exp\left(-\frac{t}{\theta_1}\right) + r_{02} \exp\left(-\frac{t}{\theta_2}\right) \tag{10.11}$$

From the curve fit of $r(t)$ (solid line—Figure 10.5C), the values $r_{01} = 0.23$, $r_{02} = 0.09$, $\theta_1 = 6.59$ ns, and $\theta_2 = 0.05$ ns were obtained. Once again for visualization, the time-dependent anisotropy data (open circles) and corresponding fitted curve (solid line) was scaled logarithmically (Figure 10.5D).

The equivalent hydrodynamic radius (r) of a spherical protein would be described by the equation:

$$r = \left(\frac{3kT\theta}{4\pi\eta} \right)^{1/3}$$ (10.12)

where
  T is the temperature
  k is the Boltzmann constant (1.3807×10^{-16} erg K^{-1})
  θ is the rotational correlation time
  η is the viscosity in poise (P) which, for water, can vary with temperature: 0.0069 P
    at 37°C to 0.01 P at 20°C [58]

The hydrodynamic radius can also be estimated from the hydrated protein volume,

$$r = \left[\left(\frac{3M}{4\pi N_0} \right) (V_2 + V_1) \right]^{1/3}$$ (10.13)

where
  M is the protein mass in Daltons
  V_1 is the volume of bound water related to the hydration as

$$V_1 = \delta/\rho$$ (10.14)

where the hydration (δ) is expressed as g of H_2O per gram protein and the density of water (ρ) varies between approximately 0.997 g cm^{-3} at 25°C to 0.993 g cm^{-3} at 37°C, V_2 is the partial specific volume per gram of protein and it is typically 0.73 cm^3 g^{-1}, and N_0 is Avogadro's number (6.022×10^{23} mol^{-1}) [14,51]. As an example, the hydrodynamic radius of the 10,000 Da ACBP may be estimated. Using the density of water at 25°C (298 K), $\rho = 0.997$, and the hydration, $\delta = 0.23$ g H_2O per g protein, such that $V_1 = \delta/\rho = 0.23$ cm^{-3} g^{-1} protein, the spherical radius is calculated by substituting these parameters in the equation to find

$$r = \left[\left(\frac{3M}{4\pi N_0} \right) (V_2 + V_1) \right]^{1/3} = \left[\left(\frac{3(10,000)}{4\pi(6.022 \times 10^{23})} \right) (0.72 + 0.23) \right]^{1/3} = 15.6\,\text{Å}$$

for hydrated ACBP. This can be compared to the radius calculated from the rotational correlation time. For ACBP, the rotational correlation time was found to be $\theta = 3.07$ ns as measured from the 313 nm emission using 275 nm excitation [58]. Assuming a viscosity of 0.00891 P (at 298 K), an approximate spherical radius was calculated,

$$r = \left(\frac{3kT\theta}{4\pi\eta} \right)^{1/3} = \left(\frac{3(1.3807 \times 10^{-16})(298)(3.07 \times 10^{-9})}{4\pi(0.00891)} \right)^{1/3} = 15.1\,\text{Å}$$

Fluorescence lifetime and time-resolved anisotropy measurements of SCP-2 showed that the addition of oleoyl-CoA or oleic acid to SCP-2 significantly altered both the long lifetime component as well as the overall rotational rate; confirming SCP-2 binding to these ligands [51]. Furthermore, the apo-SCP-2 is strongly ellipsoidal, and ligand binding results in protein conformational changes [51]. Similar experiments with I-FABP and L-FABP not only confirmed oleic acid and oleoyl-CoA binding, but also resolved the orientation of fatty acid ligands within each of the binding pockets [14,22,49]. Similarly, apo-ACBP was found to be an ellipsoidal protein, whose conformation was significantly altered upon oleoyl-CoA binding [58]. While confirming some binding assay results, these experiments also suggested that limitations existed for assays utilizing direct binding of fluorescent lipids, since such experiments were unable to demonstrate I-FABP binding to *cis*-parinaroyl-CoA [52]. These experiments also provide evidence as to why direct binding assays with I-FABP and *cis*-parinaroyl-CoA were unable to obtain significant results—presumably due to the orientation of the fluorescent moiety of *cis*-parinaroyl-CoA being partially exposed to the aqueous buffer, rather than completely buried within the hydrophobic ligand binding pocket as for *cis*-parinaric acid [22].

10.3.5   FLUORESCENCE RESONANCE ENERGY TRANSFER

FRET allows for confirmation of ligand binding as well as determination of the distance between molecules. This method requires the presence of fluorescent amino acids within the protein that are in close proximity to the ligand upon binding and a modest degree of overlap between the absorbance spectrum of the acceptor and the emission spectra of the donor [49,60]. Energy is transferred from the donor (i.e., intrinsically fluorescent amino acids of the protein) to the acceptor (i.e., *cis*-parinaroyl-CoA) through the conjugated double bonds, resulting in a decrease in donor fluorescence and an increase in acceptor fluorescence [49,50,60]. Since FRET varies with the sixth root of the intermolecular distance, the donor and the acceptor residues must be in very close proximity (typically 10–100 Å) for efficient FRET to occur [49,61]. In practice, depending on the specific donor/acceptor pairs used, this method is useful for determining intermolecular distances in the range of 1–100 Å. Protein is titrated with increasing quantities of ligand, that is, the energy acceptor, and emission spectra of both the donor and the acceptor are obtained by a spectrofluorometer upon excitation of the donor. Maximal fluorescence intensities are recorded after background correction for both the donor and the acceptor. The intermolecular distance between the donor and acceptor molecules is determined by the Forster equation,

$$E = \frac{R_0^6}{R_0^6 + R_{2/3}^6}$$ (10.15)

where
  E is the FRET efficiency
  R_0 is the Forster distance or critical distance for 50% efficiency
  $R_{2/3}$ is the actual distance between donor and acceptor [49,61]

The energy transfer efficiency is first calculated by quenching of donor aromatic amino acid fluorescence according to

$$E = 1 - \frac{F_{DA}}{F_D} \tag{10.16}$$

where F_{DA} and F_D are the fluorescence emission intensities of the aromatic amino acids in the presence or absence of added ligand. Alternately, the energy transfer efficiency E can also be calculated by the intrinsic increase of acceptor ligand fluorescence according to

$$E = \left[\left(\frac{F_{AD}(\lambda_A^{em})}{F_A(\lambda_A^{em})} \right) - 1 \right] \times \left(\frac{\varepsilon_A(\lambda_D^{ex})}{\varepsilon_D(\lambda_D^{ex})} \right) \tag{10.17}$$

where

F_{AD} and F_A are the maximal fluorescence intensities of the acceptor emission in the presence and absence of a donor, respectively

ε_A and ε_D are the extinction coefficients (at the excitation wavelength of the donor) of the acceptor and donor, respectively

The critical distance for 50% energy transfer efficiency (R_0) in angstroms can be calculated according to the equation

$$R_0 = 9.79 \times 10^3 \left[\kappa^2 \eta^{-4} Q_D J \right]^{1/6} \tag{10.18}$$

where the orientation factor κ^2 is assumed to be 2/3, the refractive index η is 1.4 for proteins in solution, the quantum yield of the donor in the absence of acceptor Q_D is calculated based upon the amino acid sequence of the protein, and the overlap integral J is expressed in mL M^{-1} or cm^3 M^{-1} [49,60].

The modest degree of overlap between the emission spectrum of the intrinsically fluorescent moieties of proteins and the absorbance spectrum of the parinaric acids allows for Forster energy transfer to occur from aromatic residues to parinaric acid. Titration of I-FABP with *cis*- or *trans*-parinaric acid results in quenching of I-FABP tryptophan fluorescence, which is not noted for the binding of nonfluorescent fatty acids and a concurrent increase in parinaric acid fluorescence [49]. Although the addition of nonfluorescent ligands to L-FABP results in a slight decrease of L-FABP tyrosine fluorescence, suggesting a conformational change upon ligand binding, stronger quenching is noted for the addition of *cis*- or *trans*-parinaric acid [49]. Titration of SCP-2 with *cis*- and *trans*-parinaric acid shows a similar energy transfer from tryptophan to the parinaric acid, with a calculated distance between these fluorescent moieties of 40 Å [50]. These experiments show that fatty acid binding by these lipid binding proteins is specific and that the fluorescent moieties involved in the energy transfer are located in close proximity.

For example, the quantum yield of L-FABP was previously determined to be 0.0658, and the overlap integral for L-FABP and *cis*-parinaric acid was estimated to be 1.20×10^{-14} mL M^{-1} [49], so the Forster distance would be equal to 9.79×10^3 ((2/3) $(1.4)^{-4}$ (0.0658) $(1.20 \times 10^{-14}$ mL $M^{-1}))^{1/6}$, or $9790 \times (1.37 \times 10^{-16})^{1/6}$ or approximately 22.2 Å. Upon excitation of L-FABP at 280–285 nm, the maximal peak of fluorescence emission occurs around 310 nm. After background subtraction of buffer, protein, and acceptor at each concentration, the fluorescence intensity values at this point (approximately 310 nm) can be utilized to calculate the intermolecular distance between the tyrosine molecules of L-FABP and the *cis*-parinaric acid. If the fluorescence intensity at 310 nm of the donor (L-FABP) in the absence of acceptor (*cis*-parinaric acid) was 5000 a.u. and the fluorescence intensity at 310 nm after maximal binding of acceptor was 1250 a.u., then the maximal energy transfer efficiency E would be equal to $1-(1250/5000)=0.75$, or 75%. Rearranging Equation 10.15, $E = R_0^6/(R_0^6 + R_{2/3}^6)$, gives the new equation

$$R_{2/3} = \left[\left(\frac{R_0^6}{E}\right) - R_0^6\right]^{1/6} \qquad (10.19)$$

By using $E=0.75$ and $R_0=22.2$ Å, $R_{2/3}=((22.2^6/0.75) - (22.2^6))^{1/6}$ or 18.5 Å.

10.3.6 PROS AND CONS

These techniques offer many advantages over nonfluorescent techniques, such as microcalorimetry, radioactive binding assays, radioactive competition assays, or cell-based techniques. First, fluorescent techniques allow for very low (nanomolar) concentrations of both protein and ligand. This is essential for examining ligand effects at levels below the critical micelle concentration (CMC) of the ligands and low protein concentrations required for proteins that tend to self-aggregate. Since both fatty acids and fatty acyl-CoA are molecules with limited solubility in aqueous buffer, with low CMC in the typical range of 1 μM [62,63] and 60–200 μM [62], respectively, this benefit is especially important for these ligands. Second, these assays do not require separation of bound from unbound ligand; thereby decreasing the likelihood of innate errors incurred upon disruption of binding equilibriums. Third, no extraneous substances are required (i.e., Lipidex, charcoal, or other competing matrix as used for radioligand assays). Fourth, the use of purified proteins (rather than cell-based assays) ensures that ligand binding is due to a direct interaction with the protein of interest rather than through a complex of multiple proteins.

Although there are many benefits to using fluorescent techniques to examine ligand binding, each assay has its own internal limitations, and therefore, a combination of techniques is usually used to ensure meaningful results. For example, the ligand of interest is not always available in a fluorescent form. Furthermore, the addition of a fluorescent moiety to a ligand may alter a protein's affinity for that ligand. Displacement assays overcome these issues by examining the effect of natural ligands on displacing the fluorescent ligand; however, for displacement to occur, the natural ligand must have a higher affinity than the fluorescent ligand. Alteration of protein fluorescence intensity or lifetimes upon ligand binding is the most direct method since it utilizes no synthetic compounds or moieties, yet, this method is not available for all proteins. The protein must first contain intrinsically fluorescent residues (tryptophan or tyrosine) for this method to be a viable option. If the lipid binding protein tyrosine or tryptophan residues produce a strong fluorescent signal at concentrations sufficiently low for ligand binding assays (1 μM or less for lipid binding proteins), and if the protein undergoes a conformational change upon ligand binding, which alters the aqueous exposure of one or more of the tyrosine or tryptophan residues, then ligand binding of nonfluorescent lipids can be analyzed directly by quenching of intrinsic protein fluorescence as a function of ligand concentration [32,49]. However, due to the low number of intrinsically fluorescent amino acids in most of the lipid binding proteins, this method has not been as useful for lipid binding proteins as it has been for nuclear receptors. As such, this method is described in Section 10.5. Alternately, whether in the presence or absence of a conformational change, if the ligand alters the hydrophobicity of the microenvironment of the fluorescent amino acids, then at least one of the lifetime components is likely to be altered upon ligand binding. However, the latter method also requires the ability to detect miniscule changes in each lifetime component, and it may not be sensitive enough to detect subtle changes in lifetime measurements necessary to determine ligand affinity. FRET assays, similarly, require the presence of intrinsically fluorescent amino acids within the ligand binding pocket to be useful as well as the addition of a fluorescent ligand that can act as an energy acceptor (subject to the issues listed above).

10.4   TRANSPORT OF LCFA AND LCFA-CoA INTO THE NUCLEUS

10.4.1   INTRODUCTION

In order for lipids to affect gene expression, they must be transported into the nucleus. Since LCFA and LCFA-CoA have high affinities for membrane lipids and membrane proteins with hydrophobic binding sites for these ligands, cells appear to have evolved special cytoplasmic LCFA and LCFA-CoA binding proteins for this purpose. Modifications of the LSCM and MPLSM techniques described above for intracellular lipid uptake have been used to study whether the ligands alone enter the nucleus or whether these ligands enter the nucleus bound by cytoplasmic binding proteins. These techniques allow for fluorescently labeled lipids or proteins to be examined in real time as they enter and/or move throughout the cell. The advent of fluorescently labeled organelle markers has assisted in determining cellular localization through colocalization of fluorescent signals in both fixed and live cells.

10.4.2   NUCLEIC ACID STAINS

In recent years, several fluorescent dyes that stain nucleic acids have been used in living or fixed cells to determine nuclear localization. Although, most of these dyes preferentially stain double-stranded DNA, most will stain all nucleic acids to some degree. Chemical modification of such compounds has resulted in a variety of similar nucleic acid stains with shifted optimal excitations and emissions, allowing for a set of nucleic acid dyes which span the entire visible spectrum (Table 10.5). Similarly, more sensitive probes have been designed with increased nucleic acid affinity and which are essentially nonfluorescent in the absence of nucleic acids. A variety of nucleic acid stains exist, which differ in cell permeability, nucleic acid affinity, fluorescence enhancement upon DNA binding, extinction coefficients, fluorescence quantum yields, and fluorescence spectra, thereby allowing for a broader range of applications. These nucleic acid stains can be used for LSCM or MPLSM in conjunction with a fluorescently labeled protein or lipid. A partial list of commercially available nucleic acid stains is presented in Table 10.5, with their maximal values for absorbance, emission, extinction coefficient, and quantum yield.

10.4.3   NUCLEAR LOCALIZATION OF LCFA AND LCFA-CoA

By combining the LSCM/MPLSM single-cell imaging methods described in the section on cellular LCFA/LCFA-CoA uptake with the above described nucleic acid stains, real-time nuclear localization of LCFA/LCFA-CoA can be obtained in living cells. This technique is useful for determining the overall amount of lipids that enter the nucleus, the rate of nuclear LCFA or LCFA-CoA uptake, and whether or

---

TABLE 10.5
Partial List of Commercially Available Dyes for Nuclear Localization

Name	Abs (nm)	E_m (nm)	ϵ_{max} (cm^{-1} M^{-1})	QY	FI Increase
DAPI	358	461	24,000	0.34	20
Hoechst 33342	350	461	40,000	0.42	30
Hoechst 33258	352	461	45,000	0.38	10
YOYO-1	491	509	98,900	0.52	100–1000
TOTO-1	514	533	117,000	0.34	100–1000
JOJO-1	529	545	171,400	0.44	100–1000
YOYO-3	612	631	167,000	0.15	100–1000
TOTO-3	642	660	154,100	0.06	100–1000
SYTO 11	508	527	75,000	0.49	ND
SYTO 16	488	518	42,000	0.65	ND
SYTO 42	433	460	31,000	ND	370
SYTO 45	455	484	43,000	ND	4660
SYTO 59	622	645	112,000	0.18	ND
SYTO 62	652	676	76,000	0.27	ND
SYTO 63	657	673	119,000	0.17	ND
SYTO 64	599	619	84,000	0.39	ND
SYTO 83	543	559	68,000	ND	ND
SYTO 84	567	582	95,000	ND	ND

Note: Abs, absorbance maxima; E_m, emission maxima; ϵ_{max}, extinction coefficient maxima; QY, quantum yields; FI, fold increase in fluorescence intensity upon DNA staining; ND, not determined.

not cytoplasmic lipid binding proteins affect these processes. Again, both cultured fibroblast cells overexpressing lipid binding proteins as well as cultured primary hepatocytes from mice lacking one or more of the cytoplasmic lipid binding proteins can be examined.

A monolayer of cells grown on chambered cover glass slides is first incubated with a nucleic acid stain (often in the range of 0.1–1 μM) for 20–40 min to allow the dye to reach the nucleus [30] and background fluorescence images are acquired. A nuclear stain should be chosen, whose excitation wavelength does not overlap the excitation wavelength of the fluorescent lipid; this decreases background effects and helps to differentiate which fluorescence signal is from which fluorophore. The amount of stain and time for incubation needs to be determined experimentally as it will vary, depending upon the cell line, confluency of cells, and the particular nucleic acid stain being used. Next, cells are incubated with a nonsaturating level of fluorescent lipid, preferably one which is not metabolized, and multiple digital images of the fluorescent signals are acquired over a given time frame with a laser scanning system [29,30,41,42]. Signals from each fluorophore are collected in separate channels, and the images are later merged for colocalization determination (Figure 10.6).

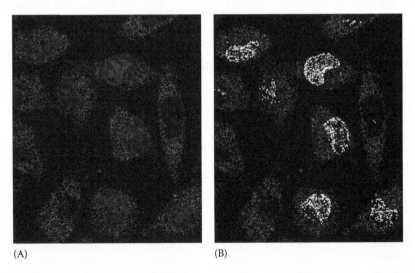

(A)   (B)

FIGURE 10.6 (See color insert following page 108.) Fatty acid nuclear localization in cultured L-cells. L-cells were seeded onto chambered cover glass, culture media was replaced with PBS containing 1.25 μM of SYTO59 nucleic acid stain, and cells were incubated at room temperature for 30 min. Background images were obtained, followed by the addition of 100 nM BODIPY C-16, and images were acquired 30 min postincubation on 0.3 μm confocal slices with a X63 Plan-Fluor oil immersion objective, N.A.1.45, an Axiovert 135 microscope (Zeiss, Carl Zeiss Inc., Thornwood, New York), and MRC-1024 fluorescence imaging system (Bio-Rad, Hercules, California). Syto59 and BODIPY-C16 probes were excited at 488/568 lines with a krypton-argon laser (Coherent, Sunnyvale, California), and emission was recorded by a photomultiplier after passing through a 522/D35 or 680/32 emission filter, respectively. (A) Merged images of BODIPY C16 (green) and SYTO59 (red) fluorescence. (B) Colocalized pixels representing BODIPY C-16 nuclear localization are shown in white.

The amount of LCFA or LCFA-CoA in the nucleus is determined by the percent of fluorescent LCFA/LCFA-CoA pixels colocalizing with pixels from the nuclear stain, and several programs (i.e., LaserSharp, Bio-Rad, Hercules, California; MetaMorph Image Analysis, Advanced Scientific Imaging, Meraux, Louisiana; ImageJ, NIH, Bethesda, Maryland) are available to determine these values [30,64]. Fluorescence intensity expressed in gray scale units is then used to plot the increase in intensity over time. Linear regression analysis is used to define the linear portion of the uptake curves and to obtain the rate of uptake [29,30].

10.4.4 Nuclear Localization of Cytoplasmic Lipid Binding Proteins

Three distinct fluorescent methods have been utilized to track nuclear transport of cytoplasmic lipid binding proteins. These include: (1) fusion of a fluorescent protein (e.g., green fluorescence protein, GFP; yellow fluorescence protein, YFP) to the protein of interest through DNA conjugation and genetic modification of cells or organisms, (2) addition of recombinant fluorescently labeled proteins to cells, and (3) immunofluorescent labeling of the protein of interest in fixed cells or tissues. Within each of these methods, several techniques or variations, as well as fluorescent moieties, may be used. In each case, fluorescence microscopy is needed to observe the fluorescence localization, and many of these techniques are used in conjunction with a nuclear stain. A list of techniques used to determine the nuclear localization of cytoplasmic lipid binding proteins, the lipid binding protein observed, the fluorophore used, and the effect of adding ligands is provided in Table 10.6.

10.4.4.1 Fusion of Fluorescent Expression Proteins

Perhaps one of the most commonly used techniques for determining protein localization has been to fuse the protein of interest to a second protein, which is fluorescent, such as the GFP from *Aequorea victoria* (jellyfish). The cDNA encoding this protein was originally cloned 15 years ago [65,66]. Since that time, numerous variants have been developed, resulting in a wide spectrum of excitation and emission maximums, species-specific codon bias, increased fluorescence stability, a variety of pH sensitivities, increased protein solubility, and decreased photobleaching. A plethora of these GFP variants are commercially available through Clontech in predesigned mammalian expression vectors. In most instances, the gene of interest is cloned through standard molecular biology techniques and inserted into the commercially available GFP construct in the proper orientation to ensure an open reading frame through both the protein of interest and the GFP gene, thus ensuring a fusion of the two proteins upon transcription [67]. This new construct can then be transfected into cells or used in the creation of transgenic animals and examined through fluorescence microscopy with a microscope fitted with an FITC filter [67] or through LSCM/MPLSM as previously described herein. LSCM/MPLSM allows for the use of a nuclear stain in conjunction with the fusion protein and determination of colocalization between the two fluorophores. Such techniques have been used to examine nuclear localization of A-FABP [68,69] and K-FABP [69] in cultured cells, both in the presence and absence of FABP ligands.

TABLE 10.6

Techniques Used to Determine Nuclear Localization of Cytoplasmic Lipid Binding Proteins and the Effect of Adding Ligands

Technique	Protein	Fluorophore	Effect of Ligands	Ref.
GFP variants	A-FABP	GFP	Increase protein in nucleus	[68,69]
GFP variants	K-FABP	GFP	Increase protein in nucleus	[69]
Chemical labeling with fluorescent tag	ACBP	Rhodamine derivative	Not determined	[70]
Chemical labeling with fluorescent tag	ACBP	Fluorescein and rhodamine derivative	Not determined	[71]
Chemical labeling with fluorescent tag	ACBP	Cy5	Not determined	[83]
Chemical labeling with fluorescent tag	L-FABP	Fluorescein	Increases protein association with nuclei	[5]
Immunofluorescence	ACBP	Fluorescein	Not determined	[85]
Immunofluorescence	ACBP	FITC	Not determined	[7,86]
Immunofluorescence	A-FABP	AF488	Strongly increased protein in nucleus Almost exclusively nuclear	[87]
Immunofluorescence	ALBP	FITC	TTA no effect	[86]
Immunofluorescence	KLBP	FITC	TTA no effect	[86]
Immunofluorescence	L-FABP	FITC	Not determined	[30]
Immunofluorescence	L-FABP	FITC	No effect when expressed in adipocytes	[76]

10.4.4.2 Recombinant Proteins Covalently Labeled with Small Fluorescent Tags

More recently, smaller fluorescent tags have been utilized. In this case, the purified recombinant protein is directly conjugated to a fluorescent dye, usually less than 1 kDa in size. Several such dyes, such as the Cy3/Cy5 dyes (Amersham Biosciences, Piscataway, New Jersey; Invitrogen, Carlsbad, California) and the AlexaFluor® dyes (Invitrogen), are commercially available in a protein labeling kit. The Cy3/Cy5 protein labeling kits supplied by Amersham utilize a bisfunctional NHS-ester to label free amine groups on the protein, while the AlexaFluor dyes have either a succinimidyl (NHS) ester or TFP ester moiety that reacts with the primary amines of the protein. Recently, work has been done to create new fluorescent dyes for protein labeling and nuclear imaging [70,71]. With each of these fluorophores, the protein to be labeled needs to be at a concentration of approximately 1 mg mL^{-1} in 50 mM phosphate buffer, pH 7.0–7.3. For best labeling results, the protein buffer should not contain any primary amine groups, such as TRIS or glycine, or carrier proteins, as these will inhibit the labeling efficiency. The protein and dye are incubated in a weak basic buffer such as sodium carbonate, pH 9.3, at room temperature for 30 min to 2 h, and the reactive dye molecules allowed to interact with the protein. The actual time required for the labeling reaction depends on the kit used and the size of the protein to be labeled. Size exclusion gel chromatography is used to separate unbound dye from protein-bound dye, and the dye-to-protein ratio is determined by absorbance and mass spectroscopy [64,70,71]. Cells are seeded onto chambered cover glass slides, and the labeled proteins are then added to the cells in culture media in the presence of Pep-1, a small 21-amino acid peptide, commercially available under the name Chariot® (Active Motif, Carlsbad, California), which aids in the internalization of proteins [72,73], and the cells are allowed to take up the protein at 37°C for up to 2 h [71,74]. Prior to confocal microscopy, cells are washed and incubated with a nucleic acid stain as described above. Fluorescent images are obtained and analyzed as described for colocalization of LCFA/LCFA-CoA and nuclear dyes [30,64,74,75].

10.4.4.3 Immunofluorescence Labeling of Fixed Cells or Tissues

Although the use of fluorescently tagged proteins allows for real-time visualization in live cells, both through fusion proteins and tags on recombinant proteins, both of these techniques require the use of artificially labeled proteins. The use of immunofluorescence in fixed cells allows for visualization of native proteins and confirmation that the genetically or chemically tagged proteins distribute similar to the respective native proteins in the cell. Cells are grown on chambered glass slides under the conditions of the assay, i.e., in the presence or absence of endogenous ligands. Cells are fixed by one of several protocols; for example, with an acetone:ethanol mixture [7], a paraformaldehyde solution [30,74,76], or a methanol solution [30]. For the acetone:ethanol fixation method, cells are incubated in a 70% acetone and 30% ethanol mixture for 30 min at −20°C, which both fixes and permeabilizes the cells in a single step. The other two methods require an additional permeabilization step following the fixation step to allow the antibodies to pass into the cells. For the other two methods, cells can be fixed by incubation in a 4% formaldehyde solution in phosphate buffered saline (PBS) at room temperature for 15–30 min, or the cells can be fixed in methanol for 5 min at −20°C and rehydrated by 57 mM borate buffer, pH 8.2, prior to subsequent steps. In some instances, inclusion of a small amount of glutaraldehyde may be helpful for fixation. Following fixation, cells are usually permeabilized with a weak detergent solution, such as 0.1% Triton-X100 in PBS, at room temperature for 5 min. Additional time in the permeabilization solution can lead to overpermeabilization of the cells and nonspecific binding of antibodies. Regardless of the fixation method used, after fixation and permeabilization, cells are washed several times with PBS and incubated in a blocking solution from 1 h to overnight to decrease the amount of nonspecific protein binding. A common blocking solution is a mixture of 5%–10% fetal bovine serum or albumin in PBS. After blocking, the cells are incubated with the primary antibody or antibodies diluted in a weak (0.5%–1%) blocking solution. Although the appropriate concentration of antibodies and the time required for the incubation are dependent upon the antibody titration and upon the amount of protein in the sample, the primary antibodies used for immunofluorescence are often used at a higher concentration than those for Western blot. Many antibody manufacturers supply a recommended dilution factor for immunofluorescence with their antibodies. In general, incubations are overnight at 4°C, 2 h at room temperature, or 1 h at 37°C. Next, cells are washed several times with PBS to remove unbound antibody and then incubated with a secondary antibody, which has been conjugated to a fluorophore (FITC, AlexaFluor, etc.), many of which are commercially available. Often the secondary antibody is diluted in the weak blocking solution, as used for the primary antibody, protected from the dark, and incubated at room temperature for an hour. The concentration of the secondary antibody must be determined experimentally, such that adequate labeling is noted, with negligible background fluorescence; this is dependent upon both the fluorophore and the antibody. Cells are washed several times with PBS to remove unbound secondary antibody. Prior to imaging, cells can also be stained with a nucleic acid stain to enable nuclear colocalization as described above [30]. Once cells have been stained, a solution such as SlowFade® (Invitrogen, Carlsbad, California) is added to preserve the fluorescent signal, and a cover slip is mounted over cells. The SlowFade medium can also be purchased to contain DAPI, a fluorescent nuclear marker, such that nuclear staining and fluorescence preservation can be accomplished in a single step. Fluorescent images are obtained and analyzed as described for colocalization of LCFA/LCFA-CoA and nuclear dyes [30,64].

A variety of organelle markers exist, and several of these markers have been used to examine the intracellular distribution of ACBP in fixed rat hepatoma cells (Figure 10.7). Since rat hepatoma cells produce high amounts of ACBP, this cell line allows examination of endogenous ACBP in a physiological environment. Cells were prepared as described above and double fluorescence immunolabeled with antibodies to ACBP in combination with antibodies to each of the following organelle markers: cholera toxin B marker (PM caveolae/lipid raft marker, panel A), wheat germ agglutinin (WGA, Golgi marker, panel B), cathepsin D (lysosomal marker, panel C), mitochondrial heat shock protein 70 (HSP70, mitochondrial marker, panel D), concanavalin A (ER marker, panel E), and Hoechst 33342 (nuclear marker, panel F). LSCM and colocalization analysis showed that ACBP was specifically distributed throughout cells, with the highest density in the perinuclear cytoplasm and lower but distinct distribution within nuclei and plasma membranes. Similar methods have been used to examine the intracellular distribution of SCP-2 and pro-SCP-2 in fixed and live cells [74,75].

10.4.5 Pros and Cons

Although each of these techniques has its own internal flaws, these methods provide the opportunity to observe live cells in real time as they uptake lipids and/or cytoplasmic lipid binding proteins and transport them to the nucleus. Through colocalization techniques, nuclear localization of both lipids and proteins can be determined

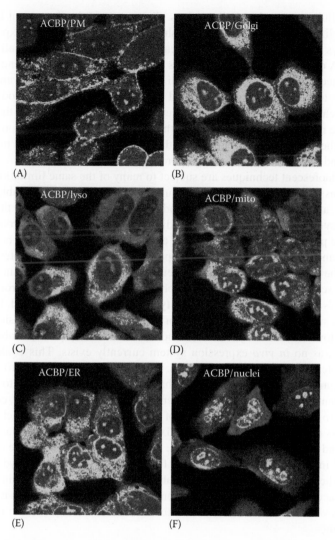

(A)   (B)

(C)   (D)

(E)   (F)

FIGURE 10.7 (See color insert following page 108.) Intracellular distribution of endogenous ACBP in fixed rat hepatoma cells by indirect immunofluorescence confocal microscopy. T-7 rat hepatoma cells were seeded onto chambered slides, the culture media was removed, and cells were fixed with 4% glutaraldehyde at 37°C for 30 min. Residual aldehyde was quenched with ammonium chloride, and nonspecific binding was blocked by incubation in 5% FBS in Hank's solution. Primary antibodies raised against ACBP were produced locally in rabbit or rat, while all other primary and secondary antibodies were commercially purchased. Images were acquired with an X63 Plan-Fluor oil immersion objective, N.A.1.45, an Axiovert 135 microscope (Zeiss, Carl Zeiss, Inc., Thornwood, New York), and MRC-1024 fluorescence imaging system (Bio-Rad, Hercules, California). Colocalization of ACBP with: (A) cholera toxin B (PM caveolae/lipid raft marker), (B) wheat germ agglutinin (Golgi marker), (C) cathepsin D (lysosomal marker), (D) mitochondrial heat sock protein 70 (mitochondrial marker), (E) concanavalin A (ER marker), (F) Hoechst 33342 (nuclear marker). Yellow pixels represent colocalized pixels.

without cellular or membrane disruptions. Since most of the fluorescent lipids are nonmetabolizable, especially within the short time frame of imaging experiments, these assays also distinguish LCFA nuclear internalization from internalization of LCFA-CoA or other fatty acid metabolites. Further, these assays differentiate binding to the nuclear membrane from internalization into the nuclear matrix and resolve effects due to ligand binding by the lipid binding proteins. Finally, modifications of these techniques can be utilized to determine whether ligands enter the nucleus through the nuclear pore as a complex with lipid binding proteins or through dissociation of the ligand from the complex at the nuclear envelope, followed by passage of the ligand alone through the nuclear pore to the nuclear receptor.

These fluorescent techniques are subject to many of the same limitations previously discussed. There are only a small number of commercially available fluorescent lipids, and although they resemble endogenous ligands, most of the fluorescent lipids are synthetic. Furthermore, the addition of a fluorophore to a protein could prevent the protein from folding properly and may cause the protein to behave differently than the native protein. This is especially an issue for small proteins. Since many of the cytosolic lipid binding proteins are small (e.g., 15 kDa), the addition of a fluorescent coexpressed protein, such as GFP (27 kDa), could strongly alter the protein's expression level, functionality, or ability to be transported. For small proteins, a smaller tag such as the Cy3 or Cy5 dyes may be more physiologically relevant. However, currently, cells must be forced to take up proteins labeled with these dyes, as no *in vivo* expression system currently exists. This could lead to altered protein levels and/or localizations. Thus, labeled proteins need to be examined for ligand affinity, conformational changes, and their localization to the same compartments as unlabeled proteins. When using a fixed cell procedure, conditions will need to be optimized for each set of antibodies. Although a variety of fixation buffers and procedures exist, some fixation procedures (e.g., methanol) allow proteins and lipids to move after fixation. Thus, while potentially more disruptive to the cell integrity, fixatives containing formaldehyde or glutaraldehyde are better since the fixed proteins are more stable (i.e., unable to move) than with the methanol/acetone fixative. By utilizing one of the live cell methods with labeled protein, as well as confirmation by the fixed cell method, many of these issues can be overcome.

10.5 LCFA/LCFA-CoA INTERACTIONS WITH NUCLEAR RECEPTORS

10.5.1 INTRODUCTION

Once ligands enter the nucleus, they must interact with nuclear receptors to affect gene regulation. To begin to understand such interactions, the affinity of nuclear receptors for LCFA and LCFA-CoA were initially examined *in vitro*. Recombinant protein techniques similar to those described for cytoplasmic binding have been used to determine that nuclear receptors bind these lipids with high affinity. Such assays include: (1) direct binding of fluorescent fatty acids or fatty acyl-CoA, (2) displacement of fluorescent fatty acids or fatty acyl-CoA by nonfluorescent fatty acids or fatty acyl-CoA, (3) quenching of intrinsic protein fluorescence upon ligand binding, and (4) FRET from the intrinsically fluorescent donor protein to a fluorescent ligand acceptor. Again, the fluorescent moiety is excited and the resulting emission spectra are obtained in the presence and absence of protein or ligand.

10.5.2 Direct Binding Assays of Fluorescent Lipid Analogues

A variety of fluorescent lipids have been examined for their ability to directly bind to nuclear receptors. These include both natural and synthetic compounds, as shown in Table 10.7. These assays are conducted much the same as those for direct binding assays of fluorescent lipid analogues with the cytoplasmic lipid binding proteins. Briefly, a very small amount (100 nM or less) of recombinant protein is titrated with increasing concentrations of fluorescent ligand, and emission spectra of the fluorophore are measured upon excitation. Maximal fluorescence intensities are used to determine the ligand binding affinity (K_d) and the number of binding sites (n) after background correction for solvent effects, protein fluorescence, and fluorescent intensities of the fluorophore in an aqueous environment [23,24,26,77,78]. Perhaps the most significant difference between these binding assays and those of the lipid binding proteins is the need to utilize small concentrations of the recombinant protein to ensure that the protein remains in solution. Further, the addition of a fluorescent moiety to LCFA or LCFA-CoA can decrease their solubility in aqueous environments (i.e., BODIPY), thus requiring such assays to be determined in a very low concentration range [30]. Although these fluorescent lipids do not necessarily bind with the same affinity as their endogenous counterparts [23], they provide a basis to determine whether LCFA or LCFA-CoA analogues can bind to the nuclear receptor.

TABLE 10.7
Ligand Binding Affinity of Several Nuclear Receptors for Fluorescent Lipid Analogues

	HNF-4α	PPARα	PPARγ
	K_d (nM)	K_d (nM)	K_d (nM)
BODIPY C-12		7.1 ± 0.4 (Figure 10.8A)	
BODIPY C-12-CoA		22.2 ± 3.2 (Figure 10.8B)	
BODIPY C-16		8.3 ± 1.2 [24]	
BODIPY C-16-CoA		19.9 ± 1.7 [88]	
Cis-parinaric acid	421–589 [77]	— [23]	669 ± 75 [89]
Cis-parinaroyl-CoA	238–759 [77]	13.3 ± 3.7 [23]	
Trans-parinaric acid		30.0 ± 6.2 [26]	49.6 ± 1.2 [69]
NBD-stearate	93 [77]	18.7 ± 3.2 [23]	

Note: Ligands that resulted in no binding in the examined range and for which a K_d could not be calculated are shown as "—."

Further, even if these fluorescent lipids bind more weakly than their endogenous nonfluorescent counterparts, they can still be used in displacement assays that do allow for the determination of the ligand affinity of endogenous lipids.

10.5.3 DISPLACEMENT ASSAYS

Although several of the fluorescent lipids from Table 10.7 have been used to examine the binding affinity of nuclear receptors for endogenous LCFA and LCFA-CoA through displacement methods (Table 10.8), the more soluble lipids have been the most useful for such techniques. Unfortunately, the low solubility of the BODIPY compounds in aqueous buffers has limited their use in displacement assays, and such assays are only possible with very low concentrations of BODIPY compound. Although there has been some debate as to the optimal concentration of nuclear receptor and fluorescent lipid to use [79], the general consensus is to keep both concentrations as low as possible to obtain a strong signal upon ligand binding and that the amount of fluorescent lipid should be equal to or less than the concentration of the protein [23,26]. Maximal fluorescence intensities are obtained from protein bound fluorescent fatty acid or fatty acyl-CoA, the resulting complex is titrated with increasing concentrations of nonfluorescent lipid until the effects plateau, and the decrease in fluorescence intensity as the fluorescent lipid is displaced is used to calculate the affinity of the nonfluorescent lipid [23,26]. The ligand affinity (K_I) of the endogenous, nonfluorescent lipid can be calculated based upon the EC_{50} of the

---

TABLE 10.8

Ligand Affinity of Nuclear Receptors for Endogenous Fatty Acids and Fatty Acyl-CoA Determined by Quenching of Intrinsically Fluorescent Amino Acids within the Protein upon Ligand Binding (K_d) and Displacement of a Fluorescent Lipid (K_I)

	HNF-4α	PPARα	
	K_d (nM)	K_d (nM)	K_I (nM)
C16:0		— [23]	— [23,26]
C18:0			Weak [26]
C18:2			4.8 ± 0.9 [26]
C20:4	742 [77]	19.7 ± 4.5 [25]	17.3 ± 3.8 [26]
C16:0-CoA	1.7 [77]	8.4 ± 0.7 [23]	1.1 ± 0.3 [23]
C18:0-CoA	3.8 [77]	14.0 ± 0.9 [23]	4.2 ± 1.1 [23]
C18:2-CoA	4.4 [77]	4.3 ± 0.3 [23]	1.0 ± 0.3 [23]
C20:4-CoA	4.0 [77]	3.5 ± 0.4 [23]	2.0 ± 0.5 [23]

Note: Ligands that resulted in no binding in the examined range and for which a K_d could not be calculated are shown as "—."

ligand obtained from the displacement assay ($EC_{50,ligand}$), the concentration of the fluorescent lipid used ([fluorescent]), and the precalculated affinity of the nuclear receptor for the fluorescent lipid ($K_{d,fluorescent}$) by the equation [23,26]:

$$EC_{50,ligand} / [\text{fluorescent}] = K_I / K_{d,fluorescent} \tag{10.20}$$

Such assays show that even if fluorescent LCFA/LCFA-CoA analogues bind with different affinity than endogenous ligands, they are still useful for determining the ligand affinity of endogenous, nonfluorescent ligands. Furthermore, values obtained for ligand affinity by displacement assays are similar to those obtained by direct methods (see Section 10.5.4), as shown in Table 10.8.

10.5.4 QUENCHING OF INTRINSICALLY FLUORESCENT AMINO ACIDS

One technique that has been especially useful in circumventing potential problems of fluorescent ligands, especially with nuclear receptors, is to determine affinities of endogenous LCFA and LCFA-CoA based on the quenching of intrinsically fluorescent amino acids in the nuclear receptor. The larger size of nuclear receptors as compared to the cytoplasmic binding proteins increases the probability of the nuclear receptor containing intrinsically fluorescent amino acids (tyrosine and tryptophan), whose fluorescence is altered upon ligand binding [80]. Both the HNF-4α and the peroxisome proliferator-activated receptor α (PPARα) contain several such intrinsically fluorescent residues. This has enabled direct examination of nuclear receptor binding to endogenous, nonfluorescent LCFA and LCFA-CoA and requires no synthetic compounds or competition assays (Table 10.8). Small concentrations of the recombinant nuclear receptor (100 nM) are titrated directly with the lipid of interest, and the resulting fluorescence emission is measured upon UV excitation of the tyrosine and/or tryptophan residues [23,25,77,78]. The data are analyzed by plotting the change in fluorescence intensity by the ligand concentration and utilizing the equations described for direct fluorescent lipid binding of cytoplasmic lipid binding proteins [23,25,77,78].

An example of this technique is shown in Figure 10.8. Due to the poor solubility of the BODIPY analogues in an aqueous buffer and the subsequent formation of micelles, direct excitation of BODIPY was complicated by the need to distinguish the proportion of increased fluorescence emission upon ligand binding from the increased fluorescence emission noted upon micelle formation. In order to circumvent these issues, tyrosine quenching of the intrinsic PPARα fluorescence was utilized to determine the binding affinity of PPARα for BODIPY C12 fatty acid (Figure 10.8A and B) and fatty-acyl CoA (Figure 10.8C and D). In each case, the emission spectrum of PPARα protein is measured from 300 to 380 nm upon excitation of the tyrosine molecules at 280 nm, both in the absence and presence of increasing concentrations of BODIPY C12 fatty acid or fatty acyl-CoA. In this example, 100 nM PPARα protein was titrated with 0, 2.5, 5, 10, 15 20, 25, 37.5 50, 62.5, 75, 100, 125, 250, and 500 nM, and emission spectra obtained. After background subtraction of buffer only, the resulting spectra were used for ligand affinity calculations. Although 15 concentrations of each

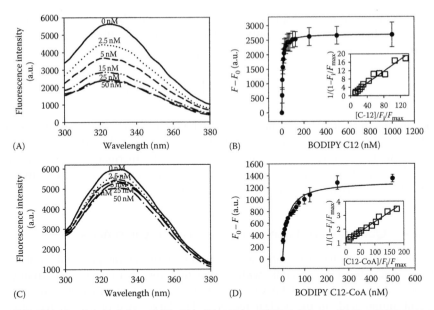

FIGURE 10.8 The nuclear receptor PPARα binds BODIPY C12 fatty acid (A, B) and fatty acyl-CoA derivative (C, D) with high affinity. Recombinant PPARα protein (100 nM) was excited at 280 nm in the absence and presence of increasing concentrations of BODIPY C12 fatty acid (A, B) or fatty acyl-CoA (C, D), and the resulting emission spectra measured from 300–380 nm in a PC1 photon counting spectrofluorometer (ISS, Inc., Champaign, Illinois). Each spectrum was corrected for background fluorescence, and a small subset of these spectra is presented for titration of the fatty acid (A) and the acyl-CoA (C). The maximal intensities (emission peak at 322 nm) for each concentration of BODIPY C12 were used to produce a graph of the change in fluorescence intensity ($F - F_0$) as a function of fatty acid (B) or fatty acyl-CoA (D) concentration, fit to the ligand binding equation for one site saturation $y = B_{max}x/(K_d + x)$ (solid line), and used to calculate the dissociation constant (K_d) shown in Table 10.7. Insets, linear plot of the binding curve from each panel.

BODIPY analogue were analyzed, only a limited number of the acquired spectra are presented in Figure 10.8A (fatty acid) and C (fatty acyl-CoA) for simplicity. Maximal fluorescence intensity (emission peak at approximately 322 nm) for each concentration shown in Figure 10.8A gives the following values: 0 nM = 5626, 2.5 nM = 4444, 5 nM = 3679, 15 nM = 2870, 25 nM = 2537, and 50 nM = 2385. Using these values, the change in fluorescence intensity ($F - F_0$) in arbitrary fluorescence units (a.u.) would be equal to: 0 nM = 0, 2.5 nM = 1182, 5 nM = 1947, 15 nM = 2756, 25 nM = 3089, and 50 nM = 3241. By plotting the change in fluorescence intensity ($F - F_0$) on the y-axis and the BODIPY C12 fatty acid concentration on the x-axis for all values and replicates, a saturable curve was obtained (filled circles—Figure 10.8B). This curve was fit (solid line—Figure 10.8B) in Sigma Plot 8.0 (SPSS, Inc., Chicago, Illinois) to the ligand binding for one site saturation equation

$$y = \frac{B_{max}\,x}{K_d + x} \tag{10.21}$$

and gave a $K_d = 7.1 \pm 1.6\,nM$. By transforming the data to a linear least-square plot of $1/(1 - \Delta F/\Delta F_{max})$ versus $C_L/(\Delta F/\Delta F_{max})$ as previously described herein (Figure 10.8B, inset), a similar binding affinity is obtained $K_d = 7.1 \pm 0.4\,nM$. Although the effect of BODIPY C12-CoA on the PPARα fluorescence spectra was weaker (Figure 10.8C) than that for the fatty acid (Figure 10.8A), fitting the data to the ligand binding equation for one site saturation (Figure 10.8D) showed that the CoA thioester was still strongly bound ($K_d = 22.4 \pm 3.2$), albeit with weaker affinity than the fatty acid.

10.5.5 FLUORESCENCE RESONANCE ENERGY TRANSFER

The higher probability of the nuclear receptors containing intrinsically fluorescent amino acids has also enabled determination of the molecular distance between such amino acids and fluorescent lipids, especially *cis*-parinaroyl-CoA, through FRET. For the most part, the technique and data analysis conducted are the same for nuclear receptors [23,77] as for the cytoplasmic binding proteins [49,61]. These values can be used to determine the ligand binding affinity (K_d) and the number of binding sites (n) under FRET conditions by using the equations described in the section on direct binding of fluorescent lipids. However, since the likelihood of the nuclear receptors undergoing a change in fluorescence emission upon ligand binding due to conformational changes is high, the values calculated for the quenching of tyrosine or tryptophan residues may be different than those calculated for the increase in sensitized emission. The FRET values can be corrected for protein conformational changes through the equation:

$$E = \left(\frac{G(\lambda_2)}{G(\lambda_1)} - \frac{\varepsilon_A(\lambda_2)}{\varepsilon_A(\lambda_1)} \right) \times \left(\frac{\varepsilon_A(\lambda_1)}{\varepsilon_D(\lambda_2)} \right) \qquad (10.22)$$

where

E is the energy of transfer

$G(\lambda)$ is the magnitude of the corrected excitation spectrum of the energy acceptor

ε_A and ε_D are the extinction coefficient of the acceptor and donor [77,81]

10.5.6 PROS AND CONS

These techniques offer many advantages over nonfluorescent techniques as discussed in the section on ligand binding of cytoplasmic lipid binding proteins. The ability to study ligand binding with very low concentrations of protein is even more important for nuclear receptors than for lipid binding proteins due to the poorer solubility of the nuclear receptors. Moreover, the quenching of intrinsic protein fluorescence is the most direct method for studying nuclear receptor binding since it requires no extraneous or artificial substances, and as such, provides the most accurate values.

As with the lipid binding proteins, there are some limitations to these assays and a combination of techniques may be used to ensure accurate results.

10.6 PROTEIN–PROTEIN INTERACTIONS: CYTOPLASMIC LCFA/LCFA-CoA BINDING PROTEINS DIRECTLY BIND NUCLEAR RECEPTORS

10.6.1 INTRODUCTION

Data demonstrating the presence of lipid binding proteins in the nucleus, as well as ligand binding data showing that both lipid binding proteins and nuclear receptors bind LCFA and LCFA-CoA, suggested that lipid binding proteins might function to shuttle ligands through the cytoplasm and into the nucleus for interaction with nuclear receptors. Such data also suggested that lipid binding proteins might directly interact with nuclear receptors. Several of the techniques described within this review have been modified to allow examination of direct protein–protein interactions *in vitro*, not only confirming findings in fixed and living cells but, in addition, yielding direct binding affinities and stoichiometries. While labeling of recombinant proteins with fluorescent moieties has allowed direct protein–protein interactions to be examined in solution, similar techniques have also been used to determine protein–protein interaction within the nuclei of cells. Furthermore, many of these techniques have been used to study a variety of protein–protein interactions, such as nuclear receptor heterodimerization [82], transferase and lipid binding protein interactions [74,75], nuclear receptor and cofactor interactions [83], as well as lipid binding protein and nuclear receptor interactions (Table 10.9).

10.6.2 FLUOROPHORE CHOICES

As described earlier in this review, small fluorescent tags can be conjugated to recombinant proteins or fixed cells can be immunofluorescently labeled. For confocal

---

TABLE 10.9
Lipid Binding Protein Interaction with Nuclear Receptors

Lipid Binding Protein	Nuclear Receptor	Fluorophore	Technique	Ref.
L-FABP	PPARα	Not given	Colocalization	[84]
ACBP	HNF-4α	Texas red and FITC Cy3 and Cy5	Colocalization, FRET, EM	[7]
L-FABP	PPARα	Not given	Colocalization	[30]
L-FABP	PPARα	Cy3 and Cy5	Spectroscopy, FRET	McIntosh et al. (2008)
L-FABP	PPARα	AF488 and AF594	Spectroscopy, FRET	Figure 10.9

microscopy and colocalization, two fluorescent moieties are chosen in which the excitation of one does not result in the excitation of both. For both spectroscopy- and microscopy-based FRET assays, two fluorophores are chosen such that there is a modest degree of overlap between the emission spectrum of the acceptor and excitation spectrum of the donor, allowing for energy transfer from the donor to excite the acceptor. The more commonly used FRET pairs have been Cy3/Cy5 and FITC/TRITC, and secondary antibodies are readily available with these fluorescent conjugations.

10.6.3 Spectroscopy FRET in Vitro

By utilizing the FRET methods previously described for confirming fluorescent LCFA-CoA binding to nuclear receptors and binding proteins, the distance between the fluorescent moieties on one protein and the fluorescent moieties on a second protein can be determined. Recombinant pure proteins can be fluorescently labeled as previously described herein. The fluorescence intensity of the donor fluorophore-labeled protein (excitation wavelength more toward the blue spectrum) is first measured in the absence of the acceptor fluorophore-labeled protein (excitation wavelength more toward the red spectrum). Next, the donor protein is titrated with increasing concentrations of acceptor-labeled protein and fluorescence spectra of each fluorophore recorded. As FRET occurs, the fluorescence intensity of the donor will quench while the acceptor will have increased sensitized emission (Figure 10.9). The fluorescence intensity of the fluorophore-labeled acceptor needs to be measured at each titration concentration in the absence of donor. These spectra can then be subtracted from the spectra of the donor titrated with acceptor to correct for any excitation of the acceptor that may be a direct result of excitation of the donor fluorophore. The affinity of these two proteins for each other, as well as the distance between them, can be calculated from both, the decrease in fluorescence intensity of the donor and the increase in fluorescence intensity of the acceptor, using the equations presented in the section on Binding in the cytoplasm and FRET [64,74,75]. However, the binding affinity and intermolecular distance between the fluorophores is often calculated using the decrease in fluorescence intensity of the donor, rather than the increase in fluorescence intensity of the acceptor, due to difficulties in obtaining accurate measurements of the extinction coefficients and the requirement of complete donor labeling [60].

For example, recombinant L-FABP protein was fluorescently labeled with the AlexaFluor488 (AF488) dye (Invitrogen, Carlsbad, California) and recombinant PPARα protein was fluorescently labeled with the AlexaFluor594 (AF594) dye (Invitrogen, Carlsbad, California). The AF488-labeled L-FABP protein served as the donor and the AF594-labeled PPARα served as the acceptor, and R_0 is equal to 60 Å (Invitrogen). Emission spectra of 25 nM AF-488-L-FABP were obtained from 505–650 nm upon excitation at 494 nm, both in the absence and presence of AF594-PPARα and corrected for background fluorescence (Figure 10.9A and C). Maximal fluorescence intensity of the donor (emission peak at 520 nm) for each concentration shown in Figure 10.9A gives the following values: 0 nM = 54,123, 4 nM = 53,505, 10 nM = 53,223, and 50 nM = 53,286. Using these values, the change

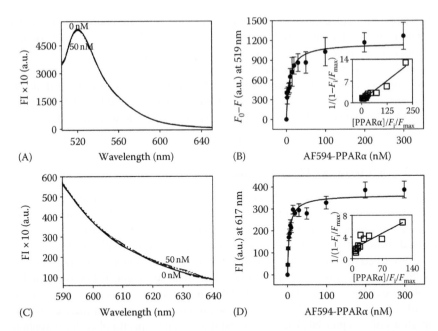

FIGURE 10.9 FRET from donor AF488-labeled L-FABP to acceptor AF594-labeled PPARα protein. AF488-labeled L-FABP (10 nM) was titrated with increasing concentrations of AF594-labeled PPARα protein, and FRET was detected as quenching of AF488 emission (A, B) and as the appearance of AF594 sensitized emission (C, D). (A) Emission spectra of AF488 upon excitation at 490 nm. (B) Plot of the average change in maximal fluorescence emission ($F_0 - F$) at 519 nm upon excitation at 490 nm as a function of AF485-labeled PPARα. Values represent the mean ± SE, $n = 4$. Inset, linear plot of the binding curve in (B). (C) Emission spectra showing a slight increase in AF495 fluorescence intensity upon excitation of AF488. (D) Plot of the average increase in fluorescence emission at 617 nm upon excitation at 490 nm as a function of AF594-labeled PPARα protein concentration. Values represent the mean ± SE, $n = 4$. Inset, linear plot of the binding curve in (D).

in fluorescence intensity ($F - F_0$) in arbitrary fluorescence units (a.u.) would be equal to: 0 nM = 0, 4 nM = 618, 10 nM = 900, and 50 nM = 837. The change in fluorescence intensity ($F - F_0$) versus the concentration of AF594-PPARα gives a saturable curve (filled circles—Figure 10.9B). This curve was fit (solid line—Figure 10.9B) to Equation 10.21 $y = ((B_{max}x)/(K_d + x))$ as previously described herein and gave a $K_d = 8.2 \pm 2.1$ nM. By transforming the data to a linear least-square plot of $1/(1 - \Delta F/\Delta F_{max})$ versus $C_L/(\Delta F/\Delta F_{max})$ as previously described herein (Figure 10.9B, inset), a similar binding affinity is obtained. Using the values above, the efficiency of energy transfer would be equal to $E = 1 - F_{DA}/F_D = 1 - (53,286/54,123) = 0.0155$ or 1.55%. Substituting this value into the equation for determining the distance between the two fluorophores, $R_{2/3} = ((R_0^6/E) - R_0^6)^{1/6} = ((60.0^6/0.0155) - (60.0^6))^{1/6}$ or 120 Å. Similarly, maximal fluorescence intensity of the acceptor (emission peak at 618 nm) for each concentration shown in Figure 10.9C gives the following values: 0 nM = 1825, 4 nM = 1987, 10 nM = 2013, and 50 nM = 2059. Using these values, the change in fluorescence intensity $(F - F_0)$ in arbitrary fluorescence units (a.u.) would be equal to $0\,nM = 0$, $4\,nM = 162$, $10\,nM = 188$, and $50\,nM = 234$. The change in fluorescence intensity $(F - F_0)$ versus the concentration of AF594-PPARα gives a saturable curve (Figure 10.9D) and a $K_d = 5.3 \pm 0.8\,nM$. The maximal $F_{AD} = 2{,}059$, $F_A = 2{,}026$, $\varepsilon_A = 73{,}000\,M^{-1}$ cm^{-1}, $\varepsilon_D = 71{,}000\,M^{-1}$ cm^{-1}, so the efficiency of energy transfer $E = ((F_{AD}/F_A) - 1)(\varepsilon_A/\varepsilon_D)$, $E = ((2{,}059/2{,}026) - 1)$ $(73{,}000\,M^{-1}$ $cm^{-1}/71{,}000\,M^{-1}$ $cm^{-1}) = 0.017$ or 1.7%. Using the sensitized emission of the acceptor to calculate the distance between the two fluorophores, $R_{2/3} = ((R_0^6/E) - R_0^6)^{1/6} = ((60.0^6/0.017) - (60.0^6))^{1/6}$ or $118\,Å$.

Although variation may occur due to the fluorescent label chosen, such assays have shown that PPARα and L-FABP directly interact with high affinity and in close molecular proximity. For example, assays in which Cy3-labeled PPARα served as the donor and Cy5-labeled L-FABP served as the FRET acceptor yielded affinities in the 50–100 nM range and a molecular distance of approximately $50\,Å$ [64]. However, when AF488-labeled L-FABP was used as the donor and AF594-labeled PPARα was the acceptor, much stronger binding affinities were obtained $(K_d = 5-8\,nM)$, but the intermolecular distance between these fluorophores was much larger, $116-117\,Å$ (Figure 10.9). Such discrepancies are most likely due to different amino acids being labeled by the respective fluorophores, differences in the amount of spectral overlap of the two fluorophores, differences in protein labeling efficiencies, or additional factors. Although both the Cy3/Cy5 dyes and the AF488/AF594 dyes label free amine groups, and both sets of protein were approximately labeled with one dye per protein, the exact amino acids labeled by the dye were not determined. Moreover, the dye to protein ratio of 1:1 is an average value of the dye labeling of all the protein molecules in solution. If in one case all the protein molecules are labeled with one dye (1:1 ratio) and in another instance half of the protein molecules are labeled with two dyes while the other half of the protein molecules have no dye (still a 1:1 ratio), then it would be expected that these two cases might result in different values. It is ideal to have one dye per protein molecule; however, this is not always going to be the case. The presence of more than one dye on a protein molecule is a more complex situation. If several dyes label one protein in a given region, such labeling might result in self-quenching. If a protein molecule has several dyes which are far apart, then each dye may act differently upon interaction with the other labeled protein (i.e., the FRET distance will be an average of the distances between these dyes and the other dye). Regardless, both assays demonstrate the direct interaction of these proteins. Similar techniques have been used to show that Cy3-labeled PEX5 binds Cy5-labeled SCP-2 and pro-SCP-2 with similar distances (67 and $72\,Å$, respectively) but with different affinities $(K_d = 26$ and $2\,nM$, respectively) [74].

10.6.4 CONFOCAL MICROSCOPY—COLOCALIZATION

To date, the most common technique utilized for the visualization of protein–protein interactions within cells has been the fixation and immunolabeling of cells with fluorescent antibodies. This technique is a variation of that described in the section on nuclear localization of lipid binding proteins, with the difference being that two antibodies are required—one for each protein. Briefly, cells grown on chambered glass slides under assay conditions are fixed and permeabilized by one of several protocols as previously described herein. Cells are blocked with serum or BSA to reduce nonspecific binding of the antibody. Next, cells are incubated with either fluorescently labeled primary antibodies or nonlabeled primary antibodies, followed by fluorescently labeled secondary antibodies. If using fluorescently labeled secondary antibodies, then it is necessary to choose primary antibodies produced in different species to ensure proper labeling with the secondary antibody [7,30]. This technique has been used to show colocalization of ACBP with HNF-4α [7] as well as colocalization of L-FABP with PPARα [30].

For example, the following demonstrates a protocol for the formaldehyde fixation and the use of fluorescently labeled secondary antibodies. Volumes for each step depend on the number of wells of the chambered slide (e.g., for a two-well chambered slide 1 mL volume is adequate to cover the cells completely). Culture media is removed from cells grown on chambered glass slides, and the cells washed several times with PBS. It is important that the washes be gentle to prevent disruption of the cells from the surface of the slide. The cells are then fixed in a 4% formaldehyde solution in PBS at room temperature for 20 min. Cells are washed briefly with PBS and permeabilized in 0.1% Triton X-100 and 0.05% SDS in PBS at room temperature for 5 min. Cells are immediately washed with PBS to prevent overpermeabilization. The cells are then incubated in 5% BSA in PBS for 2 h at room temperature. Following the blocking step, cells are incubated with both nonlabeled primary antibodies diluted 1/200 in 1% BSA in PBS at 37°C for 1 h. For this example, the primary antibodies used are goat anti-PPARα and rabbit anti-L-FABP; since these antibodies are from different species, they can be incubated together in a single step. The cells are washed 3–4 times with PBS and then incubated with both fluorescently labeled secondary antibodies diluted 1/500 in 1% BSA in PBS at 37°C for 1 h. Since our primary antibodies were produced in goat and rabbit, we can incubate both secondary antibodies together. One secondary antibody would be fluorescently conjugated anti-rabbit IgG (FITC) and the other antibody would be fluorescently conjugated anti-goat IgG (TRITC). Cells are again washed with PBS, SlowFade medium applied, and a coverslip sealed over the cells.

Once cells are immunolabeled with the appropriate antibodies, the cells are visualized by confocal microscopy, which allows for the acquisition of the individual fluorescent signals from each fluorophore simultaneously. This is achieved through the use of dichroics and multiple photomultiplier tubes. It is important to image control cells (e.g., cells fixed and incubated with antibodies which are not fluorescently labeled as well as cells immunolabeled with each fluorescently labeled antibody individually) to determine the amount of background fluorescence and the amount of fluorescent bleed-through. If a significant amount of fluorescence is noted from both fluorophores through a single channel or photomultiplier tube, then the results will indicate a higher amount of colocalization than actually occurs. Although small amounts of bleed-through or autofluorescence can be corrected for by setting the threshold value above these artifact values, careful filter selection and fluorophore choice limit the amount of fluorescence from artifact sources. Using a software program such as ImageJ (NIH, Bethesda, Maryland) or MetaMorph Image Analysis (Advanced Scientific Imaging, Meraux, Louisiana), the images can be merged and the number of pixels that colocalize determined. Such programs allow threshold values to be set manually or automatically, calculate the colocalization coefficient for each fluorophore, and indicate the number of pixels for a given image that are colocalized.

10.6.5   CONFOCAL MICROSCOPY—FRET

In order to determine the molecular distance between proteins/fluorophores inside cells, modifications of several of the above techniques are used. Cells are fixed and immunolabeled as described above, except that it is common to conjugate the fluorescent moiety directly to the primary antibody (preferably the Fab fragment which is threefold smaller than the whole IgG), rather than to use a secondary antibody. This is due to the fact that antibodies are large proteins, and the use of both a primary and secondary antibody pushes the fluorophores apart, resulting in an underestimation of the actual proximity of the proteins of interest (i.e., lipid binding protein and nuclear receptor) [7]. Qualitatively, the sensitized emission of the acceptor is measured upon excitation of the donor, both before and after photobleaching of the acceptor. This allows for determination of any bleed-through of the donor emission into the acceptor emission spectrum. Quantitatively, FRET efficiency (E) is calculated according to the equation:

$$E = (FI_{post} - FI_{pre})/FI_{post} \qquad (10.23)$$

where

FI_{post} is the acceptor fluorescence intensity after donor photobleach
FI_{pre} is the acceptor fluorescence intensity before photobleach

The fluorescence intensities can be measured with a number of computer programs (i.e., LaserSharp, Bio-Rad, Hercules, California; MetaMorph Image Analysis, Advanced Scientific Imaging, Meraux, Louisiana; ImageJ, NIH, Bethesda, Maryland). The FRET efficiency can then be used to estimate intermolecular distance with the Forster equation (Equation 10.15) [7,61]. This technique has been used to show direct interaction of HNF-4α and ACBP at a distance of 53 Å within the nuclei of fixed cells [7].

10.6.6   IMMUNOGOLD ELECTRON MICROSCOPY

Although not a fluorescent technique, it is worth mentioning the use of immunogold electron microscopy (EM) for determining interaction of lipid binding proteins and nuclear receptors within the nucleus of fixed cells. This method allows for the examination of fixed tissues and is often used to confirm FRET imaging data. However, due to the intrinsically higher resolution, the immunogold EM method yields less information about a large number of cells for a single field of view. Since the resolution is higher, less of the cell (or fewer cells) is visualized in a single microscopic field; this requires more images to be captured through multiple fields to obtain data from a similar number of cells as used for LSCM. For this technique, tissue samples are often fixed in 4% formaldehyde, 0.1% glut in 100 mM sodium phosphate buffer overnight at 4°C, dehydrated with an ethanol series, and embedded in resin at 48°C for 2 days. Ultrathin sections are placed on Formvar-coated nickel grids and immunogold stained with antibodies to nuclear receptors and cytoplasmic lipid binding proteins. Antibodies to lipid binding proteins can be linked to gold particles of one size, while a larger or smaller gold particle can be used for the nuclear receptor antibody. Sections can be stained with a solution such as uranyl acetate and Reynold's lead citrate prior to examination by EM. Immungold EM techniques have been used to determine nuclear interactions of both ACBP with HNF-4α [7] and L-FABP with PPARα [64].

10.6.7 FUNCTIONAL ASSAYS—TRANSACTIVATION ASSAYS

Another nonfluorescent technique that has been used to confirm the functional significance for lipid binding protein expression on the activation of nuclear receptors is the use of transactivation assays and luminescence. Cultured cells are transfected with expression vectors for nuclear receptor expression, a nuclear receptor response element conjugated to a reporter construct (luciferase), and an internal transfection control reporter vector (*Renilla luciferase*). The internal transfection control reporter allows for correction of such variables as differences in cell numbers, differences in transfection efficiencies, and pipetting errors. The internal transfection control needs to be a protein that can be easily measured (e.g., luminescence) and that is not dependent upon the experiment or any of the other experimental expression vectors. These cells are grown in the presence or absence of LCFA or nonmetabolizable LCFA-CoA, and the cell lysates are examined for luminescent signal in the presence of substrate. Several kits are available for the analysis of cell lysates for luminescence (e.g., Dual-luciferase reporter assay, Promega, Madison, Wisconsin). These assays can be conducted in 96-well plates and the luminescence signals determined with a plate-reading luminometer. Values are often presented as % activity of luminescence intensity after correction for the transfection efficiency (*Renilla luciferase*) [7,24]. By comparing the effects of transactivation between cells which do not express lipid binding proteins and those that do, the effect of these proteins on nuclear receptor function can be examined. Such assays have demonstrated that the presence of ACBP increases transactivation of HNF-4α [7], while the presence of L-FABP increases transactivation of PPARα [84].

10.6.8 PROS AND CONS

Although more complex than the previously described assays, these assays have provided the first evidence for direct, functional interaction of lipid binding proteins and nuclear receptors within cell nuclei. Further, several of these assays can be used in cultured cells, primary cells derived from transgenic or knock-out animals, and tissues for results *in situ*. By combining several of these techniques, one can determine the binding affinity of the respective proteins for each other *in vitro*, the intermolecular distances between the interacting proteins *in vitro* and in cells, and what effect lipid binding proteins, LCFA, or nonmetabolizable

LCFA-CoA have on nuclear receptor interactions and activity. Energy transfer techniques can also be expanded upon to study how lipids are handed off within nuclei.

These fluorescent techniques are subject to many of the same limitations previously discussed herein. Again, fluorescent-labeling of a protein can alter protein structure, function, or localization, and thus require additional testing. The use of fluorescently labeled antibodies in fixed cells for colocalization, FRET, or immunogold EM techniques may underestimate the distance between the proteins due to the size of the antibodies. Further, the fixation and permeabilization of cells may result in inappropriate fluorescence labeling and, again, additional experimentation must be done to validate the results. However, the use of several of these techniques can compensate for many of these potential limitations, and these assays have the potential to be expanded upon to answer more complex questions.

10.7   CONCLUSIONS

Overall, the use of fluorescent techniques has allowed us to begin to understand the role lipid binding proteins play in nuclear regulation. Although there are many lipid binding proteins, and only a few have been examined in detail, such assays demonstrate that some lipid binding proteins are responsible for shuttling of LCFA and LCFA-CoA to the nucleus. Further, several lipid binding proteins have been shown to interact directly with nuclear receptors, suggesting that a complex of the proteins and ligand is formed in which the binding protein passes the ligand directly to the nuclear receptor. This indicates that lipid binding proteins function in gene regulation through the nuclear receptors they choose to interact with. The role of cell context, nutrient LCFA type and status, and other variables on these interactions remain largely unexplored.

ACKNOWLEDGMENTS

This work was supported in part by the USPHS National Institutes of Health grants DK41402 (FS and ABK), NIH P20 grant "Fluorescence Probes for Multiplexed Intracellular Imaging" GM72041 (Project 2, ABK and FS), National Research Service Award DK066732 (HAH), grant DK70965 (BPA), and grant DK77573 (HAH).

LIST OF ABBREVIATIONS

ACBP	acyl-CoA binding protein
A-FABP	adipocyte fatty acid binding protein
BODIPY C-12	4a-diaza-s-indacene-dodecanoic acid
cis-parinaric acid	9Z,11E,13E,15Z-octadecatetraenoic acid
CMC	critical micelle concentration
EM	electron microscopy
FRAP	fluorescence recovery after photobleaching
FRET	fluorescence resonance energy transfer

GFP	green fluorescence protein
GR	glucocorticoid receptor
HNF-4α	hepatocyte nuclear factor-4α
HPLC	high-performance liquid chromatography
I-FABP	intestinal fatty acid binding protein
LCFA	long-chain fatty acid
LCFA-CoA	long-chain fatty acyl-CoA
L-FABP	liver fatty acid binding protein
LSCM	laser scanning confocal microscopy
LXR	liver X receptor
MPLSM	multiphoton laser scanning microscopy
NBD-stearic acid	12-(*N*-methyl)-*N*-[(7-nitrobenz-2-oxa-1,3-diazol-4-yl) amino]-octadecanoic acid
PPARα	peroxisome proliferator-activated receptor-α
RAR	retinoic acid receptor
RXR	retinoid X receptor
SCP-2	sterol carrier protein-2
SREBP	sterol regulatory element binding protein
TLC	thin layer chromatography
TR	thyroid hormone receptor
trans-parinaric acid	9E,11E,13E,15E-octadecatetraenoic acid
YFP	yellow fluorescence protein

REFERENCES

1. Stewart, J.M., Driedzic, W.R., and Berkelaar, J.A. (1991) Fatty-acid-binding protein facilitates the diffusion of oleate in a model cytosol system. *Biochem. J.* 275:569–573.
2. Veerkamp, J.H. (1995) Fatty acid transport and fatty acid-binding proteins. *Proc. Nutr. Soc.* 54:23–37.
3. Jump, D.B. et al. (1995) Effects of fatty acids on hepatic gene expression. *Prostaglandins Leukot. Essen. Fatty Acids* 52:107–111.
4. Pegorier, J.-P., May, C.L., and Girard, J. (2004) Control of gene expression by fatty acids. *J. Nutr.* 134:2444S–2449S.
5. Lawrence, J.W., Kroll, D.J., and Eacho, P.I. (2000) Ligand dependent interaction of hepatic fatty acid binding protein with the nucleus. *J. Lipid Res.* 41:1390–1401.
6. Wolfrum, C. et al. (2000) Binding of fatty acids and peroxisome proliferators to orthologous fatty acid binding proteins from human, murine, and bovine liver. *Biochemistry* 39:1469–1474.
7. Petrescu, A.D. et al. (2003) Physical and functional interaction of acyl CoA binding protein (ACBP) with hepatocyte nuclear factor-4alpha (HNF4alpha). *J. Biol. Chem.* 278:51813–51824.
8. Hertz, R. et al. (1998) Fatty acyl-CoA thioesters are ligands of hepatic nuclear factor-4α. *Nature* 392:512–516.
9. Hertz, R. et al. (2001) Suppression of hepatocyte nuclear factor 4α by acyl-CoA thioesters of hypolipidemic peroxisome proliferators. *Biochem. Pharmacol.* 61:1057–1062.
10. Forman, B.M., Chen, J., and Evans, R.M. (1999) Hypolipidemic drugs, polyunsaturated fatty acids, and eicosanoids are ligands for peroxisome proliferator-activated receptors α and δ. *Proc. Natl. Acad. Sci.* 94:4312–4317.

11. Isseman, I. and Prince, R.A. (1993) The peroxisome proliferator activated receptor: Retinoic X receptor heterodimer is activated by fatty acids and fibrate hypolipidaemic drugs. *J. Mol. Endocrinol.* 11:37–47.

12. Keller, H. et al. (1993) Fatty acids and retinoids control lipid metabolism through activation of peroxisome proliferator activated receptor-retinoid X receptor heterodimers. *Proc. Natl. Acad. Sci.* 90:2160–2164.

13. Kliewer, S.A. et al. (1997) Fatty acids and eicosanoids regulate gene expression through direct interactions with peroxisome proliferator-activated receptors α and y. *Proc. Natl. Acad. Sci.* 94:4318–4323.

14. Frolov, A. et al. (1997) Isoforms of rat liver fatty acid binding protein differ in structure and affinity for fatty acids and fatty acyl CoAs. *Biochemistry* 36:6545–6555.

15. McArthur, M.J. et al. (1999) Cellular uptake and intracellular trafficking of long chain fatty acids. *J. Lipid Res.* 40:1371–1383.

16. Rolf, B. et al. (1995) Analysis of the ligand binding properties of recombinant bovine liver-type fatty acid binding protein. *Biochim. Biophys. Acta* 1259:245–253.

17. Rasmussen, J.T., Börchers, T., and Knudsen, J. (1990) Comparison of the binding affinities of acyl-CoA-binding protein and fatty-acid-binding protein for long chain acyl-CoA esters. *Biochem. J.* 265:849–855.

18. Richieri, G.V., Ogata, R.T., and Kleinfeld, A.M. (1994) Equilibrium constants for the binding of fatty acids with fatty acid binding proteins from adipocyte, intestine, heart, and liver measured with the fluorescent probe ADIFAB. *J. Biol. Chem.* 269:23918–23930.

19. Rosendal, J., Ertbjerg, P., and Knudsen, J. (1993) Characterization of ligand binding to acyl-CoA-binding protein. *Biochem. J.* 290:321–326.

20. Norris, A.W. and Spector, A.A. (2002) Very long chain $n − 3$ and $n − 6$ polyunsaturated fatty acids bind strongly to liver fatty acid binding protein. *J. Lipid Res.* 43:646–653.

21. Faergeman, N.J. and Knudsen, J. (1997) Role of long-chain fatty acyl-CoA esters in the regulation of metabolism and in cell signalling. *Biochem. J.* 323:1–12.

22. Frolov, A.A. and Schroeder, F. (1997) Time-resolved fluorescence of intestinal and liver fatty acid binding proteins: Role of fatty acyl CoA and fatty acid. *Biochemistry* 36:505–517.

23. Hostetler, H.A. et al. (2005) Peroxisome proliferator activated receptor α interacts with high affinity and is conformationally responsive to endogenous ligands. *J. Biol. Chem.* 280:18667–18682.

24. Hostetler, H.A. et al. (2008) Glucose directly links to lipid metabolism through high-affinity interaction with peroxisome proliferator activated receptor-α. *J. Biol. Chem.* 283:2246–2254.

25. Hostetler, H.A., Kier, A.B., and Schroeder, F. (2006) Very-long-chain and branched-chain fatty acyl CoAs are high affinity ligands for the peroxisome proliferator-activated receptor α (PPARα). *Biochemistry* 45:7669–7681.

26. Lin, Q. et al. (1999) Ligand selectivity of the peroxisome proliferator-activated receptor α. *Biochemistry* 38:185–190.

27. Burnett, D.A. et al. (1979) Utilization of long chain fatty acids by rat liver: Studies of the role of fatty acid binding protein. *Gastroenterology* 77:241–249.

28. Hamilton, J.A. (1998) Fatty acid transport: Difficult or easy. *J. Lipid Res.* 39:467–481.

29. Huang, H. et al. (2002) Liver fatty acid binding protein targets fatty acids to the nucleus: Real-time confocal and multiphoton fluorescence imaging in living cells. *J. Biol. Chem.* 277:29139–29151.

30. Huang, H. et al. (2004) Liver fatty acid binding protein colocalizes with peroxisome proliferator receptor α and enhances ligand distribution to nuclei of living cells. *Biochemistry* 43:2484–2500.

31. Chao, H. et al. (1999) Microsomal long chain fatty acyl CoA transacylation: Differential effect of SCP-2. *Biochim. Biophys. Acta* 1439:371–383.
32. Jolly, C.A. et al. (1997) Fatty acid binding protein: Stimulation of microsomal phosphatidic acid formation. *Arch. Biochem. Biophys.* 341:112–121.
33. Jolly, C.A., Murphy, E.J., and Schroeder, F. (1998) Differential influence of rat liver fatty acid binding protein isoforms on phospholipid fatty acid composition: Phosphatidic acid biosynthesis and phospholipid fatty acid remodeling. *Biochim. Biophys. Acta* 1390:258–268.
34. Jolly, C.A., Wilton, D.A., and Schroeder, F. (2000) Microsomal fatty acyl CoA transacylation and hydrolysis: Fatty acyl CoA species dependent modulation by liver fatty acyl CoA binding proteins. *Biochim. Biophys. Acta* 1483:185–197.
35. Schroeder, F. et al. (1998) Fatty acid binding protein isoforms: Structure and function. *Chem. Phys. Lipids* 92:1–25.
36. Murphy, E.J. et al. (1996) *Cis*-parinaric acid uptake in L-cells. *Arch. Biochem. Biophys.* 335:267–272.
37. Murphy, E.J. et al. (1996) Liver fatty acid binding protein expression in transfected fibroblasts stimulates fatty acid uptake and metabolism. *Biochim. Biophys. Acta* 1301:191–198.
38. Prows, D.R., Murphy, E.J., and Schroeder, F. (1995) Intestinal and liver fatty acid binding proteins differentially affect fatty acid uptake and esterification in L-Cells. *Lipids* 30:907–910.
39. Nemecz, G. et al. (1991) Interaction of fatty acids with recombinant rat intestinal and liver fatty acid-binding proteins. *Arch. Biochem. Biophys.* 286:300–309.
40. Heyliger, C.E. et al. (1996) Fatty acid double orientation alters interaction with L-cell fibroblasts. *Mol. Cell. Biochem.* 155:113–119.
41. Atshaves, B.P. et al. (1998) Cellular differentiation and I-FABP protein expression modulate fatty acid uptake and diffusion. *Am. J. Physiol.* 274:C633–C644.
42. Atshaves, B.P. et al. (2004) Liver fatty acid binding protein expression enhances branched-chain fatty acid metabolism. *Mol. Cell. Biochem.* 259:115–129.
43. Atshaves, B.P. et al. (2004) Liver fatty acid binding protein gene ablation inhibits branched-chain fatty acid metabolism in cultured primary hepatocytes. *J. Biol. Chem.* 279:30954–30965.
44. Luxon, B.A. and Weisiger, R.A. (1993) Sex differences in intracellular fatty acid transport: Role of cytoplasmic binding proteins. *Am. J. Physiol.* 265:G831–G841.
45. Weisiger, R.A. (1996) Cytoplasmic transport of lipids: Role of binding proteins. *Comp. Biochem. Physiol.* 115B:319–331.
46. Weisiger, R.A. (2005) Cytosolic fatty acid binding proteins catalyze two distinct steps in intracellular transport of their ligands. *Mol. Cell. Biochem.* 239:35–42.
47. Kuklev, D.V. and Smith, W.L. (2004) Synthesis of long chain $n - 3$ and $n - 6$ fatty acids having a photoactive conjugated tetraene group. *Chem. Phys. Lipids* 130:45–158.
48. McIntosh, A.L. et al. (2005) Uptake kinetics of fluorescent long chain $n - 3$ and $n - 6$ fatty acids in intact cells. *FASEB J.* 19:A292.
49. Nemecz, G., Jefferson, J.R., and Schroeder, F. (1991) Polyene fatty acid interactions with recombinant intestinal and liver fatty acid binding proteins. *J. Biol. Chem.* 266:17112–17123.
50. Frolov, A. et al. (1997) Lipid specificity and location of the sterol carrier protein-2 fatty acid binding site: A fluorescence displacement and energy transfer study. *Lipids* 32:1201–1209.
51. Frolov, A. et al. (1996) Sterol carrier protein-2, a new fatty acyl coenzyme A-binding protein. *J. Biol. Chem.* 271:31878–31884.
52. Hubbell, T. et al. (1994) Recombinant liver fatty acid binding protein interactions with fatty acyl-coenzyme A. *Biochemistry* 33:3327–3334.

53. Richieri, G.V., Ogata, R.T., and Kleinfeld, A.M. (1996) Thermodynamic and kinetic properties of fatty acid interactions with rat liver fatty acid-binding protein. *J. Biol. Chem.* 271:31068–31074.

54. Stolowich, N.J. et al. (2002) Sterol carrier protein-2: Structure reveals function. *Cell. Mol. Life Sci.* 59:193–212.

55. Faergeman, N.J. et al. (1996) Thermodynamics of ligand binding to acyl-coenzyme A binding protein studied by titration calorimetry. *Biochemistry* 35:14118–14126.

56. Huang, H. et al. (2005) Acyl-coenzyme A binding protein expression alter liver fatty acyl-coenzyme A metabolism. *Biochemistry* 44:10282–10297.

57. Lakowicz, J.R. (2006) Frequency-domain lifetime measurements. In *Principles of Fluorescence Spectroscopy*, Lakowicz, J.R. (ed.), pp. 158–204. New York: Springer Science.

58. Frolov, A. and Schroeder, F. (1998) Acyl coenzyme A binding protein: Conformational sensitivity to long chain fatty acyl-CoA. *J. Biol. Chem.* 273:11049–11055.

59. Lakowicz, J.R. (2006) Fluorescence anisotropy. In *Principles of Fluorescence Spectroscopy*, Lakowicz, J.R. (ed.), pp. 353–382. New York: Springer Science.

60. Lakowicz, J.R. (2006) Energy transfer. In *Principles of Fluorescence Spectroscopy*, Lakowicz, J.R. (ed.), pp. 443–475. New York: Springer Science.

61. Forster, T. (1967) Mechanism of energy transfer. In *Comprehensive Biochemistry*, Florkin, M. and Statz, E.H. (eds.), pp. 61–77. New York: Elsevier Scientific Publishers.

62. Powell, G.L. et al. (1985) Fatty acyl CoA as an effector molecule in metabolism. *Fed. Proc.* 44:81–84.

63. Vorum, H. et al. (1992) Solubility of long chain fatty acid in phosphate buffer at pH 7.4. *Biochem. Biophys.* 1126:135–142.

64. Hostetler, H.A. et al. (2009) L-FABP directly interacts with PPARa in cultured primary hepatocytes. *J. Lipid. Res.* 50:1663–1675.

65. Inouye, S. and Tsuji, F.I. (1994) *Aequorea* green fluorescent protein. Expression of the gene and fluorescence characteristics of the recombinant protein. *FEBS Lett.* 341:277–280.

66. Prasher, D.C. et al. (1992) Primary structure of the *Aequorea victoria* green-fluorescent protein. *Gene* 111:229–233.

67. Amsterdam, A. and Hopkins, N. (1998) The uses of green fluorescent protein in transgenic vertebrates. In *Green Fluorescent Proteins: Properties, Applications, and Protocols*, Chalfie, M. and Kain, S. (eds.), Wiley-Liss, New York.

68. Ayers, S.D. et al. (2007) Continuous nucleocytoplasmic shuttling underlies activation of PPARγ by FABP4. *Biochemistry* 46:6744–6752.

69. Tan, N.-S. et al. (2002) Selective cooperation between fatty acid binding proteins and peroxisome proliferator activated receptors in regulating transcription. *Mol. Cell. Biol.* 22:5114–5127.

70. Bandichhor, R. et al. (2006) Synthesis of a new water soluble rhodamine derivative and application to intracellular imaging. *J. Am. Chem. Soc.* 128:10688–10689.

71. Bandichhor, R. et al. (2006) Water soluble through bond energy transfer cassettes in intracellular imaging. *Bioconj. J.* 17:1219–1225.

72. Morris, M.C. et al. (1997) A new peptide vector for efficient delivery of oligonucleotides into mammalian cells. *Nucleic Acids Res.* 25:2730–2736.

73. Morris, M.C. et al. (2001) A peptide carrier for the delivery of biologically active proteins into mammalian cells. *Nat. Biotechnol.* 19:1173–1176.

74. Parr, R.D. et al. (2007) A new N-terminal recognition domain in caveolin-1 interacts with sterol carrier protein-2 (SCP-2). *Biochemistry* 46:8301–8314.

75. Martin, G.G. et al. (2008) Structure and function of the sterol carrier protein-2 (SCP-2) N-terminal pre-sequence. *Biochemistry* 47:5915–5934.

76. Adida, A. and Spener, F. (2002) Intracellular lipid binding proteins and nuclear receptors involved in branched-chain fatty acid signaling. *Prostaglandins Leukot. Essen. Fatty Acids* 67:91–98.

77. Petrescu, A.D. et al. (2002) Ligand specificity and conformational dependence of the hepatic nuclear factor-4α (HNF-4α). *J. Biol. Chem.* 277:23988–23999.

78. Petrescu, A.D. et al. (2005) Role of regulatory F-domain in hepatocyte nuclear factor-4-alpha ligand specificity. *J. Biol. Chem.* 280:16714–16727.

79. Adamson, D.J.A. and Palmer, C.N.A. (2002) Fluorescence-based ligand binding assays for peroxisome proliferator activated receptors. *Methods Enzymol.* 357:188–197.

80. Lakowicz, J.R. (2006) Protein fluorescence. In *Principles of Fluorescence Spectroscopy*, Lakowicz, J.R. (ed.), pp. 530–577. New York: Springer Science.

81. Stryer, L. (1978) Fluorescence energy transfer as a spectroscopic ruler. *Ann. Rev. Biochem.* 47:819–844.

82. Feige, J.N. et al. (2005) Fluorescence imaging reveals the nuclear behavior of peroxisome proliferator-activated receptor/retinoid x receptor heterodimers in the absence and presence of ligand. *J. Biol. Chem.* 280:17860–17890.

83. Schroeder, F. et al. (2008) Role of fatty acid binding proteins and long chain fatty acids in modulating nuclear receptors and gene transcription. *Lipids* 43:1–17.

84. Wolfrum, C. et al. (2001) Fatty acids and hypolipidemic drugs regulate PPARα and PPARγ gene expression via L-FABP: A signaling path to the nucleus. *Proc. Natl. Acad. Sci.* 98:2323–2328.

85. Elholm, M. et al. (2000) Long-chain acyl-CoA esters and acyl-CoA binding protein are present in the nucleus of rat liver cells. *J. Lipid Res.* 41:538–545.

86. Helledie, T. et al. (2000) Lipid-binding proteins modulate ligand-dependent *trans*-activation by peroxisome proliferator-activated receptors and localize to the nucleus as well as the cytoplasm. *J. Lipid Res.* 41:1740–1751.

87. Adida, A. and Spener, F. (2006) Adipocyte-type fatty acid-binding protein as inter-compartmental shuttle for peroxisome proliferator activated receptor γ agonists in cultured cell. *Biochim. Biophys. Acta* 1761:172–181.

88. McIntosh, A.L. et al. (2009) Liver type fatty acid binding protein (L-FABP) gene ablation reduces nuclear ligand distribution and peroxisome proliferator-activated receptor-α activity in cultured primary hepatocytes. *Arch. Biochem. Biophys.* 485:160–173.

89. Palmer, C.N.A. and Wolf, C.R. (1998) *Cis*-parinaric acid is a ligand for the human peroxisome proliferator-activated receptor gamma: Development of a novel spectrophotometric assay for the discovery of PPARγ ligands. *FEBS Lett.* 431:476–480.

Part IV

Methods to Assess Fatty Acid
and Phospholipid Metabolism

11 Methods for Measuring Fatty Acids Using Chromatographic Techniques

Gwendolyn Barceló-Coblijn,
Cameron C. Murphy, and Eric J. Murphy

CONTENTS

11.1 INTRODUCTION

Lipid analytical methodology was developed during the first part of the last century and, compared to other analytical methods, it might be seen as "old fashioned." However, the interest in techniques that allow the measurement of fatty acids (FA) has reemerged with the evidence that FA have very important roles in cellular function. Initial interests were focused on FA because they were part of glycerophospholipids, critical building blocks of cellular membranes. Changes in phospholipid FA composition can lead to the alteration of the membrane biophysical properties. While this is definitely an important role for FA, it is now known that FA have significantly greater functions than merely being structural constituents of biological membranes. For instance, through their binding to different nuclear receptors, such as RXR or PPAR (see Chapters 9 and 10), FA and fatty acyl-CoA are involved in gene expression regulation [1]. FA also affect ion channels, regulating events such as heart contraction [2,3]. In addition, FA such as arachidonic acid and its metabolites (prostaglandins and eicosanoids) are involved in cell signaling pathways [4–6]. Therefore, from the basic research point of view, there are many reasons to be interested in the methodology used in FA analysis. In this chapter, traditional methods used in the analysis of FA will be demonstrated. For methods using mass spectrometry approaches, see Chapter 6.

Another important application is the analysis of biological samples, such as tissue or blood. The presence or absence of certain FA may be indicative of a pathological process, such as an increase in unesterified arachidonic acid during an inflammatory response or the FA content of milk used to assess infant nutrition. For all of these reasons, FA analysis techniques have become very popular and used worldwide. The most established methodologies use gas–liquid chromatography (GLC), an instrument that is less expensive than those used in mass spectrometry techniques and readily available to most researchers worldwide. An additional, widely used method is high-performance liquid chromatography (HPLC), yet the availability of these instruments is somewhat more limited than GLC although, as it will be mentioned later, in some applications, HPLC is more adequate. In this chapter, some of the most common procedures used to analyze FA are summarized. Additional extensive reviews on this topic can be found elsewhere, and the reader may want to consult these reviews as well [7–10].

There is no "best method" to analyze FA, and, consequently, the analyst will have to evaluate the advantages and disadvantages of each method taking into account several aspects required for analysis. For instance, the analyst needs to consider if the FA is free or esterified onto a glycerol moiety, if it is bound to a protein in an unesterified form, the type of the chemical bound, and finally the characteristics of the particular FA to be analyzed. In nature, unesterified (free) fatty acids (FFA) are usually present at a very low concentration. In fact, the presence of a high concentration of FFA in a sample is generally associated with sample deterioration [11] or a pathophysiological event, e.g., ischemia, neurotrauma [12–19]. However, FFA are common in other types of samples such as milk and milk derivatives [9]. Thus, the majority of FA are found bound to molecules through different chemical bonds. These bonds can be through ester linkage as in glycerophospholipids or cholesteryl esters (O-acyl linkage), through an amide linkage as in sphingomyelins (N-acyl linkage), or through an ether linkage as in plasmalogens. In turn, FA vary in chain length and the degree of unsaturation. FA can be divided as short chain (C_{4-12}), long chain (C_{14-22}), or very long chain ($C_{>22}$) FA and, according to their degree of unsaturation, as saturated, monounsaturated, and polyunsaturated fatty acids (PUFA). In addition, FA may contain other polar functional groups such as hydroxy-, epoxy-, and keto-groups, which may require derivatization to improve the analysis. Finally, because of the high sensitivity of these techniques, the methods were developed to analyze small quantities of samples, especially after the introduction of capillary GLC column technology. Most of the protocols can be scaled up proportionally, although for a large amount of sample, some additional considerations should be taken into account.

11.2   FATTY ACID ANALYSIS BY GAS–LIQUID CHROMATOGRAPHY

11.2.1   DERIVATIZATION BEFORE GLC ANALYSIS

One of the advantages of GLC is that it allows for the determination of a complete FA profile in a very short time. However, if the FA are not derivatized, they are difficult to analyze by GLC because of their high polarity, low volatility, and the possibility to establish hydrogen bonds with each other [7]. Thus, underivatized FA have boiling points higher than the structurally related alkanes, usually close to the temperature of decomposition. For this reason, prior to analysis, FA are derivatized to reduce their reactivity and increase their volatility. This process reduces peak tailing, thereby improving the peak configuration and increasing detector sensitivity. Nonetheless, several methods for the separation and quantitation of FA without any derivatization can be found in the literature [20–23], although, if a higher sensitivity is demanded, then the derivatization is required. Most frequently used derivatives are alkyl derivatives, such as methyl [24–26], ethyl [27], propyl [27], isopropyl [28], butyl [29], and i-butyl [30] compounds. FA analysis by HPLC is very useful during the analysis of temperature-sensitive molecules that could be damaged during the process of volatilization but more importantly is useful during studies examining FA chain elongation or shortening [31–35].

Traditionally in the past, lipid extracts were subject to a saponification reaction with sodium or potassium hydroxide to release esterified FA; for detailed procedures, consult [9,25]. However, saponification–esterification procedures require long reaction times, while direct transesterification procedures are completed within minutes. In fact, there is no need to hydrolyze lipids to obtain FFA as most lipids with esterified FA can be directly transesterified [36]. Nonetheless, derivatization steps required for HPLC analysis often require saponification (see Section 11.3.1).

Despite the large number of protocols for FA derivatization, each procedure should be optimized according to the characteristics of the sample. Some of the problems associated with ester preparation include incomplete conversion of lipids into FA esters, alteration of the original FA composition or the formation of artifacts during the reaction, GLC column damage due to residual traces of the esterification reagent, incomplete extraction of FA esters, and loss of highly volatile short-chain

FA esters [37]. Methyl esters are by far the most common derivatives synthesized, although, in some particular cases, isopropyl or butyl esters are more convenient. This section covers the commonly used methods used in esterification of FFA and for the transesterification of esterified FA.

11.2.1.1 Acid-Catalyzed Esterification

Common reagents for acid-catalyzed esterification or transesterification are methanol containing hydrochloric acid, sulfuric acid, or boron trifluoride. Because these reactions involve a reversible step, a large excess of alcohol is required to favor the desired forward reaction and to displace the equilibrium point. In addition, because water is a stronger electron donor than the aliphatic alcohols, these reactions need to take place under completely anhydrous conditions. Drying agents such as anhydrous sodium sulfate can be added to the reaction medium to achieve these conditions [38], but it is generally a better choice to operate with dry reagents and glassware [25].

Many lipid chemists consider HCl in anhydrous methanol as a mild and useful derivatization reagent [39] and, in fact, it is the most frequently cited reagent for preparation of methyl esters. This reagent consists of 5% anhydrous hydrogen chloride in methanol, prepared by bubbling dry gaseous HCl into dry methanol [25]. This method is useful for both esterification of free or unesterified fatty acids and transesterification of O-acyl lipids, but note that bubbling gaseous HCl into dry methanol is dangerous and should only be done by those skilled in the art. There is a large list of methods using alternative electrophilic catalysts [9]. For example, a solution of 1%–2% concentrated sulfuric acid in methanol is very easy to prepare and has a reaction time that is almost identical to 5% methanolic HCl solution [28,40]. In our protocol, we use toluene, rather than the more toxic benzene, in a 1:1 solution with dry methanol. By adding 2% sulfuric acid to this solution, FFA are esterified in a screw-top test tube (phenolic cap with a Teflon liner) under a nitrogen atmosphere at 95°C for 4 h with continuous shaking (see Protocol A) [41]. Note that care needs to be taken when mixing the concentrated sulfuric acid with the methanol:toluene mixture as this reaction is very exothermic and has the potential to boil if the sulfuric acid is not added slowly. Other acidic catalysts are aluminum trichloride [42] or p-toluenesulfonic acid [43], but these are rarely used.

An alternative method to esterify free (unesterified) FA is using 12%–14% boron trifluoride (BF_3) in methanol under refluxing conditions [44]. Boron trifluoride in methanol is a powerful acidic catalyst. While esterification of FFA is completed in 2 min [45], transesterification of esterified FA have longer reaction times [46]. Boron trifluoride can be used with other alcohols such as ethanol [47], propanol [48], and butanol [29,47]. This method has proved to give complete transesterification of all lipids, although the addition of butylated hydroxytoluene (BHT) is recommended to prevent the decomposition of PUFA [8,16,49]. However, this method presents some limitations such as the limited shelf-life of the reagent and significant artifacts produced from the oxidation of unsaturated FA. In addition, when used in high concentrations, boron trifluoride will cleave plasmalogens, releasing aldehydes that yield dimethylacetals, which interfere with the separation of some major FA [8,50,51]. This is especially important in plasmalogen-enriched samples such as red blood cells or central nervous system tissues.

11.2.1.2   Base-Catalyzed Transesterification

Base-catalyzed procedures are, under most circumstances, much simpler and faster than those catalyzed by acids. In fact, this method should be considered for mixed lipid samples that do not contain FFA or for single lipid classes containing ester-bound FA [25]. The mechanism of the reaction is different when compared with acid catalysis and the reaction cannot esterify an FFA. During transesterification, water needs to be avoided because its presence quenches the forward reaction and increases the presence of FFA, thereby reducing the yield. Thus, base-catalyzed transesterification, as is the case for acid-catalyzed reactions, should also occur with a large excess of alcohol and in the absolute absence of water [25].

The most useful basic agents used for transesterification are 0.5–2 M sodium or potassium methoxide in anhydrous methanol, prepared by dissolving the clean metals in anhydrous methanol, which is a strongly exothermic reaction. Potassium hydroxide in methanol at similar concentrations is occasionally used to transmethylate triglycerides [52] or glycerophospholipids [53]. The latter is a simple method and consists of adding 0.5 M KOH in anhydrous methanol to the lipid extract for a total reaction time of 2 min at 37°C [53]. The reaction is quenched by the addition of methyl formate and the reaction can be done in less than 1 min (see Protocol B). Note that sodium hydroxide will not dissolve in methanol whereas potassium hydroxide is readily soluble. While some authors [25] do not recommend this method because appreciable hydrolysis of lipids to FFA can occur if the least trace of water is present, especially if the reaction is prolonged [54,55]; using short reaction times limits this problem. An additional advantage of this protocol is that the reaction proceeds at room temperature, so there is no need to add BHT to protect PUFA. The mild conditions achieved with this method protect labile groups such as cyclopropane rings or double bonds, which will not undergo heat-induced isomerization. However, the mildness of the reaction prolongs the reaction time for cholesteryl esters, which require twice as long of a reaction time and prevent the transesterification of amide-bound FA, as in sphingolipids, or the liberation of aldehydes from plasmalogens [8]. However, when analyzing samples rich in plasmalogens, the base-catalyzed transesterification prevents the release of aldehydes that form dimethylacetals associated with the acid-catalyzed reactions [49,56,57].

Some strongly basic quaternary salts of ammonium such as trimethyl(m-trifluorotolyl)ammonium hydroxide [58], trimethylammonium hydroxide [59,60], trimethylphenylammonium hydroxide [61], and trimethylsulfonium hydroxide [62] have been used for transesterification of glycerides. After the addition of these reagents, lipid-bound FA are converted into quaternary ammonium salts which, in turn, are pyrolyzed to methyl esters in the hot injector port of the GLC. The major advantage of this method is the elimination of one step, the one that involves the extraction of the derivatized product. This can be particularly useful to analyze short-chain FA methyl esters, which are easily lost during extraction and solvent evaporation [59].

11.2.1.3   Methylation Using Diazomethane

A third group of FA derivatization methods involves diazomethane as a reagent. The latter is generally prepared as a solution in diethyl ether by the action of alkali on a nitrosamide [7]. Diazomethane reacts rapidly with FFA in the presence of a little methanol, which catalyses the reaction, to form methyl esters, but it does not induce transesterification [63]. Thus, this method allows for free FA methylation in the presence of bound FA [9], saving the analyst the time-consuming steps required for the separation of FFA by thin-layer chromatography (TLC) or HPLC [64,65].

The main disadvantage of this method is the toxicity of diazomethane as well as its potential explosiveness. For this reason different analysts have developed modified methods to minimize the hazards. Schlenk and Gellerman described a method which generates small quantities of diazomethane for immediate use [63]. This method gives few byproducts and decreases the dangerousness if the appropriate conditions are used [66]. The reaction occurs at 0°C or at room temperature and requires a short reaction time. However, this procedure is not exempt of artifacts and, during the methylation, other functional groups can be affected [9].

11.2.1.4 Derivatization of Short-Chain, Amide-Bound Fatty Acids and Nonpolar Lipids

As previously mentioned, the main problem in analyzing short-chain FA is that they are highly volatile and the short-chain methyl esters are often lost during solvent evaporation. One of the approaches to overcome this problem is the type of derivatization agent used [27,29,36,67]. For very short monocarboxylic FA such as C_1–C_3, a recommended method utilizing p-bromobenzyl, m-methylbenzyl, or p-nitrobenzyl iodide dissolved in ethanol has been described [36].

Sphingolipids that contain amide-bound FA are not easily transesterified under acidic or basic conditions. In cases in which only FA composition is required, the lipids may be refluxed with methanol containing concentrated HCl (5:1 v/v) for 5h or for 24h at 50°C [7]. Alternatively, a specific hydrolysis method can be used [51] and then the product be methylated by an appropriate procedure. However, we have found that our acid-catalyzed esterification protocol works well for liberating the FA found on sphingomyelin [68] and gives quantitative yields (see Protocol A).

Triacylglycerols and cholesteryl esters, which are nonpolar lipids, are not soluble in tranesterification reagents and consequently longer times or reaction will be needed. Using some nonpolar solvents, such as benzene, toluene, chloroform, or tetrahydrofuran [8,46,69] can be helpful. However, chloroform is not advisable because it usually contains ethanol as a stabilizer, which competes with methanol for esterification [25]. Again, we have found nearly 95%–98% recovery of triacylglycerol FA [70–72] content using our base-catalyzed procedure (see Protocol B).

11.3 PRACTICAL ASPECTS TO CONSIDER IN FATTY ACID DERIVATIZATION

11.3.1 FATTY ACID AUTOXIDATION

To reduce the risk of oxidation damage of lipids, particularly PUFA, lipids should always be handled under a nitrogen atmosphere in the test tube and, in case that the reaction needs heating, BHT can be added to the solvents. Some investigators routinely add BHT to solvents used in extraction to PUFA oxidation and useful concentrations are 0.005%–0.1% (w/v) [16,73,74]. However, it should be noted that the solvent chosen for extraction and storage can dramatically impact the need for BHT. For instance, in our laboratory, we traditionally use only hexane:2-propanol (3:2 by vol.) for storage and oftentimes for extraction when a single-phase extraction is desirable. In addition to using this solvent, keeping volumes >1 mL also helps minimize the impact of oxidation using this solvent system. The single-phase extraction methods are much more PUFA friendly than the traditional chloroform:methanol mixtures associated with two-phase extraction methods commonly referred to as the Folch method or the Bligh and Dyer method [75,76], due in part to the release of chlorine from the chloroform. Hence, two independent methods are used to prevent PUFA oxidation: (1) the use of an antioxidant such as BHT and (2) the use of PUFA-friendly solvents such as hexane:2-propanol.

11.3.2 Reaction Product Extraction after Derivatization

Once the sample has been esterified or transesterified, the reaction product has to be extracted. Long-chain FA can be extracted with alkanes such as *n*-pentane [46], *n*-hexane [8], or isooctane [77]. For a complete recovery, the extraction process should be carried out at least twice and with an adequate ratio of extraction solvent to the reaction medium [8,25,46]. Note that we recommend three washes with hexane or petroleum ether and a ratio of 2 volumes of hexane for every reaction volume. In addition, note in protocol B that after sitting overnight at −80°C, the glycerol potentially containing residual KOH will be at the bottom of the test tube. This can easily be removed by carefully aspirating the glycerol using a Pasteur pipet.

11.3.3 Esterification on Thin-Layer Chromatography Adsorbents

It is common to separate lipid extracts into different lipid classes using TLC. In our laboratory, we commonly use TLC for the separation of glycerophospholipids and of neutral lipids (see Figure 11.1). Although the conventional procedure is to elute the lipid fraction prior to derivatization, some methods proceed directly to the reaction without elution. In the case of base-catalyzed transesterification, this can lead to a poor recovery because water bound to the silica gel causes some hydrolysis in the presence of a base to form FFA rather than fatty acid methyl esters (FAME). In case of acid-catalyzed procedures, the best results are when the silica gel to lipid ratio is low. While working under a nitrogen atmosphere and by vigorously vortexing the silica with the reagent, the reaction yield may be improved, although extraction from the silica is the best solution, especially when working with TAG and cholesteryl esters.

11.3.4 Extraction from TLC Plates

Lipids that are separated by TLC can be subsequently eluted from the silica in preparation for phenacyl esterification or other esterification or transesterification procedures (see Protocol C) [33]. The desired lipid fractions are first scraped from the plate into a clean test tube. Deionized water is added to cover the silica, which facilitates the movement of the lipid from the silica because water is far more polar

TLC lipid separation

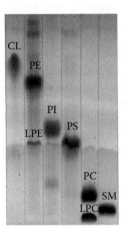
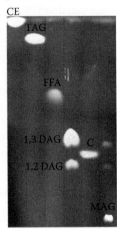

Phospholipid plate   Neutral lipid plate

FIGURE 11.1   TLC separation of phospholipid standards (Avanti Polar Lipids, Alabaster, Alabama) and neutral lipid standards (NuChek Prep, Elysian, Minnesota) on a Merck 60 plate with a preabsorbant zone. The phospholipids were separated in a solvent system containing chloroform:methanol:acetic acid:water (55:37.5:3:2 by vol.). The neutral lipids were separated in a solvent system containing petroleum ether:diethyl ether:water (75:25:1.3 by vol.). The TLC plates were activated at 110°C for 24 h prior to spotting of the samples in hexane:2-propanol (3:2 by vol.). The solvents were poured into the TLC tanks, mixed, and the TLC tank sat covered for 30–45 min. After samples were spotted, the TLC plates were heated to 110°C in an oven for 10 min, and then cooled for 30 s prior to placing in the tank containing the appropriate solvent system. Samples were visualized with iodine for the phospholipids or 6-*p*-toluidino-2-naphthalenesulfonic acid (TNS) [89] for the neutral lipids.

than the lipids and will saturate the silica's binding sites, thus freeing the lipids. Then, 3 mL of hexane-isopropanol (3:2 by vol., HIP) is added and the mixture is vortexed vigorously for at least 1 min. The lipids will preferentially dissolve in the HIP and thus are in a solution free of silica. HIP is used because it is an organic solvent that will easily dissolve neutral lipids and has just enough polarity to dissolve more polar lipid fractions such as the various phospholipids. The vortexed mixture is then centrifuged at $2500 \times g$ for 10 min to separate the organic and aqueous layers. While we often use a refrigerated centrifuge set at −10°C, because the cold temperature facilitates phase separation, cooling the samples is not necessary. If you do not have access to a refrigerated centrifuge, an alternative method to cool the samples, if cooling is desired, is to place the samples in the freezer and then subject them to centrifugation only after they are cold. The upper organic layer is then aspirated using a Pasteur pipet and transferred into another clean test tube. The remaining lower phase is re-extracted with a second aliquot of HIP (3 mL) and this mixture is then vortexed, subjected to centrifugation to facilitate phase separation, and, again, the upper phase is removed and added to the first aliquot. This process can be repeated a third time if desired, to maximize the lipid recovery. The lipid containing extract is then dried down under a stream of nitrogen and the lipids redissolved in a desired solvent for further processing or analysis.

11.4 FATTY ACID ANALYSIS BY HPLC

Methodology for FA analysis by GLC is rapid and easy, so it might be thought that there is no place for HPLC. However, there are certain circumstances when HPLC is necessary. First, the fact that HPLC runs at room temperature makes this technique suitable for the analysis of thermolabile molecules. Second, positional and conformational isomers are more easily separated by HPLC than GLC. Consequently, the separation and quantification of hydroxylated, branched-chain, trans or conjugated FA is easier by HPLC. Finally, HPLC methodology can be used for preparative scale separations of a particular FA or for studying isotopically labeled FA.

In contrast to GLC which prefers flame ionization detection, the choice of the detector for HPLC analysis determines the procedure to adopt. Several detectors are possible but the most used ones are light scattering, UV, fluorescence, and radio-activity. However, many of these detectors are not useful for detecting FA because they do not have a good chromophore or fluorophore. Some methods with UV detection (not very sensitive) at 200 nm have been described for the separation of non-derivatized FFA [21,78]. However, much greater sensitivity is achieved if phenacyl or related derivatives of FA are prepared, which allow their detection at higher wavelengths. Thus, fluorescent detection emerged as the most sensitive tool in HPLC trace analysis.

11.4.1 PHENACYL ESTERS AND RELATED DERIVATIVES

Phenacyl esters and related derivatives absorb strongly in the UV region of the spectrum (see Figure 11.2, see Protocol E). Taking advantage of this fact, FA can be separated and quantified after conversion to their corresponding phenacyl esters [79–81]. This is one of the most frequently used methods, it is easy to prepare, the reaction gives quantitative yields, and it is a convenient method with which to start. This method was used to study the ability of erucic acid (22:1n-9) to cross the blood–brain barrier and its subsequent metabolism via chain shortening in the brain [31]. First, the lipid extract is saponified at 100°C for 30 min in 2% KOH in ethanol, which is then acidified with hydrochloric acid. The released FA are extracted with hexane, and the phenacyl esters are then prepared by the addition of acetone containing 2-bromoacetophenone (10 mg/mL) and triethylamine (10 mg/mL), followed by incubation at 100°C for 5 min. After the addition of acetic acid (2 mg/mL), the samples are then incubated for another 5 min. Finally, individual FA phenacyl esters are separated by HPLC on a C-18(2) Luna column using a binary solvent system consisting of water and acetonitrile.

A broad choice of stationary phases can be used for separation, such as reverse-phase [81,82] or silver ion modes [83]. This method allows the separation of phenacyl derivatives of monounsaturated and polyunsaturated FA as well as geometric isomers [80]. The detection limit is in the nanogram ranges and the detection wavelength is 242 nm.

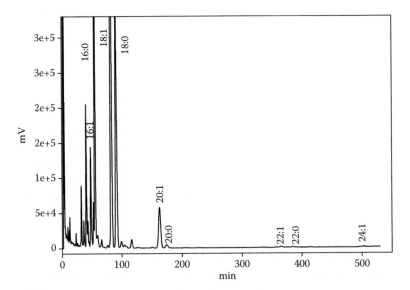

FIGURE 11.2 HPLC chromatogram of a rat liver phospholipid sample prepared as phenacyl esters. Separation is on a Luna 18(2) column (4.6×250 mm, Phenomenex, Torrance, California) using the solvent system described in the text.

11.4.2 ANTHRONYL DERIVATIVES

A highly sensitive method for the quantitative analysis of phospholipid molecular species was developed by Takamura et al. [84], which is a different approach as compared to that discussed in Chapter 6. However, this approach is very useful when examining radiotracer disposition into specific phospholipid molecular species, something that cannot be readily done using mass spectrometry techniques due to the radioactivity. Nonetheless, this protocol was successfully used to analyze brain phosphatidylethanolamine and phosphatidylcholine molecular species [85,86]. First, the individual phospholipid classes were separated by TLC. Then, after scrapping the corresponding band, the phospholipid is directly resuspended in 1 mL diisopropylether; 1 mL of 30 mM Tris–borate buffer (pH 7.5) and 20 μL phospholipase C of *Bacillus cereus* was added. This reaction needs continuous stirring and takes place overnight at room temperature, and to stop the reaction, 2 mL of hexane/isopropylether (1:1 v/v) is added. After vortexing twice, the upper phase is separated and transferred into new test tubes and evaporated at room temperature in a centrifugal evaporator for 30 min. The phospholipids are then mixed with 10 μL of *N,N*-diisopropylethylamine and 200 μL of 9-anthroyl chloride solution in a sealed test tube and heated at 60°C for 0.5 h. After cooling the tubes in ice, the reaction is stopped by adding 3 mL of 100 mM NH$_4$OH solution. The reaction product is then extracted with 3.5 mL of chloroform and washed for three times with 3 mL of 100 mM NH$_4$OH solution. The lower phase is collected into new test tubes and evaporated to dryness.

Next, the anthronyl-diacylglycerol derivatives are separated into diacyl, alkylacyl, and alkenylacyl subclasses by TLC. The solvent system used is hexane:toluene:diiso-propylether (10.0:9.0:0.8 by vol.) and the three subclasses are easily visualized under UV light. Each subclass is extracted with 2 mL of hexane/diisopropylether. The final product is redissolved in acetonitrile and injected into the HPLC for analysis. Each glycerophospholipid subclass is segregated into individual molecular species on a Supelcolsil LC-18 column using acetonitrile:propane-2-ol, 75:25 (v/v) as solvents with a flow rate of 0.2 mL [85,86].

Takamura and Kito [87] also described a method to quantitatively analyze the polyenoic fatty content by synthesizing dinitrobenzoyl derivatives. The dinitroben-zoyl derivatives were separated into individual molecular species by reversed HPLC using a combination of two solvent systems, acetonitrile/2-propanol (80:20, v/v) or methanol/2-propanol (95:5, v/v), and they were quantified at 254 nm.

In addition, HPLC or silica gel columns can be also used for the analysis or iso-lation of FA with polar functional groups, especially oxygenated moieties such as hydroperoxides. With care, the separation of isomers differing in the position of hydroperoxy or hydroxy groups on an aliphatic chain can be achieved [10]. A variety of commercial chiral stationary phases are available, and they can be used to resolve FA containing enantiomeric functional groups and hydroperoxy or hydroxy groups [88]. Silver-ion chromatography is a good choice to quantitatively separate cis- and trans-FA.

11.5   CONCLUSIONS

We live in a time where there is a revolution in techniques used to measure lipids as the reliability and availability of mass spectrometry instruments has improved. This current era is similar to what was seen in the early 1960s, with the onset of GLC analysis and then again in the mid-to-late 1970s, with the proliferation of HPLC techniques for lipid analysis. However, just as lipid analytical data from the 1960s is still valid today, so is the use of these "older technologies" to address lipid sepa-ration and analysis. We have found that many of these standard techniques are as valuable today as they were 20 years ago. In fact, when assessing FA conversion into longer or shorter chains, the greater sensitivity of radiotracer methods cannot be ignored, yet the use of radioactive samples in a mass spectrometer is not appropri-ate. Clearly, there is a need to maintain and improve our more classical approaches and use them hand-in-hand with methods such as mass spectrometry. Hence, the importance of this chapter is to introduce analysts to this "older technology" and to provide detailed protocols that the reader may find useful in addressing their lipid analytical needs.

ACKNOWLEDGMENTS

This work was supported by grants 1R21 NS060141 to EJM from the NIH and by a project on a 1P20 RRR17699. We thank Cindy Murphy for the typed preparation of this chapter.

APPENDIX

Protocol A: Acid-catalyzed esterification

Solutions

1. Toluene:anhydrous methanol 1:1 + 2% H_2SO_4 *freshly prepared and it must be clear!

Or

2. Toluene:anhydrous methanol 1:1 + 4% H_2SO_4 *freshly prepared and it must be clear!
3. Toluene:anhydrous methanol 1:1 *freshly prepared and it must be clear!
4. Hexane
5. dH_2O (distilled = d)
6. 2,2,4 trimethylpentane (isooctane)

Samples from TLC plates to be stored overnight

1. TLC scrapings containing FFA are placed in large screw-top test tubes (16 × 125 mm) and 1 mL fresh toluene:methanol 1:1 is added per tube.
2. Flush tube with N_2 gas, tightly cap and store overnight at −80°C.
3. To esterify this sample, add 1 mL of toluene:methanol 1:1 + 4% H_2SO_4 for a final H_2SO_4 of 2%.
4. Place in a shaking water bath at 70°C for 4 h.
5. After 4 h, add 1 mL of dH_2O to each tube to quench the reaction.
6. Add internal standard of choice (we use 17:0 methyl ester).
7. Add 5 mL of hexane and vortex vigorously.
8. Chill in −80°C freezer for 15–30 min to facilitate phase separation.
9. Centrifuge at 600–800 × g for 10 min.
10. Remove upper phase with a Pasteur pipet (you need not get 100% of the upper phase at this step) and put it into a clean screw-top test tube (16 × 125 mm).
11. Rinse remaining lower phase with 3 mL of hexane and repeat steps 8–10, but add to the original tube (repeat this step again, i.e., two 3 mL rinses).
12. Chill sample overnight at −80°C to facilitate phase separation of any transferred lower phase, remove with a Pasteur pipet.
13. Prepare sample for transfer to microvial with limited volume insert (see below).

Samples from TLC

1. Place samples containing silica into a screw-top tube (16 × 125 mm).
2. Add 1 mL of toluene:methanol + 2% H_2SO_4 to bring final to 2% H_2SO_4.
3. Flush with nitrogen and tighten cap (Teflon lined phenolic cap is ideal).
4. Place in a shaking water bath at 70°C for 4 h.
5. After 4 h, add 1 mL of dH_2O to each tube to quench the reaction.
6. Add internal standard of choice (we use 17:0 methyl ester).

7. Add 5 mL of hexane and vortex vigorously.
8. Chill in −80°C freezer for 15–30 min to facilitate phase separation.
9. Centrifuge at 600–800×g for 10 min.
10. Remove upper phase with a Pasteur pipet (you need not get 100% of the upper phase at this step) and put it into a clean screw-top test tube (16 × 125 mm).
11. Rinse remaining lower phase with 3 mL of hexane and repeat steps 8–10, but add to the original tube (repeat this step again, i.e., two 3 mL rinses).
12. Chill sample overnight at −80°C to facilitate phase separation of any transferred lower phase, remove with a Pasteur pipet.
13. Prepare sample for transfer to microvial with limited volume insert (see below).

Sample transfer

1. Remove all of the hexane in the tube under a stream of nitrogen (**remember to heat the tube in a water bath or dry bath; about 45°C will work**).
2. Rinse the sides of the tube with approximately 0.25 mL of hexane (repeat twice).
3. Working with one tube at a time, add ~125 µL of hexane to the tube using a pipetor, mix, aspirate, and place into an insert in microvial.
4. Then add ~80 µL hexane to the sample tube, mix, and repeat transfer to the insert in the microvial (repeat two more times).
5. When all microvials are ready, remove the solvent using a Speedvac set on medium heat for 15–20 min.
6. Repeat the above procedure three times and then discard original tubes.
7. Cap and store the microvials containing the last rinse of hexane at −80°C until analysis.

Preparation for GLC

1. Remove the hexane in the microvial using a Speedvac set on medium heat for 15–20 min.
2. Rinse around top of each microvial with 30–40 µL hexane and remove this hexane using the Speedvac.
3. To each outer vial, add 200 µL of isooctane (2,2,4 trimethylpentane), and to each insert add the same volume as the 17:0 added previously.
4. Cap vials securely and run on the GLC.

Protocol B: Base-catalyzed transesterification

Solutions

1. 0.5 M KOH in anhydrous methanol ***freshly prepared!**

| Potassium hydroxide | 2.806 g | 3.51 g | 4.21 g |
| Anhydrous methanol | 100 mL | 125 mL | 150 mL |

Or

2. 1 M KOH in anhydrous methanol ***freshly prepared!**

Potassium hydroxide	5.611 g
Anhydrous methanol	100 mL

3. Methyl formate
4. Hexane
5. dH$_2$O (distilled = d)
6. 2,2,4 trimethylpentane (isooctane)

Preparation

1. Silica scrapings from TLC plates are placed in large screw-top test tubes (16 × 125 mm), and 1 mL anhydrous methanol is added per tube, which can be stored overnight at −80°C.
   Or
2. Samples separated by HPLC are collected in screw-top test tubes and the solvent removed by evaporation under a stream of nitrogen and transesterified immediately or samples can be stored in hexane:2-propanol (3:2 by vol.) at −80°C.
   Or
3. Silica scraping from TLC are placed in large screw-top test tubes (16 × 125 mm) and transesterified.

Procedure

1. For silica samples stored in methanol, add 1 mL of 1 M KOH/MEOH per tube for a final concentration of 0.5 M KOH.
   Or
2. For fresh silica samples, add 2 mL of 0.5 M KOH/MEOH per tube.
   Or
3. For HPLC samples, evaporate all of the solvent under a stream of nitrogen and add 2 mL of 0.5 M KOH/MEOH per tube.

Note: Final KOH concentration is always 0.5 M KOH

4. Flush the tube with nitrogen gas.
5. Cap, vortex, and in 1 min quench the reaction by adding 200 µL of methyl formate and mix by vortexing.
6. Add the desired amount of internal standard (we use 17:0 methyl ester).
7. Add 3 mL of hexane and vortex for about 30 s per tube.
8. Chill in −80°C freezer for 15–30 min to facilitate phase separation.
9. Centrifuge at 600–800 × g for 10 min.
10. Remove upper phase with a Pasteur pipet (you need not get 100% of the upper phase at this step), and put it into a clean screw-top test tube (16 × 125 mm).

11. Rinse remaining lower phase with 2 mL of hexane and repeat steps 8–10, but add to the original tube (repeat this step again, i.e., two 2 mL rinses).
12. Chill sample overnight at −80°C to facilitate phase separation of any transferred lower phase, remove any remaining lower phase with a Pasteur pipet (note: This will contain glycerol and KOH, neither of which is good for GLC columns).
13. Prepare sample for transfer to microvial with limited volume insert (see below).

Sample transfer

1. Remove all of the hexane in the tube under a stream of nitrogen (**remember to heat the tube in a water bath or dry bath, about 45°C will work**).
2. Rinse the sides of the tube with approximately 0.25 mL of hexane (repeat twice).
3. Working with one tube at a time, add ~125 µL of hexane to tube using a pipetor, mix, aspirate, and place into an insert in microvial.
4. Then add ~80 µL hexane to the sample tube, mix, and repeat transfer to the insert in the microvial (repeat two more times).
5. When all microvials are ready, remove the solvent using a Speedvac set on medium heat for 15–20 min.
6. Repeat the above procedure three times and then discard original tubes.
7. Cap and store the microvials containing the last rinse of hexane at −80°C until analysis.

Preparation for GLC

1. Remove the hexane in the microvial using a Speedvac set on medium heat for 15–20 min.
2. Rinse around top of each microvial with 30–40 µL hexane and remove this hexane using the Speedvac.
3. To each outer vial add 200 µL of isooctane (2,2,4 trimethylpentane), and to each insert add the same volume as the 17:0 added previously.
4. Cap vials securely and run on the GLC.

Protocol C: Elution of lipids from silica

Solutions

1. Hexane: 2-propanol (3:2 by vol.)
2. dH$_2$O (distilled = d)

Procedure

1. Scrape each lipid fraction from the TLC plates and place into a clean test tube.
2. Add enough water to cover the silica (typically 1 to 1.5 mL) and then add 3 mL of hexane-isopropanol (HIP 3:2 by vol.).

3. Vortex for at least 1 min.
4. Centrifuge at 600–$800 \times g$ for 10 min.
5. Transfer upper organic layer into another clean test tube using a Pasteur pipet and save this test tube as the lipids extracted from the silica are here.
6. Add another 3 mL of HIP to the silica and repeat the vortexing and centrifugation and add this organic layer to that transferred in Step 5.

Protocol D: Extraction of human red blood cells

Solutions

1. Hexane: 2-propanol (3:2 by vol.)
2. dH$_2$O (distilled = d)
3. Hexane
4. 2-propanol

Procedures

1. Into large screw top tubes, pipet 500 μL packed cells, add 500 μL H$_2$O, and vortex well, then hold on ice for 15 min.
2. Then add 4 mL 2-propanol per tube and vortex well, then vortex frequently over the next hour.
3. To the 2-propanol, add 6 mL of hexane per tube, vortex well, and then vortex frequently over the next hour.
4. Pellet the debris by centrifugation at 600–$800 \times g$ for 10 min.
5. Transfer the extract as cleanly as possible into a clean large screw-top tube (16×125 mm).
6. Remove the solvent under a stream of nitrogen and rinse the sides of the tube with multiple rinses of ~250 μL of hexane:2-propanol (3:2 by vol., HIP).
7. After rinsing the tube, add ~200 μL of HIP, mix, and then transfer the lipids into a clean tube using a Pasteur pipet (repeat this procedure three times).

Note: This separates the lipid extract from residual blood proteins and other debris not removed by the first centrifugation step.

8. This sample can then be used for further processing by TLC (Figure 11.1) or direct transesterification to assess FA changes in RBC phospholipids using the steps outlined in Protocol B.

Protocol E: Phenacyl derivatives

Solutions

1. Ethanol containing 2% KOH (2 g per 100 mL)
2. Concentrated HCl
3. dH$_2$O (distilled = d)
4. Acetone containing 1% 2-bromoacetophenone (1 g per 100 mL)

5. Acetone containing 1% triethylamine (1 mL in 100 mL)
6. Acetone containing 0.2% acetic acid (0.2 mL in 100 mL)
7. Acetonitrile

Procedures

1. Dry lipid-containing samples down under a stream of nitrogen.
2. Immediately after drying down, add 2 mL of 2% KOH in ethanol (2 g in 100 mL of ethanol).
3. Heat the samples in a boiling water bath for 30 min.
4. Allow the samples to cool to room temperature.
5. Add 1 mL of H_2O and 0.1 mL of concentrated HCl.
6. Then add 3 mL of hexane and vortex for at least 1 min.
7. Facilitate phase separation by centrifugation at $600-800 \times g$ and remove the upper organic phase, which contains the FA.
8. Repeat the hexane extraction to maximize recovery.
9. Dry down the extracted samples under a stream of nitrogen.
10. Immediately after drying, add 0.1 mL of 1% 2-bromoacetophenone in acetone and 0.1 mL of 1% triethylamine in acetone.
11. Vortex the samples for at least 1 min.
12. Heat in boiling water bath for 5 min.
13. Allow the samples to cool to room temperature.
14. Then add 0.16 mL of 0.2% acetic acid in acetone.
15. Vortex for at least 1 min.
16. Heat the samples in a boiling water bath for 5 min.
17. Then dry the samples down under nitrogen and dissolve them in acetonitrile for HPLC analysis.

LIST OF ABBREVIATIONS

BHT	butylated hydroxytoluene
CoA	coenzyme A
FA	fatty acids
FAME	fatty acid methyl esters
FFA	free or unesterified fatty acids
GLC	gas–liquid chromatography
HPLC	high-performance liquid chromatography
PUFA	polyunsaturated fatty acids
TLC	thin-layer chromatography

REFERENCES

1. Jump, D.B. (2002) Dietary polyunsaturated fatty acids and regulation of gene transcription. *Curr. Opin. Lipidol.* 13:155–164.
2. Xiao, Y.F., Sigg, D.C., and Leaf, A. (2005) The antiarrhythmic effect of n-3 polyunsaturated fatty acids: Modulation of cardiac ion channels as a potential mechanism. *J. Membr. Biol.* 206:141–154.

3. Guizy, M. et al. (2008) Modulation of the atrial specific Kv1.5 channel by the n-3 poly-unsaturated fatty acid, alpha-linolenic acid. *J. Mol. Cell. Cardiol.* 44:323–335.

4. Szekeres, C.K. et al. (2000) Eicosanoid activation of extracellular signal-regulated kinase 1/2 in human epidermoid carcinoma cells. *J. Biol. Chem.* 275:38831–38841.

5. Fang, K.M. et al. (2008) Arachidonic acid induces both Na$^+$ and Ca^{2+} entry resulting in apoptosis. *J. Neurochem.* 104:1177–1189.

6. Wolfe, L.S. (1982) Eicosanoids: Prostaglandins, thromboxanes, leukotrienes, and other derivatives of carbon-20 unsaturated fatty acids. *J. Neurochem.* 38:1–14.

7. Christie, W.W. (1989) The preparation of derivatives of fatty acids, in *Gas Chromatography and Lipids: A Practical Guide*, Christie, W.W. (ed.), The Oily Press Ltd., Dundee, Scotland, pp. 36–47.

8. Eder, K., Reichlmayr-Lais, A.M., and Kirchgessner, M. (1992) Studies on the metha-nolysis of small amounts of purified phospholipids for gas chromatographic analysis of fatty acid methyl esters. *J. Chromatogr.* 607:55–67.

9. Brondz, I. (2002) Development of fatty acid analysis by high-performance liquid chro-matography, gas chromatography, and related techniques. *Anal. Chim. Acta* 465:1–37.

10. Christie, W.W. (2003) Preparation of derivatives of fatty acids, in *Lipid Analysis*, Christie, W.W. (ed.), 3rd edn., The Oily Press, Bridgwater, U.K., pp. 205–224.

11. Christie, W.W. (1982) *Lipid Analysis*, Pergamon Press, Oxford, U.K.

12. Bazan, N.G. (1970) Effects of ischemia and electroconvulsive shock on free fatty acid pool in the brain. *Biochim. Biophys. Acta* 218:1–10.

13. Bazan, N.G. (1971) Changes in free fatty acids of brain by drug-induced convulsions, electroshock and anesthesia. *J. Neurochem.* 18:1379–1385.

14. Bazan, N.G. (2003) Synaptic lipid signaling: Significance of polyunsaturated fatty acid and platelet-activating factor. *J. Lipid Res.* 44:2221–2233.

15. Murphy, E.J. and Horrocks, L.A. (1993) Mechanisms of hypoxic and ischemic injury: Use of cell culture models. *Mol. Chem. Neuropathol.* 19:95–106.

16. Murphy, E.J. et al. (1994) Lipid alterations following impact spinal cord trauma in the rat. *Mol. Chem. Neuropathol.* 23:13–26.

17. Golovko, M.Y. and Murphy, E.J. (2008) Brain prostaglandin formation is increased by α-synuclein gene-ablation during global ischemia. *Neurosci. Lett.* 432:243–247.

18. Saunders, R.D. et al. (1987) Effects of methylprednisolone and the combination of α-tocopherol and selenium on arachidonic acid metabolism and lipid peroxidation in traumatized spinal cord tissue. *J. Neurochem.* 49:24–31.

19. Saunders, R. and Horrocks, L.A. (1987) Eicosanoids, plasma membranes, and molecu-lar mechanisms of spinal cord injury. *Neurochem. Pathol.* 7:1–22.

20. James, A.T. and Martin, A.J. (1952) Gas-liquid partition chromatography: The separa-tion and micro-estimation of volatile fatty acids from formic acid to dodecanoic acid. *Biochem. J.* 50:679–690.

21. Ackman, R. and Burgher, R.D. (1963) Quantitative gas liquid chromatographic estima-tion of volatile fatty acids in aqueous media. *Anal. Chem.* 35:647–652.

22. Brondz, I., Olsen, I., and Greibrokk, T. (1983) Lipids separation and analysis. *J. Chromatogr.* 274:299–304.

23. Ceccon, L. (1990) Quantitative determination of free volatile fatty acids from dairy products on a Nukal capillary column. *J. Chromatogr.* 519:369–378.

24. Firestone, D.H.W. (1979) IUPAC gas chromatographic method for determination of fatty acid composition: Collaborative study. *J. Assoc. Off. Anal. Chem.* 62:709–721.

25. Christie, W.W. (1993) Preparation of ester derivatives of fatty acids for chromatographic analysis, in *Advances in Lipid Methodology-Two*, Christie, W.W. (ed.), The Oily Press Ltd., Dundee, Scotland, pp. 69–111.

26. Liu, K.-S. (1994) Preparation of fatty acid methyl esters for gas-chromatographic analy-sis of lipids in biological samples. *J. Am. Oil Chem. Soc.* 71:1179–1187.

27. Brondz, I., Olsen, I., and Sjøøstrøm, M. (1989) Gas chromatographic assessment of alcoholyzed fatty acids from yeasts: A new chemotaxonomic method. *J. Clin. Microbiol.* 27:2815–2819.
28. Peuchant, E.W.R., Salles, C., and Jensen, R. (1989) One step extraction of human erythrocyte lipids allowing rapid determination of fatty acid composition. *Anal. Biochem.* 181:341–349.
29. Iverson, J. and Sheppard, A.J. (1977) Butyl ester preparation for gas-liquid chromatographic determination of fatty acids in butter. *J. Assoc. Off. Anal. Chem.* 60:284–288.
30. Molnár-Perl, I. and Pintér-Szakács, M. (1983) Gas chromatographic analysis of C1-C20 carboxylic acids esterified in aqueous solutions as isobutyl esters. *Chromatographia* 17:493–500.
31. Golovko, M.Y. and Murphy, E.J. (2006) Uptake and metabolism of plasma derived erucic acid by rat brain. *J. Lipid Res.* 47:1289–1297.
32. Sprecher, H., Chen, Q., and Yin, F.Q. (1999) Regulation of the biosynthesis of 22:5 n-6 and 22:6 n-3: A complex intracellular process. *Lipids* 34:S153–S156.
33. Murphy, C.C., Murphy, E.J., and Golovko, M.Y. (2008) Erucic acid is differentially taken up and metabolized in rat liver and heart. *Lipids* 43:391–400.
34. Moore, S.A. et al. (1989) Brain microvessels produce 12-hydroxyeicosatetraenoic acid. *J. Neurochem.* 53:376–382.
35. Moore, S.A. et al. (1991) Astrocytes, not neurons produce docosahexaenoic acid (22:6 omega-3) and arachidonic acid (20:4 omega-6). *J. Neurochem.* 56:518–524.
36. Craig, B., Tulloch, A.P., and Murty, N.L. (1963) Quantitative analysis of short chain fatty acids using gas liquid chromatography. *J. Am. Oil Chem. Soc.* 40:61–63.
37. Eder, K. (1995) Gas chromatographic analysis of fatty acid methyl esters. *J. Chromatogr. B Biomed. Appl.* 671:113–131.
38. Molnár-Perl, I. and Pintér-Szakács, M. (1986) Modifications in the chemical derivatization of carboxylic acids for their gas chromatographic analysis. *J. Chromatogr.* 365:171–182.
39. Brondz, I. and Olsen, I. (1986) Chemotaxonomy of selected species of the *Actinobacillus-Haemophilus-Pasteurella* group by means of gas chromatography, gas chromatography-mass spectrometry and bioenzymatic methods. *J. Chromatogr.* 380:1–17.
40. Gander, G., Jensen, R.G., and Sampugna, J. (1962) Analysis of milk fatty acids by gas-liquid chromatography. *J. Dairy Sci.* 45:323–328.
41. Akesson, B., Elovsson, J., and Arvidsson, G. (1970) Initial incorporation into rat liver glycerolipids of intraportally injected [³H] glycerol. *Biochim. Biophys. Acta* 210:15–27.
42. Segura, R. (1988) Preparation of fatty acid methyl esters by direct transesterification of lipids with aluminum chloride-methanol. *J. Chromatogr.* 441:99–113.
43. Isaiah, N., Subbarao, R., and Aggarwal, J.S. (1969) Analysis of phenyl esters of fatty acids, hydroxyphenyl alkyl ketones and long chain alkyl by thin-layer chromatographic techniques. *J. Chromatogr.* 43:519–522.
44. Shantha, N.C. and Napolitano, G.E. (1992) Gas chromatography of fatty acids. *J. Chromatogr.* 624:37–51.
45. Metcalfe, L. and Schmitz, A.A. (1961) The rapid preparation of fatty acid esters for gas chromatographic analysis. *Anal. Chem.* 33:363–364.
46. Morrison, W.R. and Smith, L.M. (1964) Preparation of fatty acid methyl esters and dimethylacetals from lipids with boron fluoride-methanol. *J. Lipid Res.* 5:600–608.
47. Jones, E. and Davison, V.L. (1965) Quantitative determination of double bond positions in unsaturated fatty acids after oxidative cleavage. *J. Am. Oil Chem. Soc.* 42:121–126.
48. Salwin, H. and Bond, J.F. (1969) Quantitative determination of lactic acid and succinate acid in foods by gas chromatography. *J. Assoc. Off. Anal. Chem.* 52:41–47.

49. Murphy, E.J. and Horrocks, L.A. (1993) Effects of differentiation on the phospholipid and phospholipid fatty acid compositions of N1E-115 neuroblastoma cells. *Biochim. Biophys. Acta* 1167:131–136.

50. Lough, A.K. (1964) The production of methoxy-substituted fatty acids as artifacts during the esterification of unsaturated fatty acids with methanol containing boron trifluoride. *Biochem. J.* 90:4C–5C.

51. Alexander, L.R., Justice, J.B. Jr., and Madden, J. (1985) Fatty acid composition of human erythrocyte membranes by capillary gas chromatography-mass spectrometry. *J. Chromatogr.* 342:1–12.

52. Christopherson, S. and Glass, R.L. (1969) Preparation of milk fat methyl esters by alcoholysis in an essentially nonalcoholic solution. *J. Dairy Sci.* 52:1289–1290.

53. Brockerhoff, H. (1975) Determination of the positional distribution of fatty acids in glycerolipids. *Methods Enzymol.* 35:315–325.

54. Hubscher, G., Hawthorne, J.N., and Kemp, P. (1960) The analysis of tissue phospholipids: Hydrolysis procedure and results with pig liver. *J. Lipid Res.* 1:433–438.

55. Glass, R. (1971) Alcholysis, saponification and the preparation of fatty acid methyl esters. *Lipids* 6:919–925.

56. Ansell, G.B. and Spanner, S. (1963) The alkaline hydrolysis of the ethanolamine plasmalogen in brain tissue. *J. Neurochem.* 10:941–945.

57. Murphy, E.J., Anderson, D.K., and Horrocks, L.A. (1993) Phospholipid and phospholipid fatty acid composition of mixed murine spinal cord neuronal cultures. *J. Neurosci. Res.* 34:472–477.

58. McCreary, D. et al. (1978) A novel and rapid method for the preparation of methyl esters for gas chromatography: Application to the determination of the fatty acids of edible fats and oils. *J. Chromatogr. Sci.* 16:329–331.

59. Metcalfe, L. and Wang, C.N. (1981) Rapid preparation of fatty acid methyl esters using organic base-catalyzed transesterification. *J. Chromatogr. Sci.* 19:530–535.

60. Misir, R., Laarveld, B., and Blair, R. (1985) Evaluation of a rapid method for preparation of fatty acid methyl esters for analysis by gas-liquid chromatography. *J. Chromatogr.* 331:141–148.

61. Williams, M. and MacGee, J. (1983) Rapid determination of free fatty acids in vegetable oils by gas liquid chromatography. *J. Am. Oil Chem. Soc.* 60:1507–1509.

62. Butte, W. (1983) Rapid method for the determination of fatty acid profiles from fats and oils using trimethylsulphonium hydroxide for transesterification. *J. Chromatogr.* 261:142–145.

63. Schlenk, H. and Gellerman, J.L. (1960) Esterification of fatty acids with diazomethane on a small scale. *Anal. Chem.* 32:1412–1414.

64. Stein, R.A., Slawson, V., and Mead, J.F. (1967) Gas-liquid chromatography of fatty acids and derivatives, in *Liquid Chromatographic Analysis*, Marinetti, G.V. (ed.), Marcel Dekker, Inc., New York, vol 1, p. 364.

65. Prasad, M. et al. (1988) Analysis of tissue free fatty acids isolated by aminopropyl bonded-phase columns. *J. Chromatogr.* 428:221–228.

66. Drucker, D. (1981) Detection of microorganisms by gas chromatography, in *Microbiological Applications of Gas Chromatography*, Drucker, D. (ed.), Cambridge University Press, London, U.K., pp. 166–296.

67. Biondi, P.A. and Cagnasso, M. (1975) A procedure for boron trifluoride-catalyzed esterification suitable for use in gas chromatographic analysis. *J. Chromatogr.* 109:389–392.

68. Golovko, M.Y. et al. (2006) Acyl-CoA synthetase activity links wild-type but not mutant α-synuclein to brain arachidonate metabolism. *Biochemistry* 45:6956–6966.

69. Lepage, G. and Roy, C.C. (1986) Direct transesterification of all classes of lipids in a one-step reaction. *J. Lipid Res.* 27:114–120.

70. Murphy, E.J. et al. (2000) Phospholipid and phospholipid fatty acid composition of L-cell fibroblast: Effect of intestinal and liver fatty acid binding proteins. *Lipids* 35:729–738.

71. Prows, D.R. et al. (1996) Intestinal fatty acid-binding protein expression stimulates fibroblast fatty acid esterification. *Chem. Phys. Lipids* 84:47–56.
72. Murphy, E.J. et al. (1996) Liver fatty acid binding protein expression in transfected fibroblasts stimulates fatty acid uptake and metabolism. *Biochim. Biophys. Acta* 1301:191–196.
73. Yang, P. et al. (2006) Determination of endogenous tissue inflammation profiles by LC/MS/MS: COX- and LOX-derived bioactive lipids. *Prostaglandins Leukot. Essent. Fatty Acids* 75:385–395.
74. Golovko, M.Y. and Murphy, E.J. (2008) An improved LC-MS/MS procedure for brain prostanoid analysis using brain fixation with head-focused microwave irradiation and liquid-liquid extraction. *J. Lipid Res.* 49:893–904.
75. Folch, J., Lees, M., and Sloan Stanley, G.H. (1957) A simple method for the isolation and purification of total lipids from animal tissues. *J. Biol. Chem.* 226:497–509.
76. Bligh, E.G. and Dyer, W.J. (1959) A rapid method of total lipid extraction and purification. *Can. J. Biochem. Physiol.* 37:911–917.
77. Brown, J. and Ault, W.C. (1930) A comparison of the highly unsaturated acids of beef, hog, and sheep brain. *J. Biol. Chem.* 89:167–171.
78. VanRollins, M. et al. (1982) High pressure liquid chromatography of underivatized fatty acids, hydroxy acids, and prostanoids having different chain lengths and double bond positions, in *Methods in Enzymology*, Lands, W.E.M. and Smith, W.L. (eds.), vol. 86, Academic Press, New York, pp. 518–543.
79. Durst, H.D. et al. (1975) Phenacyl esters of fatty acids via crown ether catalysts for enhanced ultraviolet detection in liquid chromatography. *Anal. Chem.* 47:1797–1801.
80. Wood, R. and Lee, T. (1983) High-performance liquid chromatography of fatty acids: Quantitative analysis of saturated, monoenoic, polyenoic and geometrical isomers. *J. Chromatogr.* 254:237–246.
81. Chen, H. and Anderson, R.E. (1992) Quantitation of phenacyl esters of retinal fatty acids by high-performance liquid chromatography. *J. Chromatogr.* 578:124–129.
82. Engelmann, G.J. et al. (1988) Rapid method for the analysis of red blood cell fatty acids by reversed-phase high-performance liquid chromatography. *J. Chromatogr.* 432:29–36.
83. Christie, W.W. (1989) Silver ion chromatography using solid-phase extraction columns packed with a bonded-sulfonic acid phase. *J. Lipid Res.* 30:1471–1473.
84. Takamura, H. et al. (1986) Quantitative analysis of polyenoic phospholipid molecular species by high performance liquid chromatography. *Lipids* 21:356–361.
85. Barceló-Coblijn, G. et al. (2003) Modification by docosahexaenoic acid of age-induced alterations in gene expression and molecular composition of rat brain phospholipids. *Proc. Natl. Acad. Sci. USA* 100:11321–11326.
86. Barceló-Coblijn, G. et al. (2003) Gene expression and molecular composition of phospholipids in rat brain in relation to dietary n-6 and n-3 fatty acid ratio. *Biochim. Biophys. Acta* 1632:72–79.
87. Takamura, H. and Kito, M. (1991) A highly sensitive method for quantitative analysis of phospholipids molecular species by high-performance liquid chromatography. *J. Biochem.* 109:436–439.
88. Christie, W.W. (1992) The chromatographic resolution of chiral lipids, in *Advances in Lipid Methodology-One*, Christie, W.W. (ed.), The Oily Press Ltd., Dundee, Scotland, pp. 121–148.
89. Jones, M., Keenan, R.W., and Horowitz, P. (1982) Use of 6-p-toluidino-2-naphthalenesulfonic acid to quantitate lipids after thin-layer chromatography. *J. Chromatogr.* 237:522–524.

12 Methods for Measuring Fatty Acid Uptake and Targeting in Cultured Cells and *in Vivo*

Paul G. Millner, Serena M. Lackman,
and Eric J. Murphy

CONTENTS

12.1 INTRODUCTION

Why study fatty acid uptake and why is it so important? To the average lipid biochemist, this of course is a question that will have them rattling off a long list of reasons and then moving over to their soap box to further illuminate the questioner about why granting agencies should care as well. So why is this question so important? First, we do not truly comprehend how this process occurs and only now have we begun to appreciate the multiple influences by which many systems regulate this process. Second, many individuals just squirt some fatty acid into some cells, generally arachidonic acid (ARA, not the commonly seen AA), and sometime later stimulate them with a particular agonist and then measure ARA release. What they do not know may ultimately significantly impact their experimental outcomes and lead to more confusion in the literature. Third, concentrations matter. Some people figure that if a little is good, a whole lot more will be better. With fatty acids, this influences where the fatty acid goes in the cell, to a membrane or to a lipid droplet or, in more metabolic terms, to be used in a phospholipid as a membrane constituent or to be stored in a triacylglycerol (TAG) because the cell cannot figure out what else to do with all of this fatty acid. Fourth, fatty acids have really become important of late as we now know that fatty acids are intimately involved in gene expression (see Chapters 9 and 10). Hence, fatty acids coming into a cell or tissue influence what genes are turned on, which in turn regulate downstream cellular function. So, in short, there are a lot of reasons to understand the bases for fatty acid uptake and trafficking in cells. Even if you are not a lipid biochemist but a person really interested in looking at lipid-mediated signal transduction, you need to understand some important aspects of fatty acid uptake and hopefully you will find what follows useful.

12.1.1 Importance of Fatty Acids in Signaling

Fatty acids have an important and crucial role in lipid-mediated signal transduction in virtually every biological system. ARA (20:4n-6), in particular, is a well-known signaling molecule that exerts its effects primarily through its metabolism to eicosanoids and other products. Its use as a signaling molecule occurs in virtually every organ system including the immune system. For instance, ARA is critical for the maintenance of synaptic vesicle pools and cholinergic and serotonergic neurotransmission in *Caenorhabditis elegans* [1]. In higher organisms, ARA regulates alpha-amino-3-hydroxy-5-methylisoxazole-4-propionate (AMPA)-mediated excitatory postsynaptic currents, demonstrating an important role in postsynaptic modulation of CA1 neurotransmission in the hippocampus [2]. In addition, stimulation of striatal and hippocampal neurons with N-methyl-D-aspartate (NMDA) activates phospholipase A_2-mediated ARA release and subsequent downstream ARA processing [3–5]. In the hippocampus, ARA release is critical for long-term potentiation (LTP) [4–9], a process that is thought to be important for the formation of memory. In aged rats, hippocampal ARA levels and LTP are reduced with increasing age [10]; however, restoration of these levels to those found in younger rats restores LTP [11]. Further, ARA and its products prolong synaptic exposure to glutamate and GABA through a reduction in glial-mediated uptake of these neurotransmitters [12,13] as well as through stimulating an increase in presynaptic glutamate release [10,13].

ARA is an important part of the heart's lipid-mediated signaling cascades. The heart takes up fatty acids at least an order of magnitude better than the brain [14,15], primarily because the heart uses fatty acids, such as palmitic acid (PAM), for a primary source of metabolic energy via β-oxidation [14,16,17]. However, the heart also takes up polyunsaturated fatty acids such as ARA [14,18], where its role in heart lipid-mediated signal transduction is becoming more appreciated [19–26]. It is now accepted that phospholipases A$_2$ (PLA$_2$, see Chapter 1) have a key role in releasing ARA in the heart [27]. Thrombin stimulates ARA release from choline plasmalogen (PlsCho) by increasing the activity of the plasmalogen-selective iPLA$_2$ [24]. In cultured rat ventricular myocytes, IL-1β works through a receptor-mediated process to activate a membrane-associated, plasmalogen-selective, calcium-independent phospholipase A$_2$ (iPLA$_2$), releasing ARA [19]. On the other hand, in rat ventricular myocytes, TNF-α mediates ARA release through activating a cytosolic PLA$_2$ [20], which is associated with both its negative and positive effects on heart function [22]. Angiotensin II also stimulates ARA release as well as the formation of inositol phosphates (see Chapter 2) through the activation of multiple receptor subtypes coupled to PLA$_2$ and phospholipase C (PLC) activation [23]. β$_2$-Adrenergic receptor stimulation also increases ARA release through the activation of a PLA$_2$ [25] and sympathetic denervation of the heart results in a marked reduction in ARA turnover [21], consistent with a reduction in heart PLA$_2$ activity. Collectively, these studies indicate a key role for ARA in heart lipid-mediated signal transduction and a downstream physiological impact on heart ionotropic and chronotropic events [26].

However, other fatty acids also have important physiological impact in the heart and other systems as well. Docosahexaenoic acid (DHA, 22:6n-3) and other PUFA are implicated in reducing cardiac sudden death in dogs [28], presumably via their interaction with specific potassium channels in the heart [29]. DHA is also a well-known and important fatty acid in the brain and the recent discovery of the resolvins or docosanoids, bioactive hydroxyl-containing DHA that are formed via the action of lipoxygenases [30], only adds additional interest in the downstream effects of DHA in different organ systems. Because these unique molecules attenuate the inflammatory response in the brain [31] and are found in blood [32], there is considerable interest in how DHA may provide additional protection against inflammatory conditions via the formation of docosanoids [30–32]. Hence, the uptake and trafficking of DHA to specific lipid pools within different cell systems or cultured cells may have additional significance relative to merely understanding in greater detail the processes by which this uptake occurs.

12.1.2 FACTORS REGULATING FATTY ACID UPTAKE AND TARGETING

But what about targeting of fatty acids to specific lipid pools? This of course is important to understand for a number of reasons ranging from understanding targeting of specific fatty acids to pools used in signal transduction (see Chapters 1 through 5) or to specific membranes where a certain fatty acid may be involved in eliciting gene transcription (see Chapters 9 and 10). Fatty acid uptake is regulated in part by the formation of the activated form of the fatty acid by the condensation of the fatty acid carboxylic acid with the thiol group on coenzyme A (CoA) to form a thioester or what is commonly termed an acyl-CoA. This process is driven by acyl-CoA synthetases (Ascl), which are the primary enzyme family that forms acyl-CoA, and this process is energy consuming as the formation of the acyl-CoA requires the hydrolysis of an ATP. For this family of enzymes, expression of the different family members is tissue dependent and may influence how a particular tissue processes fatty acids. In mammals, there are five Acsl isoforms, Acsl 1, Acsl 3, Ascl 4, Acsl 5, and Acsl 6, although Acsl 6 has two splice variants [33].

Acsl are capable of targeting fatty acids to particular lipid pools, such as Acsl 1 targeting fatty acids to form TAG [34,35], while other family members, such as Acsl 3 or 4, target fatty acids for esterification into phospholipid pools and TAG pools [36–39]. Acsl 1, 3, and 4 are all inhibited by triacsin C [38,39], which provides a useful tool to examine their influence on fatty acid targeting. On the other hand, Acsl 6 targets fatty acid primarily to phospholipids. In PC-12 cells, Acsl 6 expression increases ARA and DHA uptake nearly by 50%, targeting these fatty acids to phospholipid pools [37], which functionally results in increased neurite outgrowth [36]. Hence, it is readily apparent how Acsl expression can impact the mass of fatty acid entering the cell as well as directing the fatty acid to particular lipid pools within the cell. This is highly suggestive that fatty acid utilization drives fatty acid uptake into cells or, in other words, metabolism drives uptake.

Similar to maintaining a strong inward gradient for glucose by the rapid phosphorylation of glucose to glucose-6-P, the plasma membrane–localized fatty acid transport proteins (FATP), a bifunctional enzyme family (FATP1-6) that has Ascl activity [40], may have a similar role. These proteins have the catalytic capacity to form the acyl-CoA and are thought to be involved in vectoral fatty acid transport into the cell or into particular organelles [40]. If so, this rapid formation of the acyl-CoA would make fatty acid uptake unidirectional in cells containing FATP and maintain a steep concentration gradient into the cell. The importance of this step is that a non-membrane permeable fatty acyl-CoA is formed, thereby maintaining a fatty acid gradient into the cell or organelle and preventing the reverse transport of the lipophilic fatty acid across the membrane. In combination, these enzymes convert the nonreactive fatty acid into a fatty acid thioester with CoA, thereby activating the fatty acid for various anabolic and catabolic reactions.

12.1.3 WHY CONCENTRATION MATTERS

For years, physiologists have known that saturated and monounsaturated fatty acids are rapidly esterified into heart TAG pools and used almost exclusively for β-oxidation [16,21,41,42]. This of course was also thought to be the case for ARA because studies in isolated hearts and in isolated myocytes indicate that ARA is predominantly esterified into TAG pools [18,43]. This led to the dogma that ARA was similar to PAM and merely an energy source for the heart. However, at physiologic plasma concentrations, 20:4n-6 is targeted to phospholipid pools [14] and at nonphysiologic concentrations, it is targeted to TAG pools [14,18,43]. This points to how concentration is incredibly important and how using the inappropriate concentration can cause significant confusion. In the end, we now know that there is a divergent processing of ARA and PAM, resulting in targeting to two vastly different heart lipid pools and that ARA targeting to the phospholipid pools is consistent with its use in heart lipid-mediated signal transduction.

Another example is when a good friend and colleague of mine incubated human embryonic kidney (HEK) cells with an enormous amount of oleic acid. As a cell biologist he was amazed at the formation of lipid droplets, whereas as a lipid biochemist I was amazed at his amazement. Cells just do not know what to do when presented with a tremendous concentration of fatty acids in a culture system. Like myself at a dinner buffet, the cells just keep taking in the fatty acid and because unesterified fatty acid is detrimental to the cells, they have to make TAG, which at some point may become useful. (There may be an additional personal corollary here.) While the point of his paper was that alpha-synuclein modulates the turnover of lipid droplets [44], it is also a fine example of cells gorging themselves on fatty acids and then having to literally stick them somewhere. How this may influence other processes such as glycolytic flux is another important point to consider because glycolysis is a critical provider of glycerol required for TAG synthesis. In the end, concentration is an important factor not only in relationship to what pools fatty acids are targeted, but also because it may lead to an inappropriate use of cellular energy for TAG synthesis, thereby affecting other processes.

12.1.4 Cell Cultures vs. *in Vivo*

In the past, we have examined fatty acid uptake in cells [45–48] and in intact animals [15,49–52]. Despite the decreased metabolic demand of cells in culture, the results are amazingly similar. For instance, we have examined the influence of alpha-synuclein on brain fatty acid uptake and have found very similar changes in cultured cortical astrocytes [45] to what we have observed in brain [15,50]. In addition, lipid mass changes observed in alpha-synuclein deficient astrocytes and lipid trafficking changes in these astrocytes [45] are very similar to what is observed in brains from intact mice [15,49,50,53]. A similar story has emerged with fatty acid binding protein (FABP); in the Schroeder laboratory we demonstrated that liver (L)-FABP and intestinal (I)-FABP differentially influence fatty acid uptake and targeting when expressed in L-cell fibroblasts [46–48,54]. When my laboratory examined the influence of heart (H)-FABP deletion on fatty acid uptake and trafficking in the heart [51] and brain [52] of intact mice, we observed a reduction in fatty acid uptake and targeting in a manner that we predicted based upon the previous culture experiments expressing L- and I-FABP [46–48]. In addition, the changes we observed in phospholipid mass in heart and brain [51,52] are consistent with what we have observed in the culture system [55]. These two examples demonstrate how cell cultures, when used appropriately, can produce similar results to experiments in intact animal models. Thus, while there can be tremendous concern over some aspects of using cultured cells to model events that occur *in vivo*, our experience is that, when used properly, these two systems can be very complimentary. Indeed, when one is examining the influence of a particular agonist on lipid-mediated signal transduction, cell cultures provide an ideal system in which various inhibitors and agonists can be used to further our understanding of how these systems work together to elicit a response. A similar situation is much more difficult in the intact animal, again demonstrating that these two systems can and should be used in parallel to further our understanding of lipid-mediated signal transduction.

12.2 FATTY ACID UPTAKE IN CULTURED CELLS

Cell cultures offer a very useful system to study lipid metabolism using fatty acid tracers or to load a cell with a particular tracer and assess its release by various agonists. In addition, there may be situations where the experimental paradigm is to test whether the metabolism of fatty acids or fatty acid uptake is impacted by the presence of particular proteins. For instance, L-FABP expression in L-cell fibroblasts causes a marked increase in fatty acid uptake [46], whereas I-FABP expression does not alter fatty acid uptake [47,48]. However, both FABP impact fatty acid targeting, with L-FABP putting the fatty acid primarily into the phospholipids while I-FABP targeting fatty acids to the TAG fraction [46–48]. The importance of using a fatty acid tracer over the fluorescent fatty acid techniques discussed in Chapter 10 is that fluorescent fatty acids are poorly esterified [46,54], hence they only measure fatty acid uptake into the cell, but not the impact of fatty acid metabolism on uptake. However, with L-FABP-expressing cells, the increase in fatty acid uptake is similar between fluorescent and radiotracer fatty acids, but only through using the radiotracer fatty acids is a functional impact of each FABP on fatty acid targeting illuminated. A similar situation is seen when assessing the impact of alpha-synuclein on astrocyte fatty acid uptake and targeting using a loss of function model [45]. In these experiments, there is a slight reduction in PAM and ARA uptake, but no impact on DHA uptake, very similar to what is observed in the intact brain [49,50,56]. However, there is a marked alteration in fatty acid targeting that is demonstrated by calculating the fractional distribution of each tracer. Using this technique, it is clear that in the absence of alpha-synuclein, more fatty acid tracer is targeted to the cholesteryl ester pools, demonstrating that in these cells alpha-synuclein influences the lipid pool to which fatty acids are targeted. These are two separate, yet very important, examples of how fatty acid uptake and targeting experiments can be used in elucidating functional consequences of protein expression.

12.2.1 CELL INCUBATION WITH FATTY ACIDS AND CELL LIPID EXTRACTION

Fatty acids can be added directly to the medium or added in a complex with bovine serum albumin (BSA). In my laboratory, I prefer to use a defined fatty acid concentration, generally $2\,\mu M$, and add the fatty acids to the cells from the stock solution of tracer and cold fatty acid (1 mM) dissolved in ethanol. This limits the amount of ethanol introduced to the cultured cells to 0.2% for 10 mL of medium. This protocol can be used for continuous fatty acid incubations [45] or in pulsed incubations [46] (see Section 12.2.2).

For continuous exposure experiments, cells are labeled with the radiotracer fatty acid of interest that is generally diluted with the same cold (nonradioactive) fatty acid. When making up the tracer, it is important to first determine how much radioactivity you want for your experiment. We generally use $50\,\mu Ci$ of tracer, which at a specific activity of 53 mCi/mmol (for [1-^{14}C]20:4n-6 (Moravek Biochemical, Brea, California)) is 943 nmol of fatty acid. For a 10 mL solution at a concentration of 1 mM, we need an additional 9.057 µmol of cold 20:4n-6. After combining the tracer and cold fatty acid, we have a total of 10 µmol of ARA in the 10 mL solution. Two aliquots are taken to assess the amount of radioactivity in disintegrations per minute (DPM), which is then converted to µCi by dividing the number of DPM by 2.22×10^6 DPM/µCi. At this point, we use the specific activity of the tracer to calculate the actual number of moles of fatty acid that enter any given compartment within the cell. By using the specific activity, we can easily calculate rates or kinetics because we have a mass number as opposed to DPM, which will ultimately be more useful information.

While we use a fixed concentration of 2 µM, additional concentrations can be used, keeping in mind that each fatty acid has a different critical micelle concentration (CMC). The CMC is the concentration at which a lipid will spontaneously form a micelle in an aqueous solution. These values are hard to find in the literature and we use an assay with cis-parinaric acid to make these measurements. Using this assay, a solution of cis-parinaric acid in PBS is put into a cuvette in the fluorometer at which point the cis-parinaric acid remains in solution and its fluorescence is quenched in the aqueous environment (see Chapter 10). A concentrated fatty acid in ethanol (note that a concentrated fatty acid is used to keep the concentration of ethanol low) is then titrated into the system and the concentration at which there is a sharp increase in fluorescence is noted. This is the point at which cis-parinaric acid has entered into the newly formed micelles, hence the concentration at which micelles are formed or the CMC. For instance, PAM has a CMC of about 5–6 µM, while ARA has a CMC of about 80 µM. Phospholipids, because of their limited solubility, have an even lower CMC. It is important to not exceed the CMC as you may have the fusion of micelles with the cells, adding a significant artifact.

To start the experiment, the medium on the cells is removed and carefully replaced with serum-free medium. After a short period of acclimation, 30 min or so, the fatty acid (20 µL for 10 mL of medium) is carefully added and mixed using a gentle rocking motion. After the desired incubation time, the medium is aspirated from the cells to remove the input function (the tracer) and rinsed with ice-cold phosphate buffered saline (PBS). The PBS is aspirated and the cells are flash frozen by floating them on liquid nitrogen to fix all metabolism [57]. We then use a modified single-phase extraction protocol [45,55,58–60]. In this protocol, 2-propanol (2 mL) is added directly to the frozen cells that are then removed by scraping in at least two different dimensions with a Teflon cell scrapper. The cells can thaw slightly to aid in removal once the 2-propanol is in place. The 2-propanol is removed with a glass Pasteur pipette and placed into a tube containing 6 mL of hexane. It is important to remove all of the residual protein that will often be denatured. Rinse the cell plate with another 2 mL of 2-propanol and remove the rinse and add to the tube containing the first extract. Vortex the tube and look for a potential lower phase, which is often residual PBS. At this point you have a classic Raden extract of hexane:2-propanol (3:2 by vol. [61,62]). However, if there is a lower phase after vortexing, add an additional 2 mL of 2-propanol, which will generally facilitate the formation of a single phase (see Protocol A).

Other methods use BSA that is mixed with the fatty acid tracer of interest in a 5:1 to a 10:1 molar ratio of BSA to fatty acid [63–66]. This is a popular method, but it is complicated by the on rate and off rate of the fatty acid from the BSA as well as the fact that BSA can back extract fatty acids from the cells. If choosing to use this protocol, we recommend making a fatty acid film on a 20 mL glass scintillation vial and then adding 5 mM HEPES (pH 7.4) buffer containing "essentially fatty acid free" BSA (50 mg/mL, Sigma Chemical Co., St. Louis, Missouri) and complex the fatty acid with the BSA by sonicating the sample in a bath sonicator for 60 min at 40°C.

12.2.2 Pulse Experiments and Why Time Is Important

Pulse labeling experiments are very useful for loading cells with a particular fatty acid tracer or for examining the metabolism of a tracer over time. These experiments are also done when one wants to load a cell with [1-^{14}C]ARA to examine agonist-induced release of ARA into the medium. Whatever the reason for a pulse experiment, there are two important points to consider. First, fatty acid tracers move from one lipid pool to another. This is best illustrated in Figure 12.1. In this figure, we show results from labeling primary mouse cortical astrocytes with 2 µM [1-^{14}C] ARA for 10 min and then removing the medium containing the tracer and replacing

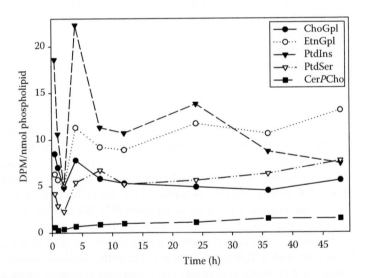

FIGURE 12.1 Primary mouse cortical astrocytes were labeled with [1-^{14}C]ARA for 10 min at a concentration of 2 µM. Cells were extracted after 0.5, 1, 2, 4, 8, 12, 24, 36, and 48 h. Phospholipids and neutral lipids were separated using a silicic acid column and then the phospholipids then separated by TLC. Lipid mass was determined by measuring individual phospholipid mass using a phosphorus assay [78]. Radioactivity was determined using liquid scintillation counting. Specific activity was calculated by dividing DPM by the phospholipid mass. Abbreviations: ChoGpl, choline glycerophospholipid; EtnGpl, ethanolamine glycerophospholipids; PtdIns, phosphatidylinositol; PtdSer, phosphatidylserine; CerPCho, sphingomyelin.

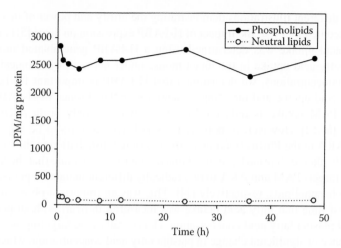

FIGURE 12.2 The neutral lipid and phospholipid fraction radioactivity is presented from the fractions obtained from the silicic acid columns as described above.

this medium with that containing 10% fetal bovine serum (FBS). At 0.5, 1, 2, 4, 8, 12, 24, 36, and 48 h, the cells were extracted as described in Section 12.2.1. It is important to note that we are presenting specific activity calculated as the DPM in a given phospholipid fraction divided by the mass (nmol) of that phospholipid. Herein, we see that sphingomyelin (CerPCho) slowly increases its specific activity, while phosphatidylinositol (PtdIns) has several sharp peaks, followed by a much smaller peak, indicating that the tracer is going into and out of the PtdIns fraction. The real take home message here is that the tracer does not stay in one given lipid fraction, but goes into and out of different phospholipids over time. Hence, if the experiment is looking at agonist-induced ARA release, the cells need to be labeled for the same amount of time to ensure that the release is from the same phospholipid fraction. While there are nominal changes in the total radioactivity in the phospholipid fraction, for the most part the radioactivity remains constant (Figure 12.2). Note, however, that very little ARA is found associated with the neutral lipid fraction, which is separated from the phospholipid fraction using silicic acid columns (see Protocol B). Hence, it is important to understand where your fatty acid of interest is at a given time to design experiments accordingly and to recall that the fatty acid will move with time, so particular attention has to be placed on monitoring labeling times prior to experiments.

12.3 FATTY ACID UPTAKE IN TISSUES

Whole animals offer a much more physiologically relevant system to study fatty acid uptake. To do this, we use a high specific activity tracer infused through the femoral vein and a well-described steady-state fatty acid kinetic model [67]. While the details of this model are discussed in Chapter 13, herein I will focus on how this model is useful to assess fatty acid uptake and metabolism.

Several different illustrations demonstrating the utility and power of this method are presented herein. First, the impact of H-FABP expression on heart [51] and brain [52] fatty acid uptake was determined using a H-FABP gene-ablated mouse. By assessing fatty acid uptake in the intact mouse under normal physiological plasma fatty acid concentrations, we determined that H-FABP is important for heart and brain fatty acid uptake and targeting. Essentially, in the absence of H-FABP, heart ARA and PAM uptake is reduced 40%–60%, whereas only brain ARA uptake is reduced [51,52]. However, in both systems H-FABP appears to be important in directing ARA to the PtdIns fraction, as the fractional distribution to this fraction is dramatically altered. Similarly, we used this model to demonstrate that the heart differentially targets PAM and ARA to two radically different metabolic pools, namely TAG and phospholipids, respectively [14]. This was an important observation as it was the first time that fatty acid uptake was measured in heart without perturbing the normal plasma fatty acid concentration. Hence, fatty acid targeting is measured in the absence of significant change in plasma fatty acid concentration, which is now known to affect ARA targeting. Another example is our work demonstrating that erucic acid enters the brain and is rapidly chain shortened [68]. This effort combined with techniques discussed in Chapter 11, illuminates how [14-^{14}C]22:1n-9 (erucic acid) is used to study the metabolism of this fatty acid in intact brain. The metabolism of this tracer was also examined in heart and liver, where we showed that this tracer underwent rapid β-oxidation in the liver and that the carbons are recycled and used by the liver to form stearic acid (18:0) [69]. Hence, the power of this technique *in vivo* is to infuse a radiotracer into an animal and assess its uptake, metabolism, and targeting in a particular organ system in a manner that does not perturb the normal uptake or metabolism parameters of that fatty acid.

12.3.1 Tracer Preparation, Surgery, and Infusion

Fatty acid radiotracer is evaporated under a constant stream of N_2 at 40°C and then the tracer is solubilized in 5 mM HEPES (pH 7.4) buffer containing "essentially fatty acid free" BSA (50 mg/mL, Sigma Chemical Co., St. Louis, Missouri). Solubilization is facilitated by sonication in a bath sonicator for 60 min at 40°C. The appropriate amount of radiotracer is prepared for each mouse or rat using the animal's weight based upon the infusion parameters of 170 μCi/kg for a [1-^{14}C]fatty acid or 1.7 mCi/kg for a [^{3}H]fatty acid [67,70].

The animal surgery and tracer infusion are performed as previously described [15,21,49,51,69]. Briefly, fasted animals are anesthetized with halothane (1%–3%) and PE-10 catheters (mice) or PE-50 catheters (rats) are inserted into the femoral artery and vein. Using an infusion pump (BS-8000, Braintree Scientific, Inc., Braintree, Massachusetts), awake (3–4 postoperative hours) animals are infused with 170 μCi/kg of [1-^{14}C]fatty acid into the femoral vein over 10 min at a constant rate of 30 μL/min (mice) or 0.4 mL/min (rats) to achieve steady-state plasma radioactivity. Prior to and during the infusion, arterial blood samples (~20 μL) are taken to determine plasma radioactivity. Following infusion, each animal is killed using pentobarbital (100 mg/kg, i.v.) and tissue rapidly removed and flash frozen in liquid nitrogen. If the experiment is assessing brain fatty acid metabolism, the animal is immediately subjected to head-focused microwave irradiation (2.8 kW, 1.35 s; Cober Electronics, Inc., Norwalk, Connecticut) to heat-denature enzymes *in situ* following the injection of pentobarbital. We routinely freeze the tissue in liquid nitrogen and then pulverize the tissue under liquid nitrogen temperatures to a fine powder using a stainless-steel mortar and pestle. This permits long-term storage and retrieval of tissue for many different analyses as needed. We normally use 100 mg of wet weight tissue for lipid analysis.

12.3.2 Tissue and Plasma Extraction

Arterial blood samples, taken at fixed times during the infusion period, are stored for up to 10 min on ice before separating the plasma by centrifugation with a Beckman microfuge (Fullerton, California). Plasma lipids are then extracted by transferring a 100 μL aliquot of plasma into a tube containing 3 mL of chloroform:methanol (2:1 v/v), then mixing it by vortexing [71]. The addition of 0.63 mL of 0.9% KCl to these tubes results in two phases. These phases are thoroughly mixed, then separated overnight in a −20°C freezer. The upper phase is removed and the lipid-containing lower phase is rinsed with 0.63 mL of theoretical upper phase to remove any aqueous soluble contaminants [71]. Phase separation is facilitated by centrifugation at $2800 \times g$ (0°C) in a refrigerated Beckman Allegra 15-R centrifuge (Beckman Instruments, Fullerton, California) for 20 min. The upper phase is discarded and a portion of the lower phase is dried and its radioactivity quantified using a Beckman 6500 liquid scintillation counter (Beckman Instruments, Fullerton, California).

In this model, whole blood is taken immediately prior to pentobarbital injection to correct for residual blood levels in the tissue. For brain, residual blood accounts for about 2% of the brain weight [15,49]. Residual blood in heart and liver is estimated to be 24% and 17%, respectively, based upon values in the literature [72,73]. By determining the amount of radioactivity in whole blood combined with the percentage of mass of a given tissue that is contributed by residual blood, the actual fatty acid uptake into any given tissue can be calculated by subtracting the amount of radioactivity contributed by the residual blood in the tissue. This then is the actual fatty acid uptake and has subtracted the contribution of the residual blood to the tissue radioactivity.

Lipids are extracted from the tissue in a Tenbroeck tissue homogenizer using a two-phase system [71]. Briefly, the tissue mass (g) is converted to volumes by multiplying with a correction factor of 1.28 [61]. This value represents 1 vol. The pulverized tissue (100 mg) is weighed directly into the homogenizer and 17 vol. of chloroform:methanol (2:1 v/v) is then added. The tissue is homogenized until there is a fine particulate-like powder. The solvent is removed and the homogenizer is rinsed with 3 vol. of chloroform:methanol (2:1 v/v). The rinse is added to the original sample and 4 vol. of 0.9% KCl solution is added to this combined lipid extract. After vigorous mixing, phase separation is facilitated by centrifugation as described above. The upper phase and proteinaceous interface is removed and saved in a 20 mL glass scintillation vial to assess β-oxidation products. The lower organic phase is washed twice with 2 mL of theoretical upper phase, which contains chloroform:methanol:water (3:48:47 by vol.). This is an important step as it aids in removing organic soluble proteins and other products, thereby yielding a better estimate of β-oxidation. Each wash is followed by centrifugation at 0°C to facilitate phase separation. While this can be done at room temperature, by cooling the sample the phase separation is much more complete. The washes are removed and combined with the previously removed upper phase and the radioactivity determined by liquid scintillation counting. The washed lower phase is dried under a stream of nitrogen and the lipids redissolved in 3 mL of *n*-hexane:2-propanol (3:2 v/v) containing 5.5% H_2O.

Following extraction, individual lipid fractions are separated using a variety of techniques such as thin layer chromatography (TLC) (Figure 11.1). By separating the lipid fractions, the fractional distribution of each fatty acid is determined. While there is minimal metabolism of ARA or DHA to longer or shorter chain products, other fatty acids, such as erucic acid, undergo extensive metabolism. To assess this metabolism, we use phenacyl derivatives as described in Chapter 11. Using this method, fatty acid chain elongation or shortening can easily be determined.

12.3.3 ASSESSING β-OXIDATION

We routinely measure products of β-oxidation using the two-phase tissue extraction procedure outlined in Section 12.3.2. While this is not an absolute accounting for β-oxidation because radioactive CO_2 is not measured, it is an excellent approximation and will yield very useful data. However, it is important to note that washing of the lower phase several times with theoretical upper phase is crucial to achieving more accurate accounting of β-oxidation as well as a much cleaner lower phase. This step, although time consuming, is highly recommended.

12.3.4 ASSESSING FATTY ACID UPTAKE AND TARGETING

Fatty acid uptake is assessed by counting the total organic fraction and aqueous fractions and combining these values. Of course, this is done after correcting for the contribution of residual blood to the tissue radioactivity. Following correction, this value will give you the total amount of fatty acid taken up. This value can be converted to mass if the specific activity of radiotracer in blood is calculated, which is a standard procedure when calculating kinetic values (see Chapter 13 for details).

Fatty acid targeting is determined by separating the organic fraction into various components. We routinely use silicic acid columns to separate the neutral lipid and phospholipid fractions (see Protocol B). This separation will provide a much better TLC separation of the individual neutral lipid and phospholipid fractions. We routinely use Merck 60 silica plates (20 cm × 20 cm, 250 μm) with a preabsorbant zone that are activated at 110°C for at least 24 h. Tissue phospholipids are separated using a chloroform:methanol:acetic acid:water (50:37.5:3.5:2 by vol.) solvent system [74] and visualized using iodine vapor. Tissue neutral lipids are separated using a petroleum ether:diethyl ether:acetic acid (70:30:1.3 by vol.) solvent system that resolves cholesterol, cholesteryl esters, diacylglycerols, nonesterified fatty acids, and TAG [75]. Neutral lipids are visualized with 6-*p*-toluidino-2-naphthalenesulfonic acid (TNS) [76]. Bands corresponding to lipid fractions are identified using commercially prepared standards (NuChek Prep, Elysian, Minnesota; Avanti Polar Lipid, Alabaster, Alabama). Following removal by scrapping, the fractions are transferred into 20 mL glass liquid scintillation vials and 0.5 mL of water is added to facilitate the transfer of the lipid off the silica. Then 10 mL of scintillation cocktail is added and the samples vigorously vortexed for 30 s. The samples are not counted until 30 min has passed, to permit the silica gel time to settle.

Because we have found over the last 25 years that some products work better than others for our protocols, we are happy to recommend the following solvents and chemicals for all of our analytical procedures:

Fisher ScintiVerse BD cocktail (Fisher catalog number SX18-4)
EMD chloroform (VWR catalog number CX1055-6)
Mallenkrodt ether (VWR catalog number MK285408)
EMD hexane (VWR catalog number HX0296-1)
EMD isopropanol (VWR catalog number PX1834-1)
EMD methanol anhydrous (VWR catalog number MX0487-6)
EMD methanol (VWR catalog number MX0488-6)
EMD petroleum ether (VWR catalog number PX0424-6)
Moravek biochemical for fatty acid tracers (Brea, California)
Avanti polar lipids for all phospholipids (Alabaster, Alabama)
NuChek Prep for all fatty acids (Elysian, Minnesota)
Unisil silicic acid (Clarkson Chemical Co. Williamsport, Pennsylvania; note that this is the best around but is a small company).

12.4   CONCLUSIONS

In this chapter, the use of radiotracers to study fatty acid metabolism or to label cultured cells for use in experiments assessing agonist-stimulated fatty acid release have been discussed. There are many nuances to studying fatty acid uptake and trafficking, many of which involve strict attention to concentrations and an understanding that lipid metabolism is heavily influenced by the concentration of fatty acids used. While it was not discussed above in detail, I do want to point out that fatty acid metabolism occurs rapidly, as illustrated by the complete β-oxidation of erucic acid in rat liver within 10 min and the synthesis of saturated fatty acids from these carbon units in this time frame [69]. The rapidity of fatty acid metabolism is often overlooked by many in the field, despite that fact that we know how rapid lipid-mediated signaling events occur. So, it is important for each and every reader to appreciate the rapid nature by which lipids are metabolized and to keep that in mind when designing experiments.

ACKNOWLEDGMENTS

This work was supported by NIH grants R21 NS043697 and R21 MS060141 to EJM and a project on P20 RR17699. We thank Cindy Murphy for the typed preparation of this manuscript.

APPENDIX

Protocol A: Cell extraction of lipids and proteins

Solutions

1. PBS pH 7.3
   8.0 g NaCl
   0.2 g KCl
   1.44 g Na_2HPO_4
   0.24 g KH_2PO_4
   Dissolve into 800 mL distilled water and q.s. to 1000 mL.
   Adjust pH accordingly using HCl or NaOH.

Preparation

1. Keep container of liquid nitrogen available to freeze the cell culture plate.
2. Label one set of large screw-top tubes (16 × 125 mm) and set aside.
3. Label one set of large screw-top tubes (16 × 125 mm) to be filled with hexane (6 mL).

Procedure

Test tube preparation

1. Fill large, screw-top test tubes with 6 mL of hexane.

Removal of cells from cell culture plate

1. Open the cell culture plate and aspirate the medium in the cell culture plate. Remove the medium from the opposite side of the cells.
2. Add ~4 mL of PBS solution using a long Pasteur pipette. Add it to the opposite side of the cell culture plate, away from the cells, so that they do not get washed away.
3. Swirl the PBS solution around the cells.
4. Aspirate the PBS solution from the cell culture plate.
5. Place the cell culture plate, cell-side down, on the liquid nitrogen. Make sure there is enough nitrogen so that the cell culture plate can float on it.
6. Leave the plate on the liquid nitrogen for ~30 s. Make sure the cell culture plate is frozen on the bottom.
7. If using a T-75 flask, use pliers to remove the top side (no cells) of the cell culture flask and remove the pieces of plastic until the bottom of the cell culture plate is completely accessible. If using a 100 mm plate or a six-well plate, you will not have to pry off the top.
8. Add 2 mL of isopropanol (2-propanol) to the bottom of the cell culture plate and swirl.
9. Scrape the cells from the cell culture plate by scraping in two different directions and gather into a corner.

10. Using a clean pipette, aspirate the cells and the isopropanol and transfer them to a large, screw-top test tube containing 6 mL of hexane.
11. Add another 2 mL of isopropanol to the cell culture plate and swirl.
12. Scrape any remaining cells into a corner.
13. Aspirate the cells from the cell culture plate and add to the same test tube as the first rinse.
14. Cap the test tube with a Teflon line phenolic cap and gently vortex it.

Separation of lipid and protein content

1. Dry down the samples using nitrogen until ~1–2 mL remains.
2. Centrifuge $3000 \times g$, at 4°C for 20 min.
3. Decant the lipid (liquid) portion into new screw-top test tubes using a long pipette and be careful not to disturb the protein pellet.
4. Samples are now ready to be separated into the NL and PL portions or analyzed by other techniques. Use the silica column separation protocol to separate the cells into NL and PL fractions if desired.

Protein preparation

1. Cover the protein portion and allow to dry overnight to remove trace levels of solvent.
2. Once the proteins are dried, add 1.0 mL of 0.2 M KOH to each sample.
3. Cap the samples with a Teflon lined phenolic cap and heat the samples at 65°C overnight.
4. Refrigerate. Protein analysis may be done using a modified Bradford assay [77].

Protocol B: Silica columns for separation of neutral lipids and phospholipids

Pipette preparation

1. Using another long pipette, place a small amount of fine glass wool (about 0.5 cm) into a long Pasteur pipette (9″).

Silicic acid preparation and addition

1. Pour enough activated Unisil silicic acid (200–325 mesh, Clarkson Chemical Company, Williamsport, Pennsylvania) into a 20 mL glass scintillation vial to fill about one-third (1/3) of the vial with the powder.
2. Fill the rest of the vial with chloroform ($CHCl_3$).
3. Shake solution until the powder is evenly distributed and the solution is clear in color.
4. Fill the stuffed pipettes with the silicic acid solution to a height of about 1.5–2 cm, collecting the eluent in a 20 mL scintillation vial.

5. Wash with the column with ~1 mL chloroform until it is clear.
6. Beware of bubbles, if there are bubbles in the silicic acid column, agitate the silicic acid with a pipette containing CHCl$_3$ to remove the bubbles.

Sample preparation

1. Dry down samples under a stream of N$_2$ gas and rinse the tubes 3×0.25 mL of hexane:2-propanol (3:2 by vol.).
2. Evaporate to dryness.
3. Reconstitute the samples with ~200 μL of chloroform.

Neutral lipid (NL) collection

1. Place NL lipid collection test tubes (16×125 mm screw topped) under poured columns and apply the sample to the column.
2. Rinse the sample tube with ~200 μL chloroform and transfer to the column.
3. Rinse three or four more times with 200 μL chloroform and transfer to the column each time.
4. Wash the NL through with chloroform:methanol (58:1) until ~10 mL is collected.
5. Dry down under N$_2$ gas and store in HIP (3:2) or use for the desired assay.

(Note that this protocol will quantitatively elute all neutral lipids including cholesterol, cholesteryl esters, TAG, monoacylglycerols, and free fatty acids, but the phospholipids will be retained.)

Phospholipid (PL) collection

1. Place new tubes (large, screw topped) under the pipettes.
2. Wash the PL through with methanol until ~10 mL is collected.
   Notice that upon rinsing with methanol the silica changes from clear to a milky white color.
3. Dry down under N$_2$ gas and store in HIP (3:2) or use for the desired assay.

Protocol C: Folch extraction: Two-phase extraction (radioactive preparation)

Solutions

1. Chloroform:methanol 2:1 by vol. (Folch)
2. 0.9% KCl
   Potassium chloride 900 mg
   dH$_2$O q.s. to 100 mL
3. FUL = theoretical upper phase
   C:M: H$_2$O 3:48:47 by vol.

Preparation

1. Pulverized tissue **must be maintained on dry ice in plastic scintillation vials**
2. Keep tools used for transfer of tissue samples cooled in liquid N_2

Procedure

1. Homogenizing ***Take care to contain and properly dispose of all supplies used; clean thoroughly all areas used***.
   a. Support Tenbroeck homogenizer in flask on balance pan; tare. Place tissue into homogenizer to desired weight.
   b. Multiply mass by correction factor of 1.28 and use this number to determine volumes.
   c. Add 17 volumes of Folch and homogenize.
   d. Using a glass pipette, aspirate (transfer) to a small glass screw-top test tube. While transferring, be sure to rinse the homogenizer.
   e. Add an additional 3 volumes of Folch, homogenize remaining tissue and transfer to the same test tube. Place in wet ice.
   f. Rinse homogenizer with small amount of Folch and discard rinse.
2. Extracting
   a. Add 4 volumes of 0.9% KCl to screw-top test tube, vortex, chill, centrifuge at $2800 \times g$ at 0°C.
   b. Aspirate upper (aqueous) phase to scintillation vial.
   c. Add 4 volumes of FUL to screw-top tube, vortex, chill, centrifuge.
   d. Aspirate upper phase to the scintillation vial, getting as much of the solid tissue as possible. Remaining lower phase is organic phase.
   e. Repeat with 4 volumes of FUL as above.
3. Aqueous phase
   Add 15 mL scintillant, shake well and count.
4. Organic phase
   Dry down.
   Add 1 mL HIP.
   Pipet 25 µL (100 µL for mice) to scintillation vial, add 10 mL scintillant and count.
Save remaining organic phase in −80°C freezer.

LIST OF ABBREVIATIONS

ARA	arachidonic acid
Ascl	acyl-CoA synthetase
BSA	bovine serum albumin
Cer*P*Cho	sphingomyelin
ChoGpl	choline glycerophospholipids
CMC	critical micelle concentration
CoA	coenzyme A

DHA   docosahexaenoic acid
DPM   disintegrations per minute
EtnGpl   ethanolamine glycerophospholipids
FABP   fatty acid binding protein
FATP   fatty acid transport protein
LTP   long-term potentiation
PAM   palmitic acid
PBS   phosphate buffered saline
PLA_2 phospholipases A_2
PLC   phospholipase C
PtdIns   phosphatidylinositol
PtdSer   phosphatidylserine
TAG   triacylglycerol
TLC   thin layer chromatography

REFERENCES

1. Lesa, G.M. et al. (2003) Long chain polyunsaturated fatty acids are required for efficient neurotransmission in *C. elegans. J. Cell Sci.* 116:4965–4975.
2. St-Gelais, F. et al. (2004) Postsynaptic injection of calcium-independent phospholipase A_2 inhibitors selectively increases AMPA receptor-mediated synaptic transmission. *Hippocampus* 14:319–325.
3. Dumuis, A. et al. (1988) NMDA receptors activate the arachidonic acid cascade system in striatal neurons. *Nature* 336:68–70.
4. Miller, B. et al. (1992) Potentiation of NMDA receptor currents by arachidonic acid. *Nature* 355:722–725.
5. Wolf, M.J. et al. (1995) Long-term potentiation requires activation of calcium-independent phospholipase A_2. *FEBS Lett.* 377:358–362.
6. Massicotte, G. et al. (1991) Modulation of a DL-α-amino-3-hydroxy-5-methyl-4-isoxazolepropionic acid/quisqualate receptors by phospholipase A_2: A necessary step in long-term potentiation. *Proc. Natl. Acad. Sci. USA* 88:1893–1897.
7. Massicotte, G. et al. (1990) Effect of bromophenacyl bromide, a phospholipase A_2 inhibitor, on the induction and maintenance of LTP in hippocampal slices. *Brain Res.* 537:49–53.
8. Clements, M.P., Bliss, T.V.P., and Lynch, M.A. (1991) Increase in arachidonic acid concentration in a postsynaptic membrane fraction following the induction of long-term potentiation in the dentate gyrus. *Neuroscience* 45:379–389.
9. Williams, J.H. et al. (1989) Arachidonic acid induces a long-term activity-dependent enhancement of synaptic transmission in the hippocampus. *Nature* 341:739–742.
10. Lynch, M.A. and Voss, K.L. (1994) Membrane arachidonic acid concentration correlates with age and induction of long-term potentiation in the dentate gyrus in the rat. *Eur. J. Neurosci.* 6:1008–1014.
11. McGahon, B., Clements, M.P., and Lynch, M.A. (1997) The ability of aged rats to sustain long-term potentiation is restored when the age-related decrease in membrane arachidonic acid concentration is reversed. *Neuroscience* 81:9–16.
12. Barbour, B. et al. (1989) Arachidonic acid induces a prolonged inhibition of glutamate uptake into glial cells. *Nature* 342:918–920.
13. Breukel, A.I. et al. (1997) Arachidonic acid inhibits uptake of amino acids and potentiates PKC effects on glutamate, but not GABA, exocytosis in isolated hippocampal nerve terminals. *Brain Res.* 773:90–97.

14. Murphy, E.J. et al. (2000) Intravenously injected [1-^{14}C]arachidonic acid targets phospholipids, and [1-^{14}C]palmitic acid targets neutral lipids in hearts of awake rats. *Lipids* 35:891–898.

15. Golovko, M.Y. et al. (2006) Acyl-CoA synthetase activity links wild-type but not mutant α-synuclein to brain arachidonate metabolism. *Biochemistry* 45:6956–6966.

16. Klein, M.S. et al. (1979) External assessment of myocardial metabolism with [^{11}C]palmitate in rabbit hearts. *Am. J. Physiol.* 237(1):H51–H57.

17. DeGrella, R.F. and Light, R.J. (1980) Uptake and metabolism of fatty acids by dispersed adult rat heart myocytes. I. Kinetics of homologous fatty acids. *J. Biol. Chem.* 255(2):9731–9738.

18. Hohl, C.M. and Rosen, P. (1987) The role of arachidonic acid in rat heart cell metabolism. *Biochim. Biophys. Acta* 921:356–363.

19. McHowat, J. and Liu, S. (1997) Interleukin-1β stimulates phospholipase A$_2$ activity in adult rat ventricular myocytes. *Am. J. Physiol.* 272:C450–C456.

20. Liu, S.J. and McHowat, J. (1998) Stimulation of different phospholipase A$_2$ isoforms by TNF-α and IL-1β in adult rat ventricular myocytes. *Am. J. Physiol.* 275:H1462–H1472.

21. Patrick, C.B. et al. (2005) Arachidonic acid incorporation and turnover is decreased in sympathetically denervated rat heart. *Am. J. Physiol.* 288:2611–2619.

22. Amadou, A. et al. (2002) Arachidonic acid mediates dual effect of TNF-α on Ca^{2+} transients and contraction of adult rat cardiomyocytes. *Am. J. Physiol.* 282:C1339–C1347.

23. Lokuta, A.J. et al. (1994) Angiotensin II stimulates the release of phospholipid-derived second messengers through multiple receptor subtypes in heart cells. *J. Biol. Chem.* 269:4832–4838.

24. McHowat, J. and Creer, M.H. (2000) Selective plasmalogen substrate utilization by thrombin-stimulated Ca^{2+}-independent PLA$_2$ in cardiomyocytes. *Am. J. Physiol.* 278:H1933–H1940.

25. Pavoine, C. et al. (1999) Evidence for a β2-adrenergic/arachidonic acid pathway in ventricular cardiomyocytes. Regulation by the β1-adrenergic/camp pathway. *J. Biol. Chem.* 274:628–637.

26. Steinberg, S.F. (1999) The molecular basis for distinct β-adrenergic receptor subtype actions in cardiomyocytes. *Circ. Res.* 85:1101–1111.

27. McHowat, J. and Creer, M.H. (2001) Comparative roles of phospholipase A$_2$ isoforms in cardiovascular pathophysiology. *Cardiovasc. Toxicol.* 1:253–265.

28. Billman, G.E., Kang, J.X., and Leaf, A. (1997) Prevention of ischemia-induced cardiac sudden death by n-3 polyunsaturated fatty acids in dogs. *Lipids* 32:1161–1168.

29. Kang, J.X., Xiao, Y.-F., and Leaf, A. (1995) Free, long-chain, polyunsaturated fatty acids reduce membrane electrical excitability in neonatal rat cardiac myocytes. *Proc. Natl. Acad. Sci. USA* 92:3997–4001.

30. Serhan, C.N. et al. (2002) Resolvins: A family of bioactive products of omega-3 fatty acid transformation circuits initiated by aspirin treatment that counter proinflammation signals. *J. Exp. Med.* 196:1025–1037.

31. Marcheselli, V.L. et al. (2003) Novel docosanoids inhibit brain ischemia-reperfusion-mediated leukocyte infiltration and pro-inflammatory gene expression. *J. Biol. Chem.* 278:43807–43817.

32. Hong, S. et al. (2003) Novel docosatrienes and 17S-resolvins generated from docosahexaenoic acid in murine brain, human blood, and glial cells. *J. Biol. Chem.* 278:14677–14687.

33. Van Horn, C.G. et al. (2005) Characterization of recombinant long-chain rat acyl-CoA synthetase isoforms 3 and 6: Identification of a novel variant of isoform 6. *Biochemistry* 44:1635–1642.

34. Li, L.O. et al. (2006) Overexpression of rat long chain acyl-CoA synthetase 1 alters fatty acid metabolism in rat primary hepatocytes. *J. Biol. Chem.* 281:37246–37255.

35. Mashek, D.G. et al. (2006) Rat long chain acyl-CoA synthetase 5 increases fatty acid uptake and partitioning of cellular triacylglycerol in McArdle-RH7777 cells. *J. Biol. Chem.* 281:945–950.

36. Marszalek, J.R. et al. (2004) Acyl-CoA synthetase 2 overexpression enhances fatty acid internalization and neurite outgrowth. *J. Biol. Chem.* 279:23882–23891.

37. Marszalek, J.R. et al. (2005) Long-chain acyl-CoA synthetase 6 preferentially promotes DHA metabolism. *J. Biol. Chem.* 280:10817–10826.

38. Igal, R.A., Wang, P., and Coleman, R.A. (1997) Triacsin C blocks *de novo* synthesis of glycerolipids and cholesterol esters but not recycling of fatty acid into phospholipid: Evidence for functionally separate pools of acyl-CoA. *Biochem. J.* 324:529–534.

39. Muoio, D.M. et al. (2000) Acyl-CoAs are functionally channeled in liver: Potential role of acyl-CoA synthetase. *Am. J. Physiol. Endocrinol. Metab.* 279:E1366–E1373.

40. Watkins, P.A. (2008) Very-long-chain acyl-CoA synthesis. *J. Biol. Chem.* 283:1773–1777.

41. DeGrella, R.F. and Light, R.J. (1980) Uptake and metabolism of fatty acids by dispersed adult rat heart myocytes. II. Inhibition of albumin and fatty acid homologues, and the effect of temperature and metabolic reagents. *J. Biol. Chem.* 255(20):9739–9745.

42. Tamboli, A. et al. (1983) Comparative metabolism of free and esterified fatty acids by the perfused rat heart and rat cardiac myocytes. *Biochim. Biophys. Acta* 750:404–410.

43. Saddik, M. and Lopaschuk, G.D. (1991) The fate of arachidonic acid and linoleic acid in isolated working rat hearts containing normal or elevated levels of coenzyme A. *Biochim. Biophys. Acta* 1086:217–224.

44. Cole, N.B. et al. (2002) Lipid droplet binding and oligomerization properties of the Parkinson's disease protein α-synuclein. *J. Biol. Chem.* 277:6344–6352.

45. Castagnet, P.I. et al. (2005) Fatty acid incorporation is decreased in astrocytes cultured from α-synuclein gene-ablated mice. *J. Neurochem.* 94:839–849.

46. Murphy, E.J. et al. (1996) Liver fatty acid binding protein expression in transfected fibroblasts stimulates fatty acid uptake and metabolism. *Biochim. Biophys. Acta* 1301:191–196.

47. Prows, D.R., Murphy, E.J., and Schroeder, F. (1995) Intestinal and liver fatty acid binding proteins differentially affect fatty acid uptake and esterification in L-cell fibroblasts. *Lipids* 30:907–910.

48. Prows, D.R. et al. (1996) Intestinal fatty acid-binding protein expression stimulates fibroblast fatty acid esterification. *Chem. Phys. Lipids* 84:47–56.

49. Golovko, M.Y. et al. (2005) α-Synuclein gene-deletion decreases brain palmitate uptake and alters the palmitate metabolism in the absence of α-synuclein palmitate binding. *Biochemistry* 44:8251–8259.

50. Golovko, M.Y. et al. (2007) α-Synuclein gene ablation increases docosahexaenoic acid incorporation and turnover in brain phospholipids. *J. Neurochem.* 101:201–211.

51. Murphy, E.J. et al. (2004) Heart fatty acid uptake is decreased in heart fatty acid binding protein gene-ablated mice. *J. Biol. Chem.* 279:34481–34488.

52. Murphy, E.J. et al. (2005) Brain arachidonic acid incorporation is decreased in heart-fatty acid binding protein gene-ablated mice. *Biochemistry* 44:6350–6360.

53. Barceló-Coblijn, G. et al. (2007) Brain neutral lipids mass is increased in α-synuclein gene-ablated mice. *J. Neurochem.* 101:132–141.

54. Murphy, E.J. (1998) Fatty acid binding protein expression increases NBD-stearate uptake and cytoplasmic diffusion in L cells. *Am. J. Physiol.* 275:244–249.

55. Murphy, E.J. et al. (2000) Phospholipid and phospholipid fatty acid composition of L-cell fibroblast: Effect of intestinal and liver fatty acid binding proteins. *Lipids* 35:729–738.

56. Alling, C., Liljequist, S., and Engel, J. (1982) The effect of chronic ethanol administration on lipids and fatty acids in subcellular fractions of rat brain. *Med. Biol.* 60:149–154.

57. Demediuk, P. et al. (1985) Mechanical damage to murine neuronal-enriched cultures during harvesting: Effects on free fatty acids, diglycerides, Na+,K+-ATPase, and lipid peroxidation. *In Vitro Cell. Dev. Biol.* 21:569–574.

58. Murphy, E.J. et al. (2000) Phospholipid composition and levels are not altered in fibroblasts bearing presenilin-1 mutations. *Brain Res. Bull.* 52:207–212.

59. Murphy, E.J. and Horrocks, L.A. (1994) A model of compression trauma: Pressure induced injury in cell cultures. *J. Neurotrauma* 10:431–444.

60. Murphy, E.J., Rosenberger, T.A., and Horrocks, L.A. (1997) Effects of maturation on the phospholipid and phospholipid fatty acid compositions in primary rat cortical astrocytes. *Neurochem. Res.* 22:1205–1213.

61. Radin, N.S. (1988) Lipid extraction, in *Neuromethods 7 Lipids and Related Compounds*, Boulton, A.A., Baker, G.B., and Horrocks, L.A. (eds.), Humana Press, Clifton, NJ, pp. 1–62.

62. Hara, A. and Radin, N.S. (1978) Lipid extraction of tissues with a low-toxicity solvent. *Anal. Biochem.* 90:420–426.

63. Swendsen, C.L. et al. (1987) Human neutrophils incorporate arachidonic acid and saturated fatty acids into separate molecular species of phospholipids. *Biochim. Biophys. Acta* 919:79–89.

64. MacDonald, J.I.S. and Sprecher, H. (1989) Studies on the incorporation and transacylation of various fatty acids in choline and ethanolamine-containing phosphoacylglycerol subclasses in human neutrophils. *Biochim. Biophys. Acta* 1004:151–157.

65. Tranchant, T. et al. (1997) Mechanisms and kinetics of α-linolenic acid uptake in Caco-2 clone. *Biochim. Biophys. Acta* 1345:151–161.

66. Chen, W. et al. (2007) Inhibition of cytokine signaling in human retinal endothelial cells through modification of caveolae/lipid rafts by docosahexaenoic acid. *Invest. Ophthalmol. Vis. Sci.* 48:18–26.

67. Robinson, P.J. et al. (1992) A quantitative method for measuring regional in vivo fatty acid incorporation into and turnover within brain phospholipids: Review and critical analysis. *Brain Res. Rev.* 17:187–214.

68. Golovko, M.Y. and Murphy, E.J. (2006) Uptake and metabolism of plasma derived erucic acid by rat brain. *J. Lipid Res.* 47:1289–1297.

69. Murphy, C.C., Murphy, E.J., and Golovko, M.Y. (2008) Eurcic acid is differentially taken up and metabolized in rat liver and heart. *Lipids* 43:391–400.

70. Freed, L.M. et al. (1994) Effect of inhibition of β-oxidation on incorporation of [U-14C] palmitate and [1-14C]arachidonate into brain lipids. *Brain Res.* 645:41–48.

71. Folch, J., Lees, M., and Sloan Stanley, G.H. (1957) A simple method for the isolation and purification of total lipids from animal tissues. *J. Biol. Chem.* 226:497–509.

72. Smith, B.S.W. (1970) A comparison of 125I and 51Cr for measurement of total blood volume and residual blood content of tissues in the rat; evidence for accumulation of 51Cr by tissues. *Clin. Chim. Acta* 27:105–108.

73. Regoeczi, E. and Taylor, P. (1978) The net weight of the rat liver. *Growth* 42:451–456.

74. Jolly, C.A. et al. (1997) Fatty acid binding protein: Stimulation of microsomal phosphatidic acid formation. *Arch. Biochem. Biophys.* 341:112–121.

75. Marcheselli, V.L. et al. (1988) Quantitative analysis of acyl group composition of brain phospholipids, neutral lipids, and free fatty acids, in *Neuromethods 7 Lipids and Related Compounds*, Boulton, A.A., Baker, G.B., and Horrocks, L.A. (eds.), Humana Press, Clifton, NJ, pp. 83–110.

76. Jones, M., Keenan, R.W., and Horowitz, P. (1982) Use of 6-p-toluidino-2-naphthale-nesulfonic acid to quantitate lipids after thin-layer chromatography. *J. Chromatogr.* 237:522–524.

77. Bradford, M. (1976) A rapid and sensitive method for the quantitation of microgram quantities of protein utilizing the principle of protein-dye binding. *Anal. Biochem.* 72:248–254.

78. Rouser, G., Siakotos, A., and Fleischer, S. (1966) Quantitative analysis of phospholipids by thin layer chromatography and phosphorus analysis of spots. *Lipids* 1:85–86.

13 Methods for Kinetic Analysis of Fatty Acid Incorporation and Turnover *in Vivo*: A Steady-State Kinetic Radiotracer Approach

Dhaval P. Bhatt and Thad A. Rosenberger

CONTENTS

13.1   INTRODUCTION

For many years, fatty acids and phospholipids were thought to be of little biological importance, serving only as a source of energy and a primary component of the cellular membrane. These concepts were largely supported by early kinetic analyses measuring the turnover of lipid in cellular membranes in which the effluxes of radiolabeled components were measured *in vivo* [1,2]. These analyses suggested that the rates of turnover of fatty acids and phospholipids in biological membranes occurred on the order of days to weeks. The seemingly long half-lives of membrane components lead to a paradox, which suggested that biological membranes were metabolically inactive despite containing known precursors to important signaling molecules. At this time, known lipid-derived signaling components included the polyphosphoinositides that coupled to extracellular signaling receptors [3], arachidonic acid (20:4n-6) that was shown to be the biosynthetic precursor of prostaglandins whose formation is involved in multiple inflammatory processes [4], and platelet-activating factor, the first phospholipid metabolite to demonstrate known biological activity [5]. Since then, it is greatly appreciated that lipid metabolism is closely related to multiple signaling pathways distinct from its function either as a source of energy or its role as a simple constituent of the unit membrane [6].

In 1992, a model used to measure fatty acid and phospholipid turnover *in vivo* based on steady-state radiotracer incorporation kinetics was introduced [7]. This model was based on the rapid equilibration of circulating fatty acids and the ability to achieve a steady-state level of plasma radioactivity in fatty acid form. The method was also the first to account for the recycling of lipid components and the necessity of long-chain fatty acyl-CoA formation prior to incorporation of the fatty acid into phospholipid (Figure 13.1). Based on these principles, kinetic equations were developed to measure incorporation and turnover rates of fatty acids *in vivo*. Therefore, application of this model reflects those enzymatic pathways necessary to support signaling mechanisms, turnover of fatty acids in individual phospholipids, and the recycling of fatty acids that predominate in most tissues (Figure 13.1). This is in stark contrast to efflux kinetic analysis that reflects the loss of radiolabeled lipid components from the membrane due to metabolism [8–10].

Because fatty acids and phospholipids are utilized for diverse cellular processes, understanding how their metabolism is altered by drug treatment or in disease pathology can provide important information regarding their mechanism of action. Pathologically increased or decreased intracellular fatty acid and phospholipid concentrations can cause cellular apoptosis and have been linked to various disorders associated with a disruption in lipid metabolism. Therefore, measuring fatty acid and phospholipid turnover following drug treatment or during pathology is of great importance in biomedical research. The remainder of this chapter focuses on detailing how steady-state radiotracer incorporation kinetic analysis is applied *in vivo* with an emphasis on measuring brain lipid metabolism.

13.2   STEADY-STATE INCORPORATION KINETICS

Fatty acids and phospholipids are structural components found in all cellular membranes that participate in signal transduction secondary to cell surface receptor

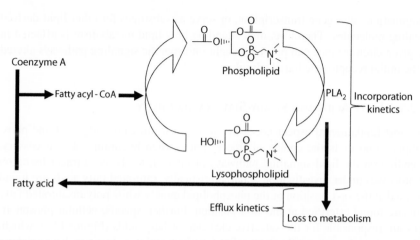

FIGURE 13.1 Schematic diagram demonstrating the PLA$_2$-mediated incorporation and turnover of fatty acid and phospholipids in mammalian cell. The solid arrows demonstrate the condensation of a fatty acid into the activated fatty acyl-CoA molecule and its loss to metabolic processes. The open arrows demonstrate the short-term turnover of a phospholipid molecule. At steady-state conditions two must be maintained. One is that the concentration of phospholipid and fatty acid remain constant. The second condition is that fatty acid incorporated into a phospholipid moiety is equal to that being hydrolyzed out (vacancy hypothesis). The brackets indicate that incorporation kinetic analysis measures both the incorporation of fatty acid and the turnover of phospholipid, while efflux kinetics measures only the loss of fatty acid due to metabolism.

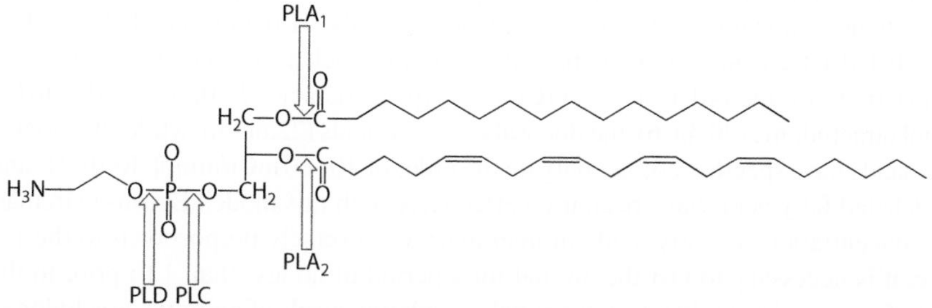

FIGURE 13.2 Schematic diagram demonstrating the chemical bonds on the phospholipid molecule cleaved by the different phospholipase classes. For demonstration purposes, the phospholipid depicted is phosphatidylethanolamine (PtdEtn) (1-palmitoyl-2-arachidonoyl-*sn*-glycerol-3-phosphoethanolamine). Abbreviations: PLA$_1$, phospholipase A$_1$; PLA$_2$, phospholipase A$_2$; PLC, phospholipase C; and PLD, phospholipase D. The specifics of each enzyme class are outlined in Chapters 1 through 3.

activation. Receptor activation can couple to intracellular signaling pathways that stimulate effector enzymes, including phospholipases A$_2$, C, and D (see Chapters 1 through 3), which hydrolyze the phospholipid molecule at specific sites (Figure 13.2). Activation of these effector enzymes typically alter the release of lipid-derived signaling molecules and influence normal cellular ionic homeostasis, propagate inflammation, alter gene transcription, or serve as substrates for other lipid-derived signaling molecules. Therefore, understanding how lipid metabolism is affected in any given circumstance can provide insight into specific signaling pathways altered by the initial receptor-mediated stimulus.

13.2.1 FATTY ACIDS AND STEADY-STATE RADIOACTIVITY

The most fundamental aspect necessary to begin to measure fatty acid and phospholipid turnover is the choice of tracer. Thus, when beginning, it is necessary, depending on the lipid-signaling pathways of interest, to choose a tracer based on the processes under investigation. Characteristically, saturated fatty acids are found esterified at the *sn*-1 position of the phospholipid moiety while polyunsaturated fatty acids are found esterified at the *sn*-2 position. Further, specific cellular phospholipases are responsible for the selective cleavage of fatty acids (Figure 13.2), which, as demonstrated using steady-state radiotracer incorporation kinetics, have selectivity for the molecular species of fatty acids esterified on the phospholipid molecule. Therefore, the choice of fatty acid provides a means to discriminate between different intracellular signaling processes.

Other aspects of steady-state incorporation kinetics that are important to consider before the start of the experiment include the radiotracer-specific radioactivity and how circulating steady-state levels of radioactivity are to be maintained. The first is to choose a radiotracer molecule that has a specific radioactivity high enough so as to not significantly increase the circulating plasma concentration of the fatty acid during infusion. The second is to control the rate of infusion of this fatty acid so that a constant level of circulating radioactivity can be achieved rapidly and maintained throughout a known period of infusion. Commercially ^{3}H-(tritium) and ^{14}C (carbon 14)-labeled fatty acids are available. Most tritium-labeled fatty acids have specific radioactivity on the order of 30–60 Ci/mmol (palmitic acid, 16:0) and 180–240 Ci/mmol (arachidonic (20:4n-6) and docosahexaenoic acids (22:6n-3)), while ^{14}C-labeled fatty acids have specific radioactivity on the order of 40–60 mCi/mmol. Both ^{3}H- and ^{14}C-labeled fatty acids have been used effectively with this model. Because circulating concentrations of fatty acids in mammals are inversely proportional to the fed state, it is necessary to fast the animal for a period of no less than 12 h prior to the start of an experiment. In a fasted animal, circulating levels of palmitic, arachidonic, and docosahexaenoic acids are approximately 150, 26, and 16 nmol/mL, respectively [11]. Based on the above stated tracer-specific radioactivity, the quantity of fatty acids infused into an animal range between 30–130 (^{3}H) and 3–4 (^{14}C) μmol/kg body weight over a defined period of time. Therefore, the infused concentration of radiolabeled fatty acids is low enough so as to not appreciably disturb steady-state levels of circulating fatty acids.

The second consideration is maintaining steady-state levels of plasma radioactivity during the infusion process. For starting the experiment, a known amount of radiolabeled fatty acid is made by sonicating dry radiolabeled fatty acid into radiotracer infusion solution (see below). The typical radiotracer infusion dose based on tracer-specific radioactivity is 7.5 mCi/kg body weight for ^{3}H-labeled fatty acids and 175 μCi/kg body weight for ^{14}C-labeled fatty acids. This solution is placed in a syringe and infused into the animal using a syringe pump via the femoral vein (see below). Two methods have been used in the past to infuse radiolabeled fatty acids. The first is to infuse the radiotracer at a constant flow rate of 0.4 mL/min for a period ranging between 8 and 10 min. Because circulating plasma half-life of fatty acids is very short (0.5–0.75 min), constant infusion will provide steady-state levels of plasma radioactivity. With the constant infusion method, steady-state levels will not be achieved until after 1 min following the start of infusion.

The second method is to use a computer-controlled variable-rate infusion pump that will allow the radiotracer to equilibrate more rapidly so that steady-state levels of plasma radioactivity can be achieved quickly. The basis of the variable rate infusion pump is that at the start of the infusion, the infusion pump runs more quickly and slows over a period of time so that the radiolabel is rapidly dispersed into the animal, and steady-state levels of radioactivity are achieved faster. This method allows one to infuse the radiotracer for a shorter period of time and to control steady-state levels of radiotracer more accurately.

An example of the variable radiotracer infusion protocol has been included in Table 13.1. This program was written using MS BASIC and can be adapted to control the infusion rate of most commercially available infusion pumps that allow external computer control. The salient feature to this program is the equation listed on line 7 of the code that sets the infusion rate of the pump based on a multivariable equation, all of which is normalized to an average-sized rat weighing 185 g (line 23). Thus, when using this protocol, several parameters need to be known before the start of the infusion. These parameters are infusion time (t), final pump rate ($a1$), the lag time (lag), four variables ($a2$, $a3$, $b1$, $b2$) based on the plasma half-life of radiolabeled fatty acids, and the average size of the rat included in the study. The procedure necessary to calculate the four variables is to give a rat a bolus injection of radiolabeled fatty acid and measuring its plasma half-life as a function of time. Once this is accomplished, a regression analysis of the plasma half-life is performed using a 4 variable regression equation to calculate the variables necessary for accurate pump operation. Most commercially available graphing or statistical software will allow one to make this calculation easily. After the plasma half-life variables are calculated, the lag time variable (lag) is determined experimentally. This parameter will hold the infusion pump at the initial calculated rate for a period defined by the equation outlined on line 7. Therefore, the lag function will provide a momentary rapid infusion to quickly achieve a plasma level of circulating radioactivity close to steady-state levels, and then allows the pump to slow the final pump rate ($a1$) as determined by the plasma half-life of circulating fatty acid. The lag time variable typically ranges from 0.5 to 3 s depending on the size and blood volume of the animals being used. The benefits of using a variable rate infusion method are that steady-state radioactivity can be achieved quickly, it can be maintained more accurately between animals, and it can provide more control during the infusion.

13.2.2 SURGICAL CONSIDERATIONS AND CANNULA PLACEMENT

When planning a radiotracer infusion experiment, several considerations must be made prior to initiating the study. The first and foremost is to gain approval from your

TABLE 13.1

Variable Rate Infusion Program

```
1.    from math import exp
2.    from sys import stdin
3.    from time import sleep
4.    import sys
5.    import Infusion Pump
6.    def rate_func(t, lag, a1, a2, a3, b1, b2, correction):
7.    return (a1+(a2 * exp(-b1 * (t - lag)))+(a3 * exp(-b2 *
       (t - lag)))) * correction
8.    def loop(pump):
9.    """
10.   Handle one pass of the constant plasma level program
11.   """
12.   pump.send_cmd('STP') # Stop
13.   pump.send_cmd('CLT') # Clear target volume to zero
14.   pump.send_cmd('MLT 10')
15.   print 'enter range (mlm, mlh, ulm, ulh):'
16.   range=stdin.readline().strip().upper()
17.   if range not in ['MLM', 'MLH', 'ULM', 'ULH']:
      a. print 'invalid range'
      b. return
18.   print 'enter coefficients:'
19.   coeff=stdin.readline().strip().split(',')
20.   a1, a2, a3, b1, b2=[float(x) for x in coeff] # convert
       list of strings to list of floats
21.   print 'rat weight in grams:'
22.   weight=float(stdin.readline().strip())
23.   correction=weight/185.0
24.   print 'CORRECTION FACTOR=%4.4f' % correction
25.   print 'enter lag time in seconds:'
26.   lag=float(stdin.readline().strip())
27.   rate=rate_func(0, lag, a1, a2, a3, b1, b2, correction)
28.   print 'initial rate=%4.4f' % rate
29.   response=pump.send_cmd('%s %4.4f' % (range, rate))
30.   if response:
      a. print response, '(out of range?)'
      b. return
31.   pump_range=pump.send_cmd('RNG')
32.   print 'enter final time (sec):'
33.   final_time=float(stdin.readline().strip())
34.   print 'press stop/start on pump to interrupt'
35.   pump.send_cmd('RUN')
36.   current_time=0
```

TABLE 13.1 (continued)
Variable Rate Infusion Program

```
37.    while current_time < final_time:
       a. if pump.prompt !=' > ':
            i. print 'interrupted'
           ii. break
       b. sleep(1)
       c. current_time += 1
       d. new_rate = rate_func(current_time, lag, a1, a2, a3,
          b1, b2, correction)
       e. print 't=%6d %4.3f %s' % (current_time, new_rate,
          pump_range)
       f. if new_rate !=rate:
            i. rate = new_rate
           ii. pump.send_cmd('%s %3.4f' % (range, rate))
       g. else:
            i. pump.send_cmd('VOL')
       h. pump.send_cmd('key')
```

institution's Animal Care and Use Committee. Thus, while the procedures outlined in this chapter result in the euthanasia of the animal at the end of the experiment, a surgical procedure is involved that is considered a survival surgery categorized as USDA pain category D, requiring that pain or distress is appropriately relieved with anesthetics, analgesics, and/or tranquilizing drugs. Further, because these protocols require the infusion of radiolabeled fatty acids, other permission must be acquired before initiating the studies to gain approval for the use of radiotracers and disposal of radiotracers. Because phospholipid metabolism is rapidly attenuated following anesthesia, infusion of tracer must be performed in the awake animal [12,13]. Therefore, a period of restraint must also be included in all animal protocols when applying for use access so that all animals can recover completely from anesthesia without causing harm to the catheters. The cannula placement procedures outlined in this section are those previously reported in the rat [7,14]. Although similar procedures have been reported in the mouse, these surgical procedures can also be applied to larger animals if necessary.

13.2.2.1 Surgical Placement of Catheters

To begin the catheter placement rats are anesthetized with 2%–3% Halothane™ or isoflurane in a Plexiglas chamber, then they are secured on a plastic dissecting tray equipped with a constant volatile anesthetic delivery and scavenging system to expose the abdomen. The right inner thigh and abdomen are shaved and cleaned, and an incision is made to expose the right femoral artery and vein. Polyethylene catheters (PE 50) filled with 100 IU of sodium heparin is placed in the right femoral artery and vein. The catheters are secured in place with sutures and the incision closed using wound clips. A 1.0% solution of lidocaine is applied locally to the incision to lessen postsurgical pain. The animals are removed from the surgical platform, then loosely wrapped in fast setting plaster body casts, and taped to wooden blocks. Care must be taken at this step to ensure that the animals are allowed free upper body movement, yet secured tightly enough to prevent injury due to catheter damage. The animals are then allowed to completely recover from the anesthesia for at least 3h before tracer infusion. During the recovery period, body temperature must be maintained to prevent hypothermia and to speed recovery. This can be accomplished using isothermal heating pads or by maintaining the body temperature in a controlled chamber equipped with a rectal feedback heating device.

Once the animals have completely recovered from anesthesia and the body temperature has returned to normal, the rats are infused intravenously with the solution containing the radioactive fatty acid (see above). During the infusion period, timed arterial blood samples (100 μL) are collected to determine plasma levels of radiotracer, which are used to calculate the plasma-specific activity of the radiotracer and to construct plasma curves necessary to calculate the input function used in the kinetic analysis (see below). It is necessary to collect enough blood samples during the infusion period to accurately measure plasma radioactivity. This includes one sample prior to the start of infusion and samples collected at minimum 1 min intervals during the infusion period. Lastly, one 50 μL whole blood sample is taken 30 s before the end of the infusion to determine whole-blood radioactivity and blood fatty acid composition. Immediately following the last blood sample collection, the rat is euthanized using an overdose of sodium pentobarbital (100 mg/kg, i.v.). Animals designated for fatty acid turnover measurements are subject to head-focused microwave irradiation (5.5 kW, 3.4 s) to stop the metabolic breakdown of membrane phospholipids. Microwaved brains are excised and frozen immediately on dry ice and then stored at −80°C until use. Because fatty acid metabolism occurs rapidly, it is necessary to minimize the postmortem interval between euthanasia and microwave fixation.

13.2.3 PLASMA CURVES AND INPUT FUNCTIONS

The significance of maintaining steady-state plasma radioactivity during the infusion period is the primary simplifying factor associated with steady-state radiotracer incorporation kinetics. Because circulating fatty acids are in rapid equilibration with surrounding tissue (see Chapter 12), achieving and maintaining a constant level of plasma radioactivity allows end-point analysis of the tissues rather than sequential analysis of multiple tissues exposed to multiple levels of radioactivity as would be the case if using a pulse-chase kinetic paradigm [8–10]. Therefore, once steady-state plasma radioactivity has been achieved, radiotracer flow and, in turn, incorporation of radiotracer into stable phospholipids will depend upon specific radioactivity gradients found between the plasma and the stable tissue phospholipids. These intermediate pools include the circulating nonesterified fatty acids, tissue nonesterified fatty acids, and the long-chain fatty acyl-CoA. Therefore, under steady-state conditions, the amount of radioactivity that is found esterified in the stable phospholipids at the end of infusion will depend on the enzymatic turnover of membrane phospholipid and the exposure period to radioactivity defined by the length of infusion.

Thus, by plotting the plasma radioactivity as a function of time during the infusion period provides a basis to measure the level of exposure each tissue has to circulating radioactivity. This data is used to calculate, using trapezoidal integration (SigmaPlot, SPSS Science, Chicago, Illinois), the area under the curve or the input function of radiotracer into a tissue in units of $nCi \times s/mL$. This value is used to normalize tissue radioactivity as described in Equation 13.1 (see below) to the amount of tracer infused over a given period of time [14].

13.2.4 ANALYSIS OF NONESTERIFED FATTY ACID, ACYL-CoA, AND INDIVIDUAL PHOSPHOLIPID COMPARTMENTS

The quantitative analysis of plasma and tissue levels of fatty acids and fatty acid radioactivity is of paramount importance for the accurate application of steady-state radiotracer incorporation kinetics. This analysis involves a combination of classic lipid analytical techniques coupled with scintillation counting to quantify radioactivity. In this regard, it is important to appreciate that for each analysis being performed there are at least two measurements being performed: One, to quantify the fatty acid or long-chain fatty acyl-CoA and the second, to quantify the amount of radioactivity found in the fraction. Further, because there is unavoidable variation in animal studies, it is important to organize your analysis and data so that the calculations outlined below are performed on matched data sets from individual animals. Because most of the topics listed in this section have been covered in depth in other chapters in this book, the procedures outlined below are only described in brief followed by specific protocols used in our laboratory.

13.2.4.1 Brain and Plasma Lipid Extraction and Chromatography

Initially microwaved brains are divided equally by cutting along the midline. Half of the brain is used for analysis of total lipids and the second half is used for the analysis of long-chain fatty acid CoA. Total lipids from the microwaved brain hemispheres are weighed and extracted using n-hexane/2-propanol (3:2, by vol.) in a glass Tenbroeck homogenizer [15] (see below). Plasma lipids are extracted in chloroform/methanol (2:1, by vol.) and partitioned with 0.9% KCl. The total plasma lipids in the chloroform extract are washed once with 0.2 volumes of 0.9% KCl to remove nonlipid contaminants prior to analysis [16] (see below). Once the plasma and brains have been extracted, the organic extracts containing the lipids are concentrated to zero under a steady-stream of N_2 at 50°C, then resolvated in a known amount of n-hexane/2-propanol (3:2, by vol.) to form working stocks of each solution. At this point, the total amount of stock standards is recorded, and records are kept so that as the sample is used all results can be normalized to the wet weight of tissue or volume of plasma used.

Standards and brain lipids are isolated on Whatman silica gel 60A LK6 TLC plates using a solvent system of chloroform/methanol/acetic acid/H_2O (50:37.5:3:2, by vol.) [17]. Neutral lipids from plasma and brain extracts are separated on silica gel 60 plates using the solvent system of heptane/diethyl ether/acetic acid (60:40:4, by vol.) [18]. Gas chromatography is used to quantify esterified and nonesterified fatty acid levels from plasma and microwaved brain tissue. Liquid scintillation counting is used to measure the radioactivity found in the isolated phospholipid and nonesterified fatty acid fractions. All extracts should be stored in *n*-hexane/2-propanol (3:2, by vol.) under N_2 at $-80°C$ to prevent oxidation of the lipid components during storage.

13.2.4.2 Methylation of Esterified and Nonesterified Acids

Esterified fatty acids found in the different phospholipid classes are hydrolyzed off the phospholipid moiety and methylated using a 5% sodium methoxide solution at 37°C for 30 min as outlined below. The reaction is stopped with methyl formate and the fatty acid methyl esters are extracted with *n*-hexane. The nonesterified brain and plasma fatty acids are methylated using 2% sulfuric acid in toluene/methanol (1:1, by vol.) at 65°C for 4 h. This reaction was terminated with H_2O and extracted with petroleum ether [19]. It is important to note that the yield of these methylation procedures is highly dependent on the purity of the solvents used and both are inhibited by water. Therefore, to avoid variation in the yield, radioactivity of the esterified and nonesterified fatty acids should be measured on that sample remaining following GLC analysis and performed in triplicate. Because GLC analysis is sensitive, only a small portion of sample is necessary to quantify fatty acid content, leaving ample sample for scintillation counting.

13.2.4.3 Gas Chromatography of Fatty Acid Methyl Esters

Fatty acid methyl esters are quantified using a gas chromatograph equipped with a capillary column (SP 2330; 30 m × 0.32 mm i.d., Supelco, Bellefonte, Pennsylvania) and a flame ionization detector. Typical separation is initiated at 90°C with a temperature gradient to 230°C over 20 min and requires only a small proportion of the total sample. Because the fatty acid methyl ester samples are radioactive, it is necessary to limit the amount of sample used to quantify fatty acid content so that contamination of the detector by radioactivity is minimal. Prior to analysis, the gas chromatograph is standardized with known amounts of fatty acid methyl ester standards to establish relative retention times and response factors. The internal standard, methyl heptadecanoate, and the individual fatty acids are quantified by peak area analysis, and the detector response is used to calculate concentration. All standards require correlation coefficients of 0.998 or greater within the sample concentration range.

13.2.4.4 Long-Chain Fatty Acyl-CoA Analysis

Quantification of labeled and unlabeled brain acyl-CoA is accomplished using an extraction procedure that involves the coprecipitation of the long-chain acyl-CoA esters with acetonitrile and potassium phosphate buffer (KH_2PO_4, pH 4.5). A step-by-step laboratory protocol detailing the application of this protocol and the purification of standards is outlined in the Appendix section, which is a modification of that outlined in [20]. Extraction using this method is attained by suspending small fragments of microwaved tissue (~400 mg) with an internal standard, heptadecanoyl-CoA (10 μg, 9.8 nmol), in 2.0 mL 100 mM KH_2PO_4, pH 4.5, and homogenized. An equal volume of 2-propanol (2.0 mL) is added to the tissue homogenate, sonicated, and 0.25 mL of saturated ammonium sulfate $((NH_4)_2SO_4)$ and 4.0 mL acetonitrile are added to form an emulsion. The emulsion is vortexed and then centrifuged for

5 min at $1900 \times g$. The supernatant containing the acyl-CoA esters is transferred to a second tube and diluted with 10 mL 25 mM KH_2PO_4, pH 4.9. The diluted acyl-CoA esters are filtered using a syringe through a prewashed ABI oligonucleotide purification cartridge (OPC, Applied Biosystems, Foster City, California) and then washed with 2.0 mL 25 mM KH_2PO_4 buffer, pH 4.9. The bound long-chain acyl-CoA esters are eluted from the column with 0.2 mL 80% 2-propanol in 1 mM acetic acid. The isolated long-chain fatty acyl-CoA esters at this point are isolated and quantified using HPLC [20]. The specific radioactivity found in the long-chain fatty acyl-CoA esters is measured by collecting peaks. They are eluted off the HPLC, the radioactivity quantified with scintillation counting, then the number normalized by the concentration of the fatty acid CoA ester and the amount of tissue analyzed.

The yield of extraction using the acetonitrile-buffer method is higher than that found with other extraction protocol because long-chain fatty acyl-CoA hydroxylases are inactive below a pH of 6.0 and inhibited by the addition of saturated $(NH_4)_2SO_4$ [21]. Other means of homogenization can be used depending on the amount of connective tissue present in the sample; however, this procedure is not amenable to the parallel analysis of tissue lipids. Therefore, parallel analysis of single brain samples must be performed for the lipid analysis as detailed in this section. Because the radiotracer is distributed uniformly between the brain hemispheres, microwaved brains are bisected along the midline and half is used for lipid analysis and the other is used for long-chain fatty acyl-CoA analysis. The use of solid phase isolation of long-chain acyl-CoA esters using an OPC avoids the time-consuming steps of solvent removal found with the chloroform/methanol-based extraction. This method has been employed with great success in extracting acyl-CoA esters from brain [20] and tissue culture [22,23] and allows for the isolation of the polyunsaturated acyl-CoA esters, arachidonoyl-CoA, docosahexaenoyl-CoA, and linoleoyl-CoA [24].

13.3　CALCULATING RATES OF INCORPORATION AND FATTY ACID TURNOVER

The radioactivity of a brain phospholipid i of interest, $c_{br,i}^*(T)$ (nCi/g), is calculated by correcting its net brain radioactivity for its intravascular radioactivity [25] as outlined above. Blood samples taken at the time of death (T) after starting tracer infusion are extracted and analyzed to make this correction. Unidirectional incorporation coefficients, k_i^* (mL/s×g), of radiolabeled fatty acid from plasma into phospholipids i are calculated as described in Equation 13.1:

$$k_i^* = \frac{c_{br,i}^*(T)}{\displaystyle\int_0^T c_{pl}^* \, dt} \tag{13.1}$$

where

　　t is time after beginning of infusion
　　c_{pl}^* (nCi×s/mL) is the plasma concentration of radiolabeled fatty acid during infusion

Rates of incorporation of nonesterified fatty acid from plasma into individual brain phospholipids (i, $J_{in,i}$) and from brain acyl-CoA pools into brain phospholipid (i, $J_{FA,i}$) are calculated as described in Equations 13.2 and 13.3:

$$J_{in,i} = k_i^* c_{pl} \tag{13.2}$$

$$J_{FA} = \frac{c_{br,i}^*(T) c_{pl}}{\lambda_{acyl\text{-}CoA} \int_0^T c_{pl}^* / c_{pl} \, dt} = \frac{J_{in,i}}{\lambda_{acyl\text{-}CoA}} \tag{13.3}$$

It is important to understand that c_{pl} (nmol/mL) is the concentration of unlabeled, nonesterified fatty acid found in plasma and that lambda acyl-CoA represents the steady-state specific radioactivity of the long-chain fatty acyl-CoA relative to that of plasma during tracer infusion:

$$\lambda_{acyl\text{-}CoA} = \frac{c_{acyl\text{-}CoA}^* / c_{acyl\text{-}CoA}}{c_{pl}^* / c_{pl}} \tag{13.4}$$

In Equation 13.4, the numerator is the specific radioactivity of brain long-chain fatty acyl-CoA and the numerator is the specific radioactivity of plasma fatty acid. The fractional turnover rate of fatty acid within the individual phospholipid i, $F_{FA,i}$ (% per hour), is defined as

$$F_{FA,i} = \frac{J_{FA,i}}{c_{br,i}} \tag{13.5}$$

The half-life of the FA in i is defined as

$$\text{Half-life} = 0.693 / F_{FA,i} \tag{13.6}$$

A list of reported rates of incorporation (Equation 13.3) and fractional turnover rates (Equation 13.5) measured using this model are provided in Table 13.2.

13.4 CONCLUSIONS

Alterations in fatty acid and phospholipid metabolism are of vital importance with regard to understanding the biochemical changes that precede pathology and to fully understand the effect drug treatment has on intracellular signaling. For these reasons, a steady-state model of radiotracer incorporation has been developed and used to monitor the changes in signal transduction associated with alteration in fatty acid incorporation and phospholipid metabolism. This method has been used successfully to demonstrate the role of arachidonic acid signaling in a rat model of

TABLE 13.2
Reported Incorporation and Fractional Turnover Rates for Palmitic, Arachidonic, and Docosahexaenoic Acids in Rat Brain, Heart, and Liver

	Palmitate	Arachidonate	Docosahexaenate	Ref.
Brain				
Incorporation rate (nmol/g/s, $\times 10^{-2}$)				
EtnGpl	5.2	0.5–2.8	5.6–6.4	[11,12,27–30]
PtdIns	4.3	1.5–3.9	1.6	
PtdSer	1.3	0.6–1.3	0.7–0.9	
ChoGpl	19.0	2.0–9.6	3.5–4.1	
Fractional turnover rate (% per hour)				
EtnGpl	4.1–4.2	1.1–2.4	1.0–2.9	[11,12,27–30]
PtdIns	29.1–39.2	5.4–15.3	17.7–26.7	
PtdSer	9.4–16.1	6.8–12.1	0.2–1.1	
ChoGpl	5.0–7.0	9.4–18.3	3.1–8.9	
Heart				
Incorporation rate (nmol/g/s, $\times 10^{-2}$)				
EtnGpl	7.8	8.8		[31,32]
PtdIns/PtdSer	10.6	6.8		
ChoGpl	55.1	72.0		
Fractional turnover rate (% per hour)				
EtnGpl	378	18.0	NR	[31,32]
PtdIns/PtdSer	7523	207		
ChoGpl	142	145		
Liver				
Incorporation rate (nmol/g/s, $\times 10^{-2}$)				
EtnGpl	4.8	NR	NR	[33]
PtdIns	1.8			
PtdSer	1.5			
ChoGpl	63.4			
Fractional turnover rate (% per hour)				
EtnGpl	0.2	NR	NR	[33]
PtdIns	0.5			
PtdSer	0.4			
ChoGpl	0.4			

Abbreviations: ChoGpl, choline glycerophospholipid; EtnGpl, ethanolamine glycerophospholipid; PtdIns, phosphatidylinositol; PtdSer, phosphatidylserine; and NR, not reported.

neuroinflammation, the effect of drug treatment on arachidonic acid turnover, and selectivity in the metabolism of different fatty acids, and has been instrumental in documenting alterations in the utilization of fatty acid in several knockout animals. Therefore, based on the complexity of mammalian lipid metabolism, understanding the variations in lipid signaling is of immense value when trying to understand the roles lipid-mediated signaling has in pathology. Thus, the application or variation of this methodology can be applied to many experimental paradigms to identify the role that changes in fatty acid or phospholipid metabolism have in disease process and to pursue potential pharmacological therapies.

ACKNOWLEDGMENT

The work was supported in part by Grant Number 2P20RR017699-06 from the National Center for Research Resources (NCRR), a component of the National Institutes of Health (NIH).

APPENDIX: COMMONLY USED REAGENTS AND PROTOCOLS

Radiotracer infusion solution

Reagent	Concentration	Amount per 500 mL
HEPES Buffer, pH 7.4	5.0 Mm	596 mg
Sodium chloride	0.9%	4.5 g
BSA (fatty acid free)	0.05 g/mL	25 g

Add the HEPES buffer, sodium chloride, and BSA to 800 mL of dH_2O while stirring. Once all reagents are in solution, adjust the pH to 7.4 (at 37°C). QS to 1 L in a volumetric flask, divide into 15 mL aliquots, and store at −20°C until use. Abbreviations: HEPES Buffer, N-2-hydroxypiperazine-N'-2-ethanesulfonic acid; BSA (fatty acid free), essentially fatty acid free bovine serum albumin lyophilized powder (~0.005% fatty acid content).

Tissue extraction using n-hexane/2-propanol (3:2, by vol.)

The extraction method listed below is a single phase extraction protocol developed by Radin [15].

1. Frozen tissue is homogenized in 18 vol. n-hexane:2-propanol (3:2, by vol.) (HIP 3:2) per gram wet weight tissue in a glass Tenbroeck homogenizer.
2. The homogenate is transferred to a screw-top conical 50 mL glass centrifuge tube.
3. The homogenizer is washed twice with 2 mL n-hexane:2-propanol (3:2, by vol.).
4. The washes and homogenate are combined and then centrifuged at $1000 \times g$ to pellet the nonsoluble material.

5. The organic supernatant is transferred to a second 50 mL conical centrifuge tube and the pellet is washed twice with 2 mL n-hexane:2-propanol (3:2, by vol.).
6. The organic supernatant and washes are concentrated to zero under N_2 at 45°C and stored in 3 mL n-hexane:2-propanol (3:2, by vol.) under N_2 at −80°C.

Plasma extraction
The protocol listed below is a modified method developed by Folch et al. [16].

1. Add 10–100 μL plasma to 3 mL chloroform/methanol (2:1 by vol.) and vortex.
2. Add 600 μL 0.15 M KCl and vortex.
3. Centrifuge at $1000 \times g$ for 5 min.
4. Transfer lower organic phase to new tube.
5. Wash upper phase with 2.0 mL chloroform and vortex.
6. Centrifuge at $1000 \times g$ for 5 min and combine lower phases.
7. Concentrate organic phase and bring up in known volume.
8. Measure radioactivity in desired volume of organic phase.

Long-chain acyl-CoA extraction
The method described below is that of Deutsch et al. [20].

1. Weigh brain and place in a glass Tenbroeck homogenizer. Record weight (half hemisphere ~0.65g).
2. Add internal standard (~10 μg) (see below).
3. Add 2 mL 75 mM KH_2PO_4 pH 5.0.
4. Homogenize, then add 2 mL 2-propanol, and homogenize again.
5. Transfer homogenate to a 14 mL centrifuge tube; then immediately add 0.25 mL saturated $(NH_4)_2SO_4$. Mix sample gently by hand.
6. Add 2 mL acetonitrile to the homogenizer, homogenize, and then transfer to the 14 mL centrifuge tube. Repeat with another 2 mL acetonitrile and combine rinse with initial extract.
7. Vortex sample rapidly for 5 min (sample will form small particles while leaving the solution clear).
8. Centrifuge for 5 min at $1800 \times g$ at 4°C for 7 min.
9. Transfer supernatant into 50 mL centrifuge tube, add 10 mL 75 mM KH_2PO_4, and mix gently by hand.
10. Re-extract protein pellet with the above solutions (2 mL 75 mM KH_2PO_4, 2 mL 2-propanol, 0.25 mL saturated $(NH_4)_2SO_4$, 4 mL acetonitrile), centrifuge, and combine with above initial extract. Add another 10 mL 75 mM KH_2PO_4 and mix gently by hand.
11. Place solution in a 12 mL syringe and filter three times through an activated OPC (Applied Biosystems, Foster City, California) at a rate of 0.1 mL · min⁻¹ or 10–15 drops · min⁻¹.

12. Wash the column with 2 mL 25 mM KH_2PO_4 followed by a 1 mL wash with water and elute the long-chain fatty acyl-CoA off the column with 1 mL 2-propanol/1 mM glacial acetic acid (3:1, by vol.).

13. Concentrate the sample under a steady-stream of N_2 at ~50°C.

14. Resolvate in 2-propanol/1 mM glacial acetic acid (3:1, by vol.) and analyze sample by HPLC.

Long-chain fatty acyl-CoA standard preparation

1. Weigh 1–5 mg acyl-CoA standard and dissolve in 500 μL acetonitrile (CH_3CN). Slowly add 300 μL 25 mM KH_2PO_4. Add more KH_2PO_4 (450 μL) to attain a 40:60 CH_3CN/KH_2PO_4 ratio.

2. Solution is placed in syringe and filtered through an activated OPC Cartridge ($0.1 mL \cdot min^{-1}$ or 40–50 drops $\cdot min^{-1}$, if doing it by hand).

3. The cartridge is rinsed with 2.0 mL 25 mM KH_2PO_4 buffer and dried by pushing air through cartridge.

4. Elute cartridge with 1.0 mL 2-propanol containing 1 mM glacial acetic acid (3:1, by vol.).

5. Run portion of the standard on HPLC to determine purity (inject ~0.5 μg).

6. Take 10 μL of purified long-chain fatty acyl-CoA, and add 990 μL 2-propanol containing 1 mM glacial acetic acid (3:1, by vol.). Read absorbance at 260 nm.

$$N = (Abs._{260} / 15,400 \text{ mol} \cdot L^{-1}) \times 100 \text{ (dilution factor)}$$

$$\mu g \cdot \mu L^{-1} = N \times MWt.$$

7. Dilute purified standard in 2-propanol containing 1mM glacial acetic acid (3:1, by vol.) to a working stock concentration (0.1 μg/μL). The standard is good for up to 3 months when stored at 4°C, although the purity and concentration should be checked periodically.

5% Sodium methoxide solution (used for methylation of esterified fatty acids)

Reagent	Concentration (%)	Amount per 100 mL
Sodium methoxide (25%)	5	20 mL
Anhydrous methanol	95	80 mL

Both the sodium methoxide solution and anhydrous methanol must be of the highest purity possible. All reagents should be stored in sealed bottles under a positive pressure nitrogen atmosphere so that none of the solutions become contaminated with water. The 5% sodium methoxide solution can be stored at room temperature for up to 7 days.

2% Sulfuric acid solution (used for methylation of free fatty acids)

Reagent	Concentration	Amount per 100 mL
Sulfuric acid (18 M)	2% by vol.	2 mL
Toluene	50% by vol.	49 mL
Anhydrous methanol	50% by vol.	49 mL

Both the toluene and anhydrous methanol must be of the highest purity possible. All reagents should be stored in sealed bottles under a positive pressure nitrogen atmosphere so that none of the solutions become contaminated with water. A 1 to 1 mixture of toluene and methanol solution is made by combining equal volumes (50 mL) of each in a sealed vial. A 2 mL aliquot of the mixture is replaced with 2.0 mL concentrated sulfuric acid. The solution can be stored at room temperature for up to 2 weeks.

Base-catalyzed methylation of esterified fatty acids

The method listed below is a modification of that provided by Brockerhoff [26] and is used to form the fatty acid methyl ester from esterified fatty acids.

1. Transfer phospholipid from either TLC or HPLC into a 16 × 125 mm screw-top test tube.
2. Add a known amount of internal standard (triheptadecanoin, Nu-Chek Prep, Inc., Elysian, Minnesota). The concentration of internal standard used will determine the final volume of sample prior to GLC.
3. Concentrate the samples to zero using the nitrogen evaporator at 45°C (when solvents are present) or in oven at 85°C for 15 min for TLC.
4. To dry sample, add 2.0 mL of 5% sodium methoxide (see above), vortex to solvate the entire sample, then incubate at 40°C for 30 min.
5. Stop the reaction with the addition of 1.0 mL methyl formate.
6. Extract the fatty acid methyl esters using 3 × 3.0 mL washes with *n*-hexane.
7. Concentrate *n*-hexane extracts to zero and transfer to μvials.
8. Concentrate samples in the μvials to zero and resolvate in a known amount of *n*-hexane.
9. Measure the fatty acid concentration via GLC.

Acid-catalyzed methylation of nonesterified fatty acids and sphingomyelin

The method listed below is a modification of that outlined [19] and is used to form the fatty acid methyl ester from nonesterified fatty acids and fatty acids found in sphingomyelin.

1. Transfer phospholipid from either TLC or HPLC into a 16 × 125 mm screw-top test tube.
2. Add a known amount of internal standard (methyl heptadecanoate, Nu-Chek Prep, Inc. Elysian, Minnesota). The concentration of internal standard used will determine the final volume of sample prior to GLC.

3. Concentrate the samples to zero using the nitrogen evaporator at 45°C (when solvents are present) or in oven at 85°C for 15 min for TLC.
4. To the dry sample add 2.0 mL of 2% H_2SO_4 in toluene/methanol (1:1, by vol.) (see above), vortex to solvate the entire sample, and then incubate at 65°C for 2 h.
5. Stop the reaction with the addition of 1.0 mL H_2O.
6. Extract the fatty acid methyl esters using 3×3.0 mL washes with *n*-hexane.
7. Concentrate *n*-hexane extracts to zero and transfer to μvials.
8. Concentrate samples in the μvials to zero and resolvate in a known volume of *n*-hexane.
9. Measure the fatty acid concentration via GLC.

LIST OF ABBREVIATIONS

16:0	palmitic acid
20:4n-6	arachidonic acid
22:6n-3	docosahexaenoic acid
^{14}C	carbon 14
^{3}H	tritium
CerPCho	sphingomyelin
ChoGpl	choline glycerophospholipid
EtnGpl	ethanolamine glycerophospholipid
PakCho	1-*O*-alkyl-2-acylglycerol-3-phosphocholine
PakEtn	1-*O*-alkyl-2-acylglycerol-3-phosphoethanolamine
PlsCho	1-*O*-alk-1′-enyl-2-acylglycerol-3-phosphocholine (choline plasmalogen)
PlsEtn	1-*O*-alk-1′-enyl-2-acylglycerol-3-phosphoethanolamine (ethanolamine plasmalogen)
PtdCho	phosphatidylcholine
PtdEtn	phosphatidylethanolamine
PtdIns	phosphatidylinositol
PtdSer	phosphatidylserine

REFERENCES

1. Freysz, L. et al. (1969) Kinetics of the biosynthesis of phospholipids in neurons and glial cells isolated from rat brain cortex. *J Neurochem* 16:1417–1424.
2. Miller, S. L. et al. (1977) Metabolism of glycerophospholipids of myelin and microsomes in rat brain. Reutilization of precursors. *J Biol Chem* 252:4025–4037.
3. Hokin, M. R. and Hokin, L. E. (1953) Enzyme secretion and the incorporation of P[32] into phospholipides of pancreas slices. *J Biol Chem* 203:967–977.
4. Bergstroem, S. et al. (1964) The enzymatic formation of prostaglandin E2 from arachidonic acid prostaglandins and related factors 32. *Biochim Biophys Acta* 90:207–210.
5. Demopoulos, C. A. et al. (1979) Platelet-activating factor. Evidence for 1-*O*-alkyl-2-acetyl-sn-glyceryl-3-phosphorylcholine as the active component (a new class of lipid chemical mediators). *J Biol Chem* 254:9355–9358.

6. Eyster, K. M. (2007) The membrane and lipids as integral participants in signal transduction: Lipid signal transduction for the non-lipid biochemist. *Adv Physiol Educ* 31:5–16.

7. Robinson, P. J. et al. (1992) A quantitative method for measuring regional *in vivo* fatty-acid incorporation into and turnover within brain phospholipids: Review and critical analysis. *Brain Res Brain Res Rev* 17:187–214.

8. Zilversmit, D. B. et al. (1942) On the calculation of "turnover time" and "turnover rate" from experiments involving the use of labeling agents. *J Gen Physiol* 26:325–331.

9. Reiner, J. M. (1953) The study of metabolic turnover rates by means of isotopic tracers. I. Fundamental relations. *Arch Biochem Biophys* 46:53–79.

10. Reiner, J. M. (1953) The study of metabolic turnover rates by means of isotopic tracers. II. Turnover in a simple reaction system. *Arch Biochem Biophys* 46:80–99.

11. Chang, M. C. et al. (2001) Chronic valproate treatment decreases the in vivo turnover of arachidonic acid in brain phospholipids: A possible common effect of mood stabilizers. *J Neurochem* 77:796–803.

12. Contreras, M. A. et al. (1999) Reduced palmitate turnover in brain phospholipids of pentobarbital-anesthetized rats. *Neurochem Res* 24:833–841.

13. Kimes, A. S. et al. (1985) Brain palmitate incorporation in awake and anesthetized rats. *Brain Res* 341:164–170.

14. Washizaki, K. et al. (1994) Brain arachidonic acid incorporation and precursor pool specific activity during intravenous infusion of unesterified [^{3}H]arachidonate in the anesthetized rat. *J Neurochem* 63:727–736.

15. Radin, N. S. (1981) Extraction of tissue lipids with a solvent of low toxicity. *Methods Enzymol* 72:5–7.

16. Folch, J. et al. (1957) A simple method for the isolation and purification of total lipids from animal tissue. *J Biol Chem* 226:497–509.

17. Jolly, C. A. et al. (1997) Fatty acid binding protein: Stimulation of microsomal phosphatidic acid formation. *Arch Biochem Biophys* 341:112–121.

18. Breckenridge, W. C. and Kuksis, A. (1968) Specific distribution of short-chain fatty acids in molecular distillates of bovine milk fat. *J Lipid Res* 9:388–393.

19. Akesson, B. et al. (1970) Initial incorporation into rat liver glycerolipids of intraportally injected (^{3}H)glycerol. *Biochim Biophys Acta* 210:15–27.

20. Deutsch, J. et al. (1994) Isolation and quantitation of long-chain acyl-coenzyme A esters in brain tissue by solid-phase extraction. *Anal Biochem* 220:321–323.

21. Kurooka, S. et al. (1972) Some properties of long fatty acyl-coenzyme A thioesterase in rat organs. *J Biochem (Tokyo)* 71:625–634.

22. Chen, Q. et al. (1998) Analysis of the acyl-CoAs that accumulate during the peroxisomal beta-oxidation of arachidonic acid and 6,9,12-octadecatrienoic acid. *Arch Biochem Biophys* 349:371–375.

23. Luthria, D. L. et al. (1997) Metabolites produced during the peroxisomal beta-oxidation of linoleate and arachidonate move to microsomes for conversion back to linoleate. *Biochem Biophys Res Commun* 233:438–441.

24. Deutsch, J. et al. (1997) Relation between free fatty acid and acyl-CoA concentrations in rat brain following decapitation. *Neurochem Res* 22:759–765.

25. Grange, E. et al. (1995) Specific activity of brain palmitoyl-CoA pool provides rates of incorporation of palmitate in brain phospholipids in awake rats. *J Neurochem* 65:2290–2298.

26. Brockerhoff, H. (1975) Determination of the positional distribution of fatty acids in glycerolipids. *Methods Enzymol* 35:315–325.

27. Contreras, M. A. et al. (2001) Chronic nutritional deprivation of n-3 alpha-linolenic acid does not affect n-6 arachidonic acid recycling within brain phospholipids of awake rats. *J Neurochem* 79:1090–1099.

28. Rosenberger, T. A. et al. (2002) Rapid synthesis and turnover of brain microsomal ether phospholipids in the adult rat. *J Lipid Res* 43:59–68.

29. Bazinet, R. P. et al. (2006) Chronic carbamazepine decreases the incorporation rate and turnover of arachidonic acid but not docosahexaenoic acid in brain phospholipids of the unanesthetized rat: Relevance to bipolar disorder. *Biol Psychiatry* 59:401–407.

30. Bazinet, R. P. et al. (2005) Chronic valproate does not alter the kinetics of docosahexaenoic acid within brain phospholipids of the unanesthetized rat. *Psychopharmacology (Berlin)* 182:180–185.

31. Murphy, E. J. et al. (2000) Intravenously injected [1-¹⁴C]arachidonic acid targets phospholipids, and [1-¹⁴C]palmitic acid targets neutral lipids in hearts of awake rats. *Lipids* 35:891–898.

32. Patrick, C. B. et al. (2005) Arachidonic acid incorporation and turnover is decreased in sympathetically denervated rat heart. *Am J Physiol Heart Circ Physiol* 288:H2611–2619.

33. Golovko, M. Y. et al. (2005) Alpha-synuclein gene deletion decreases brain palmitate uptake and alters the palmitate metabolism in the absence of alpha-synuclein palmitate binding. *Biochemistry* 44:8251–8259.

Index

Printed and bound by CPI Group (UK) Ltd, Croydon, CR0 4YY

18/10/2024

01776208-0013